Packaging technology

Related titles:

Emerging food packaging technologies
(ISBN 978-1-84569-809-6)
Emerging food packaging technologies reviews advances in packaging materials, the design and implementation of smart packaging techniques, and developments in response to growing concerns about packaging sustainability. Part I of *Emerging food packaging technologies* focuses on developments in active packaging, reviewing controlled release packaging, active antimicrobial and nanocomposites in packaging, and edible chitosan coatings. Part II goes on to consider intelligent packaging and how advances in the consumer/packaging interface can improve food safety and quality. Developments in packaging material are analysed in Part III, while Part IV explores the use of eco-design and life cycle assessment in the production of smarter, environmentally compatible packaging.

Advances in meat, poultry and seafood packaging
(ISBN 978-1-84569-751-8)
Advances in meat, poultry and seafood packaging provides a comprehensive review of both current and emerging technologies for the effective packaging of muscle foods. Part I provides a comprehensive overview of key issues concerning the safety and quality of packaged meat, poultry and seafood. Part II goes on to investigate developments in vacuum and modified atmosphere packaging. Other packaging methods are the focus of Part III, with the packaging of processed, frozen, ready-to-serve and retail-ready meat, seafood and poultry products all reviewed. Finally, Part IV explores emerging labelling and packaging techniques.

Environmentally compatible food packaging
(ISBN 978-1-84569-194-3)
Food packaging performs an essential function, but packaging materials can have a negative impact on the environment. This collection reviews bio-based, biodegradable and recycled materials and their current and potential applications for food protection and preservation. The first part of the book focuses on environmentally compatible food packaging materials. The second part discusses drivers for using alternative packaging materials, such as legislation and consumer preference, environmental assessment of food packaging and food packaging eco-design.

Details of these books and a complete list of titles from Woodhead Publishing can be obtained by:

- visiting our web site at www.woodheadpublishing.com
- contacting Customer Services (e-mail: sales@woodheadpublishing.com; fax: +44 (0) 1223 832819; tel.: +44 (0) 1223 499140 ext. 130; address: Woodhead Publishing Limited, 80 High Street, Sawston, Cambridge CB22 3HJ, UK)
- in North America, contacting our US office (e-mail: usmarketing@woodheadpublishing.com; tel.: (215) 928 9112; address: Woodhead Publishing, 1518 Walnut Street, Suite 1100, Philadelphia, PA 19102-3406, USA)

If you would like e-versions of our content, please visit our online platform: www.woodheadpublishingonline.com. Please recommend it to your librarian so that everyone in your institution can benefit from the wealth of content on the site.

Packaging technology
Fundamentals, materials and processes

Edited by

Anne Emblem and Henry Emblem

Recommended by the IOM^3 Training Academy
and endorsed by The Packaging Society

Oxford   Cambridge   Philadelphia   New Delhi

© Woodhead Publishing Limited, 2012

Published by Woodhead Publishing Limited,
80 High Street, Sawston, Cambridge CB22 3HJ, UK
www.woodheadpublishing.com
www.woodheadpublishingonline.com

Woodhead Publishing, 1518 Walnut Street, Suite 1100, Philadelphia, PA 19102-3406, USA

Woodhead Publishing India Private Limited, G-2, Vardaan House, 7/28 Ansari Road, Daryaganj, New Delhi – 110002, India
www.woodheadpublishingindia.com

First published 2012, Woodhead Publishing Limited
© Woodhead Publishing Limited, 2012; except Chapter 2 © Richard Inns, 2012. Note: the publisher has made every effort to ensure that permission for copyright material has been obtained by authors wishing to use such material. The authors and the publisher will be glad to hear from any copyright holder it has not been possible to contact.
The authors have asserted their moral rights.

This book contains information obtained from authentic and highly regarded sources. Reprinted material is quoted with permission, and sources are indicated. Reasonable efforts have been made to publish reliable data and information, but the authors and the publisher cannot assume responsibility for the validity of all materials. Neither the authors nor the publisher, nor anyone else associated with this publication, shall be liable for any loss, damage or liability directly or indirectly caused or alleged to be caused by this book.

Neither this book nor any part may be reproduced or transmitted in any form or by any means, electronic or mechanical, including photocopying, microfilming and recording, or by any information storage or retrieval system, without permission in writing from Woodhead Publishing Limited.

The consent of Woodhead Publishing Limited does not extend to copying for general distribution, for promotion, for creating new works, or for resale. Specific permission must be obtained in writing from Woodhead Publishing Limited for such copying.

Trademark notice: Product or corporate names may be trademarks or registered trademarks, and are used only for identification and explanation, without intent to infringe.

British Library Cataloguing in Publication Data
A catalogue record for this book is available from the British Library.

Library of Congress Control Number: 2012944516

ISBN 978-1-84569-665-8 (print)
ISBN 978-0-85709-570-1 (online)

The publisher's policy is to use permanent paper from mills that operate a sustainable forestry policy, and which has been manufactured from pulp which is processed using acid-free and elemental chlorine-free practices. Furthermore, the publisher ensures that the text paper and cover board used have met acceptable environmental accreditation standards.

Typeset by Replika Press Pvt Ltd, India
Printed by Lightning Source

Contents

Contributor contact details		*xiii*
Note about the editors		*xvii*
Preface		*xix*

Part I Packaging fundamentals   1

1 Packaging and society   3
A. EMBLEM, London College of Fashion, UK

1.1	Introduction: packaging from a historical perspective	3
1.2	Social developments: the changing patterns of consumption and their impact on packaging	4
1.3	Business developments: the effects of globalisation and modern retailing	5
1.4	The different levels of packaging: primary, secondary and tertiary	6
1.5	Packaging materials usage and development	7
1.6	The environmental perspective: responsible use of resources	9

2 The packaging supply chain   10
G. R. INNS, PEC Partnership Limited, UK

2.1	Introduction	10
2.2	Development, structure and inter-dependence of the segments of the global packaging supply chain	12
2.3	Packaging as a means of delivering cost effective solutions	14
2.4	Challenges of the supply chain	19
2.5	The importance of training	22
2.6	Sources of further information and advice	23
2.7	Bibliography	23

3 Packaging functions   24
A. EMBLEM, London College of Fashion, UK

3.1	Introduction	24
3.2	Containment	25
3.3	Protection	26

3.4	Preservation	41
3.5	Using packaging to provide convenience	46
3.6	Packaging as the source of product information	47
3.7	Packaging to sell the product	48
3.8	Conclusion	48
3.9	Sources of further information and advice	49
3.10	References	49

4 Packaging legislation — 50
G. CROMPTON, The Packaging Society, UK

4.1	Introduction	50
4.2	Legislation relevant to packaging	51
4.3	Legislation relating to product quality and safety during manufacture, distribution, storage and use	52
4.4	Legislation concerning honesty in trade	56
4.5	Legislation concerning protection of the environment	59
4.6	Legal considerations in international trading	60
4.7	Sources of further information	61
4.8	The role of trade associations	62
4.9	What is legally required and what is good practice?	62
4.10	Methods of enabling consistency of compliance	63
4.11	The consequences of failure to comply with legislation	63
4.12	Sources of further information and advice	64
4.13	Disclaimer	64

5 Packaging and environmental sustainability — 65
H. J. EMBLEM, Environmental Packaging Solutions, UK

5.1	Introduction: why bother?	65
5.2	Key definitions	66
5.3	Waste	67
5.4	Compliance with the law	68
5.5	Packaging re-use and recovery	75
5.6	Environmentally responsible packaging	79
5.7	Compliance with voluntary agreements	80
5.8	Climate disruption	82
5.9	Sources of further information and advice	85

6 Packaging and marketing — 87
N. FARMER, Neil Farmer Associates, UK

6.1	Introduction	87
6.2	Defining marketing	87
6.3	The role of marketing	89
6.4	Branding and the impact of packaging on product promotion and advertising	93

6.5	Branding and the marketing mix	95
6.6	Promoting the brand	97
6.7	The importance of consistency of communication	98
6.8	Use of market research tools and techniques to identify customer needs	100
6.9	Sources of further information and advice	105

Part II Packaging materials and components 107

7 Glass packaging 109
P. GRAYHURST, British Glass, UK

7.1	Introduction	109
7.2	Advantages and disadvantages of glass as a packaging material	110
7.3	Glass manufacture	112
7.4	Forming glass containers	113
7.5	Annealing	115
7.6	Surface coating	115
7.7	Inspection	115
7.8	Tolerances	116
7.9	Container design	117
7.10	Decoration and labelling of glass containers	120
7.11	Other glass-making processes	121
7.12	Sources of further information and advice	121

8 Rigid metal packaging 122
B. PAGE, Packaging Consultant, UK

8.1	Introduction to metal packaging	122
8.2	Raw materials	127
8.3	Manufacture of rigid metal containers	133
8.4	Metal closures	153
8.5	Cost/performance comparison: raw materials and forming processes	155
8.6	Container specifications	157
8.7	Decorating processes	159
8.8	Environmental overview	160
8.9	References and further reading	162

9 Aluminium foil packaging 163
J. KERRY, University College Cork, Ireland

9.1	Introduction	163
9.2	Aluminium processing	165
9.3	Refining	167
9.4	Smelting	167

9.5	Production of aluminium foil	168
9.6	Foil finishes, coatings and lacquers	170
9.7	Printing and embossing	172
9.8	Using aluminium foil as a laminate	172
9.9	Aluminium metallised films	174
9.10	Conclusion	176
9.11	Acknowledgements	176
9.12	Sources of further information and advice	177

10	**Paper and paperboard packaging**	178
	A. Riley, Arthur Riley Packaging Consultant International, UK	
10.1	Introduction	178
10.2	Properties of paper and paperboard	179
10.3	Raw materials	190
10.4	The pulping process	191
10.5	Post-pulping treatment of fibres to improve performance	195
10.6	The manufacture of paper and paperboard	200
10.7	Conversion processes for paper (<250 gsm)	208
10.8	Conversion processes for paperboard (>250 gsm)	222
10.9	Sources of further information and advice	239

11	**Corrugated board packaging**	240
	T. Watkins, UK	
11.1	Introduction	240
11.2	Materials for corrugated board	242
11.3	Manufacturing processes for corrugated board	245
11.4	Different types of corrugated board container design	249
11.5	Decoration and printing of corrugated board containers	253
11.6	Special board treatments	256
11.7	Testing corrugated board materials and containers	257
11.8	Sources of further information and advice	261

12	**Basics of polymer chemistry for packaging materials**	262
	A. Riley, Arthur Riley Packaging Consultant International, UK	
12.1	Introduction	262
12.2	The basic principles of polymerisation	264
12.3	Addition (chain growth or coordination) polymerisation of polymers	265
12.4	Condensation (step) polymerisation	269
12.5	Copolymerisation and crosslinking polymerisation	271
12.6	Factors affecting the characteristics of polymers	273
12.7	Sources of further information and advice	286

13	Plastics properties for packaging materials	287
	A. EMBLEM, London College of Fashion, UK	
13.1	Introduction	287
13.2	Market overview	287
13.3	Key properties for packaging applications	289
13.4	The common packaging plastics	292
13.5	Specialist polymers used in packaging	301
13.6	Bio-based polymers	306
13.7	Conclusion	308
13.8	Sources of further information and advice	308
14	Plastics manufacturing processes for packaging materials	310
	A. RILEY, Arthur Riley Packaging Consultant International, UK	
14.1	Introduction	310
14.2	The plasticating extruder	311
14.3	Sheet and film extrusion	314
14.4	Film treatments after forming	327
14.5	Thermoforming process for making plastic packaging	335
14.6	Injection moulding	339
14.7	Multi-injection moulding	345
14.8	Comparing injection moulding and thermoforming	348
14.9	Blow moulding	350
14.10	Environmental considerations in plastic packaging	359
14.11	Sources of further information and advice	360
15	Packaging closures	361
	A. EMBLEM, London College of Fashion, UK	
15.1	Introduction: the role of packaging closures	361
15.2	Types of packaging closure	362
15.3	Push-fit closures	362
15.4	Screw-threaded closures	365
15.5	Lug closures	371
15.6	Crimped crown cork closures	372
15.7	Peelable seal lids	373
15.8	Tamper evidence	373
15.9	Child-resistance	376
15.10	Dispensing and metering closures	377
15.11	Testing closure performance	379
15.12	Bibliography and sources of further information	380
16	Adhesives for packaging	381
	A. EMBLEM, London College of Fashion, UK and M. HARDWIDGE, MHA Marketing Communications, UK	
16.1	Introduction	381
16.2	Adhesives in packaging	381

16.3	Theories of adhesion	382
16.4	Adhesive types	384
16.5	Selecting the right adhesive	389
16.6	Adhesive application methods	390
16.7	Evaluating adhesive performance	391
16.8	Troubleshooting adhesive problems	391
16.9	Common adhesive terminology	392
16.10	Sources of further information and advice	393
17	**Labels for packaging**	**395**
	A. R. WHITE, AWA Consulting, UK	
17.1	Introduction	395
17.2	Trends in label types	396
17.3	Self-adhesive (pressure sensitive) labels	396
17.4	How a label manufacturer approaches a job	398
17.5	Wet glue (gummed labels)	400
17.6	In-mould labels	401
17.7	Sleeves	402
17.8	The choice of printing process	403
17.9	Label specifications	404
17.10	What can go wrong?	405
17.11	The label market	405
17.12	The digital revolution	406
17.13	Conclusion and future trends	406
17.14	Sources of further information and advice	406

Part III Packaging processes — 409

18	**Packaging design and development**	**411**
	B. STEWART, Sheffield Hallam University, UK	
18.1	Introduction	411
18.2	Research	415
18.3	Conceptual design	423
18.4	Case study: yoghurt for children	431
18.5	Conclusion	439
18.6	Sources of further information and advice	439
18.7	References	440
19	**Printing for packaging**	**441**
	R. MUMBY, Chesapeake Pharmaceutical and Healthcare Packaging, UK	
19.1	Introduction	441
19.2	Light and colour	442
19.3	The description of colour	442
19.4	Colour vision	443

19.5	Additive and subtractive colour mixing	444
19.6	Other factors affecting colour	445
19.7	Colour printing	447
19.8	Graphic design, reprographics and pre-press	455
19.9	Proofing options and approval processes	461
19.10	Technological aspects of printing processes	463
19.11	Other processing techniques	481
19.12	Quality control in packaging	484
19.13	References	488
20	**Packaging machinery and line operations**	490
	G. CROMPTON, The Packaging Society, UK	
20.1	Introduction	490
20.2	The packaging line	490
20.3	Unscramblers	492
20.4	Fillers and filling	493
20.5	Closing and sealing of containers	506
20.6	Labelling	509
20.7	Cartonning	513
20.8	Form, fill and seal (FFS) packaging operations	514
20.9	Direct product shrink-wrapping and stretch-wrapping	519
20.10	Modified atmosphere packaging	520
20.11	Miscellaneous wrappers	521
20.12	Coding systems	521
20.13	End-of-line equipment	522
20.14	Quality and efficiency aspects of packaging operations	526
20.15	Problem-solving on the packaging line	536
20.16	Sources of further information and advice	537
21	**Hazard and risk management in packaging**	538
	M. EWART, Authenta Consulting, UK	
21.1	Introduction	538
21.2	Packaging life-cycles in the supply chain	539
21.3	Prerequisite systems and controls	544
21.4	Hazard identification and risk assessment	547
21.5	Industry technical standards	554
21.6	References	559
	Index	*560*

Contributor contact details

(* = main contact)

Editors

Anne Emblem
School of Management and Science
London College of Fashion
UK

E-mail: packaging@btinternet.com

Henry Emblem
Environmental Packaging Solutions
UK

E-mail: enviroguru@btinternet.com

Chapters 1, 3, 13 and 15

Anne Emblem
School of Management and Science
London College of Fashion
UK

E-mail: packaging@btinternet.com

Chapter 2

G. Richard Inns
PEC Partnership Limited
PO Box 160
Ashtead KT21 9BN
UK

E-mail: richard.inns@pecpartnership.com

Chapters 4 and 20

Gordon Crompton
The Packaging Society
Education Department
143A Ings Road
Hull HU8 0LX
UK

E-mail: pack-it-ed@live.co.uk

Chapter 5

Henry Emblem
Environmental Packaging Solutions
UK

E-mail: enviroguro@btinternet.com

Chapter 6

Neil Farmer
Neil Farmer Associates
Westbridge House
19 Church Lane
Alveston
Stratford upon Avon CV37 7QJ
UK

E-mail: neilfarmer.48@btinternet.com

Chapter 7

Peter Grayhurst
British Glass
UK

Chapter 8

Bev Page
Packaging Consultant
121 Nottingham Road
Ravenshead
Nottingham NG15 9HJ
UK

E-mail: bevpageuk@aol.com

Chapter 9

Dr Joe Kerry
Food Packaging Group
School of Food and Nutritional
 Sciences
University College Cork
Cork City
Co. Cork
Ireland

E-mail: Joe.Kerry@ucc.ie

Chapters 10, 12 and 14

Arthur Riley
Arthur Riley Packaging Consultant
 International
Sabena House
6 Cottam Crescent
Marple Bridge
Stockport SK6 5BJ
UK

E-mail: riley.arthur@ntlworld.com

Chapter 11

Trevor Watkins
28a Kingsdown Park
Tankerton
Whitstable CT5 2DF
UK

E-mail: trevor.watkins@talktalk.net

Chapter 16

Anne Emblem
School of Management and Science
London College of Fashion
UK

E-mail: packaging@btinternet.com

Martin Hardwidge*
MHA Marketing Communications
8 Raglan Drive
Nottingham NG4 2RF
UK

E-mail: martinh@mhamarketing.co.uk

Chapter 17

Anthony R. White
AWA Consulting
Hillview Cottage
Upper Street
Oakley
Diss IP21 4AT
UK

E-mail: tony.white@awaconsulting.co.uk

Chapter 18

B. Stewart
Sheffield Hallam University
Furnival Building
Sheffield S1 2NU
UK

E-mail: b.r.stewart@shu.ac.uk

Chapter 19

Dr Richard Mumby
Chesapeake Pharmaceutical and
    Healthcare Packaging
Millennium Way West
Phoenix Centre
Nottingham NG8 6AW
UK

E-mail: richard.mumby@chesapeakecorp.com

Chapter 21

Matt Ewart
Authenta Consulting
Damson Rest
Church Lane
Codsall WV8 1EF
UK

E-mail: matt@authenta.co.uk

Note about the editors

Anne and Henry Emblem are packaging professionals of long standing, well-informed in making packaging decisions based on technical, commercial, environmental and legislative requirements. They have first-hand experience across the packaging industry, covering packaging manufacturers (Alcan, British Aluminium, National Adhesives, W R Grace, Morplan), packer/fillers (Merck, Sharp & Dohme Pharmaceuticals, Avon Cosmetics, British Sugar) and the high street retailer Marks and Spencer. They now operate a packaging consultancy, offering advice on all facets of packaging development, including the very important aspect of the effects of environmental legislation on manufacturers, users and sellers of packaging.

They are both Fellows of the former Institute of Packaging (now the Packaging Society part of IOM^3) and Chartered Scientists. Henry is also a Chartered Environmentalist, Chartered Waste Manager and a Fellow of the Chartered Management Institute. He works as a consultant in the field of packaging and the environment, specialising in Producer Responsibility (packaging, electrical and electronic equipment and batteries). He is one of the original authors of the BRC/IoP Global Food Packaging Hygiene Standard, a previous National Chairman of the Institute of Packaging and is a member of the Packaging Society Board. He is also actively involved in local environmental groups such as the Town Centre Initiative and Cambridgeshire Community Re-use and Recycling Network.

Anne is a Senior Lecturer in Cosmetic Science at the London College of Fashion, having moved into the academic world in 2004. Education and training have always been areas of strong interest, and she was chair of the Institute of Packaging Board of Examiners for many years. More recently, she has chaired the Packaging Industry Awarding Body (PIABC) and, when not engaged in Cosmetic Science, continues to develop and deliver in-house training courses in packaging, tailored to company needs. She is also an examiner for the Society of Cosmetic Scientists' (SCS) Diploma and a speaker at SCS events.

As well as editing *Fundamentals of Packaging Technology*, and this text, Anne and Henry have written *Packaging Prototypes: Closures* and have contributed chapters to several other texts.

Preface

This is the second major text we have edited, aimed at providing a learning resource for those studying the Diploma in Packaging Technology. The first, *Fundamentals of Packaging Technology* (FOPT to those in the know) was published in 1996 and was a 'translation' of Walter Soroka's excellent text of the same title, with additional sections to give a European focus. With the development of a new syllabus for the Diploma, clearly a new text is now required.

Finding subject specialists who are both willing and able to write chapters is never easy, but we feel we have a good combination of experiences in the text and are grateful to all the contributors for their hard work. In many cases the chapter contents go beyond the Diploma syllabus, and this is to be commended, as the book is intended to be a useful resource for experienced as well as new practitioners in the industry.

'Packaging' is a diverse and complex subject, as all of the contributors to this text will no doubt confirm. Just when you think you've got it all covered, along comes another aspect, another facet which should be considered; but at some point a line has to be drawn, otherwise the text would never be ready for publication. In any event, It would be impossible to include all areas of the subject in a single text and that was never the intention. As far as possible, references have been given at the end of each chapter to additional sources of information.

One important point for Diploma students is that, although this is a key resource, it must not be the only text you use: regurgitating its contents is not a guarantee of examination success. You must range both widely and deeply across the many resources available to you, and – again – the reference sections at the end of each chapter give you some guidance here.

Finally, we feel sure that all the authors have done their best to present accurate information, based on their own knowledge and experience. However, any decisions you make must be based on your own careful evaluation after extensive trials, and authors cannot be held responsible for your decisions and their consequences.

Anne and Henry Emblem

Part I
Packaging fundamentals

1
Packaging and society

A. EMBLEM, London College of Fashion, UK

Abstract: This chapter aims to provide a brief introduction to packaging and an overview of the industry and its development. It places packaging in its context in society and explores some of the major changes that have contributed to the growth in packaging usage. It does not aim to provide a complete view, but to lead the reader to the detailed chapters in the rest of the text.

Key words: history, ancient packaging, Industrial Revolution, brand, ready meals, microwave ovens, supermarket, primary packaging, secondary packaging, tertiary packaging, packaging usage, glass, steel, aluminium, paper, Tetra Pak, jute, cork, wood, pallets.

1.1 Introduction: packaging from a historical perspective

Packaging has been used in some form or other since the first humans began making use of tools. Animal skins and hollowed-out fruit husks were used to carry water, and grasses were woven into baskets and panniers to provide a useful way of keeping together and carrying goods. Probably one of the first examples of 'packaging' to preserve foods was the use of leaves to wrap meat when the tribe was on the move and the source of the next meal was unknown. As tribes became less nomadic and settled to farm the land, there was a need to store the produce. Clay pots met this need and archaeological evidence dating to 8000 BC shows large wide-mouthed jars being used for grains, salt, olives, oils, etc. The discovery that sand could be fused at high temperatures and made into bottles and jars increased the possibilities for storing and preserving liquids such as oils and perfumes. Both clay pots and glass containers were also used for their decorative qualities, as in the painted amphora given as prizes in the early Olympic Games from 700 BC.

As townships and cities developed and men and women became skilled in crafts beyond immediate needs, trade between cities, countries and continents developed, no doubt spurred on by the spirit of exploration which we still see today. Animals were harnessed to carry goods across the trade routes using an assortment of woven grass panniers, wooden barrels and casks and the same types of pack were used in the local markets. Thus the concept of using packaging as a convenient means of transporting goods, and to some extent in protecting and displaying them, was established, albeit that this was at the bulk level rather than with any apparent consideration of what the final consumer wanted.

1.2 Social developments: the changing patterns of consumption and their impact on packaging

1.2.1 The Industrial Revolution

A major influence in moving packaging from this bulk level to addressing the individual's needs was the Industrial Revolution, which began in England in the late seventeenth century. The shift from individual crafts at home or in small groups to mass production in factories brought large-scale migration of workers and their families to towns and cities. Foods and basic commodities previously produced and readily available at home, now had to be transported to shops in the cities to be bought by the workers using their hard-earned wages. This increased the demand for barrels, boxes and bags to bring in supplies on a larger scale than had previously been known, and it also brought a need to supply goods in the small quantities now demanded by the workers. These new 'consumers' lived in relatively cramped surroundings and did not have the large storage facilities previously available on the farms. Thus they needed to make frequent purchases and to carry their goods home, keeping them in acceptable condition as they did so. Goods were often measured out into the purchaser's own container, but gradually this changed to the shopkeeper pre-packing items such as medicines, cosmetics and tea, and having them available for sale in measured quantities, thus offering the buyer some assurance as to the quality and quantity of the goods. Eventually this pre-packing moved back a further stage from the buyer, to the situation we know today, where most goods are packed at the point of production rather than sale.

Whilst we have scant evidence to support this, it is reasonable to assume that there was a limited choice of goods available to the new consumer and little information about who had supplied the goods. If butter was wanted and butter was available in the shop, the consumer bought it, with no options from which to choose and no knowledge of its provenance. It was not until the nineteenth century that we started to see the rise of the 'brand name' used as a mark of quality by producers who wanted to make sure the buyer knew which product they were buying and were not misled by inferior goods. The word 'brand' comes from the identifying mark farmers burned into the hides of their cattle, as a stamp of ownership to deter others from stealing. The same burning process began to be used on wooden barrels and boxes as producers faced competition. Some of the oldest brand names are still with us today, for example, Schweppes (1792), Perrier (1863) and Quaker (1901).

1.2.2 Modern packaging

The move from packing goods at the point of sale to packing at the point of production brought about a shift from bulk to consumer packs, which had to survive the journey not just from shop to home, but, more importantly, from factory to shop, a journey which today may span countries and even continents and will include intermediate storage stages en route. It also gave producers the opportunity to develop their own style of packs to promote their own products, and this has brought us to the modern-day pack. Now, unlike our ancestors, we expect to have a range of goods from which to choose when we shop and the packaging plays a significant role in helping us to differentiate between the options available from various companies. We also expect our products to be free from damage, and in the case of foodstuffs, wholesome and safe and, again, the packaging makes a major contribution to meeting these expectations. Brand owners now expend their resources in developing packs which attract the attention of the would-be purchaser and at the same time provide the product with the protection needed. These different roles of packaging will be expanded and discussed in detail in the next chapter, although it is inevitable that reference will be made to them here.

1.2.3 Lifestyle changes and their impact on packaging

Since the middle of the twentieth century there have been significant changes in lifestyle in the developed countries and these have had, and continue to have, a major influence on how goods are packed. This applies particularly to food and drink, but also to all other fast-moving consumer goods. The following is not an exhaustive list, but presents just some of the relevant lifestyle changes.

- Reduction in the size of the family unit, due to decreased birth rates, increased number of one-parent families and increased longevity. There are now many more single- and two-person households than there were in the 1950s and 1960s and this means a requirement for smaller packs, thus more packaging per kilogram of food.
- Growth in the number of households in which all adults are in either full-or part-time work, outside of the home. This means less formal meals where everyone sits down together; meals are required at different times, and with minimum preparation. This brings a higher than ever consumer demand for convenience in terms of portion size and food which can be made ready-to-eat at short notice. Ready meals and the packaging formats in which they are presented make a key contribution to meeting this demand.
- Growth in ownership of domestic appliances such as the fridge and freezer has allowed consumers to buy larger quantities of 'fresh' foods, which are expected to remain in good condition for prolonged periods of time. The development of the low-cost domestic microwave oven brought with it a requirement for microwave-suitable packaging.
- More disposable income means more money to spend on food, especially luxury food and drink.
- More international travel and exposure to other cultures, leading to interest in 'ethnic' foods, but with minimal preparation time.

1.3 Business developments: the effects of globalisation and modern retailing

As well as the societal changes mentioned above, changes in the way businesses operate also strongly influence the need for and the types of packaging used. For example, the growth of the modern supermarket brings a highly competitive retail environment, with different versions or brands of the same product all displayed together. The time-pressed shopper relies on the subliminal cues of the packaging to make a selection, usually based on almost instant recognition of familiar features such as colour, graphics or shape.

Other ways in which business changes have influenced packaging include the following.

- The globalisation of manufacturing, with products being shipped over long distances and through different climatic conditions places strong emphasis on packaging to provide protection against likely hazards.
- Modern supermarkets demand fast stock replenishment with a minimum of manual effort. This has brought a requirement for secondary packs which can 'double up' as shelf-ready display packs, with no handling of the individual primary packs. One of the first companies to introduce this concept was Marks and Spencer in the 1970s. The 'straight-on tray', as it was then called, was initially used for food, and then extended to other goods such as toiletries.
- Modern supermarkets also demand rapid service at the checkout, and here the ubiquitous bar code provides a quick and reliable means of identifying the product and its price. Importantly, this data is also used for stock control purposes, often linked to automatic ordering to replenish supplies.

A further aspect to consider is concern about health and hygiene, on the part of the consumer and business. Sellers, producers and consumers alike want reassurance that a product, especially a food product, is fit for consumption and will not cause illness or injury. Packaging contributes to providing such reassurance, via fully sealed containers, often with some measure of tamper evidence. A related aspect is that packaging can be used to provide authenticity to a product, giving assurance that it is the genuine article and not a substandard alternative.

1.4 The different levels of packaging: primary, secondary and tertiary

Packaging exists at different levels and the following definitions are used throughout this text.

- *Primary* packaging includes not just the materials in direct contact with the product, but all of the packaging which surrounds the product when the consumer takes it home. For a multipack of crisps, for example, the primary packaging will be the individual bags and the large bag into which the separate packs are packed. A useful way to define primary packaging is to think of it as all the packaging which eventually finds its way into the domestic waste stream, once the product is used up.
- *Secondary* packaging is used to group packs together for ease of handling. In the example above of the crisps, several multipacks are packed into printed corrugated cases. The case is the secondary packaging. Other examples of secondary packaging

are shrinkwrap film, and the corrugated board and thermoformed plastic trays used for shelf-ready packaging.
- *Tertiary* packaging is used to collate secondary packs for ease of transport. One of the most common forms of tertiary packaging is the pallet, along with stretchwrap film and a label, to secure the secondary packs to the pallet and provide a ready means of identification. Roll cages and crates are also examples of tertiary packaging.

1.5 Packaging materials usage and development

Around 70% of packaging is used for food and drink, but other sectors such as healthcare, beauty products, chemicals, clothing, electrical and electronic equipment all need packaging to ensure they stay in an acceptable condition from manufacturer to consumer. Today's global packaging industry is valued at over $400 billion, roughly broken down into 36% paper and board, 34% plastics, 17% metals (steel and aluminium), and 10% glass, the remainder being made up of materials such as wood and textiles.

Of these major materials, glass is probably the oldest in its use as a packaging medium, dating back to its use for hollow vessels, in about 1500 BC. These were made by layering molten glass around a sand core and then removing the sand once the glass had cooled and solidified. Glass blowing started to develop around the first century BC and brought with it the ability to make glass containers of many different shapes and sizes, as the formability of molten glass was exploited. This property continues to be exploited in the modern, fully automatic glass-forming machines of today, demonstrated by a visit to the perfume counters of major stores where there is a vast range of intricately-shaped bottles on display. Glass remains an important packaging material and studies have shown that in the mind of today's consumer it is associated with features such as cleanliness, inertness and high clarity. Such properties engender its high-quality image and account for its continued use in market sectors such as beers, wines and spirits, perfumes and some pharmaceuticals.

Metal packaging is reported to date back to Napoleonic times and the development of metal containers for food has gone hand-in-hand with the introduction and development of food sterilisation systems such as those used in canning. Steel continues to be used for heat-sterilised cans for food and drink and for large containers such as drums and intermediate bulk containers (IBCs). Aluminium was first used for cans in the 1950s and today is widely used for drinks cans, especially for carbonated soft drinks. Both steel and aluminium are used for aerosol containers.

Papermaking is thought to date back to China in the second century AD and originally used woven strips of wet papyrus, laid down by hand and then dried. Papermaking machinery developments in the nineteenth century brought the ability to build up layers of cellulose fibre (initially obtained from rags and then from trees) into a continuous web, thus allowing a wide range of materials of different thicknesses and performance to be made. As the data shows, paper and board are the most widely used materials in today's packaging, and applications vary from labels to that workhorse of secondary packaging: the corrugated board case.

With regard to plastics, while the first materials date back to the nineteenth century, the major developments which have resulted in today's plastics packaging industry have taken place only since the 1940s. Plastics packaging has seen the most significant growth since then, of all the packaging materials, due to the development of low-cost processes and materials with a wide range of different properties, and, probably most significantly, the ability to tailor those properties to suit a range of different needs. For these reasons plastics have replaced the more traditional materials such as glass and metal in some applications, such as:

- the change from glass to polyethylene terephthalate (PET) containers for bottled water and soft drinks
- the change from glass to high-density polyethylene (HDPE) containers for milk
- the change from metal cans to flexible plastics pouches for pet food, soups and sauces.

Of course, this is not meant to imply that such changes are complete; some notable brands of bottled water retain glass for its high-quality image, and the vast majority of 'canned' foods remain in the traditional food can. What plastics have brought is the ability to offer alternatives and brand owners have seized these opportunities to differentiate their product offerings. Importantly, they have also brought reductions in pack weight, often with associated reductions in the total cost of the packed product and, not surprisingly, product manufacturers have welcomed such changes.

Another significant packaging development relying on plastics is the reel-formed 'carton' typified by the Tetra Pak and Combibloc containers. Although primarily constructed of paperboard, this type of pack relies on plastic layers for heat sealing (and thus forming leak-tight containers) and for barrier to moisture and gases (often along with aluminium foil). Despite its many detractors, this pack format is now ubiquitous in the fruit juice and milk sectors, providing a distinctive, lightweight, regular-shaped container. Most of the criticisms aimed at the pack have been related to its opening characteristics, leading to the development of a range of caps and opening devices.

Wood has been used for barrels for bulk products such as butter, and is still used for crates and boxes for fresh produce and for heavy engineering items such as machinery parts. However, the most significant use of wood in packaging is probably in the form of pallets, still the most common way of moving goods from manufacture to seller.

Other materials used in packaging include cork (wine bottles, albeit challenged by plastics) and textiles such as jute, used for sacks and bags. Jute sacks are used for agricultural products, due to their breathability, and for commodity food products such as sugar and rice. Jute bags are available as shopping bags and gift bags, often by companies wanting to project a 'green' image by using this natural fibre.

1.6 The environmental perspective: responsible use of resources

With regard to environmental impact, packaging has attracted criticism as a squanderer of valuable resources and an environmental pollutant. This criticism is rarely justified: far from wasting resources, packaging can and does deliver significant benefits in reducing product waste by containing, protecting against physical damage, and preserving against spoilage. Indeed, modern society could not function without packaging and the potential environmental damage of not using packaging is far greater than that caused by its use. It is estimated that supply chain losses are of the order of 40–60% in less developed countries, vs. around 3% in Western Europe.

Packaging reduction projects have long been important in modern manufacturing companies; no commercially sound organisation deliberately uses more packaging than is necessary for the safe delivery of the product and its acceptance by the market. To do so is not only environmentally irresponsible; it is economically disastrous and bodes ill for the financial health of the organisation. Legal compliance is a further issue, with widespread legislation concerning packaging use and disposal.

It is irresponsible to ignore the environmental impact of *any* decisions today, not just packaging decisions. Responsible manufacturers now consider how best to reduce the total amount of packaging used, whether it can be reused or recycled and how best to recover used packaging from the waste stream. There are no easy answers and no one material which can be held up as indisputably more 'environmentally friendly' than any other.

2
The packaging supply chain

G. R. INNS, PEC Partnership Limited, UK

Abstract: This chapter examines the supply chain as a total system for the delivery of goods from the packing line through to the retail environment and the consumer. The conventional approach to envisaging the supply chain is integrated with considerations of sustainability. The central theme is the use of packaging to prevent damage and the benefits to be derived from bringing all stakeholders together in the supply chain to deliver optimum results.

Key words: testing, distribution, sustainability, shrinkage, damage, design, specification, manufacturing, packing, supply chain walk, order picking, pallet, primary, secondary, tertiary, roll cages, shelf ready, retail ready.

2.1 Introduction

This chapter describes not just the supply chain for packaged goods but also shows how our conception of that supply chain has changed over the years and will continue to change into the future. Exploring this change will bring in key elements of sustainability that are becoming interwoven with our understanding of the supply chain.

Within this chapter we will be referring to a 'package', so we should first consider what is meant by that term in the context of the supply chain. The focus of the supply chain is the product; the role of the 'package' is to get that product safely through the supply chain and on to the customer's home. Given that context, it is then self-evident that within the supply chain by 'package' we must be referring not just to the primary package but all of the elements, primary, secondary and tertiary, the 'packaging system', that come together at different stages to assure the safe delivery of the product to the consumer's home.

Each element of that system of combined packages plays both distinct and synergistic roles within the supply chain, each element being of key importance at a particular phase in the supply chain. The primary pack achieves key importance at two points in the chain, firstly during the filling process and secondly at the point of purchase and use by the consumer. During the intervening processes it is playing a synergistic role both in keeping the product intact and in good condition and very often contributing to the combined strength of the packaging system.

The fundamental role of the secondary pack is to provide collation and to enable shelf loading. As order picking from pallets into roll cages has become the norm, so the secondary pack has come to take over many of the roles previously ascribed to the tertiary pack. Indeed as the conditions within a roll cage can be significantly more stressful than in a pallet and so much of the damage that occurs is generated during this phase, it can be argued that the secondary is now more important for protecting the product than the tertiary pack.

Another change, to shelf-ready or retail-ready packaging, has significantly changed the role of the secondary pack. It is now expected to open easily, facilitate easy shelf loading and provide a good shelf fill. Designing a pack that will contain and protect a number of primary packs during stressful distribution, but then provide what is sometimes called 'one touch' opening, is far from a trivial task.

Traditionally tertiary packaging was designed to get the total package system and contained product through the supply chain with minimal damage. The extent that it fulfils that role today varies widely. Clearly for long distance journeys overseas or through long road or rail journeys that fundamental role is little changed. However, modern modes of transport including containerisation and air suspension on lorries have greatly reduced those stresses.

In many instances also goods are only sent relatively short distances as pallet loads before being broken up in order-picking centres. Despite this, many tertiary packs are still designed as if long distance palletised travel direct to the customer were still the norm. It is well worth investigating to establish the reality before setting out to specify design.

Returning to the theme of synergy, designers have got ever smarter at getting these individual units to work together so that we can be focused on the performance of the whole packaging system and not just its parts. Put another way, if a unit of a packaging system is not working synergistically with the other units of packaging, it is time to carry out a design review.

Our starting point in any development must be the relationship between a package and its contents and those contents must be the focus of any activity. We have over time moved from the idea of the package as being complementary to the product to being an integral part, in effect another component of that product.

The whole development of first environmental and then sustainability thinking on packaging started with the issue of packaging as litter. This view of packaging as an isolated item needing to be disposed of properly led to the development of the cradle-to-grave approach to packaging; realisation that recovered packaging waste was potentially a valuable resource led to the cradle-to-cradle approach and that thinking renders the product and packaging supply chains as one. Sustainability thinking and the imperative within it to prevent and eliminate waste then gave prominence to the role of packaging in preventing that waste and hence finally to the concept of the combined supply chain, where product and packaging are treated as being one.

This approach to the supply chain still concentrates on what we now know as the environmental pillar of sustainability and from there it is a natural step to examine how the other two pillars relate to the supply chain. The economic pillar has, most will agree, been with us longer even than the environmental pillar, cost and material optimisation being but two sides of the same coin. The social pillar is much newer in terms of recognition as such, but many of its component elements have been recognised for a long time. Understanding and control of the supply chain gives us understanding and control of each of these pillars. Addressing each of these pillars raises its own set of challenges not least because today it is hard to conceive of a purely national supply chain. Materials markets and sources are global, product purchasing by retailers is now primarily organised on a regional or global basis and the location of packaging manufacture is in turn closely linked to that of the product source.

One consequence of this geographically dispersed supply chain is that decisions about what elements to include within the supply chain boundaries become the subject of challenging decisions of balance. Good practice says that the specifier should at least have an overview of where the significant choices within their supply chain are made and seek to influence and change those that lie at the heart of issues and problems. Inevitably the practical problems of monitoring and controlling increase rapidly with distance. On top of this, it is now the norm for responsibility for individual segments of the supply chain to be subcontracted and the specifier may have no direct link to or authority over those subcontractors. These important issues of distance, system boundaries and exercising of adequate controls bring us on to the next part of this chapter.

2.2 Development, structure and inter-dependence of the segments of the global packaging supply chain

Although we speak of the packaging supply chain as if it were a single entity, in reality it is a series of sequential processes, the managers of which (the stakeholders) all view the package in a different light. Therefore before anything can be achieved within the supply chain, any new development or optimisation attempted there needs to be a process of stakeholder engagement.

How easy that will be depends on how effectively the supply chain has been mapped. If the supply chain has been achieved through a long chain of sub-contractual obligations, it can be difficult to establish a complete overview. That difficulty will increase the further back down the chain one moves and, as packaging sourcing by definition lies close to the start of the chain, it is also likely to be hardest to view.

In many instances that remoteness of supply can be extremely off-putting. Language barriers, sheer physical distance and the cost of establishing these links can be major obstacles. In practice, many do find these barriers insuperable and, as long as essential legal, health and safety requirements are met, decide to live with the situation. There is a clear downside to this in missed opportunities; the choice is for each individual to make.

Once the supply chain has been mapped then at each stage there will be a key stakeholder who needs to be identified. Supply chains will vary to some degree from product to product, so some stakeholders may be different according to the exact product concerned. That difference in stakeholders will be most evident between those that are manufactured in-house and those manufactured by third parties.

One valuable approach to stakeholder identification is the 'supply chain walk', a concept developed by ECR Europe[1] and used in addressing issues such as 'shrinkage' (product loss through theft, damage or process failure in the supply chain). This supply chain walk was conceived as a physical process whereby a project team actually visits each part of the supply chain to learn in some detail about its functions and what its capabilities and limitations are. This same principle can be applied, at the very least as a mental discipline, to stakeholder identification.

Whilst no two companies are alike, and responsibilities can be quite different even where individuals share similar job titles, a typical stakeholder list might look as follows:

- *Marketing* – will have clear views on the product and how it is presented, merchandised and used. They should also have a clear view of sustainability issues, real and perceived, surrounding the product and how customers and consumers are prioritising those issues.
- *Product/packaging design* – by working together can ensure that the overall sustainability of the product is maximised.
- *Purchasing* – plays a major role in identifying suitable suppliers and identifying the economic parameters within which it is acceptable to operate.
- *Packaging manufacturer* – where involved as a partner will act as a valuable source of advice and provide assurance of sustainability in their own processes and back down their supply chain.
- *Raw material suppliers* – by no means all companies will have the resources to work with even the major suppliers of raw materials used in their packaging. However, a clearly written policy setting out acceptable and unacceptable practices in respect of raw materials and their sources that has been agreed formally with suppliers will make a major contribution to establishing sustainable practices.
- *Manufacturing* – packing processes, manual or automatic, impose significant stresses on packaging that limit the scope to reduce resource usage in packaging. An in-depth understanding of how packaging and filling machinery interact and the key relevant performance parameters is essential for a successful outcome of any project. Joint work to identify and moderate those requirements and stresses can do a great deal to reduce resource usage.
- *Packaging technical development* – will provide the skills required together with the supplier to create and test a sustainably optimised specification.
- *Storage and distribution logistics* – Pira International[2] in 2003 calculated that loss through damage in the European fast-moving consumer goods supply chain cost industry €3.5bn each year. In addition, up to 3% of food is lost through spoilage before it reaches the consumer, whilst for fruit and vegetables distributed loose that loss can reach 25% according to retail industry research. Taking steps, such as 'supply chain walks' to identify points where damage occurs is a key first step towards the elimination of such damage.
- *Merchandising* – the *Global Retail Theft Barometer* report for 2010[3] shows that on average theft is equivalent to 1.36% of retail sales. In Europe that costs industry $41 billion. Prevention of theft and other losses in the retail environment can make a significant contribution to sustainability.
- *Consumers* – although consumers are not involved directly as stakeholders, their interests must be recognised. Between 3 and 6% of purchased food is discarded and food waste comprises 10–20% of food consumed[4]. According to WRAP, in the UK, food that was not but could have been consumed totalled 5.3 million

14   Packaging technology

tons. Packaging cannot tackle all of this but much can be achieved through shelf life extension, portion control and the use of packaging in the home to slow deterioration of food.

2.3 Packaging as a means of delivering cost effective solutions

As already set out above, waste remains a significant issue in the supply chain. Cost has always had a profound influence on our ability to choose and specify packaging. Today financial impacts are recognised as one of the three aspects of sustainability alongside environmental and social impacts, each of which can in fact be viewed as a form of cost in its own right. Some might say that waste is a fundamental measure for all three. Waste after all involves loss of resources, unnecessary costs and avoidable environmental impacts. The social costs of waste are harder to calculate, but in a global society where many lack access to key goods, wasting products represents at the very least an opportunity cost to those peoples.

It has become common to view the environmental impact of packaging in the light of the weight of material used. In reality, as long as the packaging is being used in the best way to protect goods, that view is erroneous. None the less it persists.

Packaging weight reduction, much more constructively viewed as packaging optimisation, had long been a feature of the packaging development process long before it became fashionable to adopt it as an 'environmental' target. If we view the process as one of optimisation rather than straight reduction, then it should become clear that consistent product damage is a clear indicator that the process has been pushed too far.

Unfortunately no research exists to show how much cost would be involved in eliminating the €3.5 billion of damage mentioned above but we can get some indication of how we might benefit from a consideration of the analogous balance of resource usage and losses. From research we know that overall the packaging used to protect a product consumes only one tenth of the resource consumed in making that product.[5] Given that there is a clear relationship between resource usage and cost, that must surely imply that spending the right amount on packaging could produce a disproportionate benefit in overall cost. Whatever that balance might prove to be, it should be clear that it can never be a correct approach simply to reduce packaging as a standalone exercise without being assured that there will be no offsetting or oversetting losses elsewhere.

Concerted efforts to reduce packaging really started in the 1970s and have continued steadily ever since with packaging overall having more than halved in weight in that time.[6] Companies that have been applying packaging optimisation over that period, and that includes all major packaging users, are reaching the limit of the 'salami slicing' process that has traditionally been applied and now look to significant innovations to reduce their packaging usage. Whilst in the past a company following this approach is likely to have achieved overall around a 5% per year packaging reduction, that is now becoming increasingly difficult to achieve. One sign of that slow down is that the Courtauld 2 Commitment in the UK had a target to reduce packaging by 5%

over three years from 2009 to 2012 (See www.wrap.org.uk), a reduction that in the past might have been targeted in a single year.

Whether or not potential still exists to make significant packaging reductions can depend heavily on the source of the packaging. One example of a very significant opportunity for reducing packaging in this way was found in respect of the secondary packaging for goods imported from the developing world. Due most often to a wish to avoid damage in transit, a significant proportion of goods susceptible to breakage from such countries can be overpackaged. According to informed sources interviewed during a study in 2010, this overpackaging is up to 100%.

Identifying where other opportunities arise is a less simple process. Distribution systems report by default and hence do not have mechanisms to detect or report where a package may be overspecified. One way that professional packaging teams will guard against such eventualities is by reviewing packaging every two years against a packaging utilisation key performance indicator (KPI). As a consequence, overspecifying is unlikely in companies with such teams; on the other hand, there is a much greater possibility of overspecifying by companies that do not apply such regular checks. Packaging technology moves forward steadily so even if looking for large improvements per year may be too challenging, given the history of packaging optimisation the chance of finding improvements in packaging over a slightly longer period is still high.

Specifying performance for primary or retail packaging brings into consideration additional factors such as shelf appearance, food contact safety and consumer appeal. The choice of packaging is still influenced by price and the best balance between performance and an acceptable end cost.

There is a trend for packer fillers and retailers to augment in-house packaging teams by working collaboratively with their suppliers. However, such companies still need to have the expertise to review and challenge supplier specifications and brief those suppliers on performance requirements. Otherwise there is a tendency to employ conventional weight-based specifications because this makes the process of price comparison easier. A company that applies this weight-based approach is not in the best position to optimise performance and thereby is likely to overspecify. This latter point is why increasing numbers of people regard performance specifying as the way forward in resource optimisation.

There can be hesitancy on the part of a supplier to suggest a change of a material grade based on performance specifying to a customer, even if their expertise indicates that an alternative grade material could provide the required performance. This is because, if the packaging fails, they may run the risk of being deemed to be responsible and be charged for any product losses incurred by the customer. What can happen then is that the supplier may recommend that a customer trial an alternative material but leaving it to the customer to make the final decision.

Smaller packaging suppliers may also experience difficulties in working with performance specifications. The resource and expertise levels required to work with such specifications are not inconsiderable and unless their customer base strongly demands such an approach their supplier may not regard the investment as justified. Even for large packaging suppliers providing the necessary level of resources to support performance specifying can be an issue. Only if a customer is large enough to place significant orders may the supplier feel justified in devoting the amount of resources needed to create a performance specification.

Packaging should be assessed primarily in its role to reduce waste either through product protection or through process optimisation. There is still a significant amount of waste in the supply chain, referred to as 'shrinkage'. Chapman and Templar[7] define shrinkage as: intended sales income that was not and cannot be realised, retailers include within this 'malicious shrinkage' (external theft, internal theft, inter-company fraud) and 'non-malicious shrinkage' (spoilage, damage, data errors, pricing errors, delivery errors and scanning errors). Around 1.4% of sales disappear in this way[1] costing retailers around one-third of their profit margin. About 0.8% of goods are known to arrive damaged in the store. Only through performance specification and process optimisation can such losses be prevented.

Process optimisation involves re-examining the role of the packaging and specifying it according to its actual function. A good example can be seen in a 2009 retailer initiative to ship wine from the southern hemisphere in bulk and only pack it once it reaches the UK. During long distance shipping the container is unnecessary so its performance is zero or indeed negative as shipping it occupies space and consumes energy. The actual performance of the bottle is in retailing and consumer use and ageing in the bottle where relevant and so bottling in the country of consumption makes great sense whilst the package only plays a role where it is essential.

Given that packaging reduction on its own is not an effective means of cutting overall cost and even liable to be counterproductive because of damage increase, how do we set about delivering cost effective packaging? Clearly, as we have seen, the first step is to characterise the supply chain in all of its elements and then to decide what the governing KPIs are going to be. The next stage must be to look at the requirements of the product.

In an ideal world the supply chain will have been designed with the product in mind. However, we have to accept that the supply chain as a whole will have represented a major investment that has to be amortised over a significant period whilst products and packaging technology tend to evolve quite rapidly. In practice, most often it is the product and its packaging that have to do the adapting to the essentially fixed supply chain requirements.

For the purposes of this chapter we will assume that the primary package has been so designed as to provide the immediate protection of the contents in terms of barrier, light, etc., and focus primarily on physical damage protection and supply chain interactions. That is not to say, however, that the influence of product shelf life on supply chain efficiency is not worthy of serious consideration.

One interesting illustration of the type of challenge from product and packaging evolution faced by the supply chain comes from the growth in the sales of short shelf life, primarily chilled, products. Firstly, the product has to arrive in the store with sufficient remaining shelf life to allow for consumer selection and the consumer's own storage requirements in the home. Secondly, products near to the end of their saleable period are frequently discounted and may need to be destroyed. Such losses can approach 20%.

Given these changes, how can process and packaging optimisation be applied to reduce those losses? There are two potential process optimisation roles for packaging to play. First of all, barrier technology continues to improve and, perhaps more importantly, so-called 'active' packaging (first mooted in the 1930s) is becoming a reality. So we can see significant potential there to increase shelf life and give more time during which the product can be purchased and consumed. The second theme to this optimisation involves improving the ability to actively determine the real remaining shelf life of a product through the application of shelf life indicators. Currently shelf life determinations must necessarily be made on a conservative basis, allowing for the possibility that conditions in transport may be less than ideal. Monitoring the packaging in transit allows a much more precise determination to be made of available shelf life once the goods arrive in store.

Some monitoring devices travel with the load, recording conditions in the vehicle and individual pallets or tertiary packs. They are more often used analytically to determine weak points to be addressed in the distribution system, although they can be used to inform a receiving store what the in-store shelf life should be set to. Monitors specifically designed to be incorporated into packaging tend to take the form of labels that react either to time/temperature effects or to a buildup of substances in the package known to relate to spoilage. This approach provides staff within a store with specific information regarding the safe residual time for the product to be displayed, cutting down on unnecessary disposals.

Both of these approaches to shelf life monitoring add to packaging and distribution costs whilst enhanced barrier protection contributes to packaging costs. Clearly, therefore, those costs need to be justified. Such a justification can only really be made based on a total supply chain cost approach, illustrating once again the importance of involving all stakeholders in decisions regarding supply chain enhancement through packaging.

Other critical factors that relate to process optimisation within the supply chain are related to the physical strength and stress-resistant factors of the packaging and its constituent materials. Here we should still continue with the theme of understanding and respecting stakeholder priorities within each section of the supply chain and through that supporting the best overall performance for the whole.

We still cannot escape from the overall impact of cost controls on packaging decisions. The first consideration and the last will be the combined cost of the units of the package system within the product as a whole. Establishing an optimum overall cost for a packaging system may sound straightforward; often it is not. A primary reason for that difficulty lies with the buying process, where very often each unit of the packaging system for a product is bought separately and each is assessed separately in respect of its contribution to the overall cost, and apparent savings that may be available for each individual unit of the packaging system.

In the early days when a product is being launched such costing inconsistencies should be easy to detect within an overall budget. At this stage there should be clear balanced KPIs for the total system including costs that simplify the establishment of an optimised overall cost. As time passes, however, there is a risk that each item within a packaging system will be cost optimised individually without necessarily taking full cognisance of the cost impacts on other parts of that system. Here the same point needs to be made as was made in respect of optimising physical performance of packaging. If you do not consider the whole when changing just a part, then the risk of losing as much or more than you are gaining must exist.

We have talked about KPIs being determined for each stage of the supply chain and there are two ways that can be done; one is through factory trial and the other through technical investigation. In essence whilst a factory trial accompanied by careful observation and recording can achieve a great deal, in particular regarding where and how failure occurs, a technical investigation will reveal much more about the underlying reasons for failure and how they may be addressed. The next section will go into greater depth on the types of test that can be applied and the results that can be expected.

A technical investigation is unquestionably more resource and time consuming but it does give a much better basis for decision making and the eventual establishment of performance-based specification. On the other hand, a factory trial tends only to point to materials that worked and hence to lead to the type of specification that primarily describes a structure and combination without giving clues as to why. A performance specification based on a technical investigation reveals much more about what aspects of a package are critical and why, hence allowing much more freedom to look creatively at generating solutions for the future.

Another benefit that the analytical approach combined with performance specifying brings to packaging is that it tends to expose and thus help to eliminate preconceived notions about which elements of a package's design and specification are critical. Such preconceptions about the reasons for packaging failure and solutions are quite common in the absence of a technical analysis, but in many instances they will be found to have grown over time and in reality to be erroneous.

In addition, by enabling real underlying causes to be established, a technical analysis approach will encourage questioning about whether a packaging change is the right solution or whether a machine change would be more effective. In the absence of such technical understanding, it is normal for the inherent cause to be assumed to lie with packaging. Determining the solution from that point may well require going back to first principles to find whether the machine requirement can be met realistically and cost effectively through packaging and if not whether instead changes to machine components or settings would produce a more effective result.

Choosing the right form of technical analysis to represent what is actually happening in a process is a challenging task that requires a significant degree of expertise and experience. That may be one reason why so many specifications avoid defining performance parameters for packaging and instead focus on simple physical measurements of parameters such as the grammage or caliper of materials that factory trials have shown to work. Whilst it may be more difficult and time consuming there is no substitute for observation first hand of the process to find the critical points and then finding a test or tests that will allow accurate prediction of whether a material and construction will be optimum.

Simulated transit testing is one example of such an underutilised form of test that is an accurate indicator of performance and hence comparator of potential solutions. Standard transit tests where loads are sent on a journey as part of a load of standard products are a classic example of the act of observation changing the outcome. Such simple factors as a pallet being marked for observation or returned to base for analysis in an otherwise empty vehicle will change the result. Cost is also a consideration here; standard transit tests over intercontinental distances are prohibitively expensive and often omitted for that very reason. Simulated testing that does not involve lengthy and costly shipping can be much more cost effective and, with the use of instrumentation, much more repeatable.

Transit test simulators allow the load to be subjected to the same critical stresses they will experience during their journey but in the laboratory. Some models even allow stresses recorded during an actual journey to be played back into the test load. Thus the tests are reproducible when applied to alternative packaging systems and, where failures occur, the reasons can be observed and used to give precise feedback to designers and specifiers.

This kind of performance-related testing and specifying has the potential to reduce cost in the longer term, even when it is more expensive in the short term. Down time in the factory is expensive, eliminating the problem through trial and error is time consuming and there is no guarantee that any changes made will not have knock-on impacts elsewhere. The underlying causes of product damage will remain unknown and require a fresh solution each time without the benefit of fundamental knowledge. This difficulty of resolution will tend to lead to overspecification as a means of avoiding further problems. Purchasing options also remain constrained through being based on material rather than performance parameters.

2.4 Challenges of the supply chain

We have already seen above how significant shrinkage is to the supply chain. Were shrinkage to be eliminated, then retailer margins would be increased by 62%. Packaging can play a significant role in preventing this shrinkage. Just how has been explored in great depth by ECR Europe and the reader is referred to their report *Package Design for Shrinkage Protection*.[8]

Poor basic packaging design significantly increases the risk of shrinkage. Badly designed packs are more likely to:

- allow their contents to become damaged in transit
- cause contamination through leakage
- experience product spoilage or damage
- contribute to product loss through misidentification
- lose their contents
- become too unsightly to sell
- suffer decreased shelf life and consequent reduced sales.

The primary drivers of shrinkage other than through theft are:

- *Spoilage*: products that have reached their expiry date or gone beyond agreed temperature parameters and are no longer safe to sell to consumers or staff.

- *Damages*: refers to products that have been damaged during the journey to the shelf and in the general store environment.
- *Data errors:* errors in the recording of product details on company systems.
- *Pricing errors*: losses caused by errors in the way in which goods are priced and sold in the business.
- *Delivery errors*: losses caused by incorrect quantities booked to the store inventory but not physically delivered or transferred.
- *Scanning Errors*: errors occurring at the point of sale leading to a discrepancy in the store book stock.

ECR Europe also identified these packaging related factors in being critical in the prevention of shrinkage:

- placement and readability of bar codes
- robustness of inner and outer packaging
- marking and design of break packs
- clear identification of at-risk items
- packs that are difficult for a would-be thief to hide
- packs that cannot be covertly opened, tamper-evident packs
- packaging that is protectable by electronic article surveillance (EAS)
- barcode usage rules that minimise reading error
- product shipment, distribution and assortment controls
- use of testers and samplers
- fixture and secondary display design
- store layout, design and lighting
- returns policy.

There are also a number of solutions in respect of supply chain processes that in some degree relate to packaging:

- robust documentation systems for orders and deliveries
- secure storage, dispatch, transit, delivery, receipt and in-store storage
- shelf replenishment and range change processes
- procedures for damaged and returned merchandise.

There is a tendency to picture the supply chain between factory and customer in terms of pallet loads of goods travelling together between warehouses by road, rail, air or sea. There is a great deal of knowledge and expertise built up to support that concept regarding matters such as optimised pallet fill and package top-load strength. On that basis one can expect optimised and damage-free pallet loads to be the norm. In fact, we do indeed find that most packaging-related issues in the supply chain now occur after the palletised phase.

We have already addressed the importance of understanding problems through technical analysis prior to specification. If that is done properly, then damage should be minimised, but what do we mean by 'minimised'? Originally it would be said that damage has been minimised when the following conditions have been met:

- All legal requirements relating to the package and its contents have been satisfied.

- There are no health or safety issues emerging from the damage.
- The damage is not causing significant knock-on effects in other parts of the business.
- Performance issues in the supply chain related to packaging fall within agreed norms.
- Consumer complaints related to failures caused through packaging fall below the threshold acceptable level set by the producer.
- Levels of packaging-related loss at the commercial customer fall below the threshold level established with that customer.
- The cost of rectifying the problem is such as to make it economic in terms of the product value.

In addition to the above, we need today to consider carefully the impact of product loss on the sustainability of a business. We have already seen that on average the environmental impact of packaging is only one-tenth of that of the product it protects. So there is a question that needs always to be addressed whether the resource implications of upgrading the packaging are justified by the reduction in resource loss in the product. In practice the answer will very often be yes and the challenge then becomes, as we have seen before, justifying the cost of that upgrade.

As always it is the social aspect of loss and damage that is hardest to address. In a world where food shortages are a fact of life and many people are deprived of necessities, can it ever be right to allow damage and wastage to occur? On the other hand, should we balance that by considering that had that wastage been prevented, it is unlikely that the goods saved would somehow have reached the needy? Can the fact that more goods were produced potentially generated additional work for those in society needing it be regarded as a social benefit? There are no easy answers to questions of this type, but it is no less important to keep on addressing them.

If despite all precautions damage higher than the acceptable levels set out above is occurring, how should it be addressed? First of all the likely source needs to be established, ideally by following the supply chain walk approach set out above for packaging development and optimisation.

At the start of the chain production and palletisation line, damage such as crushing, impact, poor closure or scuffing tends to be characteristic and the source should be straightforward to identify. Beyond the production line and palletisation, the next likely cause is from truck handling of pallets. Trucks can impact loads, pierce packaging and even drop or topple pallets. Such incidents are, or should be, sporadic and a combination of often tell-tale damage characteristics and warehouse logs should eventually run these down. Of course, the further along the supply chain the incident occurred, the harder the task of identifying the cause will be.

Damage in modern lorries or containerised sea or air transport is relatively rare and most likely will have been caused by some one-off incident. Deliveries from more remote locations may well have travelled by less modern transport and over poorer roads, so the damage may be more systematic. Often this problem of poor local transport conditions will be offset by overpackaging. Whether or not, despite this overpackaging still failing, the receiver of the goods is willing and able to undertake an investigation in a far-off location is up to the individual. Rightly or wrongly often accepting this overspecification or damage may be perceived as the lesser evil.

Today a great deal of the product damage that is found occurs after the palletised section of the supply chain where individual units, primary, secondary or tertiary, travel mixed together with other products in some form of cage or trolley. In many instances loading of these cages is random and it is easy for items to be loaded at an angle, inverted or with heavy items superimposed on fragile ones. If loading of products is found to be the source of the problem, then there needs to be consideration of making the product more cage friendly in terms of its package dimensions, appearance (to make its fragile nature more evident) and protection. It is critical once an incident has happened to visit the order-picking operation when in full operation and seek out points where damage can happen. Such visits, if they have not been carried out previously, are invaluable for picking up information to support design and specification.

2.5 The importance of training

Nobody seeking to work effectively on supply chain topics can afford to be so much of a specialist that they cannot get a perspective on the priorities and goals of others. Arguably the most effective form of training for this is 'on the job'. Just as the supply chain walk is the most effective means of understanding the supply chain, so it is the most effective way of understanding the perspectives of the people who manage it.

Clearly having identified and formed an understanding with key stakeholders, the next stage is to weld them into teams to tackle specific problems. From what has been said above it should be apparent that persuading individuals with quite different and even conflicting goals to work to a common goal is a significant skill. Training in team building is important for anybody who is involved in forming a project team, even if they are not leading it.

Knowledge of specifying is of course fundamental to any packaging manager but performance specification may be less familiar. At the same time defining performance KPIs at each point of the supply chain is clearly logical, for those performance KPIs should be captured in the form of performance specification. Anybody unfamiliar with performance specifying will clearly gain an advantage from studying the technique.

Performance specifying needs to be backed up by testing. Some tests will be familiar, others much less so. There is a skill involved in selecting the correct test that most accurately reflects the real-life situation under study and it is well worth while having training to gain that skill. Overall it is probably true to say that instrumented transit package testing is the least well understood and least widely applied area of testing technology, despite its significant potential to cut cost and wastage. From a technology point of view, this is potentially one of the most fruitful areas to gain knowledge through training.

2.6 Sources of further information and advice

- *Packaging for sustainability: Packaging in the context of the product, supply chain and consumer needs*[9]
- *Effective Packaging Effective Prevention*[10]
- *Packaging Reduction – Doing More with Less*[11]
- *Shrinkage – a collaborative approach to reducing stock loss in the supply chain*[1]
- *Environmental Impact of Products*[12]
- *Shelf Ready Packaging (Retail Ready Packaging) Addressing the challenge*[13]

2.7 Bibliography

1. ECR Europe. *Shrinkage – a collaborative approach to reducing stock loss in the supply chain*. Brussels: ECR Europe, 2003.
2. Pira International. *PIRA (2003) Realising Profit Improvement through Reducing Damage in the Fast Moving Consumer Goods Supply Chain*. Leatherhead: Pira International, 2003.
3. *Global Retail Theft Barometer*. Newark: Centre for Retail Research, 2010.
4. Enviros. *Review of the Environmental Health Effects of Waste Management: Municipal Solid Wastes and Similar Wastes*. London: DEFRA, 2004.
5. Kooijman, J. *Environmental Impact of Packaging in the UK Supply Chain*. Reading: INCPEN, 1996. Available at: www.incpen.org/pages/
6. Advisory Committee on Packaging (UK). *Packaging in Perspective*. London: Advisory Committee on Packaging, 2008. Available at: www.incpen.org/pages/data/PackaginginPerspective.pdf
7. Chapman, P. and Templar, S. 'Scoping the contextual issues that influence shrinkage measurement'. *International Journal of Retail & Distribution Management*, 2006, Vol. 34, No. 11, pp. 860–72.
8. ECR Europe. *Package Design for Shrinkage Prevention*. Brussels: ECR Europe, 2010.
9. Monkhouse, C., Bowyer, C. and Farmer, A. *Packaging for Sustainability: Packaging in the context of the product, supply chain and consumer needs*. Brussels: IEEP, 2004.
10. Pro-Europe. *Effective Packaging Effective Prevention*. Brussels: Pro-Europe. Available at: www.ecoembes.com/es/documentacion/Prevencion/Documents/PREVENTION.pdf
11. INCPEN. *Packaging Reduction – Doing More with Less*. Reading: INCPEN, 1995. Available at: www.incpen.org/pages/
12. European Commission Joint Research Centre/ESTO/IPTS. *Environmental Impact of Products*. Brussels: European Commission, 2006. Available at: ec.europa.eu/environment/ipp/pdf/eipro_report.pdf
13. ECR Europe and Accenture. *Shelf Ready Packaging (Retail Ready Packaging) Addressing the challenge: a comprehensive guide for a collaborative approach*. Brussels: ECR Europe, 2006. Available at: www.ecrnet.org

3
Packaging functions

A. EMBLEM, London College of Fashion, UK

Abstract: This chapter examines in some detail the different functions of packaging. Emphasis is placed on understanding the properties of the product and the hazards it will face in the supply chain, including the demands of the consumer. There are different and sometimes conflicting requirements and expectations at each stage of the life of the product, and these must be understood before appropriate packaging solutions can be sought.

Key words: functions, containment, protection, physical damage, shock, vibration, compression, humidity, water, light, oxygen, pilferage, tampering, tamper evidence, journey hazards, preservation, biotic spoilage, abiotic spoilage, shelf life, water activity, pH, MAP, vacuum packaging, convenience features, information, selling, product cost.

3.1 Introduction

It is difficult to imagine life without packaging, and yet in many cases the consumer barely notices it is there, let alone thinks about what functions it is performing. The delivery of basic commodities such as coffee, tea, sugar and jam from factory to consumer relies on efficient use of packaging, and highly perishable foods such as fresh meat and fish, ready-prepared meals and salads would not be so readily available without the packaging which plays a key role in preventing spoilage, thus extending shelf life and minimising wastage. Electronic equipment such as televisions, personal computers and DVD players, along with domestic appliances such as irons, kettles, toasters and microwave ovens all rely on packaging to protect them against the potential damage encountered in the distribution chain. The growth in Internet shopping brings a need for packaging to be robust enough to survive the mail order delivery system, and to readily identify the product throughout the handling processes.

But packaging has to fulfil more than these functions of containment, preservation, protection and identification. It also influences the convenience in use of a product, and, just as importantly, it is instrumental in selling the product, by attracting the consumer.

Each of these functions will be discussed in the rest of this chapter. Many of the points being made will apply not only to consumer goods, but also to all other 'products' such as raw materials and ingredients for food, pharmaceuticals and cosmetics, components for the automotive and aeronautical industries, and packaging materials being shipped to their point of use. The functions of containment, protection and identification are just as important for the safe and efficient movement of such materials, and examples will be given where appropriate.

In all cases, to understand the functions of the pack and to ensure that they are adequately met, it is essential to define the product in terms of its nature, critical properties and value. This should always be the starting point; packaging does not have a 'life' of its own and without a product it has no reason to exist. The packaging designer/specifier needs to work closely alongside the product developer to gain a full understanding of the product and what can cause it to deteriorate to the point of being unacceptable. Only then can appropriate packaging be specified and sourced.

3.2 Containment

Properly designed, constructed and sealed packs provide complete containment for the contents, preventing unsightly or dangerous leakage, or loss of parts. This containment must be assured throughout the expected life of the product, including the numerous handling stages from the end of the packaging line to the final consumer use. Containment also means keeping a number of different or the same items packed together, and this applies to primary, secondary and tertiary packs. Examples include:

- packs of different varieties of crisps assembled into one bag, or the various parts needed in a hair colouring 'kit' such as colour, developer, gloves and comprehensive instruction sheet
- filled bottles of shampoo collated in a display tray
- stretch wrap film used to secure goods to a pallet.

If leakage or loss occurs, the consequences can vary from a minor inconvenience to a major and potentially catastrophic incident. A leaking bottle of bath foam is likely to result in a failure of the selling function, as no one wants the leaking pack on the shelf, and it may also have affected neighbouring packs so that they are also unsaleable. It may also fail to inform, if the type matter has become obliterated, in which case it may be illegal, and this also applies if it is underweight due to loss of contents. If, instead of a relatively innocuous bath product, the product is an aggressive or corrosive liquid, such as some household and industrial cleaning chemicals, the implications of leakage can be much more serious. If a vital part is lost, for example an instruction leaflet, the product may be rendered unusable.

3.2.1 Evaluating the containment function

A key step in evaluating the risk of leakage is to identify the factors which can affect the efficient containment of the product. Consider first the potential leakage points of a pack, including not just the obvious opening points, e.g. a screw threaded cap, but all other points where the pack could fail. Then consider how and why failure could occur at each of these points. Some examples to stimulate thought are given in Table 3.1.

Once the potential leakage-causing factors are identified, it is the role of the packaging technologist to ensure that the required performance characteristics are designed-in at the development stage, carefully specified on component and process

Table 3.1 Examples of containment failure

Pack type	Potential failure points/ mode of failure	Typical possible causes
Cartons	Glue seams split	Wrong adhesive for board and conditions of use; poor control of adhesive application conditions
	Tuck-in flaps work loose, tear	Wrong board weight for the weight of the product; poor cutting and creasing of the carton
Bottles/caps	Leakage at neck, misalignment of cap	Dimensional inaccuracies in bottle and/or cap finish; wrong cap application force; wrong wadding material
	Leakage at mould part line and/or injection point (plastic)	Poor control of moulding conditions
Sachets	Leakage in seal areas	Wrong sealing layer; poor control of sealing conditions; product in seal area
	Leakage in body of sachet	Puncture by product or by external means
Tubes	Leakage at neck, misalignment of cap	Dimensional inaccuracies in tube and/or cap finish; wrong cap application force; wrong dimension of orifice
	Leakage at base of tube	Wrong sealing layer; poor control of sealing conditions; product in seal area

specifications, and that checks are in place to ensure specifications are followed. Leak testing should be carried out during development, making sure that the likely conditions of use are taken into account. For example, a product intended to be carried around in the handbag requires a different test protocol from one which will be stored in one place for most of its useful life. Leak testing can be carried out by a variety of methods, including simple inversion of packs or by applying pressure to seals, in both cases observing the effects over time.

3.3 Protection

Protection means the prevention/reduction of physical damage to the product, during all stages of its life. This includes manufacture and packaging operations, storage and handling in warehouses, transport to the merchant, distributor or store for sale, display, and moving to the final usage point. It also includes storage and use of the product, e.g. in a kitchen, garage or garden, and any other handling operations which the final user may be reasonably expected to carry out on the product.

Damage can occur at any of these handling stages, although most physical damage happens in the warehousing and distribution environment, due to dropping (from pallets, and during order picking and transit), jolting, vibration (in vehicles), compression (when stacked in warehouses) or puncturing (often due to use of poor quality pallets). Damage can also result from environmental factors such as dust, dirt, birds, insects and rodents. See Table 3.2 for a list of typical hazards, their causes and effects.

Protection against physical damage is important for all products, not just the

Table 3.2 Typical hazards in the supply chain, their causes and effects

Hazard	Causes	Possible effects
Shock	Falls from conveyors, pallets, vehicles, possibly due to poor stacking; shunts due to irregular movement along conveyors; drops due to manual handling; impacts in transit due to driving over poor road surfaces	Breakage; deformation
Vibration	Vibration occurs naturally in all types of transport. In road transport the effects are enhanced over the rear axle of the vehicle, and by any imbalance in the load. Irregular road surfaces also increase vibration	Breakage; scuffing; product separation and/or settlement; loosening of screw caps; garments falling from hangers
Compression – static	Stacking in storage, made worse by damp conditions	Breakage; crushing; load collapse
Compression – dynamic	Clamp truck pressure; severe vibration during transport	Breakage, crushing, stack resonance
Puncture	Poor quality pallets, bad handling practices	Breakage; product spoilage; load collapse
Changes in relative humidity	Loads left outside; goods stored in damp warehouses, or where climatic conditions are not controlled; goods shipped via and to different climates	Product spoilage, e.g. corrosion; packaging failure, e.g. damp corrugated board cases
Changes in temperature	As above	Product spoilage; drying out of paper/board materials;
Exposure to light	Retail display	Fading of product and/or pack; product spoilage, e.g. rancidity
Insects, rodents, birds, dust, dirt	Goods stored in warehouses not cleaned or treated for pest control, or where doors/windows are left open or badly fitting	Product spoilage due to poor hygiene; contamination of product and pack
Pilferage and tampering	Goods exposed to uncontrolled personnel access; display on shelf	Loss of products; damaged packs and products; contamination; counterfeit products

obviously fragile such as glass and ceramics, or electronic components, but also for plastics, textiles, paper and board items, etc., whilst the unacceptability of damaged product may be obvious, aesthetic damage to the packaging, such as scuffing of labels/cartons, or scratches on plastic tubs, may also be causes of rejection.

3.3.1 Evaluating the protection function

To understand the level of protection required for a product, a simple equation to bear in mind is that the inherent ruggedness of the product plus the protection provided by the pack must be equal to the likely hazards encountered in the journey from factory to consumer and ultimate use of the product (Fiedler, 1995). The key steps to follow to decide what type of packaging will provide the product with the appropriate level of protection are:

- define the product
- define the environment to which the product is likely to be exposed
- investigate the properties (including cost) of available protective packaging materials (this step will be covered only in outline here).

Only when armed with the relevant information from each of these three steps will it be possible to propose possible packaging materials and pack formats which will match the defined characteristics. These proposals must then be tested in conditions which simulate 'real life' before packaging specifications can be agreed and the proposals implemented and monitored.

When developing solutions, remember it is the combined performance of all levels of packaging which is important. There will potentially be several different solutions, perhaps using a relatively weak primary pack which relies on its secondary pack for physical strength. Or it may be possible to design a sturdy primary pack which dispenses with the need for any secondary packaging, and is stacked directly onto a pallet. Therefore it is essential to consider not only the primary packaging when carrying out a development programme, but to combine this with an investigation of possible secondary/tertiary options. This combined approach will almost certainly result in the most cost effective solution.

3.3.2 Defining the product

As already stated, understanding the product is the key to any packaging development process and product development or research departments are likely to be the source of this information. The data may have to be developed from 'scratch' by carrying out actual tests, or there may be similar products with proven performance which can be used for guidance.

The aim is to define the product type (physical form) and how 'rugged' it is in terms of what conditions it can withstand before damage occurs, for example what level of shock, what vibration frequency, what top load, what range of temperature and humidity?

The product value will also have to be considered as it will decide the level of protection which can be afforded, and it will have a bearing on its attraction to the pilferer and/or the counterfeiter. Also at this point, any applicable legislation must be noted, e.g. legislation concerning hazardous goods.

3.3.3 Defining the environment

How the product is destined to be stored, moved, displayed and sold needs to be identified and clearly understood, as each of these stages has its own inherent hazards. This includes all operations, both internally within the premises and externally when the goods are outside of the direct control of the producer. One of the results of globalisation of manufacturing amongst large producers is that the distribution chain has become more complex, with multiple handling and exposure to a range of different climatic conditions. The more control which can be exercised, the more the risk of damage can be reduced and the less protection will be required from the packaging. Conversely, if control is poor, packaging requirements – and costs – will be high.

The common hazards will now be considered in more detail. In this section, refer again to Table 3.2 for possible causes and effects of each hazard, and also to Table 3.3 for ways of minimising the effects of each hazard. Examples will be given of possible protective packaging materials, but it is not the intention to cover materials in detail in this chapter, as they are covered in other chapters in this text.

Shock

Shock is defined as an impact brought about by a sudden and substantial change in velocity, and is usually encountered when an item lands on a stationary surface such as a floor, or it can happen during horizontal impact such as when items knock against each other on a conveyor. Key points known about the shock hazard are:

- Not surprisingly, the more manual handling there is, for example in a mail order environment, the greater the possibility of shock damage occurring. Most of these drops will be from a height of around 1 metre, i.e. hand height.
- The heavier the pack the lower the likely drop height.
- As the weight of a pack increases, and personal injury becomes an important consideration, the human being will handle the pack carefully and will eventually resort to mechanical handling. Palletised loads, moved by forklift truck are much less likely to be damaged than primary or secondary packs.
- Handholds reduce the likely drop height.
- Cautionary labelling such as 'This way up' has minimal effect.
- Damage to the packaging around a product is most likely when the pack is dropped on a corner or edge, but damage to the contents is greatest (and unseen until the pack is opened) when the pack is dropped flat onto one of its faces.

The most effective way of minimising the effect of the shock hazard is to cushion the impact. Figure 3.1 shows the shock pulses for a cushioned and an uncushioned drop. In the cushioned drop, the cushioning attenuates (weakens) the initial shock at the pack surface so that the product's response takes place over a longer period of time. The areas under the curves represent energy.

Some options for cushioning materials are listed in Table 3.3. The more resilient the material, the better its cushioning properties, so polymeric foams are more effective than corrugated board. The choice must always be made by going back to that fundamental point: what level of protection does the product need? Other factors to be considered are availability, possible contamination, surface abrasion, environmental impact and, of course, total cost.

Manufacturers of specialist cushioning materials produce dynamic cushioning curves which allow the user to calculate the thickness of material needed. This is

Table 3.3 Ways of minimising the effects of the common hazards

Hazard	Minimising the effects	
Shock	Reduce the amount of manual handling	
	Use cushioning materials, such as:	Polyethylene foam Expanded polystyrene Loose fill polystyrene chips Bubble wrap Compressed paper Moulded pulp Corrugated board
Vibration	Reduce product/pack movement:	Use tight shrink wraps Use accurate dimensions when sizing packaging Use pallet adhesives to prevent case movement
	Reduce contact points:	Good design of containers Label recess areas
	Protect surfaces:	Scuff resistant lacquers and film coatings
	Isolate the vibration	Use appropriate cushioning, or special 'air ride' vehicles, as used for susceptible products such as electronics
Compression	Good design of all levels of packaging	Design as a total pack, primary, secondary and tertiary
	Selection of pallet stacking pattern	Consider pallet stability and the likely stacking height
	Good storage conditions	Monitor temperature and humidity, especially when using paper/board
Puncture	Good pallet quality	Specify and monitor
	Good handling practices	Operator training
Changes in relative humidity	Operate good handling and storage practices throughout the supply chain Monitor relative humidity and introduce controls if necessary Use moisture-resistant materials/coatings	
Changes in temperature	Operate good handling and storage practices throughout the supply chain Monitor temperature and introduce controls if necessary Use temperature-resistant materials, e.g. expanded polymeric foams	
Exposure to light	Use opaque packaging	
Insects, rodents, birds, dust, dirt	Operate good standards of hygiene, including pest control, throughout the supply chain Carry out regular inspections to check on compliance	
Pilferage and tampering	Consider tamper evidence features in all levels of packaging Consider need for surveillance Consider anti-counterfeit measures	

important in the packaging of electronic goods and household appliances and further information can be found in the sources of information and advice at the end of this chapter.

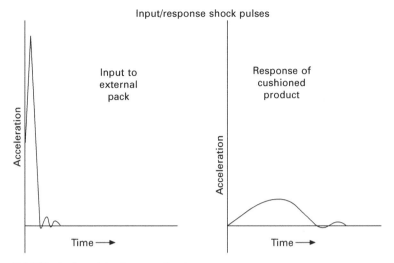

3.1 Effect of cushioning on shock.

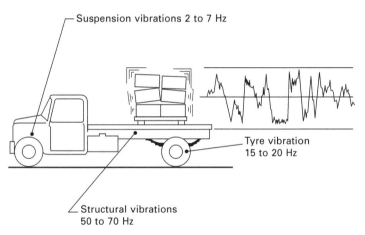

3.2 Typical sources of vibrations in road vehicles.

Vibration

Vibration refers to oscillation or movement about a fixed point. The distance moved is the amplitude and the number of oscillations per second is the frequency, measured in hertz (Hz). It is impossible to avoid vibration, as it is associated with all modes of transport. Frequencies below 30 Hz are most commonly encountered in road transport, and are of most concern. The combined effects of different frequencies can be considerable, as shown in Fig. 3.2. The greatest effect is typically experienced over the rear wheels.

Vibration resonance describes a condition in which a vibrational input is amplified, such that the output is out of all proportion to the input, sometimes resulting in severe damage. Often the only way to avoid this is to redesign the product to eliminate critical resonance points. Occasionally, loads go into a stack resonance condition, where each container goes into resonance with the previous one until the entire stack is bouncing (see Fig. 3.3). This condition can result in destruction of the product and pack. The dynamic load on the bottom container can be several orders of magnitude greater than the actual weight resting on it, which means that a corrugated case specified to withstand a calculated top load will now be totally inadequate. Also, the top container is subjected to extremes of repetitive shock and vibrations of considerable amplitude. Since this top layer is essentially weightless for short periods of time, small side movements will cause it to move. In a stretch-wrapped load this movement can skew the entire pallet load to one side.

Vibrational effects should be designed-out at the development stage, by ensuring that packs are tightly wrapped and accurately dimensioned so that movement is limited, and by removing obvious contact points. As these are likely to be low-cost or even no-cost options, they should take precedence over adding features such as scuff-resistant coatings.

Compression

Like vibration, compression is an unavoidable hazard in the distribution environment, as all products are stacked in warehouses and on vehicles. The challenge for the packaging technologist is to understand the likely compressive forces and specify appropriate packaging to limit product damage. Typically, the warehouse condition is one of static loading over time (static compression) but dynamic compression, encountered when using clamp trucks, during rail shunting and in stack resonance must also be considered. Dynamic compression describes a condition where the compressive load is applied at a rapid rate. Compression testing in the laboratory

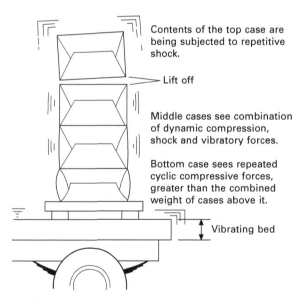

3.3 Stack resonance on a vehicle.

(e.g. test method ISO 12048:1994) is completed in a short period of time, i.e. it is a measure of dynamic compression values but the results can be correlated to indicate probable long-term warehouse stacking performance, which is likely to be of greater interest. This correlation for corrugated board is shown in Fig. 3.4. The initial part of the curve shows what the case will bear under dynamic or short-term load application. A container with a compression strength of 100 kg load clamped at an 85 kg load would be in danger of failing after about 10 minutes. The same container would fail in about 10 days if loaded to 60 kg. To last 100 days, the stack load should not exceed 55% of the dynamic compression value.

Not surprisingly, the humidity conditions encountered during warehousing and distribution can dramatically reduce the performance of corrugated board cases due to the hygroscopic nature of paper. For example, a change in relative humidity from 40% to 90% can result in a loss of about 50% of the case stacking strength. Thus corrugated cases destined for very humid conditions will have to be specified with sufficient stacking strength to allow for this inevitable loss.

As well as humidity, case stacking strength is also influenced by palletisation features such as pallet construction and pallet stacking:

- *Pallet construction.* Pallets with close-boarded decks (i.e. no gaps between the boards) allow pack weight to be evenly spread across the deck. Open-boarded pallets use less timber, but small packs can be compressed against the edges of the boards, causing excessive damage. Where palletised loads are double-stacked, reversible pallets (i.e. top and bottom decks are identical) allow even spreading of the weight of the top pallet across the top area of the lower pallet,

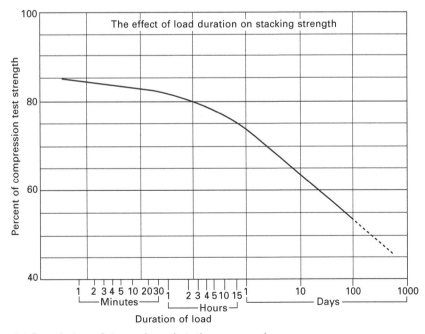

3.4 Correlation of dynamic and static compression.

thus avoiding excessive compressive forces in this area. Single-sided pallets without load-spreading perimeter boards do not allow this load spreading and can result in high damage levels (see Fig. 3.5). The greatest load per unit area is not on the bottom case, but on the top case of the lower pallet.

- *Pallet stacking.* Allowing cases to overhang the edge of a pallet leaves the load-bearing walls of the case suspended in mid-air, with an associated loss of available compression strength, as indicated in Table 3.4. Probably more controversial is the choice of column or interlock stacking pattern. Figure 3.6 shows the distribution of load-bearing ability around the perimeter of a case, clearly indicating that this is greatest at the corners. Therefore, the best possible use of container load-bearing ability is when cases are stacked directly on top of each other in a vertical column. However, this is the least stable stacking pattern

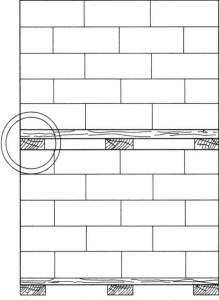

The greatest load per unit container area is not on the bottom container, but on the top container of the lower pallet.

3.5 Compressive forces when using single-side pallets without load-spreading perimeter boards.

Table 3.4 Effect of overhang on compression stack strength

Degree of overhang	Percentage loss (range)
25mm on one side	14–34
50mm on one side	22–43
25mm on one end	4–28
50mm on one end	9–46
25mm on one side and one end	27–43
50mm on one side and one end	34–46

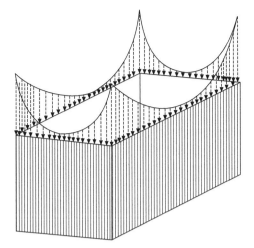

3.6 Distribution of load-bearing ability around the perimeter of a case.

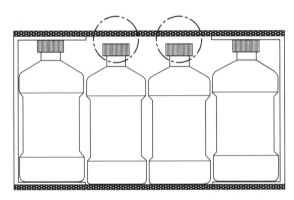

3.7 The 0201 style corrugated case, showing that not all primary packs contribute to the case compression strength.

and therefore cases are commonly interlocked (brick-style as shown in Fig. 3.5), although this format has a lower stacking strength.

Finally, the case contents, i.e. the product in its primary pack, will also influence compression strength. Rigid bottles contribute to the overall strength of the secondary pack, although when using the 0201 style case (FEFCO), in which the short flaps do not meet, not all bottles contribute, as demonstrated in Fig. 3.7. A way to address this is to use an all-flaps meeting style, such as 0204. Flexible primary packs such as bags and stand-up pouches make little or no contribution to case strength and thus the case specification will need to be more robust than for rigid primary packs. The degree of rigidity of the primary packs is important, especially when using apparently rigid containers such as plastic bottles and the known 'creep' characteristics of the plastic must be taken into account. Bottles which are expected to contribute to load-bearing should be designed avoiding sharp corners and edges, since these features act as stress concentrators and promote cracking. Circular cross sections, large neck areas to spread the load and shallow angles are all preferable, although of course such features may not meet other functions, such as ease of use and brand image.

Typically, whatever the load-bearing packaging component chosen, it should be designed to have a compression test value three to seven times greater than the maximum stacking load expected during storage. This is frequently referred to as the stacking factor or the safety factor. The maximum stacking load must take account of whether or not pallets are double or triple stacked, i.e. block-stacked on top of each other, without using racking or any other means of supporting the weight. The shorter the distribution cycle and the less handling there will be, the lower the safety factor. Table 3.5 gives guidance for starting points, across a range of different conditions, but actual factors should be calculated for each application.

Puncture

Puncture refers to the piercing of a pack, which invariably results in product leakage and/or damage. It can be caused by external agents or by the product itself. External agents include piercing by the forks of a forklift truck, with the associated catastrophic effects or, less dramatic but nevertheless still unacceptable, piercing by nails or splinters in wooden pallets. In the former case the effects can be avoided by training; indeed, it is not possible to cost effectively protect goods against forklift truck damage by packaging specification, and training out bad forklift practice is the only feasible option. With regard to damage caused by unsuitable pallets, this emphasises the importance of treating pallets just like any other packaging component and applying appropriate specifications and inspection standards.

Temperature and humidity

This section will concentrate on the direct effects of changes of temperature and humidity on the *packaging* around the product, rather than on the *product* contained therein. Product spoilage will be discussed more fully in Section 3.3, the preservation function of packaging. The two hazards are discussed together here because of their interdependence; the higher the air temperature the more moisture it can hold without condensation occurring. The dew point is the temperature at which the air is saturated with water vapour; below this, water condenses out of the moist air in the form of droplets.

Table 3.5 Guidance for stacking factors

Condition	Stacking factor
Column stack, on overhang, minimum warehousing	3.5
Column stack, on overhang, normal warehousing	4.0
Interlock stack, on overhang, normal warehousing	5.5
Column stack, overhang, normal warehousing	5.5
Column stack, no overhang, freezer storage	5.5
Interlock stack, overhang, normal warehousing	6.0
Interlock stack, extended distribution and warehousing	7.0

Temperature and humidity changes will be encountered anywhere in the supply chain and are usually, but not always, due to climatic changes. The more varied the climatic conditions to which the pack is exposed, the more severe the hazard and its effects, so shipping goods between continents is obviously more hazardous than working solely in a domestic market. However, temperature and humidity changes can also occur due to poor control of storage conditions regardless of the climate, e.g. warehouse doors/windows left open, heating left on/off, or excessive use of gas-powered forklift trucks, and this can occur anywhere, even in a relatively short supply chain.

The effects of these changes are most apparent on cellulose-based packaging, i.e. paper and board. (Metals are also affected and unless suitably treated will corrode in high humidity – see Chapter 8 for more detail.) Cellulose fibres exposed to moist air will expand across their width as they absorb the moisture, hence sheets of board will vary in dimensions according to the humidity level, and such variations can affect carton cutting and creasing accuracy. Sheet flatness can also be affected by moisture, as the different layers absorb moisture differentially, resulting in curl (See Chapter 10). Paper and board packaging components should always be allowed to acclimatise in the conditions in which they are going to be used; bringing folding cartons from a cold warehouse into a warm packaging area and expecting good performance in terms of automatic make up, folding and gluing is unrealistic, and will result in high wastage and downtime.

Almost every property of paper and board is affected by temperature and humidity, but perhaps the most significant point to consider in the protection function of packaging is that as the moisture level increases, strength characteristics can decrease very rapidly. As already mentioned, a change in relative humidity from 40% to 90% can result in a loss of about 50% of the case stacking strength. Hence corrugated board cases specified to provide a level of resistance to compression during pallet stacking in one set of humidity conditions, will fail if the humidity increases significantly. Figure 3.8 gives a chart to estimate the compression strength of corrugated board at different moisture levels. To use the chart, draw a line from the compression strength to the board moisture content (see Chapter 10). Mark the position where the line crosses the pivot line. Now project a line from the new moisture content through the pivot line mark and read off the new compression strength.

As well as taking care over storage conditions, the main way of providing protection against the effects of moisture on paper and board is to use coatings to protect the cellulose base material. With regard to protection against extremes of temperature, other than the inclusion of insulating materials such as polymeric foams, this can only be done by careful management of the conditions throughout the supply chain.

Light

Light is most likely to present a hazard when packed products are on display and exposed to retail lighting, when typical effects will be change of colour, usually fading. To the consumer a faded carton may indicate an 'old' product and thus this is likely to be discarded in favour of an apparently newer option. Colourfastness of

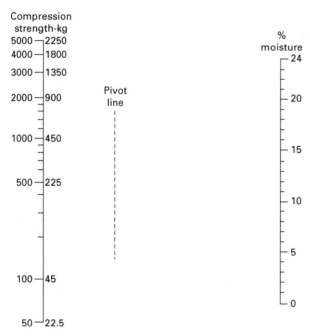

3.8 Corrugated board compression strength vs moisture content.

substrates and printing inks should be checked during the development stages, and susceptible items should be packed in opaque primary and/or secondary packaging, or UV filters can be incorporated into packaging materials. Exposure of some products to light can result in chemical changes, and these will be covered in the section on the preservation function.

Dust, dirt, insects and pests

These hazards can be encountered at any stage and their effects vary from an aesthetically unpleasing dusty pack to a serious product infestation with potential damage to human health. Prevention is the only viable approach to these hazards, operating good standards of hygiene throughout the whole supply chain. This applies just as much to the manufacture, storage and delivery of packaging components to the packer/filler as it does to the product handling stages in the chain. See Chapter 21 on hazard and risk management for further information on this aspect.

Pilferage and tampering

Pilferage and tampering by humans, which is most likely to occur in the selling environment, also comes into the category of physical damage, and concern about this has created a requirement to consider the use of tamper evident and, in some cases, anti-counterfeit packaging. Tamper evidence is defined by the US Food and Drug Administration as: 'Having an indicator or barrier to entry which, if breached or missing, can reasonably be expected to provide visible evidence to consumers that tampering has occurred'. For a tamper evident pack to be effective, the consumer must recognise that tampering has taken place, i.e. the tamper-evident feature needs to be obvious. Typical tamper-evident devices currently in use include:

- metal and plastic caps with a breakaway band
- pop-up safety buttons used on metal caps (e.g. on jars of jam)
- shrink bands and sleeves with the sleeve covering the closure
- diaphragm seals, e.g. as used on jars of coffee and pots of vitamin tablets
- permanent adhesive labels which disintegrate and/or reveal a warning message, if an attempt at entry has been made.

None of these would present the determined, malicious tamperer with an insuperable task. The skilled tamperer will not confine his/her attempts at entry to the closure, and all other potential points of access should be considered when assessing the risk of tampering.

Counterfeiting is a serious problem in products such as pharmaceuticals, spirits, tobacco and perfume, and brand owners are constantly striving to stay ahead of the counterfeiter. Anti-counterfeit measures include building in a high level of complexity to the pack, such that it will be difficult to mimic, e.g. unique, custom-designed shapes for containers and closures, holograms on labels, printing inks which show up only in special conditions, e.g. UV light, and embedded micro-chips. The spirits sector also uses a range of special non-refillable fitments inserted in the neck of the glass bottle. The flow of product is maintained but the bottle cannot be refilled once empty.

3.3.4 Completing the process of evaluating the protection function

A useful exercise is to map the journey of the packed product from production to consumption. Note the number and type of product movements and storage conditions, and against each one, list the likely hazards (i.e. shock, vibration, etc., as studied in Section 3.3.3) and their causes and effects. This will identify where the hazards occur and what types of hazards are most prevalent. See Table 3.6 for a list of typical questions to consider, and refer to Chapter 21 on hazard and risk management.

Understanding and, wherever possible, quantifying these hazards is vital in ensuring the packed product can perform satisfactorily throughout its expected shelf life. This can be done by a combination of observation, use of known data (e.g. frequency levels for different modes of transport) and measurement. By quantifying the hazards in this way, the uncertainty surrounding events taking place in the environment can be removed and packaging specifications can be tailored to meet real, rather than imaginary hazards.

Shock, vibration and climatic hazards can be measured *in situ*, by including data loggers in specific loads. These can be purchased or hired and used to record and measure actual events and the time at which each event occurs. The data obtained can be analysed to show the extent of damage-causing hazards experienced during

Table 3.6 Mapping the journey; typical data which can be obtained

Hazard	Data
Shock	What level and skill of labour is available? To what extent is manual handling used? How many times is the pack handled as: • primary pack • secondary pack • tertiary pack What is the most likely drop height?
Vibration	What methods of transport are used? What are the most likely vibration frequencies experienced?
Compression	Are pallets always racked, or is block stacking likely? What is the maximum top load during storage? Are loads evenly distributed? What is the most likely maximum dynamic load?
Pressure	What are the likely altitude and pressure changes? This is important if air freight is to be used
Environment	Will the product be stored outside? What is the range of humidity and how rapidly does it change? What is the range of temperature and how rapidly does it change? What are the likely conditions of cleanliness? Are the goods likely to be stored in poorly managed conditions and is security a potential problem?

a journey. Due to the cost involved, this approach is usually only employed when high levels of damage have been experienced, or in the early packaging development stages for high value, fragile items.

Testing should be carried out as part of the development programme and time for this will need to be built in to the development schedule (see Chapter 18). Transit testing is commonly done using actual conditions, or it may be simulated by using calibrated laboratory equipment and the known conditions identified in the process of mapping the journey. The latter is the more expensive option, although it may be more reliable and thus avoid costly mistakes. When using actual conditions, sample packs are made up, sent out on a 'typical' journey and assessed at the end. This may be done at various stages, first of all trying out a small number of packs and then building up to full pallet loads, or full vehicle loads. Provided the process is managed well, this testing can provide valuable 'real-life' information. Points to consider include:

- Trial packs should be representative of final packs and use the same materials.
- Vehicles should follow a prescribed route, over an agreed distance and class of roads.
- If the packs are palletised, the trial should continue to the end point, with secondary packs being transferred to mixed pallets if this is what will happen in reality. Primary packs should then be treated as they are during display, selection and use by the final consumer, e.g. taken home in a shopping carrier bag.
- Wherever possible, use a packed product of proven performance as a control.
- Standards must be set for what levels of damage are acceptable.

3.4 Preservation

Preservation means the prevention/reduction of changes due to biological and chemical hazards, which would lead to product spoilage. The objective of preservation is to extend the shelf life of a product. This section applies mainly, but not exclusively, to the food, drink, pharmaceutical and cosmetic industries.

3.4.1 Shelf life

According to the Institute of Food Science and Technology, 'shelf life' is defined as the time during which the product, when stored at the recommended conditions, will:

- remain safe;
- be certain to retain desired sensory, chemical, physical and microbiological characteristics;
- comply with any label declaration.

When considering the preservation function of packaging, it is important to recognise that whilst packaging can and does contribute to shelf life, it cannot overcome inherent product problems; if the product is unsafe at the point of packing, it is likely to remain unsafe inside the pack. Also, if temperature is a key factor in maintaining preservation, e.g. chilling or freezing, the packaging has only a 'supporting' role to play; if the temperature of the packed product is allowed to rise to the point where deterioration occurs, the pack will not compensate for this failure to manage storage conditions.

Within the limitations mentioned above, to determine the optimum packaging required to extend shelf life, we need to define the product in terms of what will cause it to deteriorate, i.e. what is the spoilage mechanism. We then need to understand what process (if any) will be used to prevent/delay spoilage and the extent to which this will affect the packaging used, and therefore determine its key properties. Only when we have a packaging specification which defines the required properties can we begin to investigate possible solutions.

3.4.2 Define the product-spoilage mechanisms

Defining the spoilage mechanism of a product is part of the research and development stage of the *product*, and this is an example of how product and pack development personnel must work closely together. Product development specialists should be able to provide the information needed to define the product in terms which allow the packaging technologist to specify key packaging attributes. This section provides only a broad introduction to spoilage mechanisms, which is no substitute for the detailed knowledge expected of the food, cosmetic or other product scientist.

Product spoilage, and therefore shelf life is determined by microbiological, physical or chemical factors, depending on the product, the process, the packaging and the storage conditions (Blackburn, 2000). Broadly, spoilage due to microbiological spoilage is referred to as biotic spoilage, and that due to physical and/or chemical factors is known as abiotic spoilage.

Biotic spoilage is caused by microorganisms (bacteria, moulds, yeasts) which may render a product unacceptable in appearance, taste, smell and effectiveness, or be toxic and cause sickness. Different organisms have preferred conditions for growth and adverse conditions in which they will not propagate, and this is the basis of product preservation systems. The conditions to be considered are as follows:

- *Temperature*. Microorganisms may be classified by their preferred reproduction temperature:
  - psychrophiles grow best in fairly cool conditions (10–20°C);
  - mesophiles grow best at 20–40°C;
  - thermophiles prefer temperatures in the range 40–60°C.

  Note, however, that there is considerable variation within each classification and these temperature ranges are not exact.
- *Humidity*. Microorganisms need water in which to grow and as a general rule, the lower the level of available water in a product, the less likely it is that microorganisms will propagate.
- *Acidity*. Microorganisms have an optimum pH level at which they will grow. In general, moulds and yeasts grow best in acidic environments and bacteria grow best in neutral to slightly alkaline conditions, although there are exceptions to this.
- *Presence of oxygen*. Some microorganisms need oxygen to propagate and are known as aerobes, while others cannot propagate in the presence of oxygen and are known as anaerobes. Some can propagate in either oxygen or oxygenless environments. In general, moulds and yeasts need oxygen to propagate, although some yeasts grow in anaerobic conditions.
- *Nutrient source*. All microorganisms need a nutrient source, although their needs vary greatly. Nitrates, lactates and amino acids are typical nutrient sources found in foodstuffs.

Abiotic spoilage refers to the chemical or physical changes brought about by external factors such as oxygen, moisture, light, temperature, loss/gain of volatiles, e.g.:

- oxygen, causing rancidity in fats, creams and oils
- loss of moisture causing drying out and hardening of bread, lipstick, pastes, etc.
- gain of moisture causing lumping of powders and loss of crispness of cereals and biscuits; corrosion of metal products and packs also comes into this category
- light causing colour fading or oxidation
- excessive heat causing drying out
- excessive cold causing undesirable freezing e.g. emulsions
- loss of volatiles such as some of the oils in tea, which affects its taste
- gain of volatiles which make a product taste odd, e.g. chocolate stored next to highly fragranced soaps, without an adequate barrier in the packaging will quickly pick up the volatiles in the soap and taste soapy. Unacceptable volatiles

can also be picked up from printing inks and adhesives, due to high levels of retained solvent.

3.4.3 Preservation processes

The basic principle of product preservation processes is to address the cause(s) of spoilage, and then to use appropriate packaging and storage conditions to maintain the product in its desired state. Referring back to the causes of biotic spoilage listed in the previous section, it can be seen that these can be addressed by:

- changing the temperature to destroy microorganisms, or impede their growth, using heat (pasteurisation, sterilisation) or cold (chilling, freezing).
- reducing the water activity (A_w) in a product. A_w is a measure of the amount of available water in the product and lowering this limits microorganism growth. Methods of reducing A_w include drying (which removes water) and the addition of salt or sugar (which 'ties up' the free water).
- changing the acidity level, e.g. pickling using vinegar.
- varying the oxygen level, which can be done by vacuum packaging (where the product is packed and then the air is evacuated from the pack before sealing) or by changing the gaseous mixture around the product (modified atmosphere packaging – MAP; discussed further in Chapter 20). The use of oxygen scavengers to reduce the free oxygen available may also be considered under this heading.

See Table 3.7 for a summary of these preservation methods, along with considerations for packaging requirements.

Using the same approach to reduce/prevent abiotic spoilage, it can be seen that the key property required of the packaging is an appropriate barrier to the ingress/loss of the damage-causing factor, e.g. moisture, air and light (see Table 3.8). To specify the correct level of barrier, it is necessary to know the extent to which the factor can be tolerated before spoilage takes place, e.g. knowing the extent to which a biscuit can absorb moisture before it becomes unpalatable allows the packaging technologist to design the optimum pack, which will give the required shelf life, but not be overspecified such that resources are wasted.

3.4.4 Predicting packaging characteristics for particular foodstuffs

Provided the critical values of a food product are known, i.e. at what level the environmental spoilage-causing factor of gain or loss of moisture vapour, gas (or light) becomes unacceptable, it is possible to calculate shelf life based on the relevant barrier properties of the packaging material. Conversely, knowing the desired shelf life can dictate the barrier specification of the packaging material. Assuming correct storage conditions are maintained, especially for chilled and MAP products, it is the pack's barrier to gain or loss with respect to these environmental factors which will determine shelf life. Knowing the spoilage mechanism of the product is thus the first step in predicting packaging requirements. A useful list of deterioration indices for

44 Packaging technology

Table 3.7 Summary of the common preservation methods and associated packaging requirements

Preservation	Method	Packaging properties
Cooling	Chilling	Able to withstand storage temperature without deterioration such as cracking or loss of print Odour barrier May also require light barrier Toughness, puncture resistance to withstand handling on display Pack should fit standard domestic refrigerator
	Freezing	Able to withstand blast freezing and storage temperatures without deterioration, as above Appropriate moisture barrier, to prevent freezer burn Odour barrier May also require light barrier Toughness, puncture resistance to withstand handling on display
Heating	Pasteurisation, hot filling Sterilisation (canning and retorting)	Able to withstand temperature and pressure changes during heating process Water resistance Toughness, puncture resistance to withstand physical handling during process
Drying	Air drying, heat drying, salting	Appropriate moisture barrier Resistance to chemicals used
Chemical Preservation	Pickling/other chemicals Oxygen scavengers	Resistance to chemicals used
Varying oxygen levels	Vacuum packaging MAP Oxygen scavengers	Gas barrier Puncture resistance
Irradiation		Resistance to ionising radiation

Note also the requirement for correctly sealed packs.

different classes of foods is given in Table 3.9. Note that many foods are sensitive to more than one spoilage factor.

The following is a very simple approach to calculating packaging barrier requirements, which can be used as an indicator. It is demonstrated in relation to determining the required moisture barrier, although it can be applied to gas barriers.

First of all the maximum amount of moisture allowable in the product contained in the pack, before spoilage starts to occur must be known. Assume this to be W grams. Next we need to measure the surface area of the pack as carefully as possible, allowing for changes due to handling, especially for flexible packs. Assume this to be A square metres. If the desired shelf life is T days, then we must look for a pack with a moisture vapour transmission rate (MVTR) of less than:

$$\frac{W}{T \times A}$$ grams per square metre per 24 hours

When considering barrier properties such as MVTR, packaging materials can be divided

Packaging functions 45

Table 3.8 Summary of packaging requirements to reduce/prevent abiotic spoilage

Spoilage mechanism	Packaging properties
Oxygen	Appropriate gas barrier
Loss of moisture	Appropriate moisture barrier, related to the equilibrium relative humidity of the product
Gain of moisture	Appropriate moisture barrier, related to the equilibrium relative humidity of the product May require moisture permeable pack
Light	Appropriate light barrier
Heat/cold	Insulation Importance of control of storage and handling conditions
Loss of volatiles	Appropriate gas barrier Appropriate chemical resistance
Gain of volatiles	Free from taint and odour

Note also the requirement for correctly sealed packs.

Table 3.9 Approximate order of importance of specific deterioration indices for certain foods

Foods	Microbial changes	Inherent changes	Moisture changes	Oxidation changes	Taint, etc.	Light	Physical damage
Baked goods	4	4	1	2	4	4	2
Raw and cooked meats	1	2	4	2	6	4	6
Fish	1	3	4	2	–	–	–
Shellfish	1	4	1	1	–	–	–
Potatoes	2	–	1	–	–	3	4
Green vegetables	–	–	1	2	–	–	3
Soft fruits	1	–	–	–	2	–	3
Salads	2	–	1	–	–	–	–
Breakfast cereals	–	–	1	4	3	–	2
Chocolate	5	–	1	2	3	–	4

1 = most important, 7 = least important.
Source: Paine (1992).

into two categories: those such as glass and metal, which have an absolute barrier, i.e. they are impermeable, and the permeable materials such as paper and plastics. MVTR is a measure of how quickly moisture permeates through the packaging material and is not the only consideration; if a pack is not correctly sealed, even an impermeable glass bottle will result in a permeable pack, hence the need for careful control of sealing parameters during filling, and in-process checks on the packaging line.

MVTR data for some of the common packaging plastics are given in Chapter 13 and these can be used as an initial guide to packaging material selection and to inform storage trials, thus providing a basis on which to start comparing potential alternatives and gain a preliminary idea of packaging material costs. Note that permeation is inversely proportional to thickness of the barrier and hence a 50 μ layer will be twice as good a barrier as a 25 μ layer, and that the temperature and humidity conditions to which the product is likely to be exposed in the supply chain are a vital factor in calculating the required barrier. It is essential to specify these and check that the data being quoted are applicable to the conditions expected.

The data given in Chapter 13 are generic and transmission rates for specific grades within each type of plastic are quoted by individual manufacturers and can usually be obtained from data sheets. Actual measurements of moisture, oxygen and light barriers can be carried out using laboratory testing, and references to independent laboratories carrying out such work are given at the end of this chapter. It should also be noted that the quoted data are for sheet materials in pristine condition and any creases or folds introduced during the packaging and handling processes will reduce the barrier performance. This is why the calculated barrier levels should be regarded as minimum values.

Important factors which have an effect on shelf life and barrier requirements are the size and geometry of the pack. Because moisture vapour and gas transmission rates are related to pack surface area, the smaller the pack (i.e. the higher the pack surface area to product ratio), the greater the permeation through the pack and the higher the barrier required per gram of product. This is important when developing different pack sizes and shapes and it cannot be assumed that a material which provides an acceptable barrier for one size already on the market will be suitable for another size or shape variant.

3.5 Using packaging to provide convenience

The way in which a pack is handled (and the product used) is determined by the design of the pack itself. Packaging designers have the opportunity to build in features to make handling easy, convenient and safe. If it is both intuitive and ergonomically sound how the pack should be picked up, opened and unpacked, potential damage to the contents and personnel will be minimised. This applies just as much to secondary packaging as it does to the consumer-facing primary pack.

Good pack design will determine how easy/difficult it is to dispense the desired amount of product and this is especially important for potentially 'difficult' products such as nail enamel, shoe polish, syrups, motor oil, viscous adhesives, paint, etc. If the pack offers clean and safe delivery, with no mess or loss of product, it can provide the all-important competitive edge in a crowded market. Just some examples of consumer convenience are:

- the ability to dispense and direct the required amount of product using aerosols, natural pump sprays, special nozzles on tubes, brush applicators, etc.;
- easy product access from tubes and squeezable bottles;
- easy-open features such as tear-tapes in film wraps or 'tear here' cuts in sachets;
- easy-open and reclose features such as flip-top closures on sauces and shampoos;
- packs which collate, for example, five individually-wrapped snacks into one pack which is easier to handle and store in the cupboard than single packs;
- boil-in-bag and heat-in-tray foods, which mean no dirty saucepans;

- ring-pull cans, requiring no special opening tools.

Convenience on the filling and packaging line and in distribution is also important in the quest for cost and resource-efficient solutions, e.g.

- the stability (affected by material and shape) of a container is critical in determining the running speed – lightweight plastic bottles are less stable than glass and tall/narrow shapes are less stable than short/squat shapes;
- containers which do not stand vertically for filling usually require pucks (specially made holders in which each container is supported while on the filling line), which in turn add to cost and development time;
- designing a 'punt' into a container (usually on the base) which locates it in the correct orientation for labelling;
- the use of pallets, roll cages and special trays, for ease of movement, loading and unloading;
- the use of cut-out holes in large corrugated cases for easy lifting by hand;
- the collation of products into shelf-ready packs makes them convenient to handle in the store when the shelves are being filled;
- the use of custom-made display systems for ease of recognition at the point of consumer selection;
- the use of marking and coding systems for manual or automatic recognition of the product as it is moved into and out of storage;
- assembling the components required for a production operation (e.g. car assembly) in one 'kit' which speeds up delivery to line, and quickly identifies if anything is missing.

3.6 Packaging as the source of product information

Packaging provides the ideal (and often the only) means of delivering valuable information to anyone engaged in handling the product, as well as the final consumer. There is a need for information on the identity of the product, its weight/volume, destination and handling and possibly unpacking/repacking instructions, and such information must be easy to locate and understand. For important instructions, consideration should be given to the use of pictograms, to overcome any language barriers. Information on secondary and tertiary packaging must also show information such as product, number in pack, product codes and bar codes, for ease of recording stock.

3.6.1 Types of information

For the consumer there is an increasing need to provide legal, promotional and usage data for a product, whether it is a bar of chocolate, a tube of toothpaste or an expensive perfume or skincare product. The information required is further increased when products are sold into several different geographical markets, where multi-lingual copy is needed. Legal information required includes product name, use of the product (if not obvious), weight/volume, manufacturer's or seller's (e.g. Tesco) name and address, expiry date, ingredients list, batch code (for traceability) and any relevant warning statements.

With regard to promotional and usage information, along with advertising, packaging provides the seller with a means of informing the consumer about the uses and benefits of the product, designed to encourage purchase. Once that purchase is made, it is the packaging alone which is expected to convey the definitive information on the product. This may necessitate the use of a leaflet, which must be signalled to the user (e.g. by a statement on the carton, such as 'See leaflet for instructions').

3.6.2 Information formats

Information is required in the form of printed copy, perhaps supported by illustrations, and in electronically readable form, such as bar codes and matrix codes. Whatever the form, 100% legibility every time is the goal, which means using an appropriate size of type and colour contrast, and ensuring that the printing process used is adequate for the accuracy demanded. Increasingly, the provision of information for the visually-impaired is required, using raised characters as in Braille.

3.7 Packaging to sell the product

Today, most shopping is done in large supermarkets and this applies not only to food and drink, but also to cosmetics, garden chemicals, hardware, textiles, electrical equipment, etc. The role of the specialised, trained sales assistant to help one make a choice has declined and in the generally impersonal world of the retail store, it is packaging which carries out the selling role – the 'silent salesman'. The manufacturer/seller must use the packaging, usually seen before the product, as the means of attracting the would-be purchaser. This is done by a combination of the colour, graphics, shape and size of the pack, which must combine to provide novelty, or a familiar chord of recognition, usually backed up by advertising depicting the pack and portraying the desired product image. Think about brands with which you are familiar, and the features used by the brand owner to ensure that you instantly recognise their product. For more information, see Chapter 6 on packaging and marketing and Chapter 18 on pack design and development.

3.8 Conclusion

As pointed out in Chapter 1, packaging has the potential to deliver cost effective solutions by reducing wastage and costly (and perhaps irreparable) damage to the product. The development and use of correctly specified packaging materials and formats, related directly to the demands of the goods, will ensure that excessive costs are not incurred unnecessarily. Good packaging solutions also have the potential to reduce the total cost of moving goods, by providing safe and convenient handling systems.

The cost of packaging must be related to the product's value, image and end use. For most manufacturers this means the minimum spend commensurate with meeting all the required functions of packaging. This is not the same as the lowest cost option. If a product fails to attract the attention of the consumer, due to poor packaging presentation, the packaging is not meeting its required functions.

Finally, the relative importance of each of the functions of packaging varies with the product. For example, protection against physical damage will be very important for a television set, but the selling function of the packaging will be less important, as most people buy a TV after having seen a display model in the store. To the consumer, the packaging is simply a convenient means of getting the item home safely. On the other hand, the packaging of a breakfast cereal or pet food on the supermarket shelf, amongst 25 other varieties of similar products, has a very important selling function to fulfil.

3.9 Sources of further information and advice

There are several general texts available on packaging technology, recommended for reference, such as:

- Yam, K. L. (2009) *The Wiley Encyclopaedia of Packaging Technology*, 3rd edn. Wiley, Chichester.
- Soroka, W. (2010) *Fundamentals of Packaging Technology*. Institute of Packaging Professionals, Naperville, IL (www.iopp.org).

References for pack testing:

- Instrumented Sensor Technology: www.isthq.com
- Dallas Instruments: www.dallasinstruments.com
- Pira International: www.pira.co.uk

Note: Sources used in the preparation of this chapter also include teaching and learning materials written and used by the author in the delivery of courses to a number of organisations, such as the Packaging Society, Loughborough University, University of Warwick, University of Bath and London College of Fashion (University of the Arts London).

3.10 References

Blackburn, C. de W. (2000) 'Modelling shelf-life', in Kilcast, D. and Subramanium, P. *The Stability and Shelf Life of Food*, Woodhead, Cambridge.
Fiedler, R. M. (1995) *Distribution Packaging Technology*, Institute of Packaging Professionals, Herndon, VA (out of print).
Paine, F. A. (1992) *The Packaging User's Handbook*, Blackie, London.

4
Packaging legislation

G. CROMPTON, The Packaging Society, UK

Abstract: This chapter provides an introductory summary of the main legal requirements relating specifically to packaging. These include aspects of law relating to the design of packaging and the performance of its functions, together with the legislation introduced to reduce its environmental impact. Although they have no direct statutory authority, the guides provided by the enforcing authorities, and the normal commercial requirements with the related standards are also described.

Key words: UK packaging law, food safety and labelling regulations, filling regulations, environmental regulations, including essential requirements, producer responsibility obligations.

4.1 Introduction

4.1.1 The role of law within civilised society

Whatever we are doing, our activities, and the materials used to undertake those actions, are governed by some form of legislation. Although at times this may seem inconvenient, it could be argued that it is impossible to have an effective society without some form of agreement as to what is to be acceptable and what is not. From the simple group decisions of prehistoric times, the concept of law has developed into national, regional, and international regulations. As the situations which society would wish to address have become more complex, so have the laws regulating them. Within the trading activities of society, law has functions of establishing normal expectations, and thereby making communications within business easier. The area of trade and regulation may be within the same country, in which case the applicable laws would be those of that country. However, the country may be within a legal region, such as the European Union (EU) in which compliance would be required with the legislation of the region, in addition to those of the countries involved. In international dealings there may also be legislation and standards established by recognised international authorities, such as the World Health Organisation (WHO) or the United Nations (UN).

4.1.2 How laws are made and enforced

The motive behind making new laws, or revising existing ones, is normally the recognition of a situation in which the current legislation is considered to be no longer appropriate. Within the European Union, the issue is the subject of representation by those concerned to the national government, and thence into the European Commission, which works through consultation processes to establish an appropriate piece of legislation to be submitted to the European Parliament for approval. The resulting laws may be in one of two forms. EU Regulations are directly applicable to all member states and must be implemented and followed exactly as laid down, whereas EU Directives are less prescriptive and require the enactment into national legislation before they become applicable. The development of this national legislation may lead to national variations of interpretations and applications of the Directive. Also, some Directives are optional harmonising directives and so their inclusion within national legislation is at the discretion of the national governments. An example of an optional harmonising directive is the Aerosol Dispensers Directive, which has mandatory force elsewhere in Europe, but not within the UK. As part of this process of bringing both Regulations and Directives into national legislation, each national government is required to establish the appropriate enabling legislation, including the selection of the appropriate enforcement authority. Once the enabling legislation is in force, the details are issued and in the UK this is done via Statutory Instruments (SI).

These processes are normally undertaken in consultation with advisory groups drawn from representative trade organisations. It is usual for the enforcement authority to provide guidance to those affected by the new legislation as to their interpretation, usually in cooperation with the same advisory groups they would have been consulting during the preparatory stages. Such guides clearly state that they have no force in law, but keeping records which show the appropriate guide has been followed is an important way of building a 'due diligence' defence in the UK. The basis of such a defence is to be able to present documentary evidence to show that there was the intention to comply, as all reasonable steps had been taken to seek to comply with the legislation. Eventually the first case will come to court, and begin to establish precedents. This is termed a 'test case' as it is the opportunity for the authorities' interpretation of the law to be examined and established as 'case law'. The details continue to be clarified as further cases are brought to the courts.

4.1.3 Law and trading conventions

Alongside the law, there is a series of trading conventions, which have been established by specific trade associations, or wider representative bodies. Whilst these have no direct authority in law, their influence should not be underestimated, for example in the control of bar-code systems. These conventions will be discussed later.

4.2 Legislation relevant to packaging

Before considering the legislation directly related to packaging, it is necessary to remember that there is a wide range of legislation with which companies also have to ensure compliance. Examples include: the ownership and use of premises; employment of staff; health and safety; financial transactions; taxation; contracts with suppliers and customers; control of the environmental impact of processes; and the appropriate licensing of operations and facilities. Also, there is legislation which is concerned with the product which is being packed, and so becomes applicable to

52   Packaging technology

the packaging being used. Examples include the supply of foods, and the control of hazardous substances.

4.3 Legislation relating to product quality and safety during manufacture, distribution, storage and use

The general legislation under this heading includes the provisions of consumer protection and the safe handling and use of products.

4.3.1 Legislation concerning safety of packaging for foods

As most packaging is used to supply food, the legislation relating to packaging for foods will be described more fully. Directive 89/397 on the Official Control of Foodstuffs was enabled in the UK by the Food Safety Act of 1990. As the requirements of the Directive have been updated, these have been incorporated into UK legislation by several statutory instruments, and advised by codes of guidance. Most of the requirements are concerned with the quality of the food itself, but amongst them are some which apply to packaging and others which apply to labelling. Earlier legislation concerning materials in contact with food is derived from Directive 89/109/EEC, enacted in the UK as the Materials and Articles in Contact with Food Regulations 1987, and applied as Statutory Instrument SI1523 1987. The main area of concern is the possibility that components of the materials in contact with the food could leach into the food and endanger a person's health, or adversely affect the quality and taste of the food. There is also the design of a symbol which may be included as the moulding of a plastic, or into the labelling of the product to indicate that the materials which have been used for the food contact have been tested in accordance with the various regulations and are suitable for contact with foods (see Fig. 4.1).

As the use of plastics in packaging is so widespread, it is not surprising that there is emphasis on legislation specifically directed towards the control of substances in plastics. Some materials are restricted in their use (such as vinyl chloride monomer and regenerated cellulose) and there is a series of leaching tests to establish whether a plastic will release materials into the food, simulating aqueous, acidic and fatty foodstuffs. At the time of writing, this legislation is being revised and several pieces of legislation are being compiled into an updated Directive. The latest guidance is available for reference on the Food Standards Agency website: http://www.food.gov.uk/.

The enforcing authority for food standards in the UK is the Local Authority Trading Standards Officer, who is mainly concerned about the quality and fairness of trade of the product. The enforcing authority for the hygiene of the materials, preparation premises and processes is the Local Authority Environmental Health Officer.

Other packaging materials used in contact with foods include paper and board. The general requirement for all materials in contact with foods is given by the Regulation (EC) No. 1935/2004, which states:

> Materials and articles ... shall be manufactured in compliance with good manufacturing practice so that, under normal or foreseeable conditions of use, they

4.1 The only EC-approved symbol that may be used to denote food contact materials and articles when they are not already in contact with food when sold (FSA guide, 2009).

do not transfer their constituents to food in quantities which could: endanger human health; or bring about an unacceptable change in the composition of the food; or bring about a deterioration in the organoleptic characteristics of the food.

However, with respect to paper and board the legislation has not been developed to specify the methodology by which these objectives are to be met, and because of the absorbent nature of the material, the procedures used for plastics are not suitable. Consequently the Confederation of European Paper Industries has developed an industry guideline document, based mainly upon the recommendations of the German industry, covering the permitted components of the papers and boards, and their coatings.

Although glazed pottery has been used as containers and serving crockery for many centuries, there is ongoing concern about the content of the materials used to colour the glazes, and particularly to restrict the amounts of lead and cadmium which are able to leach from them into the food. The most recent legislation introduced concerning these materials implements the amending provisions of European Directive 2005/31/EC and was introduced in the UK under the Ceramic Articles in Contact with Food (England) Regulations 2006. Further information may be found on the Food Standards Agency website.

Glass containers for foods are required to comply with the general requirements for materials in contact with foods, but there is no specific EU legislation relating to glass. The high temperatures used in its manufacture clean any organic contamination, and glass is regarded as almost inert, with a low risk of unacceptable leachate unless it has been decorated with unsuitable enamel. Any cold-end treatment should be of food contact grade. There are concerns for the potential effects of traces of alkali leaching from glass and affecting sensitive pharmaceutical ingredients, which are covered by a specification for 'Pharmaceutical grade' glass (as described in Section 4.3.2).

Metal containers for foods are also required to comply with the general requirements for materials in contact with foods. Petroleum-based lubricants used in the formation of containers have to be effectively removed. Other materials which are used in association with metals for the construction of packaging containers and closures include a range of coatings and sealing compounds. These are subject to the same general requirements concerning contact with foods, and some potentially harmful ingredients are restricted. For example, a specific limit of migration into food has been established for the use of PVC gaskets to seal the lids of glass jars.

A current example of a recent EU Regulation which has an indirect effect upon packaging is REACH, i.e. Regulation (EC) No. 1907/2006 concerning Registration, Evaluation, Authorisation and Restriction of Chemicals. The purpose of REACH is to extend the current legislation covering all chemical substances manufactured within, or imported into the EU at quantities of 1 tonne or more. The requirements of REACH are

> **R**egistration of basic information of substances to be submitted by companies, in a central database. **E**valuation of the registered information to determine hazards and risks. **A**uthorisation requirements imposed on the use of high-concern substances. This process will be used for both new and old **CH**emicals. The new regime also creates a European Chemicals Agency (ECHA) and amends current legislation.

The Food Labelling Regulations 1996 (SI 1996/1499) consolidate and replace the Food Labelling Regulations 1984 in England, and corresponding legislation elsewhere within the UK. They principally implement Council Directive 79/112/EEC on the labelling, presentation and advertising of foodstuffs (apart from the provisions of that Directive relating to net quantity: see Section 4.4.1).

The name of the food must be an accurate description, supported by the ingredients list, which is to indicate the ingredients in descending order of the amount present. Some ingredients, such as vitamins, have to be blended with other materials in order to handle them, and so are known as compound ingredients. The make-up of compound ingredients is also to be declared, as are the functions of specific groups of ingredients (such as acidity regulators). The amount of the product within the pack is required to appear in a prescribed print format (type height and location). The metric system is normally used, but derogation has recently been agreed for dual labelling with imperial units to continue to be used on some groups of products within the UK. The information on the pack is also required to advise of potential allergic reactions to the listed ingredients of food products.

The information on a pack is also required to include instructions for use and the country of origin of the product. The shelf life of the product must also be indicated adjacent to the batch number and storage conditions, as the shelf life is dependent upon the storage conditions being maintained. The form in which the shelf life is communicated depends upon the duration of the shelf life. Finally, the pack is also required to provide the name and address of the importer, manufacturer, packer or seller, to whom the customer may complain, or obtain further information should the need arise.

4.3.2 Legislation concerning safety of packaging for medicines

There are widespread requirements that medicines are to be contained within child-resistant packs. Exhaustive panel tests are required to validate that the packs cannot usually be opened by young children and that they can usually be opened by an elderly person. The details of the tests vary widely from country to country. This legislation applies not only to re-closable containers, but also to the single use blister-packs used for some tablets and capsules. Child-resistant closures are also required to be used on hazardous substances such as household bleach and garden chemicals. The design and layout of the text of the pack is controlled as part of the product licence. Amongst these is a requirement that the product name and strength appears in Braille. This may be achieved by embossing a carton, or printing a thermally raised varnish onto a label.

Pharmaceutical measuring devices such as spoons are required to be 'CE' marked, which indicates that the device has been validated and subject to ongoing process controls to ensure that it accurately delivers the required quantity of the medicine. As expected, the requirements for materials in contact with pharmaceuticals are more demanding than those for foods. For instance, it is commonly thought that glass is inert, but it does tend to leach a small amount of alkaline material. There are some pharmaceutically active compounds which are sensitive to such small traces of alkali, and so pharmaceutical grade glass has limits on the amount of alkali released under test conditions. Similarly, any plastics which may be in contact with medicines which are to be injected directly into the patient's blood stream have more stringent requirements concerning possible leachates from the plastics, in addition to those controlling the sterility of the materials.

4.3.3 Legislation concerning safety of aerosols

Aerosols are commonly used but are potentially hazardous because they are pressurised containers, often containing flammable or corrosive substances. The Aerosol Dispensers Directive has been subjected to several updates, and covers areas such as: the size of the container by material; the testing of empty cans; the testing of finished aerosol product; and the requirements for clear labelling for instructions for safe use and disposal, together with appropriate warnings. The legislation includes the provision of the 'reversed epsilon' mark ᴲ to indicate compliance with the requirements of the Directive and enable smooth transport across the borders between EU countries. Aerosols are required to carry specific UN warning markings on the cases and shipping containers.

4.3.4 Legislation concerning protection of the workforce

Legislation on the control of substances hazardous to health (COSHH) requires that the operators are fully trained in the safe handling of the materials they are using. Also the risks involved in their procedures are required to be fully assessed and eliminated, or reduced to a safe minimum. The packaging of bulk ingredients, and the storage of intermediate products, should be designed to minimise risks in handling and use. Similarly, all equipment being used is required to be constructed and guarded to prevent accidental injury. There are strict limits on the amount of noise that may be emitted in the working environment, and under normal circumstances the standard minimum temperature for the workplace is 13°C.

4.4 Legislation concerning honesty in trade

There is nothing new about problems of dishonest trade. Ancient laws called for fair measures, and the prophets preached against unjust trading practices, including balances, weights and measures. Within recent decades, a range of legislation has been introduced to cover the amount of product provided and the perception of quantity and quality conveyed by the packaging, in its size, text and illustration. Although the Consumer Protection from Unfair Trading Regulations 2008 are concerned mainly with accurate commercial representation, they have implications for the marks which might be included in the print, and aspects of the packaging. These may include packaging which is significantly over-size for its contents, or containing compartments which appear to provide product but are actually empty.

4.4.1 Legislation relating to the filling of products

The Weights and Measures Act of 1963 continues to be in force within the UK, in some areas such as those concerning the calibration of weighing equipment and the amount declared by count, which is required to be precisely accurate, unless clearly stated otherwise. However, most of its requirements were replaced by the Weights and Measures Act of 1979, subsequently updated in 1985, which implements Directive 75/106/EEC and Directive 76/211/EEC. These were amongst the first pieces of EU legislation to impact upon the packaging industry in the UK. The main change was from the concept of 'minimum net weight' to 'average fill'. It is also necessary to control the variation in the amount being filled. This is achieved by introducing a 'tolerable negative error' with statistical conditions to limit the quantities of packs at the lower end of the range. These requirements are summarised in three rules:

- The average of the contents of the packs shall not be less than the nominal quantity (the label claim).
- Less than 2.5% of the packs may contain less than the nominal quantity minus the permitted tolerable negative error.
- No packs may contain less than the nominal quantity minus twice the permitted tolerable negative error.

The filling operation is required to become more accurate, as a percentage of the label claim, as the value of the permitted tolerable negative error changes as the nominal quantity increases (see Table 4.1).

Also, the place of enforcement of this legislation moved from the retailer to the factory where the product was packed. Statistically-based investigations of the operation of the filling processes were required in the setting-up of appropriate sampling systems for routine production checks, to show that the filling process was under adequate control and this became the basis for a 'due diligence' defence. The UK legislation also introduced the concept of the 'Home Authority Trading Standards Officer' so that the Officer with whom the company dealt with routinely would be involved with any complaints or legal proceedings. The Home Authority Trading Standards Officer is the authority to whom application is made to approve the use of a specific 'e' mark on the pack, to enable smooth transportation across borders within the EU. The mark must be as specified in Fig. 4.2 (not a font similar to it) at least 3 mm high and appear in the same field of vision as the statement of quantity. The height of type required to be used to declare the statement of quantity relates to the size of the pack.

This legislation also introduced the concept of 'measuring container bottles', which have a target filling mark included within the moulding and a guarantee that if filled to that mark the product will comply with the legislation. Although they give an additional assurance to the final customer, and move the responsibility for the statistical control to the bottle manufacturer, the responsibility for overall compliance remains

Table 4.1 The permitted tolerable negative error changes with the quantity of product within the pack

Nominal quantity (Qn) g or ml	Tolerable negative error	
	as % of Qn	g or ml
5–50	9	–
50–100	–	4.5
100–200	4.5	–
200–300	–	9
300–500	3	–
500–1000	–	15
1,000–10,000	1.5	–
10,000–15,000	–	150
above 15,000	1	–

4.2 The use of the 'e' mark is authorised by the Home Authority Trading Standards Officer.

with the filler. Such bottles are more expensive because of the additional controls, and the associated waste arising from their production to a tighter specification than would normally be required by the trade association conventions. However, in the case of glass containers, they provide a useful example of where trade association conventions have been upgraded into legislative regulations.

The EU Regulation No. 1169/2011 is now implemented. This affects the provision of nutritional information, and the highlighting of potential allergens. It also affects the declaration of the country of origin, and minimum type-size of the relevant text.

4.4.2 Legislation concerning deceptive packaging and label claims

The Trades Descriptions Act 1968 has been updated by the the Consumer Protection Act 1987. Part of this Act implemented European Community (EC) Directive 85/374/EEC, the product liability Directive, by introducing a regime of strict liability for damage arising from defective products. It also created government powers to regulate the safety of consumer products through statutory instruments, and identified the criminal offence of giving a misleading price indication. Similarly the Labelling Regulations are required to accurately describe the product. For instance, the Food Labelling Regulations 1996 SI 1499 requires that it is only permitted to claim that the product is a rich source of that vitamin if the amount expected to be taken during the day provides at least half of the recommended daily intake.

Another area of close legal scrutiny is whether the claims made for the product imply medicinal properties. Within the EU only licensed medicines are allowed to make or imply curative or preventative properties for the product. If so, the product falls under the exacting demands of the Medicines Act in the UK and is required to be licensed with clinical support for the claims made, and evidence of the safety of the product. This legislation brings an additional level of enforcement through the requirement for the quality control systems to be under the supervision of a 'qualified person', who is personally responsible to the medicines inspectorate for the suitability of the production facilities, and the quality of each batch released onto the market.

4.4.3 Legislation concerning design rights, patents and copyright

In the fiercely international competitive market, it is necessary to protect against competitors introducing similar products which are 'passed off' as the original product. This may be in the physical design, or in the text or graphics of the labelling. The main aspects of legislation in this area are the Trade Marks Act, and the Copyright, Designs and Patents Act. As for all aspect of legislation, professional advice must be sought by the newcomer to a market, in order to ensure that the new product is distinctly different from competitors' products.

4.5 Legislation concerning protection of the environment

4.5.1 The EU Directive on Packaging and Packaging Waste 94/62/EC

The Directive on Packaging and Packaging Waste 94/62/EC was incorporated into UK law as the Producer Responsibility Obligations (Packaging Waste) Regulations, and the Packaging (Essential Requirements) Regulations, both having been amended several times. These Regulations are examples of variation in implementation of a Directive across the differing countries of the EU (as indicated in Section 4.1.2). The system adopted in the UK is completely different from that in other member states. There are also differences between the implementation of each of the Regulations. The Producer Responsibility Obligations (Packaging Waste) Regulations impose a threshold of £2 million turnover and handling of 50 tonnes of packaging, below which no action is required. The definition of packaging is by a decision tree (basically 'Does it contain the product?', 'Is it intended to be discarded prior to use of the product?') and the Regulations are enforced by the Environment Agency. The Packaging (Essential Requirements) Regulations apply at the point at which the product was first placed onto the market, regardless of turnover or tonnage. The Regulations endorse a series of CEN standards (published by the Comité Europeén de Normalisation therefore having authority across Europe) including the definition of packaging by a definitive list, and are enforced by the Local Authority Trading Standards Department.

The basic requirement of Producer Responsibility Obligations (Packaging Waste) Regulations is the 'polluter pays' principle. In the UK this obligation is shared across the packaging supply chain, increasing as value is added to the item, from material supplier, converter, packer-filler and retailer. It is usual for a company to perform more than one activity in the packaging supply chain. For instance, a food manufacturer would be responsible for the pack/fill obligation of the primary pack and both the pack/filling and selling obligations of a transit case used to ship the product to the retailer, as the retailer is the end user of the transit case and thus the food manufacturer is the seller as well as the pack/filler. The materials are divided across seven categories, paper/board, plastics, steel, aluminium, glass, wood and other (hessian, cork, etc.) with rules for allocating composite materials. There are additional obligations for imported materials and finished products, and exemptions for materials exported, as they do not end up in the UK waste stream. Consequently suppliers have an interest in where their materials, or packaging components, have been supplied. There are also requirements to provide information to the public to encourage recovery and recycling activities. Most companies use the services of a Packaging Compliance Scheme to handle their registration with the Environment Agency, and to assist them in updating and maintaining their data collection and reporting systems, as they may be audited to ensure they meet the requirements.

The Packaging (Essential Requirements) Regulations include limits on the content of heavy metals, and a requirement for design to enable recovery and recycling. There is also a requirement to minimise the amount of material used in the packaging. This is applied via a procedure in which it is required to be shown that the amount of material being used has been reduced until something prevented further reductions. There is a list of technical, legal, commercial and marketing considerations which might become the limiting factor.

4.5.2 Green claims and recycling information

The international standard on environmental claims ISO 14021 has been established to clarify the quality of information which the consumer may expect. It is applied in the UK through the green claims code, which is being updated. Trading Standards Officers are empowered under the Trade Descriptions Act, and the Department of Fair Trading through the control of Misleading Advertising Regulations 1998. The attributes of green claims (made in text or by the use of symbols) are that they should be: truthful; accurate; able to be substantiated; clear and relevant; in plain language; specific and unambiguous. The use of symbols is widespread but many do not have clearly understood meanings. A useful summary of these symbols and their significance is provided on the recycle-more website (www.recycle-more.co.uk). There is also a series of symbols indicating the expected availability of facilities for recycling, which is being promoted by the major retailers and recycling schemes in the UK as a means of discharging their responsibilities for providing information promoting recovery and recycling.

4.5.3 Environmental factory operations

The wider subject of the environmental impact of factory operation includes controls on effluents, emissions, categories of factory waste, resource utilisation, etc. These are also connected with the international standards ISO 14000/14001. However, this legislation is outside the scope of this chapter.

4.6 Legal considerations in international trading

With regard to trade across international borders, there are legal requirements in the country in which the product is made, the shipping of the product, and its distribution and sale in each destination country. There is also the possibility that the product may not end up in the intended market. There are shared regulations, such as the various United Nations transport regulations, affecting the specification and testing of containers and the use of international standards and symbols for the shipping of potentially hazardous substances. These are very detailed and the requirements vary with the material, the quantity, the method of shipment, and the location to which they are being sent. Alongside these may be requirements of the operating policies of insurance and shipping companies. It is not unusual to find shared objectives but differences in the details of the regulatory approach taken. Examples are given in Table 4.2. In light of these issues, it is recommended that the services of

Table 4.2 Comparison of example regulations or trade conventions between Britain/Europe, the United States and Japan

Area of application	Britain/Europe	United States	Japan
Food contact materials and food labelling	As Section 4.3.1 Food Standards Agency	Similar intentions but differ in detail and format, e.g. presentation of nutrition information. US Food and Drugs Authority (USFDA)	Similar intentions but differ in detail and format, e.g. use of both 'Best before' and 'Use by' dates. Japan Ministry of Health and Welfare
Filling control	Imperial and metric; average with limits on low variations, as Section 4.3.1	Traditional, or metric; average with limits on low variations; revised as the accuracy of filling techniques improves, by the National Institute of Standards and Technology	Metric; average with limits on low variations. Some pack sizes are retained to reflect the previous non-metric system
Child-resistant closures	As Section 4.3.1 and Chapter 15	The panel test criteria are different	The panel test criteria are different
Main bar code symbology	13 digit, begins with 5 GS 1 UK	Developed the original 12 digit, now extending to include 13 digit	13 digit, begins with 4
RFID frequencies – UHF 'tags' for international operation have to respond to frequencies from 860 to 960 MHz	865–868 MHz EPC Global	Developed VHF 13.56 MHz and then UHF at 915 MHz. RFID Journal	950–956 MHz RFID Journal

appropriately qualified and experienced agents are secured to help to navigate safely through the applicable legislation at each stage of the product's journey to the final customer.

4.7 Sources of further information

Hard copies of the various Directives and Regulations and their national implementation may be obtained from the appropriate publishers or suppliers. As indicated in the introduction, the various government agencies have taken the opportunity to provide guidance as to how they expect the legislation to be interpreted, and how a company may ensure compliance.

The advice of the appropriate enforcing authorities should be sought at the development stages of a new product or pack change. In the UK the Trading Standards Officers should operate on the 'Home Authority Principle', which makes it advisable to build an effective working relationship with the local officer. Information and advice may also be sought from the trade association or from common interest groups such as the Industry Council for Packaging and the Environment (INCPEN).

Appropriately qualified and experienced consultants are available to provide advice and assistance.

4.8 The role of trade associations

Trade associations are paid for by their members and are operated for their benefit. This includes the provision of advice to their members on regulatory requirements related to their industry. They also provide advice to legislators through the consultative procedures in the preparation and implementation of regulations, as indicated in Section 4.1.2. They also provide information to the public, and make collective arrangements for the discharge of legal responsibilities, as the compliance schemes do for the Producer Responsibility Obligations.

4.9 What is legally required and what is good practice?

Whilst it is imperative to comply with legal regulations to avoid the risk of prosecution and the subsequent penalties, it is also necessary to understand and fulfil the appropriate trading requirements in order to provide the product as the customer would expect it.

4.9.1 Industry guides

Each industry has its trade association and their appropriate guides which serve to provide information on both the legal and operational requirements. An example is the British Aerosol Manufacturer's Association (BAMA), which provides information to enable manufacturers in the UK to produce products suitable for shipment across Europe, as the appropriate Directive has not been adopted into UK law. There are other industry guides which are used as the enforcement instrument by the regulatory authorities. For example, the 'guide to good manufacturing practice' is used in enforcement by the medicines and healthcare regulatory authority.

4.9.2 Trade conventions

It is important to be aware of the appropriate trade conventions, as whilst trade conventions are unenforced unless invoked in commercial contracts, they may be taken as the default understanding if not otherwise stated. For example, within the print industry an order may be considered to be complete if the quantity of items provided is within 10% of the quantity requested. This could cause difficulties in scheduling and collecting materials together for a required production run, especially if it is for a specific order requiring a minimum quantity.

There is also a range of informal systems which may change rapidly, such as the major UK retailers' different symbols for the labelling of food contents and the provision of advice for the disposal or recycling of waste packaging. Other examples include: statistical conditions on glass container tolerances; the normal tolerances on thickness of paper and carton board; the management of bar code numbering registers and symbology, including radio frequency identification, and electronic data interchange software conventions. There is also a range of informal systems which may change rapidly, such as the major retailers' different symbols to indicate the analytical values relating to the content of foods and the provision of advice for the disposal or recycling of waste packaging.

4.10 Methods of enabling consistency of compliance

It is important to ensure that company policies and procedures are set up in such a way as to manage compliance throughout the supply chain, using the appropriate standards to demonstrate compliance. The most widely used standard is ISO 9001, which comes from a quality control background and has developed towards encouraging an ethos of continual improvement across all aspects of the business. There is a special version of ISO 9001 for suppliers to the pharmaceutical industry, including reprographics and printed materials.

As its name suggests, the 'BRC/IoP Global Standard for Packaging and Packaging Materials' focuses upon the requirements of those particular sectors of business activity. Its requirements cover areas which are particularly relevant to packaging under the headings of: senior management commitment to continual improvement; hazard and risk management; technical management; site standards; product and progress control; and personnel. As indicated in Section 4.9.2, it is necessary to also be aware of the trade conventions relating to the area of operation.

4.11 The consequences of failure to comply with legislation

The failure to comply with legislation will lead to the occurrence of circumstances which the regulations are in place to prevent, such as: failing to provide the appropriate level of preservation; failing to provide the required information; or delivering poor quality or unsafe packs into the market. These may be handled as customer complaints, but may also lead to legal proceedings. The penalties for non-compliance may be brought against different bodies. Commercial cases may be brought against the company and in some cases the directors. Some cases, such as medicines and some environmental issues, may be against the responsible individual within the company. Alongside the legal penalties are the issues of damage to consumers' confidence in the product, or in that sector of the market. If the situation led to failure to supply saleable product, commercial proceedings could be very severe, such as the sanction of the penalty clauses within a commercial contract to recover loss of profit. The trading relationship with the customer would be adversely affected. This may include the de-listing of the product by the major retailers, with negative consequences for the financial health of the supplying company.

4.12 Sources of further information and advice

Guide to Packaging and Labelling Law by Charles James, 2nd edition, 2012.
Online access to UK legal publications: Office of Public Sector Information (http://www.legislation.gov.uk/).

Online access to guidance notes – the references to the current UK regulations are within the guidance notes:

- Food Standards Agency (http://www.food.gov.uk/).
- Paper and Board Food Contact Guidelines (www.paper.org.uk/members/current_issues/food_contact/Pre_Publ_rev_bp-20090423-00012-01-E.pdf).
- Environment Agency – Packaging Waste Regulations (http://www.environment-agency.gov.uk/netregs/legislation/current/107198.aspx).
- Department for Business Innovation and Skills (BIS) Essential Requirements Guidance update (http://www.bis.gov.uk/assets/biscore/business-sectors/docs/p/11-524-packaging-regulations-government-guidance).
- Office of Fair Trading Guidance (http://www.oft.gov.uk/OFTwork/publications/publication-categories/guidance/cprregs/).
- Department for the Environment. Food and Rural Affairs (DEFRA) Producer Responsibility Pages (http://www.defra.gov.uk/environment/waste/business/packaging-producer/).

4.13 Disclaimer

The views expressed in this chapter are intended to provide a basic introduction to the wide range of legal requirements related to packaging. Any interpretations are expressions of the personal opinion of the author and should not be taken as definitive. Reference to the guidance provided by the appropriate regulatory authority is recommended in the first instance, followed, if necessary, by specialist legal advice.

5
Packaging and environmental sustainability

H. J. EMBLEM, Environmental Packaging Solutions, UK

Abstract: This chapter introduces the link between packaging and the environment. It explores the need to reduce consumption for the benefit of future generations. Various definitions and terms are explained including sustainable development and carbon footprinting. Compliance with European legislation concerning packaging waste and the environment is covered from waste reduction such as packaging minimisation, reuse, recycling through to disposal methods such as composting, energy recovery and landfill.

Key words: sustainable, sustainability, environment, environmental, aspect, impact, waste, recycling, packaging.

5.1 Introduction: why bother?

There are many reasons for considering the environment in the design, manufacture, use and disposal of packaging and packaging materials. The first and arguably the most important is the use of potentially scarce materials. Profligate and wasteful use of these resources will result in a shortage for the future. However, taken into consideration at the same time should be the following:

- Increasing cost with increasing scarcity
- Effect on future generations
- Waste produced
- Compliance with the law and voluntary agreements
- Climate change and disruption.

Increase or reduction in cost is core to any profitable enterprise and will not be covered in detail here.

In terms of the effect on future generations, the products that are best known to most members of the public are fossil fuels such as petrol, oil and gas and materials such as plastics derived from oil. There is some debate as to when the world will run out of oil as extraction of oil from certain locations such as the Falkland Islands or Alaskan shale becomes a likelihood when market prices rise and what was previously an expensive production cost is supported by a higher selling price. Increases in cost or shortage of these materials could have a catastrophic effect on the lives of our children and grandchildren.

Please note that it is not practical within this chapter to refer to how every country has approached its legislation, guidance and modes of operation with respect to packaging and environmental sustainability. Therefore the examples given are specific to the United Kingdom.

5.2 Key definitions

To assist in the understanding of the concepts of packaging and the environment, there are some definitions that must be explained.

5.2.1 What is 'packaging'?

'Packaging' shall mean all products made of any materials of any nature to be used for the containment, protection, handling, delivery and presentation of goods, from raw materials to processed goods, from the producer to the user or the consumer (EC Directive 94/62).

5.2.2 What is the 'environment'?

The surroundings in which an organisation operates, including air, water, land, natural resources, flora, fauna, humans and their interrelation (ISO 14001 clause 3.5).

5.2.3 What is an 'environmental aspect'?

An aspect of an organisation's activities, products or services that can interact with the environment (ISO 14001 clause 3.6).

5.2.4 What is an 'environmental impact'?

Any change in the environment, whether adverse or beneficial, wholly or partly resulting from an organisation's environmental aspects (ISO 14001 clause 3.7).

5.2.5 What is meant by 'sustainable development'?

There is no universally agreed definition of sustainability or sustainable development. The following are possible:

- The impact of our personal values, choices and behaviours on the wider world.
- Sustainable development must be considered in the context of scarce resources and the needs of future generations.
- There is a Native American saying: 'We don't own nature. We borrow and manage it in our lives, thinking about our descendants'.
- 'No generation has a freehold on this earth. All we have is a life tenancy with a full repairing lease' (Margaret Thatcher, 1988).
- Sustainable development is: 'development that meets the needs of the present without compromising the ability of future generations to meet their own needs' (*Our Common Future*, Oxford University Press, 1987; also known as the Brundtland Report, after the former Norwegian Prime Minister Gro Harlem Brundtland).

However, this raises a variety of questions about the extent of our own needs and those of future generations. In summary, 'sustainable development' can be thought of as a three-legged stool, one leg representing each of the social, environmental and economic aspects of a community. If any of the aspects is neglected and either withers or does not grow at the same rate, then the stool becomes unstable. Sustainability is no longer just 'nice to have' or exclusively part of corporate social responsibility but is seen as a business necessity to attract consumers and protect market share.

5.2.6 What is 'waste'?

'Any substance or object ... which the holder discards or intends or is required to discard' (EU Directive: 75/442/EEC).

5.2.7 What are 'recycling', 'recovery' and 'energy recovery'?

Recovery has a broad definition laid down in EU Directive 75/442 and includes recycling, organic recovery of waste by means of recycling, re-use or reclamation or any other process with a view to extracting secondary raw materials, or the use of waste as a source of energy.

'Recycling shall mean the reprocessing in a production process of the waste materials for the original purpose or for other purposes including organic recycling but excluding energy recovery' (EU Directive 94/62). In reality, this means the processing of waste material (such as glass containers or corrugated cardboard) through a process (such as melting for glass or re-pulping into fibres for board) and making into another item (such as a glass object, not necessarily packaging; or into a sheet of paper, again not necessarily packaging).

'Energy recovery shall mean the use of combustible packaging waste as a means to generate energy through direct incineration with or without other waste but with recovery of the heat' (EU Directive 94/62).

'Organic recycling shall mean the aerobic (composting) or anaerobic (biomethanisation) treatment, under controlled conditions and using micro-organisms, of the biodegradable parts of packaging waste, which produces stabilised organic residues or methane. Landfill shall not be considered a form of organic recycling' (EU Directive 94/62).

5.3 Waste

'An empty Coke bottle is no different in value or substances to a full one except you've removed the contents. The only thing that has changed is your attitude to it' (Gerry Gillespie, Advisor to the New South Wales Government, Australia).

In the UK we have buried our domestic, industrial and commercial waste in the ground for many years. This has been the cheapest and easiest method of waste disposal. Other European countries do not do this for a number of reasons, including:

- The water table in some parts of Europe is much nearer the surface than in the UK and any waste that is buried can easily contaminate water systems;

- There are fewer naturally occurring or man-made holes in the ground than in the UK (caused for example by gravel extraction);
- The geology is such that digging holes to bury waste is difficult.

Consequently, incineration of waste (with energy recovery) has developed much faster than in the UK and is the main disposal method in other EU member states.

Production, consumption and waste disposal patterns in the UK are currently incompatible with sustainable living. They account for a significant proportion of greenhouse gas emissions and are dependent on inputs of non-renewable resources, energy and water. Products and materials are currently disposed to landfill that could be reused, recycled or have energy recovered from them (Defra Waste Strategy Annual Progress Report 2008/09).

5.4 Compliance with the law

The first environmental legislation was introduced in England around 1388 and made the offence of dumping of animal remains, dung and garbage into rivers, ditches and streams, punishable by death. Today there are many environmental laws, many of which emanate from the EU. Examples that can affect packaging are described in Sections 5.4.1–5.4.4.

5.4.1 Landfill Directive 1999/31/EC

In England and Wales the Directive is applied under the Landfill (England and Wales) Regulations 2002 (as amended). This has set targets that are binding on EU member states for the diversion of biodegradable waste from landfill in favour of more environmentally acceptable alternatives:

- By 2010 reduce to 75% of 1995 figure
- By 2013 reduce to 50% of 1995 figure
- By 2015 reduce to 35% of 1995 figure.

For many generations the UK has dumped all its rubbish into a hole in the ground. The main reasons for this are:

- Mineral extractions such as gravel had left many holes
- Landfill was convenient and easy
- The cost of sending to landfill was low.

However, a combination of factors has led to a lower reliance on landfill:

- Realisation that decomposing waste has an impact on human health
- Release of methane gas
- Odour
- Pests
- Noise
- Contamination of groundwater
- Escalating costs (Landfill Tax)
- Lower availability of sites.

Packaging and environmental sustainability 69

In 2010, the government conducted a Review of Waste Policy with an aspiration of working towards a zero waste economy. The Review and Action Plan were published in June 2011 with many targets stretching into 2014.

In addition, the Government's Waste Strategy 2000 set the following timetable, focusing specifically on recycling and composting:

- Increase recycling/composting of household waste to 25% by 2005
- Increase recycling/composting of household waste to 30% by 2010
- Increase recycling/composting of household waste to 33% by 2015.

These were subsequently increased by the Waste Strategy for England 2007 to:

- Increase recycling/composting of household waste to at least 40% by 2010
- Increase recycling/composting of household waste to 45% by 2015
- Increase recycling/composting of household waste to 50% by 2020.

5.4.2 Packaging and Packaging Waste Directive 94/62/EC (as amended by Directive 2004/12/EC)

There are two types of Directive that can be enacted by the EU. The first is enacted under Article 100 of the Treaty of Rome and takes effect immediately and requires no transposition into member state laws. The second is enacted under Article 95 and, when published, sets dates for transposition into Member State laws. In addition, any targets are minimum requirements and can be increased by Member States. The Packaging and Packaging Waste Directive is of the second type and has been enacted in different ways in Member States of the EU. Packaging and Packaging Waste Directive 94/62/EC is transposed into UK legislation as the Producer Responsibility Obligations (Packaging Waste) Regulations and the Packaging (Essential Requirements) Regulations.

5.4.3 The UK Producer Responsibility Obligations (Packaging Waste) Regulations

These Regulations implement most of the Packaging Waste Directive 94/62/EC (as amended) and set targets for recovery and recycling of packaging placed on the market by materials – paper/board, plastics, glass, aluminium, steel, wood and other (hessian, cork etc.). The fundamental requirements of the EU Directive are:

(i) At least 60% by weight of packaging waste must be recovered or incinerated at waste incineration plants with energy recovery;
(ii) At least 55% and no more than 80% by weight of packaging waste must be recycled;
(iii) The following minimum recycling targets for materials contained in packaging waste must be attained:
   - 60% by weight for glass;
   - 60% by weight for paper and board;

- 50% by weight for metals;
- 22.5% by weight for plastics;
- 15% by weight for wood.

These targets apply to all packaging placed on the market within a Member State.

As stated, the Packaging and Packaging Waste Directive has been enacted differently in various EU Member States. Detailed information on implementation of the Packaging and Packaging Waste Directive in EU Member States can be obtained from *Understanding European and National Legislation on Packaging and the Environment*, published by Europen (The European Organisation for Packaging and the Environment).

The UK Government has taken a view that the targets shall not be imposed on small businesses and thus, in the UK, a company need take no action until its turnover exceeds £2m per year and more than 50 tonnes of packaging have been handled. To ensure the overall targets are met, the targets for companies above these thresholds are higher than those in the Directive. These have been steadily increasing year on year. Table 5.1 lists the minimum recycling targets for materials contained in packaging waste that must be attained. Of the material required to be recovered, a minimum of 92% of the recovery must be by recycling.

To comply with the Producer Responsibility Obligations (Packaging Waste) Regulations in the UK, a company exceeding both turnover and tonnage thresholds must in a calendar year:

(a) Register as a Packaging Producer with one of the Environment Agencies by 7 April and submit details of the packaging handled in the previous calendar year;
(b) Recover and recycle (or have recycling and recovery performed on its behalf) amounts of packaging equivalent to their obligation calculated from the targets above based on the packaging handled in the previous calendar year (the relevant Agency will perform this calculation) and keep evidence of this recycling and recovery;
(c) Provide a certificate of compliance to the relevant Agency by 31 January in the next calendar year.

The relevant Agencies will charge a fee for Registration. Evidence of recycling is in the form of Packaging Recovery Notes (PRNs) which can only be issued by a reprocessor of packaging waste registered with the relevant Agencies and these will also be charged for by the reprocessor(s).

Table 5.1 Minimum recycling targets in percent

	2012	2013	2014	2015	2016	2017
Paper/board	69.5	69.5	69.5	69.5	69.5	69.5
Glass	81	81	81	81	81	81
Aluminium	40	43	46	49	52	55
Steel	71	72	73	74	75	76
Plastic	32	37	42	47	52	57
Wood	22	22	22	22	22	22
Total recovery	74	75	76	77	78	79

Alternatively, an Obligated Producer can register with a Packaging Compliance Scheme approved by the relevant Environment Agencies, submit details of the packaging handled and pay fees to the Compliance Scheme. The Compliance Scheme will then arrange the recovery and recycling and certificate of compliance on behalf of its members. The relevant Agencies responsible for regulating the Producer Responsibility Obligations (Packaging Waste) Regulations in the UK are: The Environment Agency (EA) in England and Wales, The Scottish Environmental Protection Agency (SEPA) in Scotland and the Northern Ireland Environment Agency (NIEA) in Northern Ireland.

What does this really mean?

Packaging should not be sent to landfill as much of it can be recycled. In the UK the recycling loop has been active for many years for glass containers, newspaper and certain forms of aluminium and steel. The recycled content is significant and the public are accustomed to taking these materials to recycling centres. Since the Packaging Waste Regulations have come into force, many local authorities now have kerbside collections for all paper and board packaging, steel and aluminium cans and plastic bottles. Glass is also collected at the kerbside by some authorities and bottle banks are available in some areas.

There is currently a disagreement concerning the inclusion of glass in co-mingled kerbside collections. On the one hand, Local Authorities are targeted to reduce overall weights to landfill and glass, being relatively heavy compared with other packaging materials, is an attractive addition and is claimed to drive up total weights diverted from landfill. However, the process of collection and sortation of mixed materials including glass leads to contamination of these other materials with broken glass. The paper recycling industry does not like to receive paper/board contaminated with glass fragments and will pay less for such contaminated materials. Similarly, glass that is collected by the co-mingled route is of course mixed in colour and often contaminated with other waste, which again leads to a lower value of potential recyclate. Much glass collected by the co-mingled route is used for aggregate and not for recycling into glass containers.

There is also much debate about the export of packaging waste for recycling outside of the UK and outside of the EU. If packaging waste is exported to an accredited reprocessor overseas, the exporter can legitimately issue Packaging Export Recovery Notes (PERNs) that can be used as evidence of the Recovery/Recycling Obligations for Producers in the UK in stead of PRNs (see above). It is claimed that the UK would not meet the EU targets for some materials if exporting for recycling was not available. Conversely, as exporting is possible, the revenue from PERNs is benefiting recycling companies abroad and not in the UK.

However, since the Producer Responsibility Obligations (Packaging Waste) Regulations came into force in 1997, overall recycling of packaging in the UK has increased from 28% to 61% in 2008 (*Making the most of packaging – A strategy for a low-carbon economy*, Defra, 2009).

5.4.4 The UK Packaging (Essential Requirements) Regulations

These Regulations implement Articles 9 and 11 of the Packaging Waste Directive 94/62/EC as amended.

Article 9

Packaging shall be so manufactured that the packaging volume and weight be limited to the minimum adequate amount to maintain the necessary level of safety, hygiene and acceptance for the packed product and for the consumer. This is not considered to indicate a preference between material types (e.g. glass versus plastics) or packaging systems (e.g. single trip versus reusable), although consideration of the overall environmental impact of the packaging system used would be encouraged (Packaging (Essential Requirements) Regulations, Government Guidance Notes, May 2010).

This means that whilst there is a requirement to minimise the weight and volume of packaging, it does not mean that a glass bottle should be so light that it shatters in transit or on the filling line. It does not mean that the barrier properties of a plastic film are insufficient to prevent the transmission of oxygen that shortens shelf life of the contents. It does not mean that the pack has to be so small that the ingredients list is in type too small to be easily read. It does not mean that there would be a high level of rejects in production of the packaging component or that the pack is not accepted by consumers nor is so difficult to use that the consumer will not purchase the product again.

A series of seven standards in relation to packaging were published by the European Committee for Standardisation (CEN). These provide framework methodologies for considering reduction, reuse, recyclability and recovery. The CEN Packaging Standards can be used for demonstrating compliance with the essential requirements above. Compliant packaging enjoys freedom of movement across the European Community. The Standards can also help inform decisions on packaging design, for example around material specification, maximising recyclability and recovery, minimising component parts, reducing wasted space and optimising pack size.

The use of the CEN standards will carry with it the presumption of conformity of the packaging with the essential requirements in all Member States. In other words, if the standards are used, the product will be considered to meet the essential requirements unless there are grounds for suspecting otherwise. These standards are available from British Standards Institute (BSI). It should be noted that the standards represent only one means of establishing conformity with the essential requirements, and that other means may be acceptable (Packaging (Essential Requirements) Regulations, Government Guidance Notes, May 2010). In the UK, Local Authority Trading Standards Departments are responsible for regulating the Packaging (Essential Requirements) Regulations.

Options for packaging minimisation

Before considering where packaging can or should be reduced, it is important to revisit the functions of packaging, layers or 'levels' of packaging used and cost effectiveness. As already discussed in Chapter 2, the functions of packaging are to:

- Contain, protect and preserve the product
- Provide a convenient and safe way of handling and using the product
- Give information about the product
- Present and help to sell the product.

What do we expect from packaging? Packaging must:

- Match product needs, whether physical, chemical or aesthetic (if the product is sub-standard, the pack is unlikely to make up for this)
- Meet market needs, by satisfying both consumer expectations and the demands of retailer/wholesaler (if the consumer is disappointed with the pack, it influences how they feel about the product and the brand)
- Be technically and economically feasible to produce, fill, move, display, sell, use and dispose of.

Levels of packaging

Most products placed on the market that are packed have more than one 'level' of packaging around them.

Primary packaging is packaging around the product when the consumer takes it home, i.e. that which ends up in domestic waste. Examples of primary packaging could be a bottle for wine, beer or spirits, along with the closure and label(s) used; a jar, closure and label for jam or sauces; a carton for a dozen ball point pens; a metal can and label for fruit or vegetables; a label applied directly to a product, e.g. an apple, to provide product information. It must be noted that there is often more than one item of primary packaging, such as a plastic film bag containing breakfast cereals which is then contained within a cardboard carton.

Secondary packaging is packaging that is used to group or collate primary packs together. This packaging does not usually end up in the domestic waste stream. Examples of secondary packaging would be a corrugated board outer to hold a dozen bottles of wine or two dozen cans of peaches; or a tray with film overwrap (the tray may be used for shelf display). As can be seen from the tray/film example, there may be more than one item of secondary packaging.

Tertiary packaging is packaging used to hold secondary packs together and allow transport through the distribution chain. Examples of tertiary packaging could be pallets or roll cages, or stretch or shrink wrap film holding goods securely on a pallet and the pallet label.

One of the tasks of a packaging technologist is to ensure that the goods arrive at the retail outlet in the correct condition and that the goods can be taken by the customer to their home and arrive in that same condition. Where there is more than one level of packaging, any reduction in packaging weight or volume will have an effect on the amount of protection afforded to the goods by that packaging. It is wasteful to adopt the 'belt and braces' approach. If the change in amount of protection caused by a reduction in a level can be taken up by one of the other levels without overall reduction in protection, then that reduction is meaningful and genuine. However, where a reduction in the amount of protection cannot be taken up by one of the other levels, and indeed one of the other levels has to be increased to maintain the overall amount of protection, there is not a meaningful or genuine reduction.

For example, household cleaner is packed in an HDPE bottle and 12 bottles collated in a corrugated tray with an LDPE film overwrap. The secondary packs are able to be stacked up to 10 high without damage and transported through the distribution system. If the weight of the bottle could be reduced by 20% and the bottle can still deliver the household cleaner from the supermarket shelf to the home, this could be considered a packaging reduction. However, this may mean that the method of collation and transit results in bottles collapsing under the weight of a 10 high stack. This of course could be addressed by using a full corrugated case with divider to ensure the bottle reaches the supermarket in satisfactory condition, but this would mean (a) more labour involved at the supermarket to put the bottles on the shelf as they would need to be taken out of the case individually and placed on the shelf rather than previously 12 being put on the shelf together; and (b) the weight of the case will be significantly more than the tray and shrink-wrap. In this situation it is questionable if a genuine packaging reduction has been made, but on the supermarket shelf it can certainly be claimed a reduction has been made.

Article 11

(i) Hazardous Content

The sum of concentration levels of the heavy metals (cadmium, mercury, lead and hexavalent chromium) in packaging or any of its components must not exceed 100 ppm. There are some derogations such as recycled glass and recycled plastic crates and pallets.

(ii) Noxious Substances Minimisation

Packaging shall be so manufactured that the presence of noxious and other hazardous substances and materials as constituents of the packaging material or of any of the packaging components is minimised with regard to their presence in emissions, ash or leachate when packaging or residues from management operations or packaging waste are incinerated or landfilled.

Cost effective packaging

This is packaging that meets all the necessary functions of packaging at minimum overall cost. As with packaging minimisation, it is necessary to consider the combined performance of all three levels of packaging when looking for cost effectiveness. In the example given above, the overall cost may have increased.

Cost effective packaging is not just the cheapest option, but must provide the specified amount of protection and preservation and be suitable for the filling line and distribution system. This will avoid raw material, product and labour waste during production and avoid product deterioration that could lead to customer complaints, product recall or protection. So, in summary, cost effective packaging must:

- Be easy and safe to fill/handle/use
- Be compatible with product and packing line machinery
- Maximise packing line efficiency and minimise waste and its disposal cost
- Save money and valuable resources

This will then demonstrate good environmental management.

5.5 Packaging re-use and recovery

Packaging must be manufactured so as to permit reuse or recovery in accordance with specific requirements:

- Re-use
- Recovery as recycling
- Recovery as energy
- Recovery as composting or biodegradation.

We now examine each of these requirements in turn.

5.5.1 Re-use

The physical properties and characteristics of the packaging should enable a number of trips or rotations in normally predictable conditions of use:

- It must be possible to process the used packaging without contravening existing health and safety requirements for the workforce
- The requirements specific to recoverable packaging when the packaging is no longer re-used and thus becomes waste must be met.

Examples of re-use would be plastic crates/tote boxes used for distribution of goods (often but not exclusively foods) by certain high street retailers or wooden pallets used by a company or hired out by a pallet management company for use within the distribution cycle. In both of these examples, the crates or pallets are returned through the distribution system for re-use. It would be necessary for such packaging materials, especially for those used for food, to be subjected to inspection and/or cleaning before re-use. Another example would be returnable beer or mixer bottles. In these circumstances, as these are for direct food contact, the bottles would, as a matter of course undergo a rigorous inspection and cleaning process.

Certain bespoke toiletries retailers offer a refill service for liquid products, where the consumer takes the bottle back to the shop and it is refilled at the high street premises by the shop staff. Some laundry products are available as a refill pack where the consumer purchases a container which may include a dispensing device such as a trigger spray on the first occasion but subsequently may purchase a lightweight refill pouch of the liquid and refill the bottle in the home.

There are a number of aspects of re-use to take into consideration. On a commercial scale (e.g. crates/tote boxes, pallets or bottles for beer), re-use considerations include:

- Losses
- Economics of collection and washing
- Environmental impact of washing process – energy, water, cleaning agent
- Effluent produced
- Hygiene and safety
- Specification sufficient for the product to withstand multiple trips.

In the home, re-usable packaging:

- Must be cheaper for the consumer
- Must be easy to refill, without spilling
- Must be durable to withstand multiple use.

The item must also be capable of being recovered or recycled at the final end of life.

5.5.2 Recovery by recycling

A percentage by weight must be capable of recovery by recycling, i.e. must make a positive contribution to the output of the material recycling process for which it is considered suitable. Considerations here include:

- Environmental impact of recycling process – the use of recycled material does not always have a lower environmental impact than the use of virgin material.
- Economics and sortation of waste material – it may be difficult or costly to separate waste materials, resulting in a higher cost for recycled product; mixed plastics are very difficult to sort and are often contaminated with food residues.
- Market for recyclate – there may be little demand for recycled material; due to an increase in wine consumption, recycled green glass is available in quantity but there is little demand in the UK.
- Performance of recycled materials – whilst glass, steel and aluminium can be recycled almost indefinitely, both paper/board and plastics suffer some degradation with every recycling process.

5.5.3 Recovery as energy

Packaging waste processed for the purpose of energy recovery should have a minimum inferior calorific value (also known as 'minimum net calorific value') to allow optimisation of energy recovery. In the absence of harmonised standards, this is taken to mean that the packaging will make a positive contribution to the energy recovered in a waste incinerator (Packaging (Essential Requirements) Regulations, Government Guidance Notes, January 2011).

This means that once ignited, it must not be necessary to apply further heat or other energy to continue the combustion process. In some circumstances, packaging waste with a high residual calorific value such as plastics can be mixed with non-combustible waste to produce an overall mixture that will burn without the addition of further energy.

Considerations regarding energy recovery include:

- Collection and sortation. Most packaging waste that comes from the domestic waste stream is contaminated with other waste and has to be sorted by mechanical means. Dense materials such as glass and potentially higher value materials such as aluminium and steel need to be removed. Paper and plastic contaminated with food residue may not be suitable for recycling but is suitable for energy recovery.
- Capital investment of incineration facility. Such equipment is only economically viable if sufficient waste is available for continuous operations.
- Site of the facility and environmental impact of energy recovery. Whilst the benefits of energy recovery may be appreciated, the old fashioned image of the smelly smoking chimney generates local resistance to the siting of energy recovery facilities.

5.5.4 Recovery by composting or biodegradation

The conditions for composting and biodegradation are fulfilled when the packaging complies with the following:

- Packaging should be largely combustible solids; that is, the residue after incineration should be less than 50% of the packaging. This figure is taken as indicating the organic content.
- The organic materials should be inherently and ultimately biodegradable materials, that is break down to carbon dioxide, mineral salts, biomass and water or methane. Chemically unmodified materials of natural origin such as wood, wood fibre, paper pulp and jute are accepted as biodegradable for these requirements.
- The packaging should disintegrate in the waste treatment process.
- The packaging should not retard or adversely affect the waste treatment process.
- The packaging should not degrade the quality of the resulting compost (Packaging (Essential Requirements) Regulations, Government Guidance Notes, January 2011).

Biodegradable, degradable and compostable

The terms biodegradable, degradable and compostable are often used interchangeably; however, each term has a distinct meaning.

- *Biodegradable* materials are capable of undergoing physical, chemical, thermal or biological decomposition such that most of the finished compost ultimately decomposes into carbon dioxide, biomass and water.

- *Degradable* materials will in a period of time, break down into smaller particles when exposed to heat, oxygen or visible or ultraviolet light.
- *Compostable* materials biodegrade through the action of naturally occurring microorganisms and do so within a specified time under specified conditions.

To be compostable an article must comply with BS EN 13432:

- The article must break down into particles and no more than 10% must be particles larger than 2 mm in size after 12 weeks under test conditions.
- The article must break down into water, carbon dioxide and biomass (called mineralisation).
- The resulting compost must have no negative effect on plant growth.
- The concentration of heavy metals or fluorine must not exceed specified limits.

It should be noted that only articles or products can comply with BS EN 13432, but materials themselves (e.g. polymers, plastics or compounds) cannot. This is because the dimensions may mean that disintegration would not take place within the time allowed of 12 weeks or mineralisation within the time allowed of six months. For example a 15 μm film may disintegrate within 12 weeks (and comply with the other requirements) but a 20 μm sample of the same film may not fully disintegrate. Thus the material cannot be claimed to be compostable, even though the 15 μm film can disintegrate fully.

Home compostable packaging

There is a need for distinction between home compostable and industrially compostable materials. Articles that degrade in an industrial composting vessel (high temperatures and air flow) may never degrade in a home compost heap. In addition to certification to BS EN 13432, it is also possible for a product to be deemed 'home compostable'. The Association for Organics Recycling has launched a Home Composting Certification Scheme in conjunction with AIB Vincotte in Belgium.

Degradable, biodegradable and bio-based plastics

Biodegradable or compostable plastics can be bio-based or petroleum-based, but bio-based plastics are not necessarily biodegradable (including conventional polymers made from bio-based monomers). (See Chapter 13 for further definitions and properties of bio-based plastics.) Whilst biodegradable plastics have an emphasis on disposal and compliance with BS EN 13432, bio-based plastics have an emphasis on origin of carbon constituents (renewable carbon as measured by C_{14} content). Bio-based plastics are chemically identical to petroleum-based plastics (of the same formula) and can be recycled together.

Degradable and biodegradable plastics in the waste stream

Compostable, degradable and biodegradable plastics need infrastructure for collection, sorting, identification and composting or anaerobic digestion, often in industrial composting facilities. Many compostable, degradable and biodegradable plastics will contaminate sorted plastics waste streams at a very low concentration and render the batch unsuitable for recycling as the final recycled product can have indeterminate properties. For example, less than 1% of polylactic acid bottles in the PET bottle waste stream can adversely affect the transparency of the recyclate PET.

Biodegradable and compostable materials are not suitable for landfilling. Properly constructed landfills are dry and anaerobic so that biodegradable materials, if they degrade at all, will produce methane, a greenhouse gas 21 times worse than carbon dioxide. This is undesirable except in the very few cases where the landfill is designed to produce methane gas as a fuel. Compostable plastics do not hamper the composting process because they are biodegradable, disintegrable and do not affect the quality of the final compost.

5.6 Environmentally responsible packaging

There is no such thing as a good or bad pack and there is no such thing as environmentally friendly packaging. A pack should only be considered in conjunction with its contents. Every human action has the potential to cause damage to the environment (that is to say every action can have an adverse environmental impact). In many cases the product is more environmentally damaging than the packaging and both product and packaging must be considered. The pollution caused by a litre of milk spilt down a surface water drain far outweighs the damage caused by an HDPE milk container in a landfill site. However, the use of the correct packaging can prevent the spillage and resultant pollution of milk and an HDPE milk container can easily be recycled.

An environmentally responsible pack is one that gets the product from production to consumption with minimum use of materials and energy, generating the least amount of waste. In summary, environmentally responsible packaging should:

- Be resource-efficient throughout the distribution chain
- Prevent product wastage
- Optimise packaging materials and energy.

This can be achieved by:

- Designing primary, secondary and tertiary packaging as an integrated unit
- Designing to minimise number of lorries on the road
- Designing for recycling when it yields a net gain in resources.

It is important to remember 'Packaging does not have life of its own. Packaging only exists to ensure goods arrive at the point of use in the appropriate condition. If there were no goods, there would be no need for packaging' (Dick Searle, the Packaging Federation).

5.6.1 Beware of 'greenwash'

Greenwash is the use of vague or unsubstantiated or incorrect environmental claims. The use of such phrases as 'environmentally friendly', 'can be recycled/recyclable', 'good for the environment' should be avoided. Defra has published an updated guide 'Green Claims Code Guidance' in February 2011. A green claim should be:

- Truthful, accurate, and able to be substantiated
- Relevant to the product in question and the environmental issues connected with it
- Clear about what environmental issue or aspect of the product the claim refers to
- Explicit about the meaning of any symbol used in the claim – unless the symbol is required by law, or is backed up by regulations or standards, or is part of an independent certification scheme
- In plain language and in line with standard definitions.

A green claim should not:

- Be vague or ambiguous, for instance by simply trying to give a good impression about general concern for the environment
- Imply that it commands universal acceptance if there is actually some significant doubt or division of scientific opinion over the issue in question
- Imply more than it actually covers, if the claim is only about limited aspects of a product or its production, or does not deal with a significant issue for that type of product
- Make comparisons, unless the comparison is relevant, clear and specific
- Imply that a product or service is exceptional if the claim is based on what is standard practice anyway
- Use language that exaggerates the advantages of the environmental feature the claim refers to
- Imply that the product or service is endorsed or certified by another organisation when it has not been.

5.7 Compliance with voluntary agreements

In the UK it is now possible to comply with a voluntary agreement such as the Courtauld Commitment. The Courtauld Commitment is a voluntary agreement involving close cooperation between WRAP (Waste and Resources Action Programme) and the UK grocery sector. The Commitment has operated in two phases: Phase 1 from 2005 to 2009 and Phase 2 from 2010 to 2012. The Commitment has proved to be a powerful vehicle for change. Its first phase resulted in real reductions in packaging and food waste, realising significant commercial savings. Over 50 major retailers, brand owners, manufacturers and suppliers have signed the agreement since its launch in July 2005. The signatories are working closely with WRAP to develop solutions across the whole supply chain, including innovative packaging formats, reducing the weight of packaging (e.g. light-weighting bottles, cans and boxes), importing in bulk rather than small individual containers, increasing the amount of recycled content in packaging, designing for recyclability, increasing the use of concentrates, establishing refill and self-dispensing systems and collaborating on packaging design guidance. They are also working on solutions for reducing food waste through innovative packaging, in-store guidance, and the Love Food Hate Waste consumer programme.

Through Phase 1 of the Commitment (2005-2009), the growth in UK packaging waste was halted and 1.2 million tonnes reduction in packaging and food waste was delivered. Of the original targets set, two of the three targets were achieved:

1. To design out packaging waste growth (zero growth achieved in 2008) and
2. To reduce food waste by 155,000 tonnes (this target was exceeded with 270,000 tonnes less food waste arising in 2009/10 than in 2007/08).

A third target, to reduce the amount of packaging waste over the same period, was not achieved. Total packaging consistently remained at approximately 2.9 million tonnes between 2006 and 2009, mainly because of a 6.4% increase in grocery sales volumes since the agreement began in 2005 and participating retailers taking a greater proportion of the overall market for beer and wine. However, on average, across the range of groceries bought, packaging has reduced by around 4% for each product, whether through using more concentrated detergents or lightweight cans. The results of the first phase of the Courtauld Commitment showed real progress on reducing food and packaging waste, and demonstrated how effectively governments and businesses can work together through responsibility agreements.

Over the course of the Courtauld Commitment there has been an increasing shift to life-cycle thinking and the embodiment of life-cycle parameters and measurements within the framework. Initially, the Courtauld Commitment began with simple tonnage metrics for packaging reduction. Phase 2 of the Courtauld Commitment (2010-2012) looks at the life-cycle of products from manufacture to how they are used in households and ultimate disposal. In Phase 2 the targets focussed on the carbon impact of packaging, waste from household food and drink, and waste across the supply chain. Measurement of the Courtauld Commitment 2 targets was from January 2010 to December 2012 with targets set against a 2009 baseline. The three targets were:

1. Packaging target – to reduce the weight, increase recycling rates and increase the recycled content of all grocery packaging, as appropriate. Through these measures the aim is to reduce the carbon impact of this grocery packaging by 10%.
2. Household food and drink waste target – to reduce UK household food and drink waste by 4%.
3. Supply chain product and packaging waste target – to reduce traditional grocery product and packaging waste in the grocery supply chain by 5% – including both solid and liquid wastes.

First year progress results show that signatories were already half way to achieving the packaging reduction target (5.1% reduction in year 1) and three quarters of the way to reaching the household food waste objectives (3.0% reduction in year 1). The supply chain impact was significantly less at only 0.4%, but this is a new area for the Commitment and will be an area of additional focus going forward.

5.8 Climate disruption

There has been much written and spoken about the impact of the human race on the climate of the planet. However, many scientists are of the opinion that the earth's climate is changing because of increased concentrations of carbon dioxide and other greenhouse gases. These gases trap heat in the atmosphere and can contribute towards an increase in average temperature, which in turn can lead to extremes of weather conditions, more rainfall than average and/or more periods of higher or lower temperatures than average. Other scientists say that the impact of solar cycles on the temperature of the earth has not been adequately taken into account.

5.8.1 Greenhouse gases

Greenhouse gases (GHGs) are so called because they contribute towards the greenhouse effect. The greenhouse effect describes the natural phenomenon where certain gases in the atmosphere increase the Earth's surface temperature due to an ability to trap heat, similar to the way in which glass traps heat in a greenhouse (Defra Environmental KPI Reporting Guidelines).

What is the problem with GHGs?

There is scientific evidence that the increase in atmospheric concentrations of GHGs due to human-induced (anthropogenic) GHG emissions is having a noticeable effect on climate. The increase in the natural process of the greenhouse effect caused by human activities is known as the enhanced greenhouse effect and leads to global warming (Defra Environmental KPI Reporting Guidelines).

What are the GHGs and where do they come from?

The list of greenhouse gases includes the following (Defra Environmental KPI Reporting Guidelines):

- Carbon dioxide (CO_2) – burning of fossil fuels
- Methane (CH_4) – waste and agriculture
- Nitrous oxide (N_2O) – agriculture
- Hydrofluorocarbons (HFCs), perfluorocarbons (PFCs) and sulphur hexafluoride (SF_6) – air conditioning, refrigeration and industrial processes.

However, not all these GHGs have the same global warming potential and, for ease of calculation, have a conversion factor expressed as carbon dioxide equivalents or CO_{2e}. From Table 5.2, it can be seen that one gramme of sulphur hexafluoride has the same global warming potential as 23.9 kilogrammes of carbon dioxide.

The Kyoto Protocol

At an international level, the United Nations Framework Convention on Climate Change (UNFCCC) and the Kyoto Protocol have established a framework within

Table 5.2 Relative impact – 100 year global warming potential*

GHG	% of UK emissions	Global warming potential
CO_2	80	1
CH_4	10	21
N_2O	8	310
HFCs	2.1	150–11,700
PFCs	0.1	6,500–9,200
SF_6	0.2	23,900

* Reproduced from 'A Practical Guide to Carbon Footprinting' with the kind permission of the Carbon Trust. Please check the Carbon Trust website for the latest guidance: (www.carbontrust.co.uk). Please note that this material is © the Carbon Trust and cannot be reproduced without permission.

which many countries are taking action to limit or reduce GHG emissions. The Kyoto Protocol entered into force on 16 February 2005 and imposes a legally binding GHG emission reduction target on the UK of 12.5% of base year emissions between 2008 and 2012 (Defra Environmental KPI Reporting Guidelines).

UK emissions total

In 2008, UK emissions of the basket of six greenhouse gases covered by the Kyoto Protocol were provisionally estimated to be 623.8 million tonnes carbon dioxide equivalent CO_{2e}. This was 2% lower than the 2007 figure of 636.6 million tonnes (Department of Energy & Climate Change, 26th March 2009).

The Intergovernmental Panel on Climate Change (IPCC) is the leading body for the assessment of climate change, established by the United Nations Environment Programme (UNEP) and the World Meteorological Organization (WMO) to provide the world with a clear scientific view on the current state of climate change and its potential environmental and socio-economic consequences. There is however, speculation as to how much carbon dioxide and other GHGs contribute towards the phenomenon of climate change, and what the effects of climate change may be. The IPCC's 4th Assessment Report has been challenged, as it is alleged that errors have emerged after checking of the sources cited by the 2,500 scientists who produced the report. In an interview with *The Times* newspaper, Robert Watson (former Chairman of the Panel) said that all the errors exposed so far in the report resulted in overstatements of the severity of the problem.

Adaptation and mitigation

Mitigation is about reducing carbon emissions, to slow down climate change. Adaptation is about coping with the impacts of climate change.

The Stern Review

The Stern Review set out to provide a report to the UK Prime Minister and Chancellor assessing the nature of the economic challenges of climate change and how they can be met, both in the UK and globally. The Review was the most comprehensive review ever carried out on the economics of climate change. The conclusions, published in 2007, were:

- There is still time to avoid the worst impacts of climate change, if we take strong action now.
- Climate change could have very serious impacts on growth and development.
- The costs of stabilising the climate are significant but manageable; delay would be dangerous and much more costly.
- Action on climate change is required across all countries, and it need not cap the aspirations for growth of rich or poor countries.
- A range of options exists to cut emissions; strong, deliberate policy action is required to motivate their take-up.
- Climate change demands an international response, based on a shared understanding of long-term goals and agreement on frameworks for action.

However, following various United Nations Climate Change Conferences, no binding commitments have been reached to reduce greenhouse gas emissions.

5.8.2 The environmental impact of packaging and the concept of life cycle analysis for a total packed product

Both the Essential Requirements Regulations and the Courtauld Commitment focus on packaging minimisation. But what does that mean? Thinner materials could be used. But this could lead to physical product damage if there is now insufficient protection. It could lead to a reduction in shelf life and/or increased wasted product if shelf life is reduced due to higher oxygen transmission of a thinner film.

It may be possible to change from a glass container to a plastic container. But this could lead to a reduction in shelf life and/or increased wasted product if the shelf life is reduced due to a higher oxygen transmission of a plastic compared with glass. There is also the consideration that glass is easily recyclable through bottle banks, but in many circumstances plastic is less easily recycled.

It may be possible to reduce headspace in a pack, for example cereals. But this may lead to a lower production rate to allow the product to 'settle'. Whilst this may facilitate the use of a smaller carton and thus reduce pack weight, the overall cost may increase. Other techniques that will facilitate packaging reduction include:

- Use of refillable containers and refill packs (see re-use earlier).
- Reduction in water content of certain product such as liquid detergents, resulting in smaller and lighter pack for equivalent cleaning capability.
- Redesign of pack shape to give better pallet utilisation, for example a square profile of bottle with rounded corners instead of a cylindrical bottle.

There are many ways of measuring environmental impact and a method that is now commonly used is carbon footprinting. This links to climate change as the carbon footprint is expressed in grammes or kilogrammes of carbon dioxide.

5.8.3 What is a carbon footprint?*

A carbon footprint can be defined as the total greenhouse gas emissions caused directly and indirectly by an individual, organisation, event or product. This is broken down into:

- Direct emissions
- Emissions from use of electricity
- Indirect emissions.

Reasons to calculate your carbon footprint include:

- Managing emissions
- Reducing your footprint – minimising or preventing it
- Cost reduction
- Reporting
  - To customers
  - To comply with legislation
  - As part of corporate social responsibility (CSR)
  - To 'offset'.

This is where the differences between industries, products and competitors become more difficult to establish from published information. Differences include the following:

- Unit of measure: this is usually grams of CO_{2e} per unit of production, but could be expressed as per square metre of the site or per £ of sales.
- Scope: what is included? Direct and indirect, or only direct? Will electricity use be included?

The Defra Environmental KPI Reporting Guidelines provide guidance on these matters.

5.9 Sources of further information and advice

Europen is The European Organisation for Packaging and the Environment. It is an industry and trade organisation open to any company with an economic interest in packaging and packaged products. It presents the opinion of the packaging value chain on topics related to packaging and the environment. Publications include *Understanding European and National Legislation on Packaging and the Environment*, available at: www.europen.be

INCPEN (Industry Council for Packaging and the Environment) is a non-profit, research-based organisation established in 1974 dedicated to: analysing the environmental and social effects of packaging; creating a better understanding of

---

* Reproduced from 'A Practical Guide to Carbon Footprinting' with the kind permission of the Carbon Trust. Please check the Carbon Trust website for the latest guidance: www.carbontrust.co.uk. Please note that this material is © the Carbon Trust and cannot be reproduced without permission.

the role of packaging; and minimising the environmental impact of packaging. Publications include: *The Responsible Packaging Code of Practice, Packaging in Perspective* and *Pack Guide: a Guide to Packaging Eco-Design (GG908)*, produced by Envirowise and INCPEN, available at: www.incpen.org

WRAP (Waste & Resources Action Programme) works in England, Scotland, Wales and Northern Ireland to help businesses, individuals and communities reap the benefits of reducing waste, develop sustainable products and use resources in an efficient way. Useful reference material may be found on the WRAP website: www.wrap.org.uk

The Carbon Trust provides specialist support to help business and the public sector boost business returns by cutting carbon emissions, saving energy and commercialising low carbon technologies. By stimulating low carbon action, the Trust contributes to key UK goals of lower carbon emissions, the development of low carbon businesses, increased energy security and associated jobs. Publications include: *Carbon Footprinting*, available at: www.carbontrust.co.uk

The Packaging Federation is the 'over-arching' trade association for the UK packaging manufacturing industry. It is a unique representative body for companies and organisations in the UK packaging manufacturing sector and associated activities. Its website can be found at: www.packagingfedn.co.uk

The Association for Organics Recycling (AfOR) is the UK membership organisation committed to the sustainable management of biodegradable resources. It promotes the benefits of composting, digestion, and other biological treatment techniques and the use of biologically treated materials for the enhancement of the environment, business and society. Its website can be found at: www.organics-recycling.org.uk

Other useful publications include:

- Department for Business Innovation and Skills, *Packaging (Essential Requirements) Regulations Government Guidance Notes, January 2011*, available at www.bis.gov.uk
- British Standards Institute (BSI), *CEN Packaging Standards*, available at: www.bsigroup.com
- Department for the Environment, Food & Rural Affairs (Defra), *Environmental KPI Reporting Guidelines*, available at: www.defra.gov.uk
- Department for the Environment, Food & Rural Affairs (Defra), *How to make a good environmental claim, February 2011*, available at: www.defra.gov.uk

6
Packaging and marketing

N. FARMER, Neil Farmer Associates, UK

Abstract: This chapter examines the relationship between packaging and marketing. It defines marketing and marketing responsibility and examines the role of marketing in production-led, sales-led and marketing-led companies. It explains the main marketing responsibilities including identifying the need for a product, knowing the market, determining price, determining strategy for the distribution of a product and determining brand values.

Key words: marketing and packaging relationship, the marketing mix, market research tools, importance of a brand, importance of consistency of communication.

6.1 Introduction

The chapter evaluates the importance of a brand, and the various elements that make up a brand, including coherence, uniqueness, relevance, distinctiveness, market appeal and brand protection. The complete marketing mix is considered including product, price, promotion and place. The role packaging plays in product promotion and advertising is also considered. With the need for more detailed and complex information on packs, the importance of consistency of communication is evaluated with examples of how this can be achieved more effectively, including use of digital technology, global brand management and corporate identity manuals. The use of market research tools and techniques to identify customer needs including SWOT analysis, gap analysis, consumer research, demographics and socio-economic groupings, psychographics, consumer panels, test markets and competitor research are all considered, with the objective of achieving a successful new product launch.

6.2 Defining marketing

The Chartered Institute of Marketing defines marketing as: 'The management process responsible for identifying, anticipating and satisfying customer requirements profitably'. An alternative definition comes from The Dictionary of Marketing Terms, which describes marketing as 'the process of planning and executing the conception, pricing, promotion and distribution of ideas, goods and services to create exchanges that satisfy individual and organisational goals'. Whichever definition you favour, marketing encompasses everything to do with coming up with a product and service, making customers aware of it, making them want it and then selling it to them profitably. The important word here is 'profitably', because all marketing activity, whether market research, market development, promotional activities or advertising must ultimately be accountable to bottom-line profit-and-loss scrutiny. Without profit a company will simply go out of business and marketing management is but one facet of an organisation's management team. However, it is a fact that a business cannot succeed without marketing, although the precise role of marketing professionals will vary dependent on the nature of the business.

The role of the Marketing Department varies from organisation to organisation. Whatever the precise definition of the role, communication is a vital part of its mandate. The Marketing Department often has a better understanding of the market and customer needs, but should never act independently of the packaging technologist, product development or technical departments. Marketing needs to be involved wherever discussions take place regarding new pack or product development and other customer-related discussions. In recent years the introduction of digital media has brought an explosion in the number of channels marketers can use. However, the basic processes of understanding customer needs, developing appropriate products and providing fair and balanced information remain a vital part of marketers' role. Marketing is an integral part of any business, whatever the type of organisation.

6.2.1 Production-led companies

Some companies are production-led, i.e. in simple terms they sell only what they can make and innovation is not very high in the company's product strategy; an example would be a cement company whose product fulfils a basic requirement and is distributed and packed to provide only containment, protection and product information. This type of company usually has a core, commodity product. It may be constrained with regard to product and pack innovation. It may also concentrate on developing a unique or special service. It would rarely advertise or promote its products and services.

6.2.2 Sales-led companies

Some companies are sales-led, customer-driven, their packaging and marketing activity determined by customer requirements. Many packaging converters, particularly carton and label manufacturers and others supplying bespoke printed materials, come into this category. They produce to meet the customer's order and specification. Their vital need is to understand their customers' requirements and often their customers' customers. Opportunities exist in these types of companies for innovation in both product and service. It is vital for their sales forces to look, listen and be on the same wavelength as their customers' packaging technologists and product development teams.

6.2.3 Marketing-led companies

A large proportion of companies are marketing-led. This includes those selling food and drink, household products and toiletries, i.e. FMCG (fast-moving consumer goods) producers, plus companies selling other items such as clothing, textiles, household products, DIY and garden supplies, etc. Selling has been traditionally via supermarkets and stores, where the customer makes a self-selection, or via other selling methods such as printed catalogues and mail order. However, the advent of the world wide web (www) has changed much of this and it is now increasingly normal to purchase almost anything from the web.

6.3 The role of marketing

The marketing professional's role will include conducting the following activities:

- identifying the need for the product or service
- identifying the market sector
- determining the price
- determining the distribution strategy
- determining the brand values or the product and pack attributes.

6.3.1 Identifying the need for a product

It is among the marketer's responsibilities to identify the need for the product, whether it is existing or new. This is particularly crucial in new product launches because only 2% of those launched will ultimately succeed. Therefore the marketer must ask the the following questions:

- What is missing? Is there something which will give my product that USP (unique selling point). Look at the competition. What are their products like? How can I make mine different and better?
- What can be improved? If the product is an existing one, what can I do to improve it? Minor modifications, small design or functional changes can often transform something mundane and basic into a winner.
- Is there room for me in a crowded competitive market? Without a special feature or something which differentiates your product from the competition, the answer is probably 'no'! However, if you identify your target or niche sector and concentrate on that area alone, you may find you have something that can fill a gap in the market or, even better, supply a need which previously no one knew existed. However, to do this you will need to undertake market research, which will be dealt with later. All this has the ultimate objective of moving products or services from producer to consumer in a profitable way.

6.3.2 Identifying the market sector

It is essential to know and understand the market sector into which you are proposing to sell. Ask the following questions:

- Who are your potential customers? Do you have a database of names and addresses? Is the information up-to-date? You may be surprised to find that often it is out of date and therefore not useable. Companies change address and telephone number, and move location more often than you think. Do not launch until you are sure of your facts.

- How many customers are there and what is the value of the market? Be accurate and precise in your market assessment, this will determine many things including the investment to be made. Use accurate and reliable market data to undertake your market evaluation. Do not use old statistics otherwise you may make erroneous decisions.
- Where are your customers located? Sometimes there can be a distinct regional or local bias in markets. Know your market, know any regional variations and understand local needs.
- When do they want their products? Demand can be seasonal (e.g. Christmas or Easter) and a new product launch which is too early or too late can be disastrous. Timing is everything. Much of the information needed can be gathered by market research, or basic knowledge and understanding of the market.

6.3.3 Determining the price

To determine the price of any product it is essential to:

- Know the total cost of the product. No marketer can undertake pricing strategy without having a complete picture of start-up costs, manufacturing costs and other factors such as costs for a licence to produce in a particular country or territory.
- Know the development costs. The cost of research and development can be forgotten about or overlooked. This cost is a key part of any project, and can ultimately influence the final decision as to whether to proceed to a full market launch. Sometimes costs can be shared between the packaging producer and the customer to help reduce the burden of the development process, which can sometimes be protracted and difficult to define precisely in terms of time.
- Know the factors affecting cost. Things to consider include items which may be special one-off costs such as bespoke, customised tooling which may be expensive and involve long lead times. There may be special costs for artwork or origination which may make the product prohibitively expensive and therefore too costly for today's market. Whatever the extra cost is, be aware of it and build it into the whole costing model or matrix.
- What price will the market be prepared to pay? The more added-value features, the more the market will be prepared to pay for the product. The more the product is perceived as a commodity one, the less the opportunity for a higher price. It is also important to know the price of your competitors' products and pitch your price accordingly. Do not price yourself out of the market. However, a higher quality product will often require a higher price, by virtue of the extra costs involved in production. It will also require higher quality packaging with more added-value features.

6.3.4 Determining the distribution strategy

- Where will the product be sold? Traditionally sales have been through supermarkets, department stores and specialist retailers. Of course, the internet has changed much of this. New products are often ordered on-line, called off from warehouses and distributed direct to customers by fleets of lorries and delivery vans. The strategy adopted by marketers has therefore changed to reflect the speed, quick reaction and dynamism of the new channels of distribution. Out of stock is simply not acceptable today. 'Just-in-time' was the mantra for delivery of product some years ago. Today 'just-in-time' is almost too late.

- What selling methods are needed? Whereas before, catalogues and mail order were the alternative way to shop, rather than a visit to the supermarket or store, now the internet has created an unstoppable juggernaut of new sales patterns. Marketers must understand the new channels to market. Whatever method they choose, they must be ready with products and packaging which are lighter in weight, more vibrant in colour and design and more environmentally acceptable, to meet the needs of a greener and faster moving world. Often selling by a combination of all these methods will be required today.
- What size of pack? There are many factors to consider here, including the size of the original bulk pack, the secondary pack and the tertiary pack. Shelf-space is at a premium today and ultimately no marketer wants his or her product to be tucked away behind others on the shelf in the store. Success in achieving widespread distribution will depend on many things, not least of all the benefit of a multi-million-pound marketing and promotional budget. However, once in the stores, the size and shape of the pack will play an important part in the final decision on shelf location. One of the key factors in deciding pack size is perception of value. The pack must reflect the size or shape of the product contained. Too large or too small a pack in relation to the product it contains will have detrimental effects in terms of sales success. You can only add value so far by size of pack, and after that it becomes counter-productive.

  An important influence on pack size is the amount of product information that is now required to be displayed on the pack itself. Ingredients, weights, preservatives, nutritional information, quantification and explanation of terms such as 'low in sugar' are now all listed. Control of information stems from a number of UK and European labelling standards and guidelines. Use by, best before, storage conditions, business name and address, place of origin, instructions for use and batch numbers are now required. With a global market, much of this information has to be displayed in more than one language. When deciding on a pack size it is essential that all the above information is included, in a clear and concise way. The pack size must reflect the size of product, but it must also be sufficiently large to cover all compulsory and legal information requirements. However, this is only the start because a marketer needs to ensure that his or her product's brand identity, its logo, and any special promotional offers are all included on the pack in the correct way. This is a complex area and a balance has to be maintained between retaining brand identify and ticking all the boxes from a legal and statutory viewpoint.
- What are the display requirements? A new product launch or a re-launch of an existing product will benefit from point-of-sale display and special merchandising. The marketing responsibility here is to ensure the display retains the original

design features of the pack and that the brand values are maintained in terms of shape, style, logo and other features.
- Telling the consumer about the product. A new product launch will benefit from a launch event, something which is done on a large scale to tell the world about the exciting new product about to hit the market.
- Advertising media. Traditionally this took the form of national, trade, technical, local and regional press, or, if the product is a new or existing major consumer brand, national TV, cinema or radio. The advertising spend on TV has been affected in recent years by the internet, which is seen by many as a more cost effective way to reach a modern audience, who get their information via electronic means rather than from traditional media.
- Timescale. The successful launch of a new product depends much on the roll out and the ability to get the product into the store at a pre-determined time and date. It is no good having a high profile advertising and promotional campaign, letting the consumer know about a new product, only to visit a local store and find it is still not on the shelf. It is the marketer's responsibility to ensure the new pack is available in sufficient quantity to meet the demands of an eager and expectant consumer.

6.3.5 Determining the brand values

It is essential for marketers to discover what product and pack attributes will appeal to the market. For example, novelty or unusual pack features may appeal to children, added-value innovative features may appeal to consumers purchasing premium products. There is a whole array of consumer research techniques to discover which pack style or feature appeals to different groups of consumers. These will be investigated later in the chapter.

6.3.6 The relationship between packaging and marketing

Marketing and packaging are complementary functions in any organisation. This is particularly true in FMCG producers, where the link between the two can be vital, particularly in the launch of new products which involve packaging of a technical nature. The packaging technologist can play an important part in guiding the marketer through the design, product development, production and cost implications of choosing a new type of packaging. Whilst the marketer might understand the overall features of a new package, the packaging technologist can fully explain every function and ensure that the marketer makes the correct decisions, from a technical viewpoint. The relationship goes much further than this in some organisations, leading to the creation of a role within the company of Packaging/Marketing Specialist, whereby one person with sufficient knowledge of both functions undertakes a combined role. It will require a considerable amount of technical and commercial knowledge to aspire to this role, but the benefits can be considerable.

6.4 Branding and the impact of packaging on product promotion and advertising

A brand is a mark of authenticity, something which imparts an intrinsic value to a product. It is a guarantee of reliability, a mark of quality and inherent goodness. Iconic, long-lasting brands are truly special. It is fascinating to look at market leaders from 1925 and see how many are still in a leadership position. Here are some of them:

- Breakfast cereals – Kellogg's
- Chocolate – Cadburys
- Razors and shaving products – Gillette
- Soft drinks – Coca-Cola
- Tea – Liptons.

These are names which have a massive brand equity and kudos. This is reflected in the way these companies go about their marketing activities. It goes without saying that a brand should be fiercely protected. This is vital as the brand name differentiates one product from another. Arguably it is a promise, a commitment that the product being purchased is intrinsically good.

A brand consists of:

- the product itself
- the brand name
- promotion
- advertising
- overall presentation
- methods of distribution and sale.

If your product is the same as everyone else's, there is little encouragement to buy yours rather than theirs. If your packaging is the same as everyone else's you will devalue your brand and lose your brand identity. However, this issue becomes more complex when faced with the growth in retailers' own brands. Arguably these are not branded products as such, more taking their brand values from the presence and integrity of the retailer's name. In this instance the retailer is the brand and the customer's confidence in the individual product derives from the confidence the consumer has in the retailer's name and brand identity.

This has been further extended or indeed confused by the launch of retailer sub-brands, 'value' or lower cost lines, and added-value premium lines which intrinsically convey messages of superior quality, often by their packaging. In the lower cost brand market, Sainsbury's 'basics' range is a good example as is their 'Taste the Difference' in the added-value sector. Much of what has been discussed here comes down to the packaging and the product perception by virtue of the packaging style, quality or image.

The elements of a brand are:

- coherence
- uniqueness
- relevance

- distinctiveness
- market appeal
- it must be protectable.

Coherence

Consistency of communication of a brand is vital. This can be through several different strata: across advertising and packaging, across product range, and throughout product life.

Uniqueness

What makes a brand or product special? It might be the special structure/shape, form or material. What is its function and what feelings or emotions does it evoke? Are these feelings good or bad?

Relevance

What is the brand's relevance to peoples' needs? Is the brand still one which consumers wish to purchase? The brand can become the badge and often it is its package shape or design which signals the desirability of the brand. All Nescafe Coffee jars still have a distinctive shape which differentiates them from other coffee.

Distinctiveness

A brand must stand out. A consumer's product attention in the store whilst looking at the shelf will last 7 seconds at maximum. Some 68–80% of purchasing decisions are made whilst facing the product on the shelf. In a food superstore there are at least 16,000 products on display and in a traditional department store there are 30,000–40,000 items. It is therefore essential that a brand must be recognisable. Customer research has revealed that colour is often the first thing which attracts. Colour can be detected in the store shelf from as much as 10 metres away. From 4 metres the shape of the pack becomes discernible and only from 1 metre does the customer identify the brand. What the package needs therefore on the shelf is shape, colour, strong graphics and branding. However, the brand must balance the need to 'stand out', with the requirement of reassurance, recognition and customer awareness. Too much change in a brand identity is not always good.

Market appeal

A brand must continue to differentiate a product from the competition. It is essential that it achieves that vital commodity – product differentiation at the point of purchase, i.e. a uniqueness of product. A brand can do this by its packaging. A brand must also make the customer want to buy it. Of course, national brand loyalty has declined in recent years, almost to the extent that supermarket loyalty (influenced by the incentives which stores have introduced, such as loyalty cards) has become an important factor in purchasing decisions. However, a brand which still conveys a strong image and identity will provide an important incentive to make consumers want to buy it, once they are in the store. Procter and Gamble's 'Ariel' and Unilever's 'Persil' may face own brand competition but they are still iconic products, aided by their packaging and brand identify.

A brand must be protectable

It should contain all the elements of legal information, plus product constituent details and bar codes for ease of identification. These are all part of the 'brand equity'.

6.5 Branding and the marketing mix

The key elements of the marketing mix are:

- product
- price
- promotion
- place.

Product

The first impression of a product is its packaging. If the packaging conveys an aura, an impression of quality via its shape, graphics, use of colour or print, it is going a long way towards protecting its brand integrity. The product and the packaging can become intertwined and become one and the same to the consumer. In some cases the packaging can often overtake the product in terms of priority and importance to the overall concept.

Price

The price is determined by the position required by the brand. A high-quality product, demanding a high price, will also require the high-quality presentation available from high-quality packaging materials. A lower quality or commodity product will not have as high a profit margin and thus not be able to sustain the cost of expensive packaging or design. A product's price often communicates as much to the consumer as its advertising or promotion. Consumers perceive a product's value based on its price in many situations. Some forms of packaging, which were once considered to be new or innovative, over a period of time became less high value and therefore moved into the commodity market category. An example of this is PET bottles in the beverage sector, particularly soft drinks.

The profitability of a product is the difference between the selling price and the total cost of getting the product to the point-of-purchase. Packaging can make a major contribution to profitability both by influencing the price which can be commanded for the product and by its direct effect on total product cost. The cost of packaging includes not only the bought-in packaging material costs but also development costs including special tooling, artwork and origination.

Promotion

The promotion is essentially the communications strategy of a marketing plan. It refers to the activities used to get the consumer to recognise the product and be encouraged to buy it in preference to others. Promotion refers to the advertising which takes place both when the product is launched and later on to sustain and grow its sales. Promotional activities might also include:

- public relations and press
- direct marketing and mailings
- promotional events
- product and company marketing material, e.g. brochures, leaflets, etc.
- premium give-away items
- sales force promotional activity, designed to generate individual sales by each salesperson.

Other promotional activity is more direct and usually occurs at the point-of-purchase in the supermarket. This can take the form of 'money off' packs, added-value offers involving extra product for the same or special price, samples, the classic 'BOGOF' (buy one get one free) offers, or the offer of free/reduced price gifts. Packaging is often the major vehicle for promoting such activities. Another technique which helps to promote the brand and protect its authenticity is range extension, i.e. by adding other variations of the same product. For example, people have been eating Kellogg's cornflakes at breakfast time for more than 100 years. The company's 'Special K' brand – aimed at people managing their weight – has been around for 50 years. There are now at least 10 different varieties and to cater for the 'on-the-go' breakfast market there is a range of 'Special K' cereal bars. Kelloggs is trusted by consumers. It was the most trusted cereal maker in eight European countries from 2007 to 2011, according to the Reader's Digest Trusted Brands Survey. This means consumers are more resistant to moving to cheaper or own-label brands. Kelloggs are keeping their brand protected by emphasising that value can be obtained from more than cheaper prices. However, they are still at great pains to make clear in their advertisements that they have been making cereals for 100 years, reconfirming iconic brand status.

New media, particularly the Internet, has provided a new tool for marketing, communication and promotion. The communication aspects of the new media are immense, allowing brands to launch websites to reinforce messages which are also contained in their product packaging. To use the Kellogg's example again, 'Special K' is using the web to communicate its healthy lifestyle programme. Its 'My Special K' website includes a personal plan for ideas and advice in the whole area of healthy living. Brand integrity is therefore a key part of this process.

Place

The place is not only where the product is sold, but also refers to the distribution method and the way in which the product is merchandised. The strategy behind how a product is distributed and sold is a very important element of the marketing mix. For example, do you want your product to have wide distribution and be sold everywhere? Or do you want to create a niche market, in a more exclusive up-market sector? For traditional store and supermarket merchandising, the part packaging has to play here is in ensuring that the product arrives at the point-of-purchase in good condition so that it will still attract consumers to make the purchase. It must also maintain its function of protecting/preserving the product throughout its use.

6.6   Promoting the brand

The essential elements of packaging in the marketing mix are the brand name, the logo (which frequently refers to the brand name), the colour of the pack, its shape, size and texture. The Coco-Cola brand name, for example, is recognised world-wide because of its distinctive typeface and shape of the iconic Coco-Cola bottle, which gives the brand its own persona or identity. As a result of this you are never in doubt that the product is one from Coca-Cola.

Achieving the correct product persona is vital in persuading the target audience to buy the product. Deciding on the elements of that persona can only be undertaken when the needs and desires of the target audience are fully understood. This is the key role played by market research, which we will deal with later in the chapter.

Advertising is the most visible and highly remembered part of a marketing campaign, apart from the pack itself. Adverts frequently use images of the pack or incorporate features of the design to link the two together in the minds of the customer. With a distinctively shaped container and/or an attractive graphic design, the recognition potential of the pack on the store shelf is greatly enhanced and helps to rekindle the message delivered in the advertisement.

It is also necessary to identify which types of media to be used to carry any advertising message. This essentially is the media plan and can include magazines, newspapers, billboards, web banners, radio adverts, TV adverts, sponsored TV and radio programmes, cinema adverts, posters and flyers, directory listings and in-store displays. The TV advertisement has traditionally been the most effective way of reinforcing brand messages. However, this medium has suffered in recent years as the use of new media has risen. TV advertising is expensive and with the rise of digital communications and the ability of the computer to receive more complex and animated information, brand reinforcement using pack images is increasingly being made by this route. According to a survey produced by The Internet Advertising Bureau (IAB) and Pricewaterhouse Coopers in 2011, internet advertising in the UK increased by 13.5% to £2.26 billion. This is a 27% share of the total advertising market.

The recognition process is aided by the use of additional advertising material at the point-of-purchase. This may take the form of shelf-edge cards, store displays and other devices, such as references to promotional activity to attract the attention of the customer and assist the product launch. However, it must be remembered that the pack itself acts as on-going advertising medium in front of the consumer, reinforcing the need for good presentation and instant brand recognition.

6.7 The importance of consistency of communication

A brand image is a powerful commodity. How a company is perceived will influence a consumer's disposition and his or her readiness to buy the company's products. When a consumer purchases a branded product he or she is identifying with that image, i.e. with the perception of a brand. The image is essentially part of the product. Brand image, integrity and consistency of communication are therefore most important factors in the packaging and marketing interface and should never be overlooked. Consistency of communication must be maintained across all facets of advertising and packaging, over the complete product range and indeed through the entire life of a product.

6.7.1 Global brand management

The supply chain has been revolutionised in the last 10 years thanks to the Internet and digital communications. Global brands have global brand management, handling every facet of the complex process which brings packaging to the retailer and ultimately the consumer. All of this should ensure greater consistency of communication. However, in the case of packaging the need for global consistency through all the stages of design, product development, print and final conversion to finished pack has never been greater, and arguably never more delicately positioned. Get this process wrong in the minutist of ways and the consequences can be catastrophic, leading to a devaluation of brand equity.

6.7.2 Use of digital technology

During the past 10 years or more, a large number of packaging producers and FMCG companies have set up production sites in Eastern Europe, the Far East and even further afield. Some multinational companies have moved all corporate and production operations to new overseas sites where costs are much lower. The way these organisations develop global marketing and packaging strategies is diverse. However, what they have all found is that the use of digital technology is driving them to improved business efficiency. They have found that the same digital technology has enabled them to better control the consistency of communication in terms of global pack design, advertising, promotion and point-of-purchase because it can all be viewed electronically, proofed, amended and, only then when agreed by all, finally printed and converted into a finished pack for a world market. The greater benefit here is speed of response and cost savings.

6.7.3 Corporate identity manuals

Corporate identity or brand manuals (or 'Corporate Bibles' as they are sometimes called) are an ideal way to ensure that consistency of communication is adhered to, once a branded image has finally been produced in all its forms. By this method every aspect of use of brand – logo, colours, typeface, font, size of lettering, etc. – are all laid down in terms of what can and cannot be done. Every department and function in the organisation should retain a copy of the manual and use it as an important point of reference to ensure correct reproduction of identity and image at all times.

The responsibility for managing this document will normally belong to the Marketing Department. They will need to liaise closely with the packaging technologist to ensure the rules governing communication of a protected brand name or logo are being strictly adhered to, most particularly on the pack itself. With the multiplicity of printing methods and packaging substrates in which branded packaging appears today (from glass, metal, folding cartons and labels to corrugated cases, rigid plastics and flexible plastics), it is vital that a consistency of image is maintained in all the ways in which a brand appears on any one package at any one time. An inconsistency in a global brand in terms of colour of pack, style or use of logo across different packaging materials, for what should be the same consistent recognisable image, can have devastating effects for the brand owner, ultimately devaluing the brand.

6.7.4 Brand consistency and complex information

The need to be consistent in brand communication is complicated today by the need for complex and detailed information to appear on any pack. Features for more robust methods of product tracking and traceability, much of which is required by legislation and regulation, now appear. The strict legal requirements for consumer information must be clearly displayed on the outside of the package, allowing customers to check before purchasing exactly what is inside. In the case of the food market they will need to know where the goods come from and even detail about whether allergic reactions could result. Nutritional information is now listed in some cases, and more recently carbon footprint details. All of this means that consistent communication will become more difficult to achieve. On a small area of a pack a brand image and identity will require careful control and management to ensure it is accurately and precisely displayed, amongst all the other legal and mandatory information. Nevertheless, it must be adhered to at all times. The marketer and the packaging technologist will need to work closely together to ensure everything is included and the brand integrity is consistently maintained.

In the pharmaceutical market the leaflet label containing all statutory and legal information, including product contents, has proved to be a major benefit to ensure all these details are correctly displayed. The marketer will need to ensure this information is accurate, not least of all because it is normal today for many leaflets to be printed in multiple languages, for international/global markets. This all comes down again to consistency of communication and the importance of retaining pack and brand integrity.

6.7.5 The role of marketing communications agencies

Often today FMCG producers employ marketing communications or design agencies to create much of the image and brand identity which appears on packs. With multi-million-pound budgets, specialists with creative skills are often needed to convey high-profile messages to sophisticated audiences. Whilst their role is important, ultimately it is the in-house marketer's responsibility, working closely with the packaging technologist and other functions in the company, to ensure that image consistency, in all pack formats, is maintained at all times.

6.8 Use of market research tools and techniques to identify customer needs

The marketing process starts in many companies with the concept or idea of a new product. It ends when the product is ultimately bought again by the consumer. However, to get to this final stage, many processes have to be gone through. In the definition of marketing provided by the Chartered Institute of Marketing the words identifying, anticipating and satisfying customer requirements are used. The market research phase of any new product launch is the 'identifying' part of this definition, because identifying in this instance means to look at or understand the requirements of the market. This can only be effectively achieved by thorough market research.

6.8.1 Product viability

If the idea is a new one, much market research and evaluation has to be carried out to determine whether the idea or the product concept is viable, both economically and technically. Equally, if the product is a change or modification to an existing one, clear objectives have to be drawn up to establish the criteria for the change. For example:

- What are the expected benefits of the change?
- What is the timescale?
- What are the reasons for the change?

6.8.2 Assessing consumer needs

In assessing consumer needs, the marketer is assisted by market research and the many tools available to him or her. Without doubt, intuition, flair or 'gut feel' still have a role to play. Often a new idea or concept for a pack starts life just like that – a thought at an unlikely time which sparks off a discussion or debate in the company. However, today, in the highly competitive FMCG market, analytical assessment of consumer needs and desires is vital. It is a fact that the majority of new product launches fail. The figure is as low as a 2% success rate in many areas. Only products which are well researched, fit the market needs and are well packaged and promoted will ultimately be successful. Market research can be carried out at any stage of marketing activity.

It will normally start with researching a market to find out any areas of demand. It will continue with researching the segments or sectors of the market.

6.8.3 SWOT analysis

As an initial evaluation of a company's competitive position, SWOT analysis is a valuable market research tool. It stands for Strengths, Weaknesses, Opportunities and Threats. Normally strengths and weaknesses are determined by internal factors, whereas opportunities and threats are determined by external factors. Strengths can help improve your product, market share or performance. Weaknesses can contribute towards losing competitive advantage, market share or financial performance. Opportunities and threats are factors which exist in the market. One example is a competitor launching a new product which could impact significantly on sales values. Arguably this could be a threat and an opportunity, depending on how this is viewed. It could be an opportunity to launch a superior more advanced alternative product or a threat because it could take sales away from your company's product.

6.8.4 Gap analysis

Research can then continue into specific segments of the market. Gap analysis (finding out where there are pockets of demand for specific products or services not being currently met) is a very useful technique. It is first necessary to decide how to judge the gap over time, for example by market share, profit, sales and so on. Then ask two simple questions – 'where are we now?' and 'where do we want to be?'. The difference between the two is the gap. The next stage is to close the gap and decide on methods as to how this can be achieved. Strategic gap analysis and tactical gap analysis will help in this process.

6.8.5 Market segmentation

Gap analysis is followed by researching prospects in those specific segments identified, to find out details of size, price, variety, colour and so on, of a product which will be acceptable to potential customers. This will not only lead to more customers being reached who will ultimately buy the product, but also reduce the number of potential competitors to be faced. Finding a niche market can often be the key to success for small, medium and sometimes even large companies. Some niche markets will be very distinct, others will be more subtle. The product could also cross several market segments and marketing activities might need to be adjusted as a result.

6.8.6 Consumer research

During the stages of product and pack development, there will be further research to test performance in use and consumer reaction. Consumer panels are a very useful way to gauge reaction to different pack concepts. A variety of new pack designs can be presented to the panel and their feedback to each new idea is often a very important factor in deciding whether to proceed to market with a new idea or concept. A range of shapes, materials, colour and print options will be presented. Opinions or preferences already expressed in a company about a potential new pack can often be reinforced or, indeed, refuted by this method. During the launch of a new product, research will also be used to test the effectiveness of advertising and promotional campaigns to see how well the product is being received.

6.8.7 Demographics

When researching a product or a market sector there are various ways of segmenting the information for ease of analysis. One of these techniques is known as demographics. It works by dividing the population into categories. This essentially sorts the population by:

- gender
- occupation
- family size
- age
- education
- socio-economic status.

In the UK this is normally done according to job function or occupation of the head of the household or chief income earner, as shown in Table 6.1. Retired persons who have a company pension or private pension, or who have private means are graded on their previous occupation.

In addition to the standard categorisation, Mintel, the international market research group, also analyses the consumer research it undertakes in another way. 'Lifestages' are derived from analysis of Mintel's own consumer research and are split into four main groups:

- Pre-family/no family – age under 45 who are not parents.
- Family – any age with at least one child under 16 still at home.
- Third age – aged 45–64 with no children aged under 16.
- Retired – aged over 65 with no children aged under 16.

Mintel has also created special groups of consumers to typify consumer habits in

Table 6.1 UK socio-economic status classifications

Socio-economic group	Occupation of chief income earner
A	Higher managerial, administrative or professional
B	Intermediate managerial, administrative or professional
C1	Supervisory or clerical and junior managerial, administrative or professional
C2	Skilled manual workers
D	Semi and unskilled manual workers
E	All those entirely dependent on the state long term, through sickness, unemployment, old age or other reasons

the early years of the twenty-first century. Some Mintel reports also use consumer research analysed by ACORN category. This is a geo-demographic segmentation method, using census data to classify consumers according to the type of residential area in which they live. Each postcode in the country can, therefore, be allocated an ACORN category. With all these tools at their disposal, the marketer should be well armed with information to make informed decisions about potential pack success.

6.8.8 Psychographics

However, there are many other factors which influence purchasing decisions, such as personal preferences, which are not identified by demographics. This is where psychographics becomes a useful tool, because it gives information on what actually motivates people. Psychographics is a term used in marketing to describe consumer buying motivation and behaviour. Psychographics essentially segments a market for marketing purposes by classifying potential customers by their attitudes and values. It investigates what makes a consumer want to buy a product.

Even though you may have determined your demographic group, people within that group still have very different perceptions about the benefits or value of a particular product and will be motivated for different reasons. These differences are known as psychographics. The need is to determine not only who will buy your product, but what makes them want to buy it. Psychographic information such as spending patterns, whether consumers are brand conscious, what influences their buying behaviour and what promotional methods they respond to most, are all essential pieces of information.

As stated, the ultimate objective is to investigate and determine what makes the consumer want to buy a product. For example, is it because the product is:

- environmentally responsible?
- a natural proposition?
- a nutrition statement?
- a low fat, low-calorie product?

Psychographics is a valuable tool to identify motivation factors. Market research is a tool to help identify the need. It then helps to answer whether a proposed product or offering meets that need. For example:

- Does the product work and is it better in performance than previous ones?
- If it is new, does it perform better than the competition?
- Does the consumer like it and are the features appealing?
- Would the consumer buy it compared to others in the market?

6.8.9 Consumer panels and consumer research

Market research, particularly at the consumer panel/consumer research stage, can provide answers to some vital questions, in deciding whether a product or pack will succeed. Market research will reveal whether the target consumer recognises the offering. It will tell us:

- Does the product stand out on the shelf when put side-by-side with other potential new packs or the competitors' products?
- Is the product clearly identified, do the consumers know immediately what the brand is or the name of the manufacturer?
- Does it present the expected image; are the brand values, graphics, logo, print and design consistent with the brand equity of the product?

Tests may be conducted using new packs against old, new packs against the competition and new packs against variations or alternatives to each other. The 'winner' is normally the most preferred option, although there is rarely an outright winner. Such testing or research panels are conducted after researchers have identified a suitable group of people and recorded reactions to the packs.

For new product development, as well as researching the effectiveness of a design, it is essential to test that the product meets the needs of the consumer from a perception viewpoint, as well as in reality. Does it deliver all the emotional benefits which are expected? Does it provide the correct perceived value? Is it correctly positioned in the marketplace? The elements which make up this positioning can be assessed and categorised in many ways.

Research can be conducted by consumer panels, sometimes in town centres or shopping malls, or products can be given to individual consumers who are asked to try them at home and report their findings. The information derived from these tests helps the marketing professional to fine-tune the product and the packaging to give a greater guarantee of success when finally launched. All this is essential so as to avoid unnecessary cost. An unsuccessful new product launch will cost many millions, in the case of a major FMCG producer brand launch. The research will also reduce the chance of failure, because extensive research and investigation has been conducted into the viability of the new pack or concept before it is actually launched.

6.8.10 Test markets

Test markets are another way to avoid the large financial cost of failure. By this method a smaller launch, sometimes regional (such as in a particular commercial TV region), will be undertaken. The feedback from the consumer reaction in this area will help the FMCG company decide whether they can proceed to a full-blow national or international launch. If the response is negative, the full launch can be cancelled and large amounts of money saved.

6.8.11 Competitor research

Competitor research can play a vital part in any decision to launch a new product or modify an existing one. Marketing strategy must reflect what the market is looking for in terms of new packs (e.g. lightweight, environmental benefits, less material content) and what the pricing policy should be. Successful positioning of a new product in the market can only be achieved if its market position and perception have been fully evaluated against its competitors. Also any promotional campaign will need to be effectively planned and researched so that a new product's advertising is sufficiently different from that of its competitors. All of this is essential because, as we already know, most new product launches fail. Without market research the likelihood of failure is even greater.

6.9 Sources of further information and advice

- The Chartered Institute of Marketing, Moor Hall, Cookham, Maidenhead, Berkshire SL6 9QH, UK.
- Imber J and Toffler B-A, *The Dictionary of Marketing Terms*, 3rd edn, Barron's Educational Series Inc., USA.
- The Internet Advertising Bureau, 14 Macklin Street, London WC2B 5NF, UK.
- Mintel Group Ltd, 11 Pilgrim Street, London EC4V 6RN, UK.
- The Reader's Digest Trusted Brands Survey, The Reader's Digest Association Limited, 11 West Ferry Circus, Canary Wharf, London E14 1HE, UK.

Part II

Packaging materials and components

7
Glass packaging

P. GRAYHURST, British Glass, UK

Abstract: Glass is one of the oldest packaging materials and still maintains an important place today. This chapter covers the production of glass and its forming into containers for packaging, including an overview of the decoration processes used. Glass quality is discussed, including typical defects, and useful points for container design are given.

Key words: borosilicate, decolourisers, ultraviolet, soda glass, cullet, furnace, forehearth, lehr, gob, blow-and-blow, press-and-blow, narrow neck press-and-blow, parison, baffle, plunger, neck ring, neck finish, annealing, surface coating, head space, frosting, danner process, vello process, ampoules, vials.

7.1 Introduction

Although glass shares some characteristics of a supercooled liquid in which the state of flow has ceased, it is generally described as a solid below its glass transition temperature. As mentioned in Chapter 1, it is one of the oldest packaging materials. Commercial glass is made with silica (quartz) which is the principal component of sand. It requires high-purity silica sands from special deposits. For soda lime glass, as used in packaging, this is mixed with substances such as soda ash and limestone. A typical composition of soda lime glass, and key properties, are shown in Table 7.1. Mixing silica and sodium compounds produces sodium silicate, also known as 'water

Table 7.1 Typical composition and properties of soda lime glass

Composition/properties	
Chemical composition (% by weight):	
Silica (SiO_2)	70–74
Sodium oxide (Na_2O)	12–16
Calcium oxide (CaO)	5–11
Aluminium oxide (Al_2O_3)	1–4
Magnesium oxide (MgO)	1–3
Potassium oxide (K_2O)	≈0.3
Sulphur trioxide (SO_3)	≈0.2
Ferric oxide (Fe_2O_3)	≈0.04
Titanium dioxide (TiO_2)	≈0.01
Properties:	
Glass transition temperature, T_g (°C)	573
Coefficient of thermal expansion (ppm/K, ~100–300°C)	9
Density at 20°C (g cm^{-3})	2.52
Heat capacity at 20°C (kJ kg^{-1} K^{-1})	0.49

glass'. Adding calcium compounds makes this substance insoluble. Aluminium oxide (alumina) improves the durability of glass. Glass surfaces may also be treated with titanium, aluminium or zirconium compounds to increase their strength and allow thinner and lighter containers. Boron compounds (borax, boric oxide) give glass low thermal expansion and high heat-shock resistance. Other substances help reduce the time and temperature required for melting as well as the presence of gas bubbles.

In addition to these substances, decolourisers such as nickel, cobalt or selenium, are added to mask any colour produced by impurities such as iron. Lead compounds improve the clarity of the glass. In addition to white flint (clear) glass, the standard colours produced in the industry are:

- green, achieved by adding chromium compounds
- amber or brown, achieved using iron, sulphur and carbon for amber glass.

Amber glass is able to filter out ultraviolet (UV) light and is primarily used for UV-sensitive products such as beer and some pharmaceuticals. It is possible to produce glass in other colours by adding coloured frits at the forehearth, for example white opal (using fluorides), blue (using cobalt oxide) and red (using selenium, cadmium and antimony sulphide).

7.2 Advantages and disadvantages of glass as a packaging material

One of the advantages of glass is that it is inert to most chemicals. Foods do not react with it to produce hazardous or undesirable compounds which might either contaminate or taint the food. The fact that glass is impermeable to gases and water makes it ideal for long-term storage of foods or beverages vulnerable to spoilage from exposure to oxygen and moisture. Although glass is generally classed as inert, sodium and other ions will leach out over extended storage times which may be relevant for some pharmaceuticals such as injectable liquids. The European Pharmacopoeia (EP) has stipulated three types of specialist glass for pharmaceutical applications with specific limits to the level of titratable alkalis:

- Type 1: a borosilicate glass which meets the most stringent extractable standard. A disadvantage is the higher melting point of this glass type, requiring a furnace temperature of 1750°C, which increases the cost of the glass.
- Type 2: a soda lime glass formula that has been 'sulphated' at 500°C in the annealing oven (lehr) to reduce alkali solubility at the glass surface. The treatment produces a discoloured, hazy appearance.
- Type 3: a conventional soda glass that has been tested and shown to meet a specified extractives level. In North America, soda glasses not meeting type 3 qualifications are classed as USP type NP.

These glass types are used mostly for the manufacture of ampoules and vials, typically containing injectable drugs. Different countries specify different test methods and acceptable standards of glass for pharmaceutical use. Results which meet the requirements of the European Pharmacopoeia may not be sufficient for other national standards such as the United States Pharmacopoeia (USP) and/or the Japanese Pharmacopoeia. This underlines the importance of knowing the market in which a product is destined to be used.

In pristine condition, immediately after forming, glass is some 20 times stronger than steel but much of this strength is lost through the development of sub-microscopic flaws as it cools which lead to stresses and weaknesses in the glass. However, despite this, the rigidity of glass as a material means that glass containers are resistant to pressure which, for example, makes them appropriate for carbonated drinks or foods requiring a partial vacuum to prevent microbiological spoilage. The rigidity of glass also gives glass containers vertical strength which allows stacking. Glass is stable at high temperatures and is resistant to thermal shock (providing it has been correctly annealed), making it suitable for hot-fill and retortable products. Its stability means it does not degrade, making it suitable for long-term storage of foodstuffs such as preserves.

The appearance of glass is also an advantage. Its transparency is useful where product visibility is important. Because of a long tradition of decorative glassware, glass can be used to convey a high-quality image, e.g. in the perfume sector. Glass can be moulded into various shapes, though rounded, cylindrical shapes are preferred for maximum strength.

The main disadvantages of glass are its high density (2.5 g/cc), resulting in relatively heavy containers, and its brittle quality. When damaged, glass shatters. Broken glass produces sharp fragments and splinters which can present a serious safety hazard. Critical faults in the production, filling, transport and storage of glass containers include cracks and chips or bubbles which create weak container walls vulnerable to breakage. Because of this potential safety hazard, glass containers are typically inspected at various points during the manufacturing process, using automated inspection equipment to check 100% of the production.

Although the raw materials for glass are relatively low cost, glass manufacture requires significant amounts of heat. High energy costs thus affect the total cost, although the effect of this can be reduced through using recycled glass (known as cullet) in glassmaking. Glass can be recycled by simply heating it until molten and then reforming a container without loss of strength or quality, or the production of harmful by-products. It can be recycled repeatedly in this way, and of all the major packaging materials, is possibly the one most associated with recycling due to the availability of bottle banks where consumers can dispose of used glass containers which can then be collected for reprocessing.

The smooth, hard, inert surface of glass also makes it relatively easy to clean whilst its heat resistance makes it easy to sterilise. These properties are of particular relevance for the reuse of glass containers; once collected and returned to the packer filler they can be cleaned, inspected and sent out again for filling. However, the cleaning processes can have a significant environmental impact (use of energy, water and cleaning chemicals, and production of effluent) and this must be considered when deciding whether to use single-trip or multi-trip glass containers. In addition, there is the cost of transporting empty bottles back to the packer filler to consider, as well as the capital investment needed for the cleaning and inspection equipment.

7.3 Glass manufacture

Commercial glass is made in specialist furnaces able to generate the very high temperatures required for processing (1,350–1,600°C). Fuel used depends on local availability; gas-fired furnaces are common in the United Kingdom, other sources will be more appropriate elsewhere. Furnaces need to be large and to run almost continuously if they are to operate profitably. A typical furnace holds up to 500 tonnes of glass and produces about 200 to 400 tonnes in 24 hours. As a result specialist glasses (including those with more unusual colours) are more expensive to produce than mass-produced standard grades and colours.

The ingredients are weighed and mixed before being fed to the furnace. Most modern glass container plants have automated (continuous) batch plants to feed the furnaces. As has been noted, an important ingredient is cullet from previous glass containers. Cullet enhances the melting rate and so significantly reduces the energy requirements of glass production. Each 10% of cullet added to a batch reduces the energy requirement by approximately 2.5%. Cullet can be used in percentages as high as 80% of the ingredient batch, although 30–50% is more common, resulting in energy savings ranging from around 10% to 20% or more. Colouring agents may be added at this point or at a later stage via the forehearth.

Figure 7.1 shows a cross section of a typical glass furnace. The mixed ingredients are fed into the furnace by hopper. As the materials are heated, the bridgewall traps impurities which rise to the surface of the molten glass. The forehearth brings the temperature down to around 1,100°C and by now the molten glass has a consistency which means it can be gravity-fed through chutes to the bottle-forming machines. Molten glass flows through draw-off spouts with orifices ranging from 12 to 50 millimetres, depending on the container size. Water-cooled mechanical shears below the orifice, which are synchronised with the draw-off flow rate and speed of the bottle-forming machine, cut 'gobs' of molten glass. Each gob makes one container.

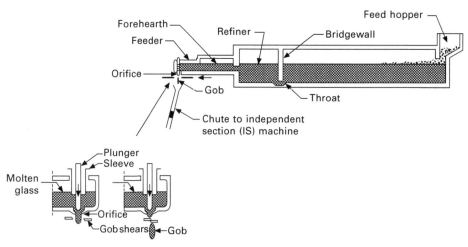

7.1 Cross section of glass furnace and gob-forming process.

Higher production speeds are achieved by the use of double, triple or even quadruple gobs to feed one forming machine.

7.4 Forming glass containers

Automatic production of glass containers usually requires a minimum run of three days for efficient and economic operation depending on the size of the forming machine. Depending on their geometry, glass containers are made by three different processes:

- blow-and-blow
- press-and-blow
- narrow neck press-and-blow.

All processes require two moulds (made from cast iron):

- a blank mould that forms an initial shape
- a blow mould to produce the final shape.

The blank mould forms the finish (the part that receives the closure), the neck and a partially formed body known as a parison. A blank mould assembly comes in a number of sections:

- a guide for inserting the gob (funnel)
- a seal for the gob opening once the gob is settled in the mould
- the blank mould cavity (made in two halves to allow parison removal)
- the finish section (neck ring assembly)
- blowing ports to supply air.

Usually between 6 and 40 identical moulds constitute a 'set' for one bottle-making machine, costing up to £50,000 or £60,000, depending upon the number of machine sections and the complexity of the containers being produced. Each mould set has a number that is formed on the bottles made by that mould. Independent section (IS) machines are the most commonly used type of forming machines throughout glass container manufacture.

In the blow-and-blow process, the container is blown in the following sequence (see Fig. 7.2):

1. The gob is dropped into the upturned blank mould through the guide.
2. The 'funnel' is replaced by a baffle (which prevents the glass from escaping) and air is blown into the mould (known as the settle blow) to force the glass into the finish section at the bottom. At this point the neck finish is formed.
3. Air is forced through the newly-formed neck orifice, expanding the glass upwards into a bottle pre-form (known as 'parison'). This is the counter blow stage.
4.–5. The parison is removed from the blank mould, using the neck ring fixture, and rotated through 180 degrees for placement into the blow mould.
6. The final blow of air in the blow mould forces the glass to conform to the

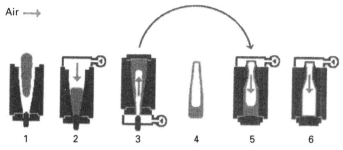

7.2 Container forming by the blow-and-blow process. 1, gob enters parison mould; 2, settle blow to form finish; 3, counter-blow to complete parison; 4, blank formed; 5, blank transferred to blow mould; 6, final shape blown.

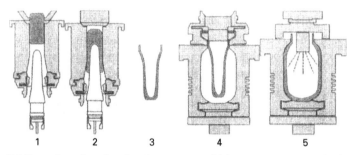

7.3 Container forming by the press-and-blow process. 1, gob drops into parison mould; 2, plunger presses parison; 3, parison completed; 4, parison transferred to blow mould; 5, final shape blown.

shape of the blow mould. The bottle is cooled so that it can be placed on the conveyor that takes it to the annealing oven.

In the press-and-blow process used to produce wide mouth jars, the process is similar except that the parison is pressed into shape with a metal plunger rather than being blown into shape (see Fig. 7.3):

1. The gob is dropped into the upturned mould.
2. The plunger presses the parison into shape.
3. The parison is completed.
4. The parison is transferred to the blow mould.
5. The final shape is blown.

For some wide mouth containers, the final blow is replaced by the application of vacuum to the wall of the mould to produce the final bottle shape.

Traditionally, the blow-and-blow process was used for narrow-necked bottles and the press-and-blow process for wide-mouthed jars. The third process, narrow neck press-and-blow (NNPB) is now being used to produce bottles with narrow neck finishes. Like press-and-blow, it uses a metal plunger, but this is much smaller. The plunger allows good control of glass distribution, resulting in containers which are significantly lighter in weight than those produced by the conventional blow-and-blow process, thus reducing cost and environmental impact.

7.5 Annealing

The moulding process relies on the glass being sufficiently hot and pliable. When the bottles leave the moulds, the temperature is about 450°C. If containers were left to cool down on their own, the low thermal conductivity of glass would mean that the interior would cool more slowly than the outside. As a result, cooling would be uneven and stresses would develop in the container walls. To avoid this, containers are transferred by conveyor to an annealing oven, known as a 'lehr' immediately after removal from the blow mould. A lehr is a belt which passes slowly through a long oven (up to 30 metres in length). The glass temperature is initially raised to 540–570°C to relieve the stresses and then gradually cooled to minimise reforming of stresses, until the containers emerge at the end of the lehr at about 60°C. They are then allowed to cool down further before packing. Improperly annealed glass has a high risk of breakage when subjected to further processing. Good annealing is also essential for thermal shock resistance.

7.6 Surface coating

Once cooled, the container surfaces are relatively smooth and pristine, though the outer surface is slightly rougher than the inner as it mimics the grain of the finely machined cast iron mould surface. Pristine glass has a comparatively high coefficient of friction. As a result, surface scratching (known as 'bruising') can occur when bottles rub together on high-speed filling lines. Scratched glass has significantly reduced breakage resistance and thus, to minimise this problem, containers are usually surface coated. Two coatings are used:

- a primer or bonding coat (e.g. tin or titanium tetrachloride), applied at the entrance to the lehr ('hot-end' coating).
- a friction-reducing coat applied at the exit to the lehr ('cold-end' coating).

Various formulations are available, depending on end use, including oleic acid, monostearates, waxes, silicones and polyethylenes. Any labelling adhesive needs to be compatible with the coating used.

7.7 Inspection

Given the potential safety hazards of broken glass, glass containers are routinely subjected to 100% inspection. Following annealing, containers are single-filed from the lehr belt by means of an unscrambler and fed through a number of inspection operations:

- Bottle spacer: this machine is pre-set to create a space between the bottles on the conveyer to avoid bottle-to-bottle contact.
- Squeeze tester: each bottle is passed between discs which exert a force to the body of the container. Any obvious weakness or crack in the bottle will cause it to fail completely with the resulting cullet being collected by a return conveyor running underneath, and recycled.

- Bore gauger: with this machine, it is possible to carry out three measurements – internal and external diameter at the neck finish entrance and the height of the bottle.
- Crack detector: bottles outside specification are automatically rejected by means of a pusher positioned downstream from the gauger. This machine can also detect uneven sealing area (land) on wide-mouthed containers.
- Wall thickness detector: by focusing a beam of light onto areas of the container where defects are known to occur from previous visual examination, any crack will reflect the light to a detector which will trigger a mechanism to reject the bottle. Typically, up to eight detectors can be found on one machine which is useful for eliminating bottles with cracks and crizzles in the finish. By using the dielectric properties of the glass, the wall thickness can be determined by means of a sensitive head which traverses the body section of the container. A trace of the wall thickness can then be obtained and bottles falling below a specified minimum will be rejected automatically.
- Hydraulic pressure tester: used for containers destined to be filled with carbonated beverages.
- Visual check: bottles are also passed in front of a viewing screen as a final inspection.

New methods of determining glass container quality are now being utilised, including automatic hot-end inspection before annealing. Advanced photoelectric methods for determining critical faults such as the very occasional glass inside or stuck plug (a piece of glass, usually very sharp, projecting inwards just inside the neck bore) are now being used. Computer methods for sensing and storing all the measurement information obtained from inspection machines are being used, with the results passed to the forming section for adjustment if necessary.

7.8 Tolerances

Tolerances for variation in any given characteristic will depend on size and container design. British Glass has agreed tolerances on glass containers and the following examples are taken from their Tec 9 booklet 'General Guidelines for the Use of Glass Containers'. Further details are available in Tec 4 'Glass Container Tolerances' and Tec 6 'Accurate Determination of Glass Container Capacity':

- Capacity (average of 12 random samples):
    Up to and including 100 ml, tolerance ± 2.7 ml
    Up to and including 200 ml, tolerance ± 3.8 ml
- Body dimensions:
    Diameter Up to and including 25 mm, tolerance ± 0.8 mm
              Up to and including 50 mm, tolerance ± 1.1 mm
    Height   Up to and including 25 mm, tolerance ± 0.7 mm
              Up to and including 100 mm, tolerance ± 1.0 mm
- Verticality (important for good performance during filling and capping):

Up to and including 120 mm in height, tolerance ± 2.2 ml
Up to and including 200 mm in height, tolerance ± 3.4 ml
(Note: tolerances for all the above vary for intermediate ranges.)

7.9 Container design

As Fig. 7.4 illustrates, containers are typically designed with cylindrical shapes and rounded edges. This is partly because viscous glass flows most easily into moulds with smooth, round shapes. Cylindrical containers are the easiest shape to manufacture since they are an expansion of the circular parison. They are also easy to handle on high-speed filling and labelling lines. They can be accurately positioned in a spot-labeller via an indexing label lug on the container exterior.

Cylindrical containers also have greater strength-to-weight ratios and better material utilisation than irregular shapes. The relative strengths of different shaped glass containers are shown in Table 7.2. Table 7.3 shows the greater weights of irregular to cylindrical container shapes. Square or angular shapes and sharp corners are difficult to form properly and more prone to weaknesses and faults. Indeed, rectangular containers typically have a round finish to minimise these problems. There are standard dimensions for cylindrical bottles which can be mass produced cheaply and ordered as required. It is also important in designing container shapes to have as low a centre of gravity and as broad a base as possible to maximise stability. Tall, narrow designs with a non-circular profile are vulnerable to tipping over whilst

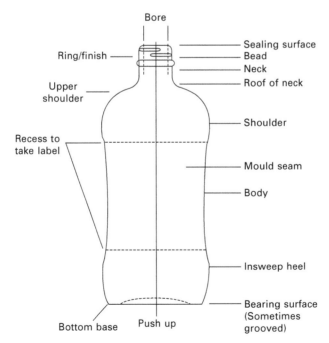

7.4 Glass container nomenclature.

Table 7.2 Relative strengths of differently shaped glass containers

Container shape	Ratio of relative strengths
Cylindrical	10
Elliptical (2:1)	5
Square with round corners	2.5
Square with sharp corners	1

Table 7.3 Relative weights of cylindrical and irregularly shaped containers

Capacity	Cylindrical	Irregular
30 ml	45 g	55 g
340 ml	225 g	285 g
455 ml	285 g	355 g
905 ml	455 g	565 g

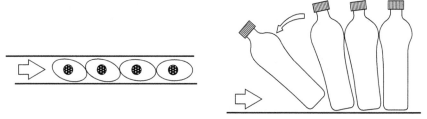

7.5 Shingling of glass containers will cause toppling and breakage on the filling line.

being moved or bunching together (a process called 'shingling' – see Fig. 7.5). These requirements also affect the design of the base (see below).

It is important to ensure the design is appropriate to the product. Issues relating to different kinds of product include the following:

- Gaseous liquids: Carbonated soft drink and beer bottles must be able to withstand internal gas pressures of 340 kPa for soft drinks and 820 kPa for beer. Bottle designs for these products are always cylindrical and have gently curving radii in order to maximise bottle strength.
- Viscous liquids: If products are dispensed by pouring, a gradually tapered and smooth neck provides good control over pouring. Some designs include a ridge on the neck to catch and hold drips which can then be wiped away. If the product is dispensed with a spoon (e.g. honey), a round, squat design is best to allow easy access.
- Semisolids: Semisolids (e.g. pastes) often need to be packed in round, squat jars to allow easy access by a spoon or knife. A simple, cylindrical design is best, avoiding corners which would be difficult to access and without a shoulder or neck to trap product. This leaves a wide 'mouth' that also makes access easier.
- Granular solids: Powdered and granular materials (e.g. coffee) usually come in wide-mouthed jars to facilitate both filling and dispensing. To protect the

contents from air and moisture, a seal often needs to be adhered to the top of the neck finish. A smooth, flat and sufficiently large area is needed to ensure an effective seal. Products such as instant coffee can vary in density but are filled to a declared weight, resulting in varying fill heights. A stippled container neck or the use of an appropriately placed label will obscure the actual fill level.
- Tablets: Since the pattern of filling is less predictable, with varying fill heights depending on how the tablets fit on top of each other, these are usually packed in simple cylindrical or rectangular containers. The design of neck needs to make it easy to dispense tablets singly if required, and to allow some margin for both varying fill heights and any wadding or sealing materials used to protect the tablets in transit.

7.9.1 The container finish

As has been noted, the container finish is designed to take the closure (e.g. screw cap). Finishes are broadly classified according to features such as diameter (expressed as the nominal inside diameter in millimetres) and sealing method. Standards for finish sizes and tolerances are established and are followed by both glass container manufacturers and closure manufacturers (see Chapter 15). Faults affecting finishes include 'overpress', where a small ridge of glass is formed on the sealing surface of the finish, or a 'choked bore', where excess glass occurs inside the finish, restricting access.

7.9.2 The container neck and shoulder

The neck design is an important part of overall container design because it has a major influence on the rate and efficiency of filling as well as the ease with which the end user can dispense the product. Filling operations are discussed in Chapter 20. A head space may be needed to facilitate filling, provide for thermal expansion or allow a vacuum necessary to control spoilage. Differences in fill level are readily visible in containers with long, narrow necks.

Blending of the neck and upper shoulder is important to container integrity and good design. The lower shoulder is a particularly vulnerable point since it is the area most likely to come into contact with other containers during filling and packing. The contact area should be as large as possible and extra thickness may be required. Container designs often include a thickened 'protruding' shoulder to protect the container and to minimise contact between containers during handling.

7.9.3 The container body

Containers are often designed with label panels that are recessed to ensure accurate placing of the label and to prevent scuffing. Cylindrical containers are easier to label than other shapes. Many apparently straight-walled containers actually have a slight (0.08 mm) 'hourglass' inward curve.

7.9.4 The container heel and base

The heel is the point at which the body of the container meets the base. Like the lower shoulder, the heel is another vulnerable area which comes into contact with other containers. To ensure maximum strength, the heel should start as high from the base as possible. In large containers, the body is blended to the heel while in small containers the heel is blended to the base. Heel tap is a manufacturing defect where excess glass is distributed into the heel.

The diameter of the base should be as large as possible to ensure stability. The centre of the base is usually domed upwards (known as the 'push-up') to create an outer ring (bearing surface) on which the container rests. This gives the container greater stability than a completely flat surface and makes it less likely to slip. It also reduces the surface area exposed to friction as the container is moved through filling and closure processes, and then packing, transport and retail display. In many designs the bearing surface is grooved for extra stability and strength. Bases may also be designed to make stacking easier, with the base fitting either into or around the cap of another container.

7.10 Decoration and labelling of glass containers

Stippling can be used for decorative effect, especially where product visibility is not an issue. Stippling can be produced by creating the design in reverse on the interior surface of the blow mould. Smooth glass surfaces are significantly weakened by surface scratches. A stippled surface pattern helps reduce this effect. A cut-glass effect can be obtained in the same way, provided the depth of the V-shaped grooves does not exceed 25% of the groove width. The cut-glass effect enhances the appearance of clear glass without impairing product visibility.

A frosted appearance can be produced by etching with hydrofluoric acid (a relatively costly process) or by sandblasting. Ceramic frosting is achieved by spraying the exterior with a ceramic paint or frit made from a ground glass and oil mixture, followed by firing. During firing, the oil evaporates and the ground glass is fused to the surface.

It is important to design a container with labelling in mind. Label panels must be large enough for the label design. As has been noted, they are often recessed to allow more accurate positioning and to protect the label from scuffing. Labelling is easier on curved surfaces, making labelling of cylindrical containers faster than flat designs, particularly if more than one face is being labelled. Wet-glue labels are widely used on glass containers, either as wraparound labels, or 'spot' labels applied to specific areas, e.g. front, back and neck. Self-adhesive labels, using paper or plastic films, are also suitable. Labels can be applied to virtually any location on a container, provided the contour is in one plane.

Screen printing can be used to apply decoration directly to the container surface. In most instances, the inks are fired on to produce an extremely durable design. There are colour limitations however, due to the temperature of the firing operation, and process printing is not possible. Metallic silver and gold effects can be achieved, albeit at an increased cost. Ceramic decals are another decorating option.

7.11 Other glass-making processes

Sealed glass containers such as vials or ampoules, which are used in the pharmaceutical industry are made from pre-formed tubing stock rather than by blowing. Ampoules are designed to be broken open at one end to dispense their contents, either by scoring at that point to weaken the glass or by coating with a ceramic paint that causes a stress concentration. Vials are small containers with a rubber seal which can be pierced by a needle to withdraw the contents. Standard ampoule and vial sizes range from 1 to 20 ml. Glass tubing is manufactured by the following processes:

- Danner process: This process produces a continuous glass tube or rod. It can make tubing of 1.6 mm to 66.5 mm diameter at a draw-off rate of up to 400 metres per minute for smaller diameters. Glass flows from a forehearth of a furnace as a hot ribbon down a sleeve onto a rotating hollow shaft or blow pipe. The ribbon is wrapped around the sleeve to form a smooth layer of glass which flows down the sleeve and over the tip of the shaft. Tubing is formed by blowing air through a blow pipe with a hollow tip.
- Vello process: Glass flows from a forehearth into a bowl or reservoir in which a hollow bell-shaped vertical mandrel rotates in a ring which allows glass to flow via the annular space to give a continuously emerging tube. Blowing air is fed via the end of the bell where there is a hollow tip. The dimensions of the tubing are controlled by the glass temperature, the rate of draw, the clearance between the bell and the ring, and the pressure of the blowing air.

To produce the ampoule or vial, the tube is cut to length then flamed to red heat, after which it can be shaped or sealed by the use of various shaping tools. Since these processes can introduce strain, an annealing process is again necessary to give reasonably strain-free containers. Tubular glass generally produces even, thin-walled containers and can be pre-gauged to give specific wall thicknesses. Specialised, low volume items such as decorative tableware and high quality glassware are manufactured by semi-automatic processes, often involving mouth-blowing and manual cutting.

7.12 Sources of further information and advice

- The British Glass Manufacturers Confederation, (British Glass), 9 Churchill Way, Sheffield S35 2PY. Tel. +44 (0)114 2901850.
- The British Bottlers Institute, P.O. Box 16, Alton, Hampshire GU34 4NZ. Tel. 01420 23632.
- Glass Packaging Institute (USA): www.gpi.org
- Owens Illinois: http://www.united-glass.co.uk/about_oi.aspx?id=1352
- GTS Ltd Report to the FSA (June 2003) A03029 – Investigation of the significant factors in elemental migration from glass in contact with food.
- CCFRA GMP Guide – CCFRA – Guide No. 18 – Safe Packaging of Food and Drink in Glass Containers: Guidelines for GMP 1998 (edited by David Rose and Robert Glaze).

8
Rigid metal packaging

B. PAGE, Packaging Consultant, UK

Abstract: This chapter discusses rigid metal packaging. After an introduction on the history of metal packaging and its main markets, it reviews raw materials and the safety and quality issues associated with them, the manufacture of rigid metal containers, closures, cost issues, container specifications and decorating processes. It concludes by looking at environmental issues affecting rigid metal containers.

Key words: metal packaging, metal can, can end, processed food, seaming, draw and wall iron (DWI), draw and redraw (DRD), metal coating, metal decorating.

8.1 Introduction to metal packaging

8.1.1 Brief history of metal packaging

Tinplate is one of the oldest packaging materials and was originally used for round, square and rectangular boxes and canisters. In the early 1800s, following the offer of a prize by Napoleon Bonaparte for the first person to develop a way of preserving foods, the first heat processed tinplate food cans in the world were made, filled, processed and sold to the public in Bermondsey, East London. This had a great impact on society and the technology soon spread to the United States where it was developed into a continuous production process.

In the early 1900s there was a major step forward when a method of mechanically attaching the ends to the food can replaced the inefficient soldering system. This improved both the speed of operation and food safety aspects. The first aerosol can was patented by Erik Rotheim (Norway) in 1929. Following this, in 1933/36 three-piece soldered side seam beer cans became available in the United States following the repeal of the Prohibition Laws and the development of suitable internal coating materials.

Shortage of tinplate in Switzerland during the Second World War led to development in the use of aluminium for creating seamless can shapes. As the result of this work, some 20 years later, in 1963, the first thin wall draw and wall ironed (DWI) aluminium beer cans were sold in the United States. Construction of similar thin wall cans in tinplate for both drinks and food followed over the next 15 years. Ernie Fraze and ALCOA developed the first easy-open can end in 1962.

In 1975 the Wire Mash (WIMA) welded side seam system was developed with an overlap small enough for three-piece food and drink cans to be made at high speed. This heralded the demise of the soldered side seam can. In 1997 shaped and embossed draw and wall ironed drinks cans were introduced. This was closely followed in 2000 by the commercialisation in Japan of the first bottle shaped thin wall can with metal reclosable screw cap.

8.1.2 Markets for metal packaging

The world market for metal containers is a little over 400 billion units. This embraces packs for food, drink, aerosols, dry and technical products. Most food cans are for wet products such as meat, fish, vegetables, fruit, rice, milk-based and recipe products, and all these need to be heat processed immediately after filling to sterilise the food for long shelf life, usually a minimum of three years. Drinks cans may contain carbonated or non-carbonated liquids. Many drinks require low level heat processing, such as pasteurisation, to ensure adequate shelf life of the product. Aerosols contain fillings ranging from personal care and toiletries through foodstuffs to household, paint and building products. Dry products include powdered foods, tea leaves, wrapped foods (candies, sweets), and non-food items. Many of these are highly decorated containers and used as promotional containers where they may be secondary packages containing, for example, a glass bottle of spirits. Cans for general line technical products are designed to hold liquids, mostly for household or industrial use. This range of containers includes tapered and parallel sided drums of up to 50 litre capacity.

As already mentioned, in 2000 the first thin wall metal bottles with reclosable screw caps were introduced for beer and soft drinks in Japan. Lower cost versions are now being developed and being marketed in both Europe and the United States. The processes for making aluminium aerosol cans are also used for forming collapsible tubes for medical products. Metal caps and closures are used for closing containers made from glass or plastic as well as those made from metal. Metal packaging is particularly suitable where a high quality of external decoration is required, as the hard surface of metal allows a very clear image to be generated.

8.1.3 Common formats for metal cans

Metal cans are mostly constructed by one of the following two basic methods, which are fully described in Section 8.3. Figure 8.1 shows these two basic systems:

- Three-piece can – comprising a cylindrical body rolled from flat rectangular sheet with the side seams overlapped and joined using electric resistance welding and two ends mechanically joined to produce a closed container.
- Two-piece can – comprising a seamless cylindrical can body with one integral end (base) shaped from a flat disc and the other end mechanically joined to produce a closed container.

The geometric relationship between can diameter and height helps to define which can format is most appropriate for which manufacturing process. This relationship may be described as:

- Tall cans   Height greater than diameter (e.g. beer can)
- Short cans   Height equal to or slightly less than diameter (e.g. tuna can)
- Shallow cans Height significantly less than diameter or width (e.g. sardine can).

124   Packaging technology

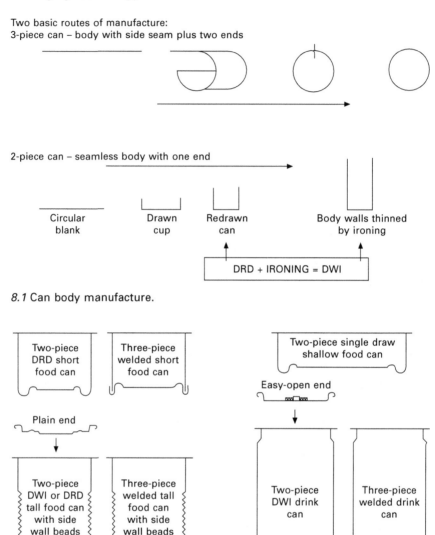

8.1 Can body manufacture.

8.2 Typical processed food and drink cans. From Coles R *et al.*, *Food packaging technology*, 2003, Blackwell. Courtesy of Blackwell Publishing.

Three-piece welded can-making systems will form tall, short or shallow cans with good metal usage efficiency. Within the two-piece seamless can-making systems, there are three variations, two (draw and wall iron, and impact extrusion) are particularly suited to making tall cans while the third (draw-redraw) uses metal more efficiently when limited to making short or shallow cans. Figure 8.2 describes typical cross sections of processed food and drink cans and the appearance of the different height

Rigid metal packaging   125

to diameter ratios. Figures 8.3, 8.4 and 8.5 describe the typical cross sections of non-round food, bottle and aerosol cans, respectively.

Most cans for food, drink or for use as aerosols or collapsible tubes have circular cross sections because these can be made, filled and ends seamed on at much higher

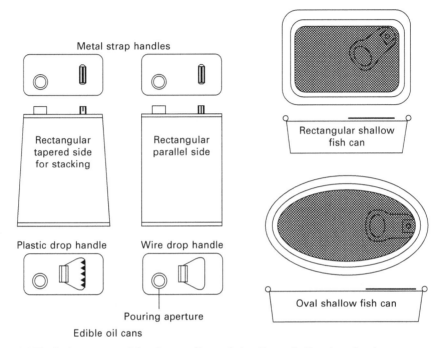

8.3 Typical non-round food cans. From Coles R *et al.*, *Food packaging technology*, 2003, Blackwell. Courtesy of Blackwell Publishing.

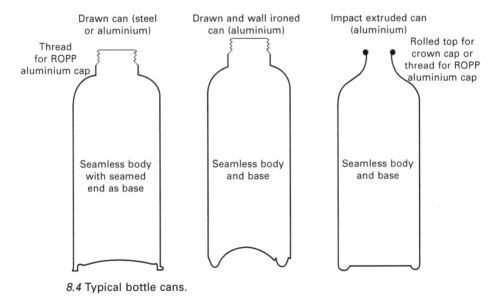

8.4 Typical bottle cans.

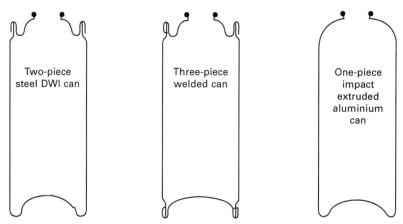

8.5 Typical aerosol cans prior to fitting of valve mechanism. Courtesy of Pira International.

speeds than those having non-round cross sections. Historically, a range of standard can diameters has been developed for each of the food, drink and aerosol markets. This was necessary in order to limit the number of different tool sets held by can makers and fillers and to limit the number of different end diameters that need to be manufactured. In the food and drink sectors, the various size ranges have been developed to produce a set of can volume capacities to provide varying portion sizes to suit the demands of the consumer markets, while minimising the number of containers having similar or nearly similar capacities.

The three-piece can-making process may be used to form any shape or size of container. Within two-piece systems, the can-making processes are less flexible in terms of making different can sizes because the investment in tooling dedicated to one can diameter and height is greater than that for three-piece can making.

A third system of forming metal containers is similar to the three-piece method but is restricted to use in making general line non-performance containers, which by definition are not required to hold liquids or pressure. These are often referred to as made-up containers and can be built from more than three pieces of metal. The main differences are that the two parts of the side seam are joined by just folding them together into a mechanical lock seam (described in detail in Section 8.3.5) and that the forming tools can be designed to make virtually any practical cross section. These types of containers do not generally fall into standard size ranges of either height or diameter.

Metal caps and closures are initially formed as shallow drawn containers before special shapes and lining/sealing materials are added to provide the necessary functions of the finished component. Because of the large quantities produced, these are generally made in standard diameters to suit the range of container neck sizes.

8.1.4 Process performance of metal cans and ends

Metal containers are manufactured from either packaging steel or aluminium. The choice will depend on the duty that the can has to perform as well as the can geometry and the method of production. Most metal containers are designed to hold liquid, or solids in liquid, at pressure, vacuum or normal atmospheric pressure, the exceptions being those designed for dry products or as decorative containers only. Information about metal selection is included in Section 8.5.

Metal cans are relatively low cost, thermally stable, strong, rigid, opaque, easy to process on high-speed lines and readily recyclable. As a packaging material, metal offers a total barrier to gas, moisture and light. These attributes make metal packaging particularly appropriate for long-term storage of perishable products at ambient conditions. Other important characteristics of metal are its stiffness, strength and ductility which permit simple forming operations and allow the containers to withstand the high mechanical loadings experienced during heat processing of foods and storage of carbonated drinks or aerosol products in conditions of high ambient temperature.

Strength is also of great importance when constructing containers to carry dangerous substances such as some organic solvents. For conveying such products across international borders by land, sea or air, the United Nations has developed a special set of regulations specifying the design of metal containers for the carriage of dangerous goods.

Metal's attractive appearance allows many graphic design possibilities and this, coupled with the hardness of the surface, makes it possible to reproduce images which are extremely sharp and detailed. Organic coatings are often applied to the metal surfaces to prevent corrosion due to chemical action with certain products or the external environment. These coatings are also used to help preserve foods safely and prevent dissolution of metallic elements into foodstuffs.

In addition to being able to store products at high pressures, aerosol containers have to be constructed to facilitate the operation of filling with both product and propellant and to allow for attachment of the valve-dispensing unit at the top of the vessel. Where product and propellant are not allowed to mix, then special arrangements such as the enclosing of the product in a flexible bag have to be incorporated.

8.2   Raw materials

The main raw materials used in metal can making are packaging steel, aluminium and organic coatings in either liquid or solid form. The coatings are used to protect the metals from chemical interaction with the products packed in the containers and, where necessary, the external environment. The coatings also help to prevent dissolution of unwelcome elements and taint from the metals into the product.

8.2.1   Metals for packaging – steel and aluminium

Steel results from the purification of liquid iron. In this process liquid iron, tapped from the base of a blast furnace into which iron ore, limestone and coke have been fed, is poured into a converter vessel. After the addition of scrap steel, oxygen is then blown into the iron to burn out all the impurities and lower the level of carbon. Whereas cast iron is a very dense and brittle material, low-carbon steel can be reshaped by mechanical rolling into very thin sections. As the steel is being rolled from a thick slab to a thin section suitable for can making its hardness and strength increase by work hardening following cold rolling steps. At times during this process the metal has to be reheated (annealed) to reduce the hardness to permit further cold rolling to take place and for the correct final combination of thickness and strength to be achieved. At this stage the metal is called blackplate, which may then be used for the manufacture of packaging where the contents will not corrode the steel. Examples of this are drums for oils, waxes and grease.

The anti-corrosion properties of steel for packaging are enhanced by the addition of tin to the surface. This is applied electrolytically (electroplated) to both surfaces of the metal while in coil form and after all shaping has been completed. An alternative metallic coating, also applied electrolytically, is electro chrome (ECCS – electro chrome coated steel) often known as tin-free steel. This material is often slightly less expensive than tinplate but must have an organic coating applied to give a complete corrosion-resistant system. It has particularly good adhesive properties.

Aluminium for light metal packaging is used in a relatively pure form with manganese and magnesium added to improve its strength properties. Aluminium is the third most abundant element and the ore from which it is extracted is bauxite. Aluminium oxide (alumina) is extracted from the other ores present in bauxite before the smelting process begins. This process requires considerable electrical energy to produce aluminium on a commercial scale. Primary and scrap aluminium are then brought together in a furnace to purify the liquid metal. It is then filtered through porous ceramic blocks for cleaning prior to casting into ingots. The faces of the ingot that will become the finished surfaces of the coil are both ground off (scalped) to remove surface imperfections and to provide clean surfaces ready for rolling. The metal is then reheated and hot rolled to an intermediate thickness. It is then allowed to cool in air for two to three days at which stage it is fully soft. Final cold rolling takes place to achieve finished thickness and hardness without further heat treatment.

8.2.2 Management of corrosion

When using steel or aluminium containers for packing wet products, some of which may also be subject to heat processing, consideration must be given to the likelihood of corrosion taking place on the inside walls of the cans due to chemical action of the product and how this may be prevented. Additionally, the external surfaces may be subject to corrosion due to excessive humidity in the atmosphere or during heat processing in steam/water systems.

For steel-based containers the presence of metallic tin on the surface always improves the corrosion resistance whether or not organic coatings are also in place. ECCS (tin-free steel) will give excellent corrosion prevention provided that an adequate layer of organic coating is in place as a seal coat. The oxide which forms naturally on the exposed surfaces of aluminium does provide some resistance to corrosion but this is insufficient for wet products, so all internal surfaces of these containers must normally be coated. Section 8.2.3 describes some exceptions to this requirement.

Organic coatings are applied to the internal container surfaces to form an inert barrier between product and metal to prevent chemical actions due to the nature of the product and to prevent dissolution of metallic elements into the product. The weight of coating and, where appropriate, the weight of tin applied to steel may be varied to suit the specific conditions in the can.

For three-piece welded side seam cans, it is necessary to leave the surfaces of tinplate forming the weld overlap free of any coating to ensure the weld is sound. After the weld has been made the previously unlacquered parts of the inside wall, including the weld itself, may then be coated with lacquer to produce a can where the internal surfaces are totally lacquered. This process is called side striping, the coating being applied by spray, roller or as a powder and then cured by a blast of hot air applied to the outside of the can body. There are some can material/product combinations where a side stripe lacquer is not necessary.

Bi-metallic corrosion can take place between aluminium and steel under certain circumstances. Where filled cans are fully heat processed after filling, it is necessary to use similar metals for both can and end to prevent this form of corrosion taking place. This applies whether the metals are coated or not. For drinks cans where only low level pasteurisation takes place, there is no problem when using aluminium ends on steel can bodies.

8.2.3 Food contact issues

When packing food or drink into metal containers, it is very important that the product packed and the container internal surfaces are compatible with one another so that no unwanted or uncontrollable chemical reactions take place between the two. This is clearly most important with wet foods and liquids. However, even for dry products, there will be situations when compatibility has to be considered. Internal lacquers on the can body and ends, as described in Section 8.2.4 below, are used to prevent basic chemical actions taking place. In addition to this, to ensure that food safety is not compromised in any way, where a lacquer protection system is in place, this must also prevent dissolution of undesirable chemical elements from either the coating itself or the metal substrate.

A number of food items, particularly meat and fish, contain sulphur. During the heat processing cycle low level reactions can take place between sulphur and tin or iron to create sulphides which are black in colour. These can take place across the surfaces of the coating if the weight is insufficient or if an incorrect coating specification has been used. While the sulphide products are not hazardous, they can impart stains to light coloured foodstuffs.

In the UK approximately 25% of cans for processed foods are supplied as plain cans, i.e. with no organic coating applied over the internal tin surface. This is done to improve the colour and flavour of white fruits and tomato-based products. As well as allowing low level dissolution of tin into the products, the presence of free tin on the surface of the tinplate helps to remove any oxygen remaining in the headspace of the can. These in turn help to retain the original colour of the products and provide a more piquant flavour. In the UK the maximum level of tin in the product is limited by law to 200 ppm. When a plain can is used in this way, the product shelf life must be set to prevent tin levels exceeding the maximum permitted level. As drink cans are designed to permit the user to consume the product directly from the can, it is critical that the external container surfaces which come into contact with consumers' lips are also considered to be food safe.

As an aid to understanding the background to current food contact regulations, the following extracts have been taken from Section 8 of *Guide to good manufacturing and hygiene practices for metal packaging in contact with food* (European Metal Packaging, 2009).

8   Food contact regulations

8.1   Introduction

Although, increasingly, metal packaging for foodstuffs and foods packed in metal are traded globally, there remains no global food contact legislative approach. The two major regulatory systems for control of materials and articles for use in contact with food are those of the EU and US (FDA), although detailed harmonised EU legislation for coated and uncoated metal packaging is still awaited. Food contact legislation is continually developing. Therefore, the information in this chapter may only remain current for a short time, and should be seen as the position as of 2008, after which this should only be used as a guide. To obtain more information regarding the current regulatory position of metal packaging, particularly as it relates to individual Member States, the appropriate European or national member state trade association (e.g. EMPAC for light metal packaging), or appropriate consultant organisations should be consulted.

8.2   Harmonised European regulations

Within the EU, materials and articles intended for use in contact with food are partly regulated at the EU level through harmonised Regulations and Directives, and partly at the member state level through their own legislation and recommendations. Wherever harmonised EU legislation exists, EU member states cannot maintain their own independent measures except in cases of specific derogation based on demonstrable risk to inhabitants of that member state. As EU harmonised legislation continues to develop, the importance of individual national member state measures will decrease.

The core of EU legislation on food contact materials and articles is the 'Framework' Regulation (EC) No. 1935/2004 which sets out in Article 3 the fundamental requirement that substances should not pass into the food at levels that may be harmful to health; or that may adversely alter the composition of the food; or that may lead to a deterioration in its organoleptic qualities. Furthermore, this regulation lays out intended means of detailed regulation of materials and articles by material type.

8.3   National regulations in the EU

The different member states within the EU have a wide range of approaches to the regulation of food contact materials and articles. As EU harmonised legislation evolves, it will replace member state legislation, but until that process is complete (and there is no certainty when that will be achieved for coated metal packaging), it is recommended that the various national metal packaging trade associations be consulted regarding the individual member state legislation in force.

8.4 USA regulations

US legislation of food contact materials and articles is managed by the Food and Drugs Administration (FDA) under the Code of Federal Regulations (CFR) Title 21 parts 170–199 in which potential migrants are considered as indirect food additives. These regulations have developed over many years and have a wider scope than the EU legislation. Under FDA, all types of food contact materials and articles are covered including coatings. Historically, the FDA regulations have been considered the most important global reference point for demonstrating the compliance and safety of packaging. With the evolution of EU legislation on food contact materials and articles, the relevance of FDA compliance in the EU is decreasing, but for materials such as coatings which are still not covered by harmonised detailed EU legislation, compliance with FDA remains an important element in the overall management of safety and compliance. Additionally, FDA compliance remains important globally.

8.2.4 Coating materials and their application

Coating materials are applied to the metal surfaces either in liquid or solid form (as powders, laminates or hot extrusions). This may be done prior to the metal forming operations, i.e. as coil or cut sheet, or after forming when the container or end has a three-dimensional shape.

Liquid coatings are organic-based materials comprising resins with combinations of solvent and water-based carriers to ensure good control of applied weight together with satisfactory wetting, adhesion and curing properties. The resin forms the hard coating that remains on the metal surface after the curing has been completed. Depending on the chemistry of the coating, curing may be achieved by the application of heat or by subjection to ultraviolet (UV) light. However, the method of curing is designed into the coating and cannot be changed later. The choice of curing method is dependent on many factors such as product packed, type of heat processing employed, the type of can and how it is used by the consumer. Historically, regulations have prevented UV coatings being used in food contact applications but some of these regulations are now being relaxed following evaluation work recently completed by the FDA. Nevertheless, for the foreseeable future, it is unlikely that UV coatings will be approved for the internal coating of heat processed food cans.

Liquid coatings may be applied by roller coating or airless spray and generally form part of the can-making operation. Figure 8.6 describes the process of roller coating the top side of a metal sheet while Fig. 8.7 shows how lacquer is applied by airless spray to the inside surfaces of a seamless can. During heat curing the coated product passes through a tunnel oven where the liquid carriers are first evaporated leaving the residual resin to be cured, by chemical cross-linking, into a hard but flexible surface.

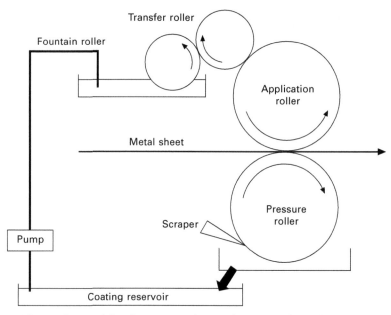

8.6 Coater for applying lacquer to sheets. Courtesy of Pira International.

- Applied by airless spray
- For food cans this is done after the last mechanical operation – wall beading
- Can lacquer is cured in a horizontal belt oven

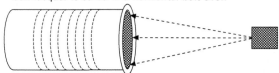

8.7 DWI food can internal lacquering.

Typically the total time for this operation is approximately 20 minutes. Ultraviolet curing is achieved by passing the wet surface of the metal sheet under a UV lamp at high speed. This creates an instant cure.

Powder coatings are 100% resin and contain no solvent or water carriers. Application is by spraying the dry powder followed by heat curing as part of the can-making operation. This process allows heavier coating weights to be applied than can be achieved by a single coat of liquid.

Polymer films, whether as laminates or direct extrusions, are normally applied by the metal manufacturer, as the most efficient way of achieving this is in a coil-to-coil operation. New systems are being developed which use pre-coated tin-free steel or aluminium for two-piece can manufacturing, either drawn or drawn and ironed. These systems may be single- or multi-layer polymers but in all cases are based on polyethylene terephthalate (PET). The main advantage of these systems is that all the processes of applying the polymers to the metal substrate take place under controlled conditions in the factories of the metal manufacturers. These then simplify the can-making process, eliminating can washing and in-line internal and external coating operations.

8.3 Manufacture of rigid metal containers

The formation of shaped can bodies from flat metal is carried out in one of two basic processes which are described as three-piece or two-piece processes. The numbers three and two refer to the number of separate pieces of metal required to construct a fully closed container. Two-piece cans may also be described as seamless cans because the base and side walls are constructed from one piece of metal without any joins in the open top container.

8.3.1 Three-piece welded cans

Three-piece welded cans for food, drink, general line decorative and industrial cans, as well as aerosols, are only constructed from steel-based materials as thin gauge aluminium cannot be welded by this process. Most of these are made from tinplate as ECCS (electro chrome coated steel), otherwise known as tin-free steel, is difficult to weld with consistency without first removing the metallic coating. Coils of tinplate, after receipt from the steel maker, are cut into sheets approximately one metre square to suit the dimensional capacity of the downstream equipment for coating and printing. After this process, which is shown in Fig. 8.8, the sheets are cut, by slitting, into rectangular blanks from which individual can bodies are made. The area in the vicinity of the weld is left without coating or print to ensure that a sound weld is made. In the welding body forming machine, each blank is rolled into a cylinder with the two longitudinal edges overlapping by approximately 0.4 mm. Using electric resistance spot welding, where alternating current passing through the metal seam heats up the material, the tinplate is softened sufficiently for the two edges to be squeezed together to form a sound joint. Each peak of electric current creates a spot of weld. As the length of the cylinder passes between the electrode rollers, a series of overlapping spots is created to form a continuous weld. If necessary, a side

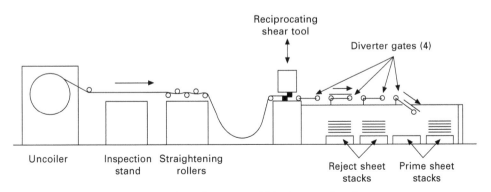

8.8 Coil cutting operation. Courtesy of Pira International.

stripe coating is applied over the weld area at this time. The forming and welding process is described in Fig. 8.9.

All three-piece cans now pass through a flanging process where both ends of the cylinder are flanged outwards to accept the can ends. Drinks and aerosol cans are usually necked in prior to flanging. This process reduces the diameter of both ends of the cylinder, before the flange is formed, which in turn allows ends to be fitted which, after seaming, are smaller in overall diameter than the can body. This in turn reduces the cost of the end and the space taken up by the seamed body. Some three-piece food cans have the bottom only necked in to permit safe stacking of one filled can on top of another.

Where food cans are going to be heat processed (retorted) after filling and where the can height is greater than its diameter, it is usually necessary to form circumferential beads in the can body wall to increase the hoop strength to resist implosion of the can during the earlier part of the process. These wall beads are shown in Fig. 8.10. The basic points of heat processing are described in Section 8.3.8.

At this point one end is now mechanically seamed onto the cylinder to create an open ended container ready for filling. This end is called the maker's end. Where easy-open ends are used, it is common for this to be fitted as the maker's end, thus allowing it to pass through the finished can testing process. The mechanical seaming process is described in Section 8.3.7. The end fitted by the packer after filling is called the canner's end. Before packing onto pallets, all cans are pressure tested to check for the presence of cracks, pinholes or weak welds. The full process sequence of making three-piece welded cans is shown in Fig. 8.11.

This three-piece can-making process will form cans of all diameters and heights provided the correct tooling is installed and the appropriate size of forming/welding machine has been employed. The welding process can only generate bodies of circular

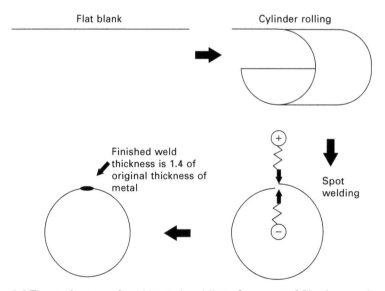

8.9 Three-piece can forming and welding. Courtesy of Pira International.

Most tall food cans are passed through a beader where the walls of the cans have circumferential beads formed in them to give added strength to resist heat processing conditions

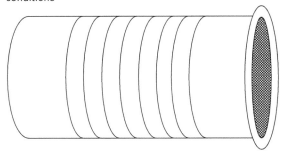

8.10 Can wall beading.

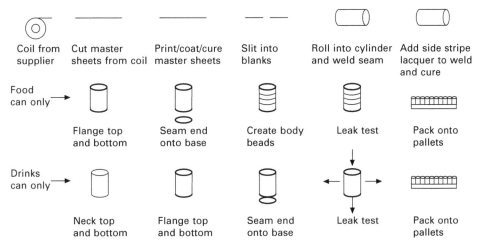

8.11 Three-piece can process flow.

cross section, therefore, where non-round, i.e. square, rectangular, etc., or slightly tapered bodies are required, the circular body must be reshaped after completion of the welding process.

Where shallow cans (height less than diameter) are to be made by this process, they can become unstable when being transported on conveyors with the can axis horizontal, which is necessary during the body forming and welding process. To overcome this, the body blank size may be increased so that multiple bodies (say 2–4, depending on the maximum blank size capacity of the welding machine) may be incorporated in the one blank. As the multiple sized blank is rolled into the cylinder it is partially scored through at the points where the formed can will be separated into the individual short bodies. After the welding (and side striping/curing) operation is complete, the long cylinder is taken into a parting machine where it is split into the individual short bodies. From this point onwards during the completion of the can-making process the bodies are conveyed with the axis in the vertical position.

8.3.2 Two-piece single drawn and multiple drawn (DRD) cans

Drawn and redrawn (DRD) cans are used for food (particularly processed fish products) and general line cans and may be made from steel or aluminium. In most cases, the metal is coated internally and externally and printed, if necessary, prior to the can-forming operation, as the presence of these materials on the metal surface provides some degree of lubrication to aid the forming. Lubricants such as waxes may need to be added to the outside surface to enhance this.

As an alternative to the use of liquid organic coatings, metal may be supplied in coil form either pre-laminated with polymer film or with polymer coatings hot extruded directly onto the metal surface. In these cases, where printing is not required, the coated coil may be fed directly into the can-forming press.

As coil printing for cans is not considered practical, coils of metal are supplied plain and cut into sheets ready for coating and printing. These processes are identical to those described above for three-piece welded cans. Because the metal blanks for DRD cans cut from the sheets will be either round or oval, metal wastage can be minimised by replacing the straight cut shear in the sheet-cutting machine with one having a castellated cut. The castellations must be designed to fit closely to the shape of the blanks across the width of the sheet. This process is called 'scroll cutting'.

Whether from coil or sheet, the metal is fed into a reciprocating press with either single or multiple forming tools. At each forming station the press first cuts a disc (blank) and, while in the same station, then draws this into a shallow can (cup) as shown in Fig. 8.12. During the drawing process, the metal is reformed from a flat sheet into a three-dimensional can while retaining the same surface area. It follows from this that the metal thickness throughout the whole of the can is the same as the starting gauge of the metal coil. During the reforming process the metal flow is controlled by a circular pressure pad to prevent wrinkles being formed in the side wall. Figure 8.13 describes the basic drawing process. The width of the coil/sheet

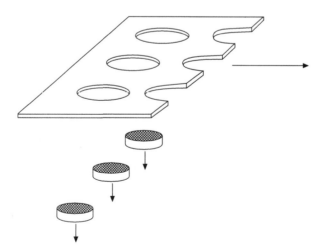

8.12 Cup making for DRD and DWI cans.

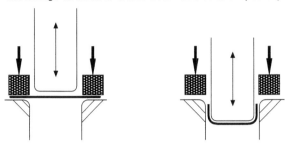

8.13 Drawing a can from flat metal.

used must match the width across the number of toolsets installed. Where multiple tooling is installed, each stroke of the press will generate a cup from every installed tool. For minimum wastage of metal the circular blanks are close packed with their centres set at 60 degrees to each other.

Alternative methods for feeding sheets are called stagger feed systems. In these, the edge of the sheet remote from the press is held by gripper fingers which are mounted onto feeding arms moving in two dimensions in the horizontal plane. With, say, a single installed tool it is then possible to feed the sheet forward and sideways through the tool so that every blank is cut from a sheet even though the width of the sheet is equal to that of multiple blanks.

The drawing tool may be configured to leave the side wall straight or to leave a flange in position. The can produced from this single draw may now be at its finished dimension, in which case a straight walled can will need to have a flange formed at the open end. Where the flange was left in position during the initial draw, it may be necessary to trim it using a circular cutting tool to ensure the flange width is consistent enough to make a satisfactory seam. For production of a deeper can, the drawing operation may be repeated (redrawing) in a second station with appropriate sized tooling; however, as for the first draw, the surface area of the deeper can and the wall thickness of wall and base will still be unchanged from that of the original disc of metal. The redrawing process sequence is shown in Fig. 8.14. The full process sequence of making DRD cans is shown in Fig. 8.15. The DRD process may also be used to form taper wall flanged cans as well as aluminium or steel taper thin wall trays to which heat sealed foil lids are applied.

For deeper cans where the height is approximately equal to the diameter, it is usually necessary to form the can in at least two steps, as described above, because there is a limit as to how far the metal can be reformed in one step. This is called the limiting draw ratio. The single drawing process is also used for making folded aluminium baking trays and takeaway containers. In these processes the metal is allowed to fold on itself as the blank is reformed into the shape of the container. Testing of all cans for minor cracks or pinholes is part of the manufacturing process and for two-piece (seamless) cans is achieved in a light-testing machine. This measures the amount of light passing across the can wall using high levels of external illumination.

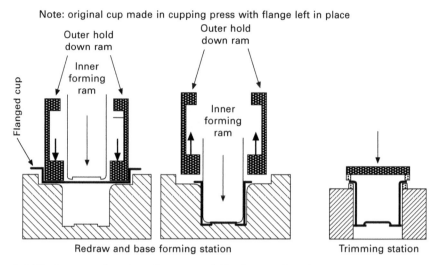

8.14 Redraw can forming. Courtesy of Pira International.

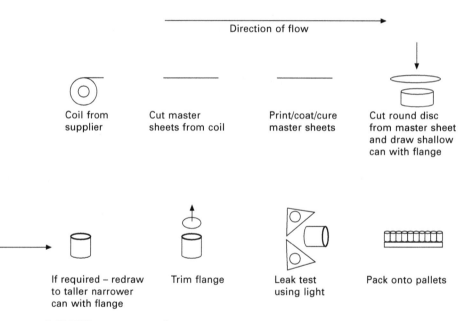

8.15 DRD can process flow.

8.3.3 Two-piece drawn and wall ironed (DWI) cans

DWI cans are used for food, drink and aerosol cans and are normally made from uncoated aluminium or tinplate. However, DWI processed food cans are only made from tinplate as thin wall aluminium cans do not have sufficient strength to withstand the external pressure imposed during the heat process cycle. In addition, DWI aerosol cans are currently only made from tinplate.

It can be seen from Fig. 8.1, which shows the basics steps in the DWI can-making process, that the first steps in forming this can, from coil to redrawn cup, are virtually identical to those used in the DRD process described above. However, the main differences are:

1. the tinplate is always fed directly from the coil into the blanking and first drawing, and
2. the coil is uncoated because subsequent wall thinning (ironing) operations are so severe that the coating would be stripped off the metal surface. Because of this, a water soluble synthetic lubricant is applied to the coil prior to the cup blanking/drawing operation. Application of coatings and print (where appropriate) later in the process are an integral part of the DWI can-making process.

The full process sequence of making DWI cans is shown in Fig. 8.16.

The shallow cups are now fed into a number of parallel body-making machines which convert them into tall thin wall cans. In this process the cup is first redrawn into a taller cup having a diameter equal to that of the finished can. Following this the taller cup is rammed through a series of ironing rings which have internal diameter marginally less than the outside diameter of the can wall but not less than the inside diameter of the can. There are usually three or four rings in the sequence each having a smaller diameter than the ring preceding it. As the cup passes through these rings the wall thickness is progressively reduced to approximately half or one third of the original starting thickness depending on the finished height of can required. Because there is no loss of metal in the process the can height is increased as the walls are thinned. This is a cold forming process but the friction generated heats up the metal. During ironing the can body is flooded with the same type of lubricant used in cup forming. As well as assisting the ironing process, the lubricant cools the can body and flushes away any metallic debris. These forming steps are shown in Fig. 8.17.

After forming the full height can body, the uneven top edge of the can is trimmed to leave a clean edge and a can of the correct overall height. Trimmed cans now pass through chemical washers to remove traces of lubricant and prepare the metal surfaces for internal and external coating (or printing in the case of drink or aerosol cans).

For DWI food cans, where paper labels are normally applied, an external clear coating is next applied by flood coating as the cans are conveyed under waterfalls of clear lacquer. This protects the external surface and base against corrosion and provides a slightly lubricated surface to improve can mobility during product filling and final packing operations. The coating is cured by heating in a hot air oven. The open end of the can is now flanged ready to accept the can end and this is followed by the addition of circumferential beads to increase the hoop strength of the can wall. The need for beading is described below in Section 8.3.8. After light testing, as described for DRD cans, an internal coating is applied using an airless spray system. This coating is cured in a hot air oven.

For DWI drink and aerosol cans, where the external surface is normally printed in line, a clear or pigmented base coat is first applied to provide a good surface for the acceptance of printing inks. A circular can printing machine is now used to apply

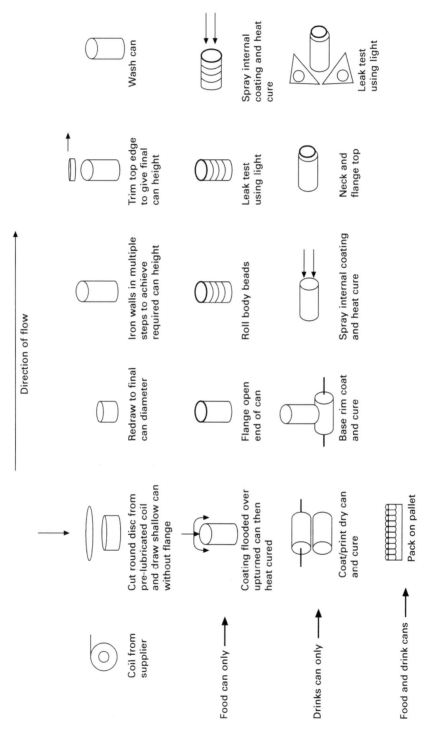

8.16 DWI can process flow.

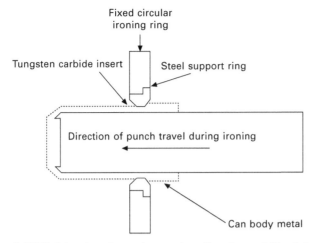

8.17 Wall ironing through one ring. Courtesy of Pira International.

the complete label design around the can body using up to eight colours. The inks are cured either in a hot air pin chain oven or by passing across UV light lamps, depending on the type of ink chemistry being used. Internal coating and curing is by airless spray and hot air oven, as for food cans. Following this, a coating is applied to the external base rim and then cured. These cans are next necked in at the open end to reduce the diameter of the end or ring (in the case of aerosol cans) fitted and then flanged. The reduced neck size allows lower cost ends to be fitted and reduces the maximum diameter of the can to that of the body. (Without necking, the diameter over the finished end seam is greater than the diameter of the body.) The finished cans are all now passed through a light-testing machine, as described for DRD cans.

8.3.4 Two-piece impact extruded cans and tubes

The process of impact extrusion is restricted to containers made from aluminium only as it is not possible to form steel cans in this way. Historically, this process has been used for aerosols, rigid and collapsible tubes. In recent years shaped bottles for drinks have been introduced where the forming process has been based on that used for aerosol production.

In this process a thick disc equal in diameter to the outside of the finished container is punched out from aluminium plate. The disc thickness is such that its mass/volume is equal to that of the untrimmed can body. The disc is placed in the bottom of a die and a reciprocating punch having maximum diameter equal to the inside diameter of the container is driven into the disc at high speed. The cold metal is forced out of the die block and flows up the side of the punch until the end of the stroke. This process is described in Fig. 8.18.

The same process is used for forming collapsible tubes. However, in this case the base of the die is modified to allow formation of the tube nozzle with or without a sealing membrane and the starting disc shape is also modified as shown in Fig. 8.19. As the forming process work hardens the tube wall, it is necessary to soften this by

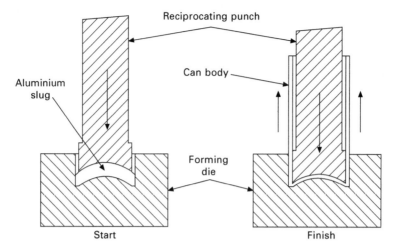

8.18 Impact extrusion process. Courtesy of Pira International.

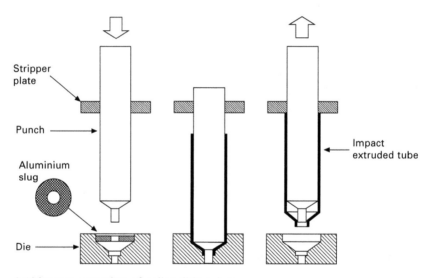

8.19 Impact extrusion of collapsible tubes.

application of heat (annealing), after forming, so that the finished tube is capable of being squeezed and rolled up during use.

Further manufacturing steps, including coating and printing, are similar to those used for DWI drinks cans. The only exception to this is that necking and flanging are replaced by a multi-step forming process, to shape the top of the container from full body diameter to that required to accommodate the ultimate can closure device, as shown in Fig. 8.20. This could be a flange on which to crimp an aerosol valve mechanism, a rolled edge to accept a crown end or a screw neck to accept a ROPP™ (roll-on pilfer proof) cap.

Rigid metal packaging

Preparation of can top to accept valve mechanism

The final process after testing is to swage the top edge of the can in approximately 15 steps to form a smooth top and roll flange to accept the valve system

8.20 Swage top of aerosol can.

8.3.5 Mechanical lock seam for general line decorative cans

The bodies of made-up general line containers are constructed from two or more pieces of pre-decorated tinplate with an almost infinite variety of cross sections available, ranging from round, oval and rectangular to triangular, hexagonal, etc. The body side joint is made by creating a lock seam where the edges of the metal are folded and pressed together. As these containers are not required to hold liquid, there is normally no cement or compound in the lock seam or the end seams. For special cases such as highly decorated cans for paint, it is possible to apply a hot melt adhesive to the seam faces prior to forming the lock. This will produce a liquid tight seam.

A typical made up container has a parallel wall body with a partly formed flange at the bottom to accept a mechanically seamed flat panel end. At the open end of the can, the top edge is hemmed or curled (internally or externally) to make safe the raw edge of the metal and provide additional strength across the section. This type of container is described in Fig. 8.21. Where the walls are to be embossed, the design is formed in the metal while the blank is still flat and before any other operations have been performed on it.

The principles used in forming and lock seaming the body are the same for all cross sections, but the tooling which forms the cross section is specific to each different can design. The edges which will form the seam must be hooked, as shown in Fig. 8.22 before the cross section of the body is formed. Provision of these ensures that the body does not spring open after the cross section shape has been formed. Figure 8.23 shows the process flow diagram for the general line lock seam can:

Step 1 A prepared metal blank is fed into the machine and placed equidistant across the expanding mandrel.
Step 2 The forming fingers close around the blank, forming it to the shape of the mandrel in the non-expanded position. At this point, the two hooks on the opposite edges of the plate loosely interlock.
Step 3 The forming fingers open up and the mandrel expands to pull the two edge hooks tightly together and give the correct internal dimensions to the can. At this point, the seam is hammered flat to provide a positive lock.

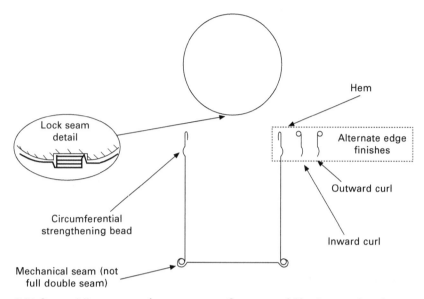

8.21 General line non-performance can. Courtesy of Pira International.

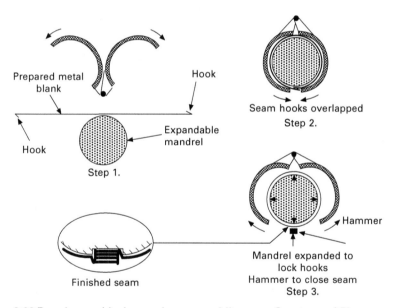

8.22 Forming and lock seaming general line can. Courtesy of Pira International.

8.3.6 End making processes

Can ends for mechanical double seaming are constructed from aluminium, tinplate or tin-free steel (TFS). Aluminium and TFS are always coated on both sides with organic lacquer or film laminate whilst the metal is still in coil or flat sheet form.

Rigid metal packaging 145

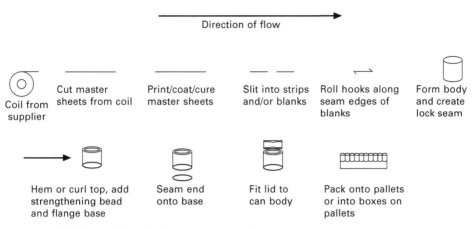

8.23 General line lock seam process flow.

For tinplate these coatings are optional, depending upon the product being packed in the container and the specified external environmental conditions.

The base of a three-piece can will always be a plain end (non-easy-open). For food cans, the top may be either plain (requiring an opening tool), full aperture easy-open or peelable membrane design. Historically, tapered rectangular solid meat cans have employed a key opening device to separate the two scored body sections; these are now gradually being replaced by containers having rectangular panel full aperture easy-open ends. For drink cans, the top is usually referred to as a stay-on-tab (SOT), enabling the opening tab and pierce-open end section to be retained on the can. The SOT end has largely superseded the traditional ring-pull end.

All ends for processed food cans have a number of circular beads in the centre panel area to provide flexibility. These allow the panel to move outwards, as internal pressure is generated in the can during the heating cycle of the process and so reduce the ultimate pressure achieved in the can. During the cooling process, this flexibility permits the centre panel to return to its original position.

Ends for beer and carbonated drink cans do not require the above feature as the can's internal pressure is always positive. The plate thickness and temper have to be appropriate to the level of carbonation of the product and, if applicable, pasteurisation treatment; otherwise excessive internal pressure may cause can ends to peak or distort.

Plain food can ends and shells for food/drink easy-open ends

The initial processes for making plain food can ends and easy-open ends for food and drink cans are the same. The body of an end that will ultimately be converted into an easy-open end is referred to as a shell. Plain ends/shells may be stamped directly from wide coils of metal or from sheets/strips cut from coils. When using coil or sheet, the metal is fed through a press that produces multiple stampings for every stroke. After removal from the forming tool, the edges of the end shells are then curled over slightly to aid in the final operation of mechanical seaming the end onto the flange of the filled can. After curling, the end shells are passed through a lining machine that applies a bead of liquid-lining compound around the inside of the curl. This process is described in Fig. 8.24. The compound lining is a resilient material that, during mechanical forming, will flow into the crevices of the double seam and thereby provide a hermetic seal. The application of this compound is shown in Fig. 8.25.

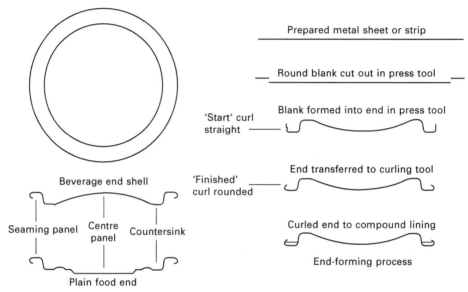

8.24 Plain food end and beverage end shell forming. From Coles R et al., Food packaging technology, 2003, Blackwell. Courtesy of Blackwell Publishing.

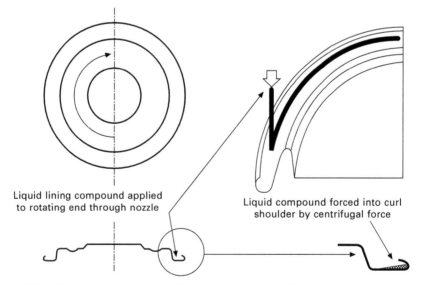

8.25 End lining compound application. Courtesy of Pira International.

Conversion of end shells into easy-open ends

The principles used in the conversion of end shells are the same for both full aperture food easy-open ends and small aperture drink easy-open ends. The conversion operations comprise scoring (partially cutting through) the perimeter of the opening panel and attaching a metal tab with which to tear open the panel. These operations are described in Fig. 8.26. Scoring is necessary to reduce the force required to open the end to an acceptable level and to determine where the break will occur.

The pull-tab is made from a narrow strip of pre-coated aluminium or steel, which is in coil form. The strip is first pierced and cut, and then the tab is formed in two further stages. At this point the tab is still attached to the strip by bridges to facilitate feeding over the rivet formed in the shell. The shells pass through a series of dies that score them and form a hollow upstanding rivet in the centre panel of the shell. The tab is then placed over the upstanding rivet on the shell, and the rivet is deformed to make a joint between the two components. The finished ends, ready for capping the filled cans, are packed into paper sleeves and palletised for shipment to the can filler.

Peelable membrane ends for food cans

A peelable membrane end is essentially a plain aluminium or steel food can end with part of the centre panel removed as a disc and replaced by a pre-cut aluminium or polymer membrane which is heat sealed to the remaining part of the centre panel. Figure 8.27 shows the assembly of this end from which it will be noted that the cut edge remaining after removal of the disc is folded back on itself so that the raw

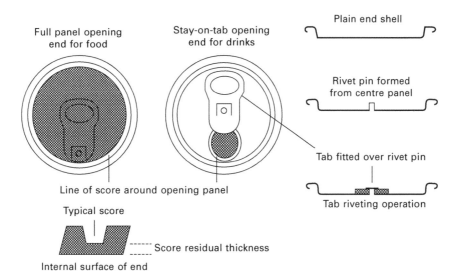

8.26 Conversion of plain end/shell into easy-open end. From Coles R et al., *Food packaging technology*, 2003, Blackwell. Courtesy of Blackwell Publishing.

148 Packaging technology

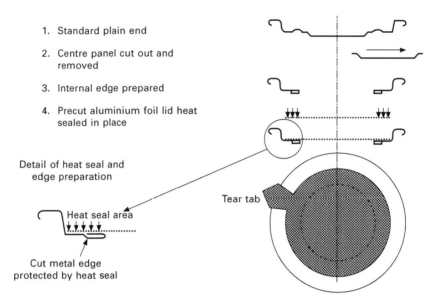

1. Standard plain end
2. Centre panel cut out and removed
3. Internal edge prepared
4. Precut aluminium foil lid heat sealed in place

Detail of heat seal and edge preparation

Heat seal area

Cut metal edge protected by heat seal

Tear tab

8.27 Peelable membrane end. Courtesy of Pira International.

metal edge is protected within the heat seal area. A recent design of this type of end is capable of withstanding full heat processing conditions without the need for overpressure in the retort to prevent rupture of the membrane due to high internal pressure.

8.3.7 Mechanical seaming of ends onto can bodies

The standards employed during the process of mechanical double seaming ends onto can bodies are extremely important in ensuring the ongoing safety of the product packed into the container as, after heat processing and cooling is completed, the internal pressure in the can is negative. As a consequence of this, any weakness in the double seam could lead to micro leaks and the introduction of non-sterile air into the container. Industry standards are designed to ensure that practical interchangeability is achieved when end and can components are purchased from different sources.

The basic system of double seaming described below, originally established at the start of the twentieth century, is the same for food, drink and aerosol cans and is applied to both round and non-round ends. The seaming tools have to follow the profile of the can/end cross section; this allows round profiles to be seamed more quickly than, say, rectangular profiles.

Figure 8.28 describes the sequence of the loose end being offered up to the can flange and the two seam roller operations to create the finished seam. The inside diameter of the end curl is just sufficient for it to drop cleanly over the flange of the can. If the end were not curled the two components would not lock together as the seam was made. During the seaming process, which is in two stages, the end is mounted on a round (or non-round) chuck which fits the external surface of the countersink wall and supports this wall during the seaming process. The first stage

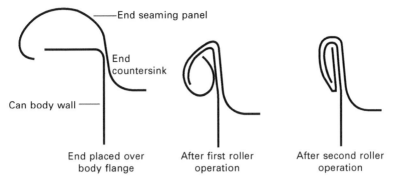

8.28 Mechanical seaming operations. From Coles R *et al.*, *Food packaging technology*, 2003, Blackwell. Courtesy of Blackwell Publishing.

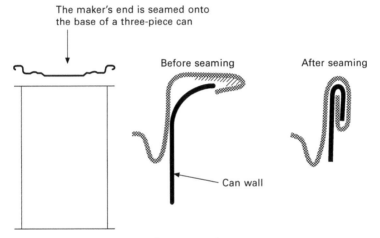

8.29 End mechanical seaming – overview.

external roll rotates the seaming panel of the end with the can flange to close them up as shown. The second stage external roll tightens up the seam to give the correct final external dimensions and produces the required hermetic seal. Both first and second stage operations are carried out in the one seaming machine. Figure 8.29 provides an overview of this process.

The cross-section view of a finished double seam is shown to the right in Fig. 8.29. To evaluate the seam, some of the dimensions can be taken from the outside of the finished seam while others can only be measured from a cross-section view. This view may be obtained either by cutting through the seam or by using X-ray technology. Some parameters cannot be measured directly from the seam cross section but require simple mathematical calculation to deduce the result.

8.3.8 Basics of heat processing of food (retorting)

For long-term storage of canned food products at ambient conditions, it is necessary to kill all living organisms within the can after filling and seaming on the closure.

This is achieved by sterilisation using heat. To complete this process so that food safety issues are completely satisfied, it is necessary to ensure that all the product in the can has received sufficient heat for the required minimum amount of time. For liquid products, convection currents within the can allow more rapid temperature increase time than, for instance, solid products where heating up the centre core of product may only be achieved by conduction.

The operating principle of the heating method used is similar to that of the domestic pressure cooker. In this process the cooker is allowed to operate at an internal pressure somewhat higher than normal atmospheric pressure. A pressure relief valve fitted to the lid controls the internal pressure and prevents it rising above this level. This elevates the boiling point temperature of the water in the cooker which in turn reduces the overall time required to heat the food to the required temperature. Industrial retort systems operate in a similar way but at higher pressures and temperatures to ensure the correct sterilisation temperatures are reached. In these systems, achieved temperatures will usually be in the range 113–132°C with processing times ranging from five minutes to periods in excess of one hour, depending on the type of product being sterilised and the can dimensions.

Industrial equipment generally falls into one of two basic types: static retorts, which are larger versions of the domestic unit or continuous retorts, where filled cans enter and exit the retort in a continuous stream. The pressures inside industrial retorts will vary between approximately one and two atmospheres above the external ambient pressure depending on the retort temperature, the higher pressure giving the higher temperature. The varying pressure and temperature experienced by the can during the heat process causes severe physical loadings on the can body and end. This needs to be understood to ensure the container is capable of withstanding these without collapse.

When the can enters the retort its contents are usually only warm, having sufficient heat to aid in conveying and filling the product into the can, and the internal pressure is atmospheric or slightly below if vacuum packed. As the full steam pressure acts on the outside of the can, the pressure imbalance has the effect of imploding the can body. This force is only resisted by providing the body with sufficient hoop strength (usually by the addition of circumferential beads) to supplement that provided by the product itself.

When the temperature inside the can ultimately reaches that of the steam on the outside, the pressure in the can has become greater than that of the retort chamber. This is because the total can internal pressure is the combination of the thermal expansion from the increasing partial pressure of the water in the product, plus that of the expanding headspace gases.

At exit from the process the contents are still relatively hot but the external retort pressure is reducing and the can is now trying to explode. At this point the can ends are temporarily deformed outwards in resistance to the internal pressure load. As the can and contents slowly cool down to room temperature, a negative pressure is induced inside (as the contents were originally warm when the end was seamed on). It follows that, for a processed food can at ambient conditions, the contents provide only a relatively low level of physical support to the can wall. This is in contrast to the high level of wall support provided by the pressurised contents inside a beverage can.

Whilst the end is primarily designed to resist the high relative internal pressure achieved towards the end of the process, it must also be sufficiently flexible to return permanently to its original profile as the internal pressure becomes slightly negative. If this does not happen, the can may have the appearance of being blown with the product wrongly being deemed unfit for use by the consumer. For very lightweight containers and those having flexible heat sealed lid systems, an over-pressure retort will allow the retort chamber pressure to be controlled so that the physical loads on the container and end are insufficient to cause damage or rupture.

8.3.9 Quality assurance of semi-finished and finished components

Quality assurance checks on semi-finished and finished components are carried out both in-line, as part of the manufacturing process, and off-line in a laboratory. Some in-line processes such as pressure or light testing for cracks and pinholes, video inspections of can internal surfaces or external decoration, are part of the manufacturing process and are carried out on every component produced. Figures 8.30 and 8.31 describe the principles of in-line air and light testing, respectively. Most other checks use predetermined sampling plans where components are removed from the line either automatically or by hand. Statistical analysis is then used to determine the performance of the various processes. This method of control is required because the line production rates are so high, being in the range 50–2500 items per minute. For dimensional checks, automatic measuring stations are often used. These stations may be situated in the production area and may be fed with sample components automatically or by hand. The equipment produces the results of each sample in a statistical format and delivers the results to the central laboratory.

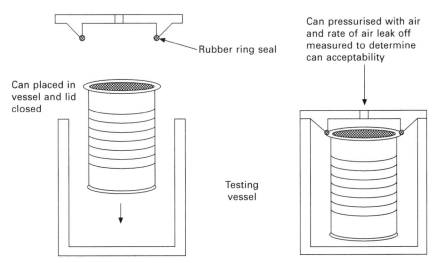

8.30 Three-piece can air testing principles. Courtesy of Pira International.

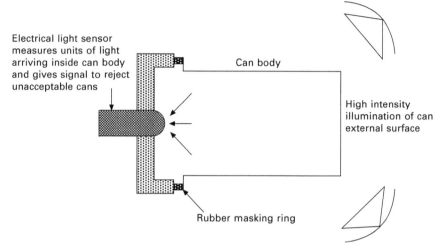

8.31 Two-piece can light testing principles. Courtesy of Pira International.

The following are examples of off-line tests usually carried out:

- Dimensions – height, diameter, wall thickness range, flange width
- Strength – axial compression, resistance to implosion (food cans), base dome buckle (drinks cans)
- Coating integrity – weight, continuity (freedom from pinholes), adhesion
- Ends (in addition to above tests), deflection due to internal pressure changes, lining compound placement and weight
- Easy-open ends, additional tests to determine 'pop' and 'pull' loads to open and tear the tab, rivet strength and integrity.

Attribute sampling is normally carried out on randomly selected pallets of finished cans or ends before these are conveyed to the warehouse for onward shipment to the customer. A sample plan is devised so that, for example, in a pallet containing 5,000 cans, six cans will be removed from the outside rows each from a different layer. These will be closely examined by eye. If more than one of these samples is deemed not fit for purpose, then the whole pallet will be quarantined for further checks. At the same time the pallets preceding and following this will also be checked in the same way until enough pallets have been held to ensure that the incidence of the problem has been contained.

Traceability of raw materials through to the finished components is assured by the recording of the coil numbers, or sheet pallet numbers where these are used, and time over which these are fed into the production line. Likewise container or batch numbers of coating materials are recorded in the same way. These references then tie up with the serial numbers of the finished pallets of components. Bar codes are generally applied to the labels on pallets of finished goods to give both fixed information such as customer name, container specification, number of items, etc., and real-time information such as pallet serial number, date, time pallet packed, etc. For drinks cans externally printed in-line, a code is embodied into the label

Rigid metal packaging   153

design, at the overlap area which indicates the makers name, factory, date and shift of manufacture.

The most common defects that may be found in metal packaging components are:

- all can bodies – low tin coating weight, badly formed flanges
- coated surfaces – pinholes, poor adhesion, undercuring, underfilm staining, cracks in coating after necking in drink cans and after curling can ends
- three-piece can bodies – poor weld strength, badly formed/incomplete weld
- two-piece draw and wall ironed can bodies – pinholes in body or flange area
- can ends (plain) – lining compound incorrect weight and bad placement
- can ends (easy-open) – broken/leaking rivets, residual score out of specification, pop and pull loads out of specification.

8.4 Metal closures

Metal closures for fitting to glass and plastic containers are highly specialised products. Because of the different strength requirements, the closures for drinks containers are made from aluminium, whilst those for processed food are made from steel. It will be seen that the threads on closures can be formed during the initial metal-forming process, or as an off-line moulding process, or during the filling/closing process or even during the (food) heat processing operation. In all cases, the dimensions (finish) of the neck of the glass/plastic container need to be specified very accurately and, for this reason, the specifications are usually set by the closure manufacturer and not the container maker. Outlines of all the closures discussed below are shown in Fig. 8.32.

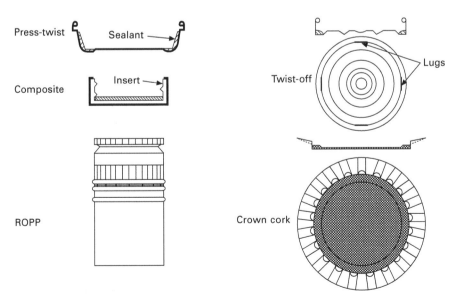

8.32 Outlines of metal closures.

8.4.1 Roll-on pilfer proof (ROPP™) caps

Roll-on closures are made only from aluminium, because it is necessary to have a soft material for the thread rolling process. This type of closure is initially produced as a printed and internally coated deep drawn two-piece can. After the addition of special mechanical features, and a wad or flowed-in liner to provide the seal to the surface of the bottle, the closure is complete and ready for shipment to the filler. When the bottle has been filled, the closure is slipped over the neck of the bottle and the threads are rolled into the closure to conform to the profile of the neck of the bottle.

The manufacturing process for the roll-on closure is the same as that for a draw/redraw can made from printed and coated sheet aluminium. If the end of the closure is to be printed, this is done whilst the metal is still flat. The closure liner is either a wad cut and inserted into the metal shell, or a flowed-in polymer applied to the inside top surface of the shell. Where the latter is used, the flowed-in material can be applied over the whole of the inside top surface or just to the circumference.

8.4.2 Composite closure

A composite closure comprises the aluminium body of a roll-on closure with a plastic moulded insert which is pressed in to give a tight fit between the two components. The low density polyethylene plastic insert has internal threads moulded into the side walls and is designed to accept a standard wad liner. This type of closure can be made with or without a tamper evident feature.

A composite closure provides two additional features to the standard ROPP™ closure.

- The thread of the closure is internal and provided by the insert, so that the external side wall of the closure has no threads in evidence after it has been sealed onto the bottle. This closure is used to improve the external appearance of the filled package.
- As the internal thread of the closure is made from moulded plastic, it is more suitable for use with products which have a high sugar content, such as liqueurs. When an all aluminium closure is used with these products, there is a tendency for the sugar, which remains on the thread after the bottle has been used for the first time, to form a sticky deposit between the aluminium and glass threads. This makes the bottle difficult to re-open. When one of the thread components is made of plastic, this problem does not occur.

8.4.3 Twist-off closures

Twist-off lug closures, which require less than one turn to apply and remove, were developed by the White Brothers in the United States in the early 1950s. For this reason, they are often referred to as 'White' caps. The product inside the glass jar is packed under vacuum, by packing the product hot and flushing the headspace with steam; the vacuum increases as the product cools to ambient temperature. The seal between closure and the top face of the glass jar is produced by a combination of the internal vacuum and the mechanical force produced when the cap is tightened in the capping machine. Many twist-off closures are now fitted with 'vacuum buttons' which indicate to the consumer, prior to opening, whether there is still vacuum inside the headspace. The internal vacuum produced during the packing operation is sufficient to pull down the vacuum button. For the light metal packaging industry, the combination of a twist-off closure fitted to a glass jar is unique, it being the only time that a fully preformed metal component is fitted to a fully preformed container made from a non-metal product.

8.4.4 Press-twist (PT) closures

The high volume market for heat processed baby food in glass jars necessitated the development of a high-speed closing system for the filled package. The PT closure, which was developed for this market, needs only to be pushed over the threads in the neck of the glass jar to make the initial closure. The heat of the processing system does the rest, by softening the sealant and allowing it to flow into the threads in the neck of the glass jar. A vacuum button is created during the metal-forming operations to indicate to the consumer, prior to opening, whether there is still vacuum inside the headspace. In the UK, a plastic moulded tamper evident ring is often incorporated into the closure. This is held in place by a curl on the wall of the closure.

8.4.5 Crown cork closures

Crown corks are used for sealing glass bottles containing carbonated or non-carbonated drinks. The cap is pre-formed during the manufacturing process and clinched onto the top of the bottle manually or automatically after the bottle has been filled. A tool is required to remove the closure from the bottle, which may be hand held or wall mounted. This closure is manufactured from tin-free steel with suitable coatings on both the inside and outside. The process is identical to that used for plain can ends until the press-forming operation. At this point, the blank is cut from the sheet in the same press station where the cap is formed. After forming, a liner is introduced into the top of the cap to make the seal across the top of the neck of the bottle.

8.5 Cost/performance comparison: raw materials and forming processes

The cost of the metal represents some 60–70% of the manufacturing cost of most metal packaging and is therefore the most significant element in total product cost. The only exception to this is where highly decorated containers are made for general line promotional markets where the decorating cost percentage is significantly more than for, say, food or drink cans. In this case the metal cost will fall to 40–50% of the manufacturing cost.

The purchase price of metal, whether steel or aluminium, is related to the amount of work (gauge reduction) undertaken by the metal manufacturer. This means that the price per tonne increases as finished thickness is reduced. While this still translates into a net reduction in price per unit of metal area purchased for each step of thickness reduction, the underlying increase in price per tonne offsets some of the potential price reduction from the reduction in gauge.

The above scenario becomes important when comparing the cost of the metal used in making three-piece welded and two-piece drawn and wall ironed (DWI) cans for processed food. In the three-piece can, for each step of gauge reduction the cost of the metal to make the body will reduce as described above, as here the metal is used in the condition received from the manufacturer with no change to thickness or properties through the can-making process. In the DWI can, which starts with heavy gauge metal to give the correct base thickness, the can-forming process does the additional work on the metal to reduce the wall to the required final thickness. A reduction in wall thickness is produced solely from a reduction in the diameter of the disc cut from the coil as the result of a change in tooling dimensions. The thickness of the metal coil purchased does not have to change for this to happen. As the disc diameter is reduced, so is the width of metal coil reduced to keep the web scrap to a minimum. This also reduces the area of metal purchased for each can. In summary, for downgauging the wall of a DWI can, the price per tonne of the purchased metal is unchanged and the reduction in wall thickness is produced solely from a change in tooling dimensions. Thus, for the same wall gauge reduction step, the DWI metal cost reduction will be greater than that for the equivalent three-piece can.

Comparing the relative prices of steel and aluminium metal for packaging is very difficult because aluminium metal is a traded commodity and often must be purchased in bulk, in advance. This can create major price variations. Steel prices, on the other hand, are generally much more stable and are normally subject to annual review to reflect cost changes. Prices of both metals are, however, subject to global political situations. On a like-for-like basis, aluminium is generally more costly than steel for packaging but the value of waste material generated from the can-making operations is significantly greater for aluminium than for steel processes. The greatest competition between these two metals is for DWI drink cans and the drink can fillers like to keep healthy competition between the two metals!

Aluminium is inherently less strong than steel, so this limits where aluminium may be economically used for the construction of metal packaging. Also, as stated in Section 8.3.1, aluminium cannot be used for making three-piece welded side seam containers. On the other hand, steel can be used to make all types of metal package except where impact extrusion is employed for forming the two-piece seamless body, because steel is not soft enough to respond to this particular forming process.

Uses and restrictions for both aluminium and steel for construction of metal packages are shown below:

- Aluminium is used for pressurised containers where the internal pressure generated by the product gives additional strength to the can wall, for products such as beer and carbonated drinks in DWI cans, and also for seamless aerosols and collapsible tubes. It is used for shallow or short drawn cans for processed foods such as fish and pâté. (Economic wall strength is not sufficient to withstand full heat process conditions in tall food cans.) Aluminium is used for roll-on pilfer

proof caps where the metal must be soft to take up the shape of the underlying bottle thread.
- Steel may be used for all metal packaging formats except those using the impact extrusion process and roll-on pilfer proof caps. ECCS (tin-free steel) cannot normally be used for welding as the chromium surface is not compatible with the welding process (special systems have been used for first removing the chrome from the steel surface but these are not very practical). ECCS can only be used for DWI cans when extruded polymers have first been applied to the metal surfaces. However, these processes are still in the development stage.

When comparing the various can-making processes and in particular three-piece welding with two-piece DWI, which together form the bulk of all cans produced, the following points need to be considered:

- The three-piece welding is a very flexible process allowing cans of different height and/or diameter to be produced from the same system, although to make different diameters there needs to be investment in appropriate tooling. In addition, facilities for coil cutting into sheets, sheet coating/printing/curing and sheet slitting need to be provided upstream of the can-forming equipment.
- Two-piece DWI can making is relatively inflexible and the process runs at its most efficient if the same size of can is made continuously on a 24 hour/day, 7 day/week basis. Lines may be constructed to make small changes in can dimensions. To suit this process, the market for the can output needs to be in the order of 600 million per annum. There is no need for coil cutting, with coating/printing (where appropriate) carried out as in-line processes. Under these conditions and assuming cans of similar dimensions and wall thicknesses, the metal cost of the tinplate DWI can should be slightly less than that of the equivalent tinplate welded can with the maker's end fitted.

8.6 Container specifications

Specifications for containers, ends and closures are necessary to ensure that the filled package is capable of fulfilling its required duty and continues to do so within acceptable ranges for all ongoing supplies until the end of a supply contract or until changes have to be made to reflect new requirements. Development of a container manufacturing specification and therefore manufacturing/delivery cost starts when the user/filler defines the performance criteria. This will contain the following information:

- type of product to be packed and shelf life expected (this will indicate whether this is a dry or wet product, what type of heat processing, if any, may be required and what filling method will be employed)
- portion size/volume contents plus headspace
- label system (printed container or paper label)
- preferred can opening method
- annual quantity required and location of customer warehouse/filling line
- system for packaging and transporting the empty containers and ends.

From the above data, it should be possible to select the container overall dimensions, the most appropriate container forming system and the type of metal to use for the container and ends. This will also dictate the label system and the empty container/end packaging system.

From the detail of the product and metal types selected, together with heat processing systems and shelf life requirement, physical strength conditions can be defined. The specifications and applied weights of organic coatings can also be derived at this stage. When the physical strength properties are known, the can maker is able to calculate the thickness and detailed specification of the metal to be purchased as well as the thickness of the can body walls and ends. Only at this point is the can maker able to calculate the cost of the metal used to make, say, 1,000 containers. Following this the manufacturing and transport costs may be computed.

Purchasing specifications for metals and organic coatings are developed from the above data. The purchasing specification developed by the filler customer will include details of quality assurance tests to be performed by the can maker together with limits of acceptance and how to handle disputes. It will also contain advice on how containers are ordered, what call-off systems are employed and whether deliveries are to warehouses or direct to filling factories. Methods for acceptance and signing off of print quality will also be included.

The basic type of information that will be included in metal container specifications is shown below:

- Can body:
  - Construction method, metal type and specification
  - Dimensions: nominal body diameter, neck diameter (if reduced from that of body), flange diameter(s), open top can height
  - Volume capacity: gross lidded volume
  - Physical performance criteria, strength
  - Surface coating specifications, application weights, performance criteria
- Can end:
  - End type, metal type and specification
  - Dimensions: nominal diameter, outside curl diameter, curl height, countersink depth
  - Physical performance criteria, strength (pop and pull opening loads for easy-open ends)
  - Surface coatings/lining compound specifications, application weights, performance criteria
- Can body and end:
  - Attribute levels of acceptance and other specific quality assurance requirements agreed with customer
  - Secondary packaging specification, barcode identification specifications.

The list of specification requirements given above are just the basic requirements. Reference may be made to meeting the requirements of national or international specifications. In the case of drink and processed food cans, specifications may include the need for achievement of practical interchangeability between, say, can ends and the flanges on can bodies to ensure ongoing satisfactory mechanical seam performance, particularly when these components are purchased from multiple supply sources.

8.7 Decorating processes

Printing on metal packaging provides decoration to attract the consumer and reinforce the image of the product brand while also giving information about the package and its contents. (See also Chapter 19 for a detailed discussion of printing for packaging.) Printing on to metal for packaging may be undertaken with the metal in flat sheet form, prior to metal forming (for three-piece, general line lock seam and drawn containers), or after the circular body has been formed in the case of DWI or impact extruded containers. Flat sheet printing allows an unlimited number of colours to be laid down, regardless of the number of printing machines/colour heads installed, because sheets can be re-fed through these machines for additional colours to be added. When printing on circular bodies the maximum number of colours is determined by the design of the machine.

The printing techniques employed for flat metal sheet are very similar to those employed for paper and board. Conventionally wet lithography has been the most common way of laying down the print on flat sheets but this is being replaced by the use of waterless plates (Toray system). This process requires good temperature control of the inks but has the benefit of being more ecologically friendly in no longer using alcohol for damping. The improved temperature control reduces the number of wasted sheets at start-up. However, once the ink has been laid down on the surface, the unique properties of metal become apparent in that metal:

- is hard and allows a very clear image to be formed
- may have small quantities of preservative oil on the surface which is not washed off before printing (or coating)
- is not absorbent, so all liquids on the surface must be removed by evaporation
- may be cold and may therefore have condensation present
- if tin plate, has a tin layer which melts at 232°C, being only some 25°C above the curing temperature of many coatings.

The other major difference is that metal sheets are relatively heavy, so handling systems in printing equipment have to be modified to cope with this.

Techniques for printing on formed circular bodies have to overcome the restriction that it is not possible to register individual colours onto a continuous surface. This issue is solved by first laying down all the colours in sequence, as a reverse dry offset image, on an intermediate rectangular surface (rubber blanket) which has the same dimensions as the external surface of the container. This surface is then brought into contact with the can and the whole image is rolled around the outside to effect the transfer. This operation is described in Fig. 8.33. Digital technology is now being developed and brought into production for circular printing of drinks cans. This will allow virtually instant change of design to take place with no set-up time or wastage of ink and containers. It will also permit topical information to be included on can

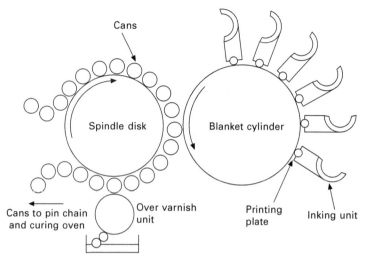

8.33 Cylindrical can decorating machine. Courtesy of Pira International.

label designs for sale of filled product within say 24 hours of the can being made. Inks for both flat sheet and circular print systems are available in heat cure or UV forms. The same restrictions of use apply as are described for organic coatings in Section 8.2.4.

For small-sized containers, it may be difficult to contain all the required information in the side wall of the container. For other containers the presence of product data on the side wall may detract from the overall image which the marketer is trying to convey. In these situations some of this information may be printed on the base of the container. For three-piece and lock seam general line containers this is straightforward. For two-piece containers, including closures, the need to have print both on the base and the side walls presents additional challenges. The only way to achieve this is to print the side walls and the base whilst the metal is still flat, prior to the first forming process. As the walls will be reformed during the drawing operation, any print placed on the flat blank outside the area of the base will become distorted. In addition to this, the further the print is placed up the side-wall, from the base of the container, the more it will be distorted from the original design. In order to overcome this problem, the design of the print for the walls of the container or closure is pre-distorted at the original design stage. Then, after the metal forming operation is complete, the print appears in its correct form.

8.8 Environmental overview

There has always been a great emphasis on reduction of metal used at source in canmaking operations because, as already mentioned in Section 8.5, the value of the metal used is generally in the order of 60–70% of the manufacturing cost. Metals for packaging are mostly purchased by area which makes it possible to closely align purchase quantities with those actually used. For the same reasons, over the years, many developments have been introduced across the industry to minimise material waste from offcuts, web scrap and finished product scrap. Lightweighting has additionally also been ongoing for many years, alongside technical process improvements. The introduction of the DWI can for drinks and later processed foods is one of the main drivers in this, together with other processes where high throughput speeds and high efficiencies can be achieved. Data recently released by EMPAC (European Metal Packaging, 2009) demonstrate that over the last 20 years the market for food and drinks cans, in steel and aluminium, has increased by 57%. Over this same period the absolute quantity of virgin metal used has reduced by 20%, and net CO_2 emissions and net energy usage have reduced by 50% and 60%, respectively. To maintain and improve on this excellent record, it will be necessary to increase the percentage of containers recycled into new metal from sources of post consumer waste.

Recycling of waste materials arising from the can-making process into new metal has been a very high priority for many years because these materials have inherent value in the economics of these processes. Metal packages are less easy to reuse than some other packaging materials, but recycling used metal packages into new metal is very straightforward. Metal is different from other packaging materials like paper and plastic because metals (steel and aluminium) are recycled on an atomic level. This means that once re-melted, recycled metal is indistinguishable from new metal smelted from ore and these metals may be re-melted an infinite number of times without loss of properties. This material property makes it a unique packaging material. During the re-melting process minor contaminants such as tin, organic ink and coatings, as well as other organic residues, do not cause any problems in the steel-making itself. Gaseous effluents arising from these are contained within the steel-making process. In some cases tin is reused in the packaging steel-making process. It may also be removed from tinplate scrap by a system of reverse electroplating. In the steel-making process commonly used for packaging steel, steel scrap from packaging and non-packaging sources is needed to control liquid iron temperature in a converter during the exothermic reaction of carbon reduction by flushing with oxygen.

Recent studies on sustainability of metal packaging confirm that an increased recycling of post consumer metal packaging waste after end of life reduces the amount of new metal that needs to be produced as virgin material. Furthermore, the use of recycled metal is infinitely more beneficial in the reduction of CO_2 emissions than the reduction of energy, transport or metal usage in container manufacture and delivery. With metals, the energy used in the mining, transportation of raw materials and smelting operations to produce new metal from ore is locked up in the metal forever, to be used in a next material life, whatever it is used for.

One of the main environmental benefits arising from fresh food packed into metal cans which are then heat processed is that the pack is then stored at ambient temperature for the remainder of its life. Metal packaging has excellent barrier properties – 100% protection from light, air and water – ensuring maximum product quality. Products packed in metal do not require additives to preserve them.

Canned food may be stored for years, no further energy being required purely to preserve the integrity of the product. In many underdeveloped countries, vast quantities of agricultural products go to waste because they have become inedible before they can be consumed. Preserving these products by packing into heat processed metal containers at source can provide a satisfactory solution to this issue. Furthermore, it does not matter how far away the final customer is located from the original source of the fresh product or how long it takes to arrive, as the contents, once opened, will still be fresh with no additional energy required to keep it so.

8.9 References and further reading

European Metal Packaging (2009) *Guide to good manufacturing and hygiene practices for metal packaging in contact with food*, EMPAC, Brussels, Belgium, www.empac.eu.

Morgan E (1985) *Tinplate and modern canmaking technology*, Pergamon, Oxford.

Page B (2006), 'Closures for metal containers', in Theobald N and Winder B, *Packaging closures and sealing systems*, Blackwell, Oxford, 68–100.

Page B (2010a) 'Packaging formats for heat-sterilised canned fish products', in Bratt L, *Fish canning handbook*, Blackwell, Oxford, 151–178.

Page B (2010b) *Metal packaging, an introduction*, 2nd edition, Pira International, Leatherhead.

Page B *et al.* (2011), 'Metal packaging', in Coles R *et al.*, *Food and beverage packaging technology*, Blackwell, Oxford, 107–121.

Turner T A (1998) *Canmaking, the technology of metal protection and decoration*, Blackie Academic & Professional, London.

9
Aluminium foil packaging

J. KERRY, University College Cork, Ireland

Abstract: This chapter reviews the manufacture and use of aluminium foil as a packaging material. It discusses processing from refining and smelting to the production of foil, printing and embossing. The chapter also describes the use of aluminium foil as a laminate as well as aluminium metallised films.

Key words: aluminium foil, aluminium laminate, aluminium metallised film.

9.1 Introduction

Aluminium is a silvery white metal belonging to the boron group of chemical elements and has an atomic number of 13. It has a number of important characteristics:

- It has a third of the density of steel, making it a lightweight metal.
- In alloy form it is strong: whilst pure aluminium has a yield strength of 7–11 MPa, aluminium alloys have yield strengths ranging from 200 to 600 MPa.
- It is resistant to corrosion due to 'passivation': the formation of a thin layer of aluminium oxide on the metal surface which prevents further oxidation.
- It is ductile and malleable at normal temperatures, making it easy to form.
- It is non-magnetic but is a good conductor of heat and electricity.

Aluminium is used widely in foil form for packaging at thicknesses of just over 6 to around 150 microns (a micron is one millionth of a metre). The key properties of aluminium as a packaging material are summarised in Table 9.1. One of the most important properties of aluminium as a packaging material is its inertness compared to most metals. As noted earlier, when exposed to air, aluminium forms a transparent oxide layer which prevents further oxidation. As well as being resistant to corrosion, aluminium is non-absorbent and thus an effective barrier against gases and liquid. Heavier gauges of foil (above 17 μm) provide a complete barrier to gases and liquids. Thinner gauges do allow some transmission: as an example, the typical water vapour transmission rate (WTVR) for foil of 9 μm thickness is 0.3 g/m^2 per 24 hours (at a temperature of 38°C and a relative humidity of 90% RH). Aluminium is also stable in cold conditions and can be used for frozen food. It is resistant to temperatures below 150°C, making it suitable, for example, for the storage and heating of chilled foods.

Aluminium does not generate toxic residues or react with most chemicals, including the majority of foods and beverages. It produces a metal with a smooth surface that is easy to clean. Aluminium does not generate sparks, making it suitable for storage of flammable and volatile materials. Its conductivity makes it useful for such applications as electrostatic shielding and induction heat sealing for packaging

Table 9.1 Key properties of aluminium as a packaging material

Property	Advantages
Appearance	Bright, reflective gloss makes for an attractive appearance
Stability	When exposed to air, aluminium forms an oxide layer that prevents further oxidation. It is also inert and does not form toxic compounds when exposed to most chemicals, including most foods and cosmetics
Barrier properties	Heavier gauges form a complete barrier to gases and water. Aluminium reflects light, making it a suitable material to protect light-sensitive products
Hygienic properties	Aluminium's smooth metallic surface is non-absorbent. It can be easily cleaned and sterilised
Formability	Aluminium's ductility makes it easy to form. It has excellent dead-fold properties. Its friability (ability to crumple) makes it useful for blister packaging
Conductivity	Aluminium conducts electricity and heat, making it useful for applications such as induction heat sealing of containers
Recyclability	Aluminium can be recycled at relatively low cost (recycling requires about 5% of the energy required to refine aluminium)

closures (see Chapter 15). Its reflectivity means that it can be used to protect light-sensitive products.

As well as these functional properties for packaging, aluminium also has a smooth, bright and reflective appearance which gives it a decorative value. In foil form, the ductility of aluminium means that it can easily be moulded into a variety of container shapes as well as be processed into foil. It also means aluminium has excellent dead-fold properties, i.e. once folded it retains the shape of the fold. Another increasingly important advantage is aluminium's recyclability. Aluminium and its alloys can be recycled at about 5% of the energy consumption required to refine the original ore. Aluminium can easily be separated from other metals for recycling as it is non-magnetic.

Aluminium does have a number of disadvantages. Pure aluminium loses significant strength at temperatures above 150°C which means a protective coating is needed if it is to be exposed to further heat processing, e.g. cooking. Whilst useful for formability, the ductility of aluminium means that aluminium foil is easily torn or punctured. A particular problem is flex cracking, i.e. the tendency of foil to split when folded or stretched. At gauges below 17 μm, the foil can also suffer from 'pinholes', minute holes caused by impurities in the metal or process variations. Pinholing allows water and gas to penetrate the metal. A protective coating or incorporation of the foil as part of a laminate (composite layer) may therefore need to be added to increase strength, prevent flex cracking and counteract pinholing. Coating and lamination are discussed in Section 9.6.

These weaknesses in aluminium foil can be an advantage in some packaging applications, for example in the production of blister packs for tablets, where the foil needs to break easily to allow access to the product. It is also important to be aware that aluminium foil is not resistant to all chemicals. Whilst the fats or mild organic acids generally found in food have little or no effect on aluminium, strong mineral acids will corrode bare foil. Mildly alkaline components (e.g. soaps and detergents) may also have a corrosive effect. Salt and other caustic agents will also corrode aluminium. In these conditions, a coating or lacquer is required to protect the metal.

9.2 Aluminium processing

Aluminium is the most abundant metal found in the Earth's crust (approximately 8%) and is the third most abundant element found on Earth, after oxygen and silicon. Due to its reactive behaviour, aluminium is never found as a pure metal in nature but combined with hundreds of minerals. The chief source of commercially manufactured aluminium today is bauxite. Bauxite is a reddish-brown clay-like deposit containing iron, silicates and aluminium oxides, the latter comprising the largest constituents. At present, bauxite is so plentiful that only deposits containing a content of aluminium oxides greater than 45% are selected to manufacture aluminium. Bauxite derives its name from a small French town called Les Baux, where bauxite was first discovered in 1821. Today, the largest bauxite mines are located in North America, the West Indies, Australia and Northern Europe.

Since bauxite occurs naturally at the surface of the Earth's crust, mining practices tend to be straightforward. Surface pits are opened using explosives to reveal bauxite beds. The bauxite ore is then excavated and loaded into trucks or rail cars for transportation to the converting or processing centre. In order to produce commercial-grade aluminium from bauxite, essentially two processes must be employed:

- the bauxite ore must be refined to remove impurities (this is called the refining step).
- aluminium metal is extracted from aluminium oxide using electrolysis (this is the smelting stage).

Following on from these initial steps two further processes are required:

- the aluminium is processed and rolled into foil
- special surface finishes are applied to enhance appearance, decorate, strengthen, protect or provide specialist functions, for example giving foil heat-sealing properties.

A flow process diagram for aluminium manufacture from bauxite is shown in Fig. 9.1.

The processes of refining and smelting require abundant electrical power and for this reason aluminium production is frequently located in areas where cheap electricity is readily available, e.g. northern Scotland and Scandinavia, where hydro-electric power is used. It is estimated that it takes 4 kg of bauxite to produce 2 kg of aluminium oxide, which, with the consumption of about 8 kW of electricity, produces 1 kg of pure aluminium. Due to the high costs associated with aluminium manufacture, metallurgists are continually investigating new approaches to the extraction of aluminium from bauxite in an attempt to reduce overall cost and environmental impact.

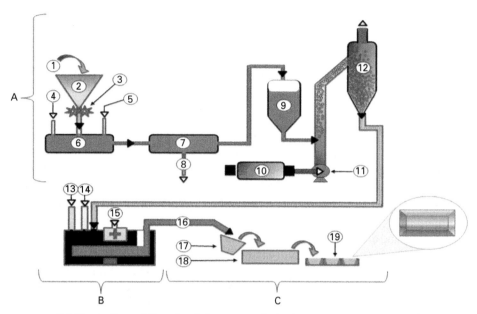

9.1 Illustration of the aluminium manufacturing process encompassing the chemical extraction process (A), electrolysis (B) and alloy casting (C) operations (1. Raw material (bauxite) is processed into pure aluminium oxide (alumina) prior to its conversion to aluminium via electrolysis. This primary step is achieved through the 'Bayer Chemical Process'. Four tonnes of bauxite are usually required in order to generate two tonnes of finished alumina which ultimately produces approximately one tonne of aluminium at the primary smelter. 2. Bauxite feed hopper. 3. Mechanical crusher employed to reduce bauxite particle size and increase surface area for chemical extraction. 4. Input chemical (sodium hydroxide). 5. Input chemical (lime). 6. Aluminium oxide is effectively released from bauxite in the presence of caustic soda solution within the primary reactor (digestion) tank. 7. The aluminium hydroxide is then precipitated from the soda solution. 8. Spent solids/tailings discard a red mud residue generated as a byproduct of the process. 9. Precipitation tank: aluminium hydroxide is precipitated from the soda solution. The soda solution is recovered and recycled within the process. 10. Drying system (air heater system). 11. Drying system (hot air blower system). 12. Drying system (cyclone fines recovery system): post calcination, the anhydrous end-product, aluminium oxide (Al_2O_3), is a fine grained free flowing, white powder. 13. Input chemical (aluminium fluoride – AlF_3). 14. Input chemical (cryolite – Na_2AlF_3). 15. Fuel source (e.g. coke, petroleum and pitch). 16. Molten aluminium: the reduction of alumina into liquid aluminium is operated at around 950°C in a fluorinated bath under high intensity electrical current. This electrolytic process (A) takes place in cells or 'pots', where carbon cathodes form the bottom of the pot and act as the negative electrode. Positive electrodes (anodes) are held at the top of the pot and are consumed during the process when they react with the oxygen generated from the alumina. Two types of industrial anodes are currently in use. All potlines built since the early 1970s use the pre-bake anode technology where anodes manufactured from a mixture of petroleum coke and coal tar pitch (acting as a binder) are pre-baked in separate anode plants. In the Soderberg technology, the carbonaceous mixture is fed directly into the top part of the pot, where self-baking anodes are produced using the heat released by

Continued

9.3 Refining

The refining step, which is also known as the Bayer process, is carried out in four stages:

- digestion
- clarification
- precipitation
- calcination.

In the digestion stage, ground bauxite is generally mixed with sodium hydroxide and pumped to heated and pressurised digestion tanks where the ore breaks down to soluble sodium aluminate and insoluble components which settle out in the tank. The clarification stage then follows. This entails passing the sodium aluminate through a series of filtered presses which are connected to tanks. The textile-based filters remove contaminants from the solution. This process can be repeated a number of times, depending on the level of contaminants present in the sodium aluminate solution. Finally this solution is then forwarded to cooling towers. From here, the aluminium oxide (alumina) is transferred to a large agitated tank or silo where the aluminium oxide fluid is seeded with hydrated aluminium crystals, thereby promoting the formation of aluminium particles. As the aluminium particles form, the hydrated and seeded aluminum crystals are attracted to each other and entrap the aluminium particles. This causes agglomeration to occur which causes clumping. As clumping develops, large agglomerates of aluminium hydrate form. As this stage in the process suggests, precipitation of aluminium hydrate occurs. The material is filtered and washed. The final stage in the refining process is calcination. This occurs when the aluminium hydrate is exposed to high temperatures in rotary kilns which drive off water and produce a white powdered material called aluminium oxide, or pure alumina, which resembles granulated sugar in appearance.

9.4 Smelting

The smelting step is employed to process the alumina. Its primary function is to separate alumina into metallic aluminium and oxygen by means of electrolysis, a procedure that was originally devised by Charles Hall and Paul-Louis-Toussaint Héroult in the late nineteenth century. The modified electrolytic method used today requires that the alumina is firstly dissolved in what is described as a smelting cell.

Continued caption fig. 9.1

the electrolytic process. 17. At regular intervals, molten aluminium tapped from the pots is transported to the cast house crucible. 18. The aluminium is alloyed in holding furnaces by the addition of other metals (according to end user needs), cleaned of oxides and gases. 19. The liquid metal is then cast into ingots. These can take the form of extrusion billets, for extruded products, or rolling ingots, for rolled products, depending on the way it is to be further processed. Aluminium mould castings are produced by foundries which use this technique to manufacture shaped components.)

A smelting cell is made from steel, lined with carbon and filled with heated cryolite. Cryolite is an aluminium-based compound that has strong conductive properties. Once the smelting cell is filled with cryolite, an electric current is passed through the cryolite and this causes a surface crust to form on the alumina. This crust is not a permanent feature and is broken regularly through further alumina additions and via stirring. As the alumina decomposes during this electrolytic process, pure molten aluminium metal falls out of solution to the bottom of the smelting cell. The oxygen gas generated as part of this process combines with the carbon lining in the cell to produce carbon dioxide.

The purified molten aluminium is now collected from the bottom of the smelting cell by siphoning it off into crucibles which are then emptied into furnaces. At this point other elements may be added to the aluminium. The addition of other elements is dependent on the characteristics required from the aluminum alloy in question. Elements added include copper, zinc, magnesium, manganese and/or chromium. Aluminium alloys containing small amounts of these elements have excellent strength properties. As an example, Alloy 3003, which contains manganese, has greater stiffness as well as improved processing properties for cans or containers for pastries and pies. Alloys 1100, 1145 and 1235 are most commonly used for reroll stock for foil. Alloy 1200 is commonly used for packaging applications. Aluminium foils are made in several tempers (i.e. degrees of hardness), dependent on their application. As an example, blister packs require the foil to be in a half-hard temper so that the foil can be easily punctured for ease of access to the product.

The modified molten aluminium is then poured into chilled casting moulds where it cools and sets to form large aluminium blocks or slabs called 'rolling ingots' which typically range in weight between 10 and 25 tons. These 'rolling ingots' are reduced to 'reroll stock' or sheets approximately 3–6 mm thick. All reroll stock is supplied to aluminium foil manufacturers in 0 temper (i.e., in its softest form) so that it can be worked easily.

9.5 Production of aluminium foil

Aluminium foil is typically less than 150 μm in thickness. Foils are available in gauges as low as 6.3 μm. Heavier foil gauges (> 17 μm) provide an absolute barrier to gases and liquids. A typical water vapour transmission rate (WVTR) for 9 μm foil is 0.3 g/m^2 per 24 hours at 38°C and 90% RH. As thickness is reduced, foil becomes more vulnerable to tearing or pinholing. Aluminium foil is produced by two basic processes:

- the traditional method of rolling aluminium slabs, ingots or thick plates into a narrow gauge aluminium web stock using heavy rolling mills (Fig. 9.2),
- by continuous casting or hot-strip casting (similar to an extrusion process) which takes place immediately after the aluminium has left the furnace (Fig. 9.3).

Because it has been established for longer, the rolling-mill method of producing reroll stock is still widely used. When being rolled to foil, ingot-rolled stock must be re-annealed (reheated) between mill passes to overcome work hardening and restore

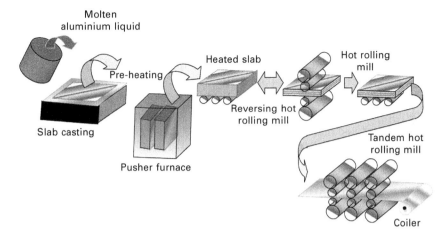

9.2 The conventional rolling-mill method of producing reroll aluminium stock and ultimately, aluminium foil.

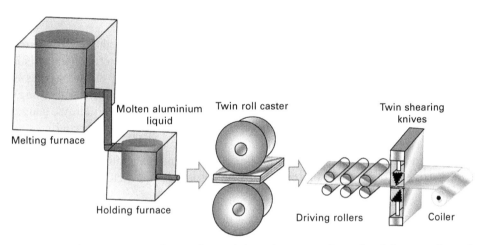

9.3 Continuous casting or 'hot-strip' casting to produce aluminium reroll stock and ultimately, aluminium foil.

workability. The most economical means of manufacturing reroll stock, however, is via continuous casting. A typical continuous-casting production line runs directly from the furnace to a winding reel. The system continuously feeds, casts, chills and coils the reroll stock. Since it is heated during production, continuous-cast reroll stock does not need to be re-annealed when being made into foil.

After the foil stock has been manufactured, it must be further processed on a rolling mill. The work rolls have finely ground and polished surfaces to ensure a flat, even foil with a bright finish. The work rolls are paired with heavier backup rolls which exert very high pressure on the work rolls to ensure stability. This pressure ensures a uniform gauge (thickness) across the resulting aluminium foil sheet (known as a web). Each time the foil stock passes through the rolling mill, it is squeezed, its thickness is reduced and its length increases, but its width remains the same. This means the required width for the final foil product must be set at the beginning of the process.

The rolling process can be viewed as a form of extrusion. Foil gauges under 25 µm are often passed through the work rolls in two webs at a time. This is done so that the two webs of foil support each other and are less likely to tear when rolled. During the rolling process, the aluminium stock requires annealing to maintain workability. The addition of lubricants to the aluminium surfaces also maintains the workability of the material. A final annealing stage removes the lubricating oil, thus making the foil receptive to printing inks and adhesives.

9.6 Foil finishes, coatings and lacquers

Rolling produces two natural finishes on foil: bright and matte. The foil surfaces in contact with the work rolls are polished to a bright and shiny finish. When a single web is run, both sides are bright. If thinner foils are rolled together, the foil-to-foil face of each web develops a satin-like matte finish. Other finishes can be produced with special patterns on the work rolls or, more commonly, by using separate or in-line mechanical finishing machines (Table 9.2).

In most packaging applications, aluminium foil is combined with other materials such as coatings, inks, papers, paperboards and plastic films. A very useful characteristic of aluminium foil is that it has the capacity to readily accept many different types of coating materials such as inks (for printing), varnishes and lacquers (for embossing), adhesives and polymers (for heat sealing, etc.). The selection of foil alloys, gauges and tempers needs to take into account the type of coating or lacquer required.

Where a coating is needed, gravure coating is used for most low-viscosity materials. Heavier coatings require some form of roll coater. Coatings generally can be classified as protective or decorative. Protective functions for coatings include:

- making the foil more heat-resistant
- increasing tensile strength
- increasing resistance to potentially corrosive agents
- enhancing the barrier properties of low gauge foil
- increasing resistance to scratching or scuffing
- increasing the UV resistance of a printed foil.

Table 9.2 Standard aluminium foil finishes and treatments used in packaging

Type of finish	Description
Bright both sides	Uniform bright specular finish, both sides
Extra-bright both sides	Uniform extra-bright specular finish, both sides
Matte one side	Diffuse reflecting finish, one side
Matte both sides	Diffuse reflecting finish, both sides
Embossed	Pattern impressed by engraved roll or plate
Annealed	Completely softened by thermal treatment
Chemically cleaned	Chemically washed to remove lubricants
Hard	Foil fully work-hardened by rolling
Intermediate temper	Foil temper between annealed and hard

Vinyl heat-seal coatings are widely used with aluminium foil. It is important that the coating is compatible with the product. It is also important to note that a normal heat-seal coating does not add significantly to a foil's bursting strength. Coatings are only likely to add to foil strength above thicknesses of 25 µm. Table 9.3 lists the chemical resistance of some coating materials. Table 9.4 lists the general properties of the various coatings applied to foil.

Decorative and other functions include giving additional gloss and depth to a decorated or printed foil, or improving the adhesive quality of the foil for other

Table 9.3 Chemical resistance of coating materials

Coating type	Acid	Alkali	Water	Solvent
Acrylics	Fair	Fair	Fair	Good
Alkyd	Fair	Fair	Good	Good
Butadiene-styrene	Excellent	Excellent	Excellent	Good
Butyrate	Fair	Fair	Fair	Fair
Cellulose acetate	Fair	Fair	Fair	Fair
Chlorinated rubber	Excellent	Excellent	Excellent	Fair
Epoxies	Excellent	Excellent	Excellent	Excellent
Ethyl cellulose	Fair	Excellent	Good	Fair
Melamine	Excellent	Excellent	Good	Good
Nitrocellulose	Good	Fair	Excellent	Good
Polyamide-epoxy	Fair	Excellent	Excellent	Good
Polyester	Good	Fair	Good	Good
Polystyrene	Excellent	Excellent	Excellent	Fair
Polyvinyl acetate	Fair	Fair	Good	Poor
Polyvinylidene chloride	Excellent	Excellent	Excellent	Fair
PVAC chloride copolymer	Excellent	Excellent	Excellent	Fair
Styrenated alkyd	Fair	Good	Good	Fair
Urea	Excellent	Excellent	Excellent	Good

Source: Aluminium Association, *Aluminium Standards and Data*.

Table 9.4 General properties and characteristics of typical aluminium foil coating compounds

Quality	Celluloses		Alkyds	Phenolic	Polyester	Epoxy Vinyls
	Ethyl	Nitro				
Abrasion resistance	1	1	2	1	2	2
Block resistance	3	1	2	1	1	3
Heat sealability	1	2	3	3	2	1
Water resistance	3	2	2	1	1	1
Alkali resistance	1	3	3	1	1	2
Fat and oil resistance	3	2	2	1	1	1
Colour stability	2	3	2	1	1	1
Film gloss	2	2	1	1	1	3
Transparency	2	2	3	1	1	1

The quality rating score equates to: 1 = best performance, 2 = satisfactory performance or frequently used and 3 = unsatisfactory performance.

coatings or printing inks. The use of transparent lacquers and varnishes gives foil a particularly bright metallic sheen. Coloured lacquers help to impart colour; for example, yellow gives aluminium a gold appearance.

9.7 Printing and embossing

If foil is to be printed, it is important to be aware that its glossy surface may make small print difficult to read. Reverse type is best avoided, unless it is large. It is sometimes necessary to print a matt white background on which black type will be more visible. Printers usually print foil on the same presses used for paper or other material. If a foil is to be printed, it is given a primer or wash coat to ensure a clean surface and to provide a foundation for the ink. Shellacs and vinyls are common primers for gravure and flexographic printing. If a thicker coating is required, e.g. for lithographic printing, vinyl copolymer or nitrocellulose may be used. A second film coat may then be applied on top of the printing to protect it from scuffing as well as to reduce surface friction.

Aluminium foil is particularly suited to embossing. This gives both a three-dimensional quality to a design and increases the number of reflective surfaces able to reflect light to create a more eye-catching effect. It also increases stiffness (see below) and allows cut pieces of foil to be easily separated, e.g. stacks of pre-cut lids. Thinner gauge foil in web (sheet) form is passed between a steel roll containing the engraving pattern and a soft matrix roll (usually paper). Heavier-gauge, coated or laminated foil is embossed with two engraved steel rolls, one with the positive image, the other the negative image. Since it requires further processing with rollers, embossing tends to improve foil stiffness and dead-fold properties.

9.8 Using aluminium foil as a laminate

Lamination involves combining sheets of different materials into a single layer, using a mixture of adhesives, pressure and sometimes temperature to bond the materials together (see also Chapter 14). Aluminium foil is laminated on web-fed rotary equipment which sometimes includes a coating unit to add further protection. Adhesives are selected on the basis of their suitability for the materials to be joined as well as such issues as any toxicity or contamination risk they might present, potential odour or colour issues, moisture and heat resistance. The main types of adhesives used are:

- heat-sealable adhesives
- water-soluble emulsions and dispersions
- thermoplastics in lacquer-type solvents
- hot melts (natural or synthetic waxes)
- extruded-film adhesives.

Four methods are used to laminate aluminium foil:

- wet bonding
- dry bonding

- extrusion bonding
- hot-melt bonding.

As the name suggests, wet bonding involves combining the various layers before the adhesive is dry. A water- or solvent-based adhesive is used and is normally applied to the foil. Further layers are then applied on top and the laminate passed through a combining or nip roll at varying drying temperatures, depending on requirements (Fig. 9.4). Materials laminated to the foil base need to be porous so that the liquid medium of the adhesive can be taken up and the adhesive properly dried, thus paper is an ideal material for wet bonding. The smoother, denser and less porous the paper, the less adhesive is required. A low paper moisture content is important to ensure good adhesion and prevent problems such as staining. It is important to combine and glue the materials quickly to ensure good adhesion and prevent materials slipping as they are fed through the machine. The use of a rapidly setting adhesive also helps to minimise the risk of air pockets or blisters in the laminate, particularly if the laminate is stressed or flexed in subsequent processing, e.g. coating operations. High drying temperatures improve bonding and water resistance, but they can over-dry and damage papers or boards.

Where materials are not porous, dry bonding is used instead of wet bonding. Dry-bond adhesives use both natural and synthetic sealing agents and can be either water or organic solvent based (see Chapter 16). Synthetic agents include vinyls, epoxies, polyesters and urethanes. The adhesive is applied to the foil and allowed to dry. The layers are then aligned and passed through a heated combiner roll which reactivates the adhesive to create the bond. This method is well suited to non-porous materials such as polyester films, which add strength and flexibility when combined with aluminium foil.

Extrusion bonding involves extrusion of one or two molten plastic films which are then combined with the aluminium foil. As the aluminium foil web approaches the combiner roll, an extruder die deposits a layer of hot extrudate across the width of the web. The laminate then passes through the chilled nip of the combiner roll, cooling the plastic layer which solidifies. No drying is required. When only the two layers are involved, the process is known as extrusion coating.

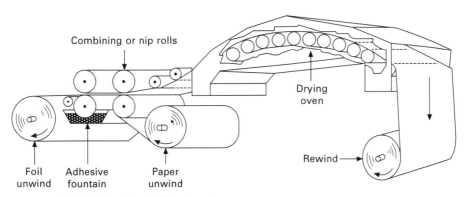

9.4 Wet-bonding foil lamination process.

The hot-melt process is used for high-speed lamination since there is no need for a drying stage. Hot melts include polymer resins, waxes and resin-wax combinations. The hot-melt can be melted at a lower temperature than extruded coatings, applied to the web and chill-set in the nip of the combiner roll. The plastic nature of the hot melt improves heat-sealability and the dead-fold characteristics of the foil.

Foil laminates intended for barrier applications should be evaluated for their barrier properties using the finished pack after all of the machining is complete, and preferably after a real or simulated shipping cycle. It is not uncommon for a prospective barrier laminate to have no measurable permeability when flat (i.e. at the point of manufacture) but to have significant permeability when formed, folded and creased into a final pack format.

9.9 Aluminium metallised films

Vacuum metallising or metallisation involves depositing a metal layer onto a substrate (e.g. paper or plastic film) in vacuum conditions. Aluminium is the only metal used for vacuum metallising in packaging applications. Initially used for decoration, metallisation is now widely used in flexible packaging since it improves gas and moisture barrier properties, heat resistance, light reflectance and electrical conductivity. Crisp packets, for example, are typically made of metallised polypropylene which provides effective protection for the required shelf life of this type of product. Metallised films are often a component in a laminate. Table 9.5 lists examples of laminates that use metallised films. Since aluminium is effective in converting microwave energy to heat, metallising is also used in 'susceptor' packaging of microwavable foods where the metallised component serves to create a local microwave energy 'hot spot'. Given the high temperatures involved, the aluminium needs to be protected by a heat-resistant layer, for example of polyester.

Batch processing is the most widely used approach to metallising. It involves a horizontal tubular chamber, up to 2 m in diameter and 3 m long (Fig. 9.5). A series of rollers carry the substrate through the chamber from which the air has been evacuated by vacuum pumps. Pure aluminium wire is fed into containers ('boats') which are electrically heated to vapourise the aluminium. (The vacuum reduces the vapourisation temperature of the aluminium.) As the vapour rises, it condenses on the underside of the substrate as it passes over a chilled drum. The thickness of the metal deposited is controlled by a combination of web speed, wire feed rate and boat

Table 9.5 Typical examples of laminates using metallised films

Product	Type of laminate
Coffee	12 μm metallised BON/50 μm LDPE
Savoury snacks	18 μm OPP/adhesive/18 μm metallised OPP
Condiments/spices	12 μm metallised PET/38 μm MDPE
Bag-in-box wine	50 μm ionomer/12 μm metallised PET/75 μm EVA
Biscuits	OPP/18 μm metallised OPP
Medical products	paper/adhesive/18 μm metallised OPP/ionomer
Cold meats	metallised PET/PE

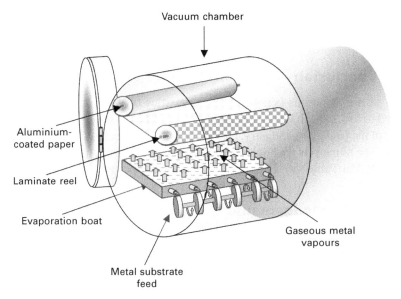

9.5 Vacuum metallising is a form of physical vapour deposition, a process of combining metal with a non-metallic substrate through evaporation. The most common metal used in vacuum metallisation is aluminium for a variety of reasons including cost, thermodynamic and reflective properties. The evaporation takes place by feeding aluminium onto heated sources or boats, which operate at approximately 1,500°C (2,700°F). The atmosphere in the vacuum metallising chamber is evacuated to a vacuum level suitable for the evaporation of the aluminium wire. Upon contact with the substrate being processed, the aluminium vapour condenses and creates a uniform layer of vacuum deposited aluminium.

temperature. A batch can handle reels of up to 20,000 m long in just under an hour. The thickness of the aluminium layer is as low as three-millionths of a centimetre. Electrical resistance and optical density are the two main methods for testing deposition and layer thickness. Metallised patterns can be achieved by metallising a pre-printed film, or using caustic solutions to selectively remove the metallic layer to create patterns or windows in the metallised film.

Metallised papers need to be high-quality virgin stock usually with a clay coating to provide a smooth surface and high reflection in the finished metallised surface. Papers also need to be lacquer coated using gravure coaters to seal the paper and ensure a good, even metal adhesion. The lacquer is dried using radiant heating, UV, electron beam or infrared curing. Developments in paper quality and lacquer technology have significantly expanded the use of metallised papers in areas such as labels and cartons for cosmetics. Since the vacuum and temperature conditions reduce the moisture content of paper substrates to below 5%, remoistening is an essential step in the manufacture of metallised paper to avoid curling. Although most plastic films can be metallised, oriented polypropylene (OPP), biaxially oriented polypropylene (BOPP), polyethylene terephthalate (PET) and biaxially oriented nylon (BON) are the most commonly metallised packaging films. Unlike paper, plastic films do not need to be sealed to be metallised.

9.10 Conclusion

Because of the numerous beneficial characteristics of the metal, the way it can be converted to foil, used in laminates or to metallise films, aluminium use is widespread throughout the packaging industry, particularly with respect to fast-moving consumer products. The numerous packaging formats and the products that it is associated with are presented in Table 9.6. Examples of container types are shown in Fig. 9.6.

9.11 Acknowledgements

I gratefully acknowledge the assistance of Mr Eddie Beatty and Mr Dave Waldron in the preparation of the photographic material presented in this chapter and to Dr John F. Kerry, Echo Ovens International Ltd, Raheen, Limerick, Co. Limerick for his assistance with the preparation of figure material for this chapter. All ownership of such material resides with The Food Packaging Group in University College Cork and with Echo Ovens International Ltd.

Table 9.6 End uses for aluminium foil as packaging material

Packaging format	Products
Rigid smooth wall containers	Meat joints for roasting, large portioned ready-meals and convenience-style food products (Fig. 9.6)
Semi-rigid wrinkle wall containers	Take-away meals, savory pies, bakery products, frozen and chilled ready-meals (Fig. 9.6)
Closure systems	Milk, beverages, instant coffee, dried powders, health foods and pharmaceutical products, cosmetic creams
Labels	Used widely from foodstuff to beverage packaging
Composite cans	Powdered drinks, snack foods, juice-based beverages, instant biscuit/cookie dough, chilled and frozen foods
Flexible packaging	Dairy products: milk, cheese, butter and ice cream
	Beverages: wines, juices, soft drinks, liquors, beer; non-beverage products: soaps, shampoos, conditioners, detergents, cosmetics
	Dessicated and powdered products: coffee, tea, cocoa powder, custard, fruit, concentrates, vegetables, dehydrated powders, soups, herbs and spices, yeast and other powdered extracts, salt, sugar, pharmaceuticals, tobacco products
	Cereal and baking foodstuffs: cake mixes, cereals, frosting mixes, pasta-based products, rice-based products, biscuits, crackers, snack foods, breads
	Confectionery: chocolate, hard and soft sweets, all products containing volatiles contributing to flavours such as mint, orange, coffee, aniseed, clove, etc.
	Muscle-based foods: meat, poultry, fish, game, casseroles, stews, soups and broths, general muscle-based retorted products, chilled and frozen products, pet foods

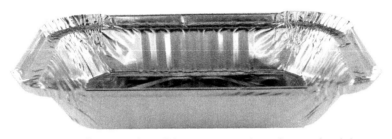

9.6. Examples of a smooth wall heavy gauge deep drawn aluminium container (above) and a wrinkle wall narrow gauge shallow drawn aluminium container (below).

9.12 Sources of further information and advice

Most general textbooks on aluminium discuss the metallurgical properties of the metal and its applications as a sheet or extruded material in construction, aerospace and automotive applications, and are thus of limited interest in the context of this textbook.

The following organisations provide useful information specific to aluminium foil in packaging:

- Aluminium Foil Container Manufacturers Association: http://www.afcma.org
- Aluminium Federation: http://www.alfed.org.uk/downloads/documents/D53ZC9P4LP_15_aluminium_packaging.pdf
- European Aluminium Association: http://www.alueurope.eu/wp-content/uploads/2011/10/Aluminium-in-packaging-brochure.pdf
- European Aluminium Foil Association: http://www.alufoil.org
- Australian Aluminium Council: http://aluminium.org.au/packaging
- The Packaging Federation: http://www.packagingfedn.co.uk
- Aluminium in Packaging: http://packaging.world-aluminium.org

10
Paper and paperboard packaging

A. RILEY, Arthur Riley Packaging Consultant International, UK

Abstract: Paper and paperboard are made from cellulose fibres, extracted from trees, combined together with additives to make a continuous matted web. This chapter covers the raw materials, processes and on- and off-line treatments used to manufacture fibrous substrates (paper and paperboard) used for the conversion into packaging components. These components include wrapping materials, bags, sacks, cartons, tubs, lids, moulded fibre packaging, and backing cards for various plastic, paper and paperboard combinations. This chapter will also discuss the different substrates, conversion methods and decorating methods.

Key words: paper, paperboard, carton, sack, bag, fibre moulding.

10.1 Introduction

'Paper' and 'paperboard' can be described as a matted or felted sheet, usually composed of plant fibre (commonly from trees or recycled paper or paperboard waste, e.g., corrugated cases, newsprint, sacks, bags and cartons). It can also be made from other fibrous materials such as linen, sugar cane, cotton and the stalks of cereal plants such as corn (commonly known as straw) (see Fig. 10.1). The terms 'paper' and 'paperboard' generally refer to the packaging they are used to make. For example, you would refer to the material to make a carton as paperboard and the material to make a corrugated case or paper sack as paper. However, this can be confusing as there is a large range of weights and thicknesses used to make a wide diversity of fibre-based packaging.

As a result of this confusion, the International Standards Organisation (ISO) decided to separate the two. 'Paper' is defined by the ISO as that substrate, made from vegetable fibres, which has a grammage (basis weight) of less than 250 grams per square metre (gsm); 'paperboard' (also known as 'cartonboard', 'cardboard', 'boxboard' or just 'board') has a grammage of 250 gsm or over. This definitive difference, however, is not widely used within the industry, different countries using varying terminology. In the United Kingdom, for example, we use the word 'card' when referring to that particular paper used to make greetings cards and we use the term 'cardboard' when referring generally to stiff paper/paperboard materials, such as corrugated or heavyweight solid paperboard.

Paper and paperboard are the most common packaging raw materials. Paper is used mainly in corrugated board manufacture, spiral tube making, and also in laminates, sack and bag manufacture, and wrapping material for such products as ream-wrap for copier paper. Paperboard is used mainly in cartons. There are many forms of paper with a variety of properties. These properties are further utilised with the help

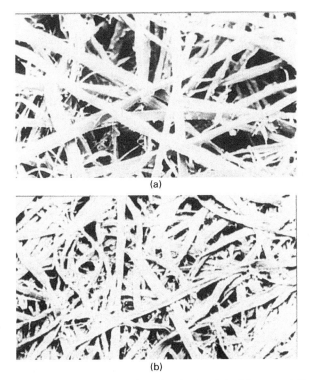

10.1 (a) Matted fibres (mechanical pulp), (b) matted fibres (chemical pulp) (courtesy of Iggesund Paperboard; www.iggesund.com).

of other materials when combined together in laminates. The other materials, apart from clay and chalk (discussed later), used in conjunction with paper are aluminium foil and plastic films. Coating with polymers also enhances the properties of paper. These coatings vary from fluorocarbon (grease resistance) to water-based and film-based barrier coatings (grease, water, water vapour and gas barrier), making the paper base a very versatile material indeed. Paperboard, though not quite as versatile on its own, can benefit from coating and lamination with polymeric materials and aluminium foil and vacuum deposition in the same way as paper.

10.2 Properties of paper and paperboard

The properties of paper and paperboard depend upon many factors and can be tailored to meet the specific needs of the packaging industry. Typical factors included in the discussion which follows throughout this chapter are:

- the source of fibre
- the type of fibre
- how the fibre is extracted from the source
- the amount of treatment the fibres are exposed to during the pulping process
- how the fibres are combined together

180 Packaging technology

- how the fibres are converted to paperboard, and the type of 'mill' used (the term given to the machine on which the paper and paperboard are made)
- the number of plies used to make up a paper or paperboard sheet
- what performance additives and process aids are used
- whether or not the paper or paperboard is bleached, coated or combined with other materials.

10.2.1 The main properties of paper and paperboard

Properties of paper and paperboard vary depending on the grade and specification used. The main properties to be considered in packaging applications are as follows:

- stiffness
- printing surface
- absorbency
- burst strength
- tensile strength
- tear resistance and compression strength
- grease resistance.

Excellent stiffness (some tissue papers excepted) and deadfold of all papers and paperboards, especially bleached and unbleached Kraft and sulphite grades, allow very lightweight papers to be used for bags ensuring their creases are sharp and they stand erect on shelf. The downside is that paper keeps its crease when deformed, detracting from its aesthetic appeal. The use of a sandwich of mechanical pulp between chemical pulps (known as folding boxboard – FBB) maximises the stiffness obtainable for a carton and provides good deadfold when creased (see Fig. 10.2).

The density, whiteness, porosity and smoothness of the bleached paper and paperboard, especially sulphite grades, portray a bright high-quality print when

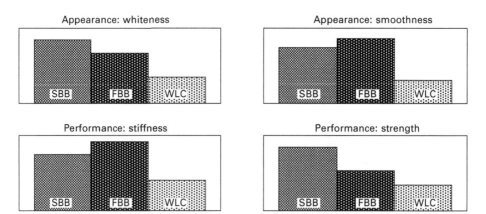

10.2 Paper properties (SBB = solid bleached board; FBB = folding boxboard; WLC = white lined chipboard) (courtesy of Iggesund Paperboard; www.iggesund.com).

decorated by the offset lithographic, flexographic, gravure or digital processes. (See Chapter 19 for detailed discussion of printing processes.)

The Cobb value (water absorbency) and porosity value of paper and paperboard can be adapted to suit the performance required (see Fig. 10.3). Water-based adhesives ideally require an absorbent surface to dry and set the adhesive by fast removal of water. This allows for efficient adhesion in bag, sack and carton making and labelling. The Cobb value can be adjusted by 'hard sizing' the paper (see later additives for paper and paperboard production) so that the fibres are protected from water, allowing use in frozen and wet environments without significantly reducing the physical properties of the paper. Where absolute water repellency is required, coatings are used, for example plastic film, water-based barrier coatings or silicone coatings.

One of the negative properties of paper and paperboard is that they absorb moisture vapour and water. If one side of the substrate is more absorbent than the other, as with clay-coated or one side polymer-coated material, it will warp. This is a very serious issue if not controlled. It can occur during or after manufacture. It is important, therefore, to control the manufacture to make a stable web and to control the storage to ensure moisture variation is kept to a minimum (see Fig. 10.4).

10.3 Cobb test for water absorbency (courtesy of Iggesund Paperboard; www.iggesund.com).

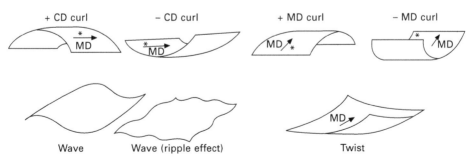

10.4 Warping of paper and paperboard (* = least absorbent side; + = greatest moisture gain; – = least moisture gain) (courtesy of Iggesund Paperboard; www.iggesund.com).

Another property is the burst strength of paper, especially 'sack Kraft'. Paper can be micro-crimped to increase the burst strength and this helps to control and balance the total energy absorption (TEA) and enables it to be used for industrial packaging such as cement sacks and 500 kg flexible intermediate bulk containers (see Fig. 10.5). Once the required total energy absorption levels are known, different combinations of paper plies and weights can be used to achieve it (see Fig. 10.6).

The tensile strength of paper and paperboard is high and their extensibility low, allowing for good constant tension to be applied when printing and laminating papers and during the manufacture of corrugated board, spiral and linear paper tube making and form fill seal (FFS) liquid packaging cartons (see Fig. 10.7).

The tear resistance of papers is variable, depending on the type and the manufacturing process, but in general a grade can be selected to meet most packaging uses. The short span compression strength of paper and paperboard can determine the resistance to compression, provided the height and dimensions of the carton footprint are known.

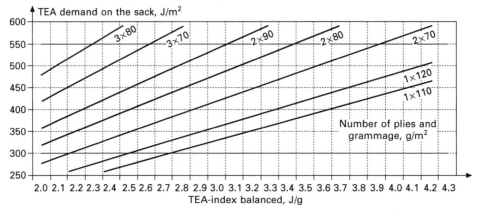

10.5 Total energy absorption (TEA) levels for different combinations of paper plies and weight.

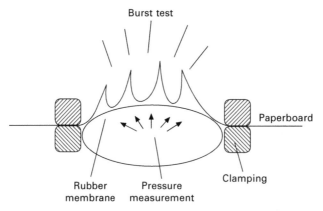

10.6 Principle of burst strength (courtesy of Iggesund Paperboard; www.iggesund.com).

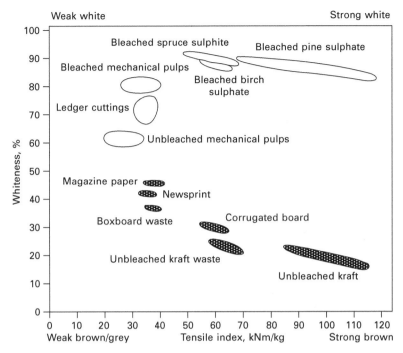

10.7 Difference in whiteness and tensile strength of primary and recycled paper and paperboard grades (courtesy of Iggesund Paperboard; www.iggesund.com).

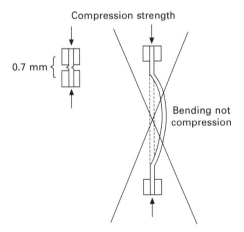

10.8 Compression strength testing (courtesy of Iggesund Paperboard; www.iggesund.com).

It is important that the distance between the jaws of the test rig is no greater than 0.7 mm, otherwise bending will affect the result (see Fig. 10.8).

Paper can be made to have varying degrees of grease resistance, either by treating the fibres and paper physically or chemically. Physical treatment of the fibres is known as beating or refining, to produce grease proof or grease resistant (GP or GR) paper. Chemical fibre treatments include adding fluorocarbon to the furnish or at the size press. GP papers can be further treated by supercalendering to produce a translucent grease resistant paper known as 'glassine'. The ultimate in grease resistance is achieved by dissolving some of the fibres in sulphuric acid to produce parchment, which also has a very low gas permeability.

10.2.2 Paperboard grades

Grades of paperboard used can be generally classified as follows.

- Solid bleached board (SBB): paperboard made from virgin bleached chemical pulp (Fig. 10.9).
- Solid unbleached board (SUB): paperboard made mainly from unbleached virgin chemical pulp. A layer of bleached fibre is sometimes added to the top to provide greater whiteness (Fig. 10.10).
- Folding boxboard (FBB): made from a layer or layers of mainly virgin mechanical pulp sandwiched between layers of virgin chemical pulp. The top layer is bleached chemical pulp and the bottom layer can be either bleached or unbleached virgin chemical pulp (Fig. 10.11).
- White lined chipboard (WLC): made from multi-layers of recycled fibres. The top layer can be made from bleached virgin chemical pulp or white deinked recycled fibres. Between the top layer and the middle layer(s) there can be a layer of chemical, mechanical or deinked recycled fibres. The bottom layer can be made from selected recycled or bleached and/or unbleached virgin fibres (Fig. 10.12).

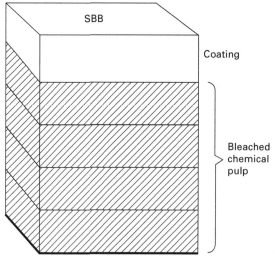

10.9 Solid bleached board (SBB).

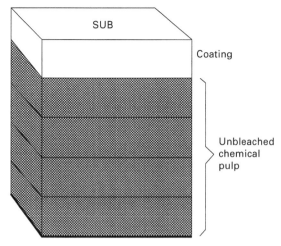

10.10 Solid unbleached board (SUB).

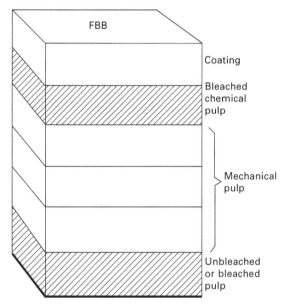

10.11 Folding boxboard (FBB).

There are other boards available which are combinations or variations of the above. Two examples are:

- CCNB – clay coated news back – news back refers to the layer(s) of pulp beneath the coating made specifically from recycled newspapers rather than mixed waste. This creates a very smooth light grey underside but, due to the short fibres, this material can only be used where the physical strength of the carton is relatively unimportant.
- CCKB–clay coated Kraft back (also known as CKB in Europe and coated kraft

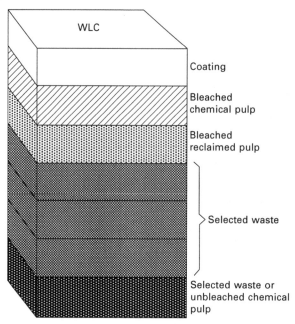

10.12 White lined chipboard (WLC).

back or CCCB – clay coated craft back in the United States). Kraft back can refer to virgin sulphate unbleached pulp or fibre recovered from used corrugated cases. This is a much stronger board than CCNB and in some instances can compete with Kraft boards.

The DIN 19303 standard for classification of paperboards is a little more refined than the traditional terminology used in the United States and the United Kingdom, and is now being adopted by most European carton manufacturers (see Table 10.1).

10.2.3 Comparison and properties of different types of paper and paperboard

As has already been mentioned, the properties of paper and paperboard vary greatly, depending on grade and specification. At this point it is worth considering some general comparisons of different types of paper and paperboard as shown in Figs 10.13–10.15. Comparisons are made taking account of bleached vs. unbleached materials, and materials made from virgin chemical pulp, virgin mechanical pulp and recycled fibres. These materials and processes are discussed in detail later (Fig. 10.13). It can be seen that bleached chemical pulps are generally weaker than unbleached pulps but both are stronger than mechanical pulps. Also, as expected, bleached pulps are whiter than unbleached pulps (see Fig. 10.14).

Mechanical pulp is stiffer than chemical pulp but not as strong. Fortunately the manufacturer is able to combine pulps for maximum overall performance, for example, folding boxboard (FBB) has inner plies of mechanical pulp and outer plies of chemical pulp. This provides a board with maximum stiffness at minimum grammage. If we

Table 10.1 DIN 19303 European classification for qualities of paperboard

Abbreviations (German terminology)	Description
GZ	Coated SBB
GGZ	Cast coated SBB
GG1	Cast coated FBB white back
GG2	Cast coated FBB manilla back
GC1	Coated FBB white back
GC2	Coated FBB manilla back
GT	Coated CB manilla or white back
GD1	Coated CB high bulk (spec. volume min 1.5 cm^3/g)
GD2	Coated CB (spec. volume min 1.4 cm^3/g)
GD3	Coated CB low bulk (spec. volume min 1.3 cm^3/g)
GN1	SUB
UZ	Uncoated SBB
UC1	Uncoated FBB white back
UC2	Uncoated FBB manilla back
UT	Uncoated CB manilla or white back
UD1	Uncoated CB top liner woodfree
UD2	Uncoated CB top liner near woodfree
UD3	Uncoated CB top liner partly mechanical pulp
Key:	
SBB	Solid bleached board
SUB	Solid unbleached board
FBB	Folding boxboard
CB	Chipboard (more often WLC = white lined chipboard)
G	Gestrichen, coated
U	Ungestrichen, uncoated
GG	Gussgestrichen, cast coated
Z	Zellulosekarton, solid boxboard
C	Chromoersatzkarton, folding boxboard
D	Duplex (CB construction)
T	Triplex (CB manilla or white back construction)

compare four properties of mechanical and chemical fibres it can be seen that

1. density increases as the amount of chemical pulp increases
2. stiffness increases by increasing the amount of mechanical pulp
3. strength increases with increased chemical pulp content
4. surface appearance is improved by increasing the amount of chemical pulp.

Figure 10.15 compares the properties of recycled fibre with the two main virgin fibres. It can be seen that, in general, recycled fibres are stronger than mechanical fibres but less strong than chemical fibres. The same comparison cannot be used for whiteness. Most recycled fibres are grey in colour. As we change from mechanical pulp to recycled pulp and on to chemical pulp, the density of the paper and paperboard produced from these pulps increases. As density increases it would be expected that strength and surface appearance increase and stiffness decreases. However, as a result of the secondary processing of recycled fibres and the contaminants contained within (clay and chalk coatings, for example) the paperboard made from recycled fibres does not usually follow this expected trend.

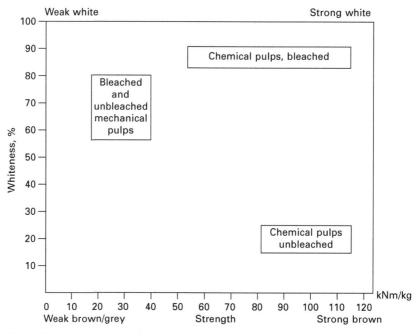

10.13 The combination of whiteness and strength for various pulps (courtesy of Iggesund Paperboard; www.iggesund.com).

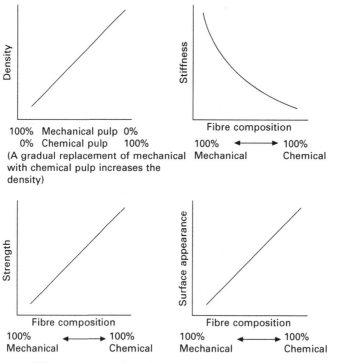

10.14 Effect on properties of changing types of pulp (courtesy of Iggesund Paperboard; www.iggesund.com).

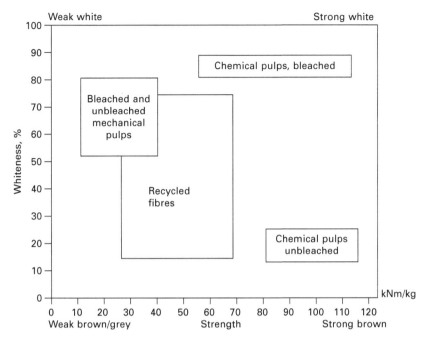

10.15 Whiteness and strength for different paperboard grades (courtesy of Iggesund Paperboard; www.iggesund.com).

10.2.4 Changing properties by combining paper with other materials

Paper and paperboard, when combined with selected polymeric film materials, can result in materials with significantly enhanced properties. Coating with polyethylene and other polymer films makes it possible to heat-seal paper-based materials, which is very useful in sack and bag manufacture where an integral seal is required, for example for dry pet foods.

Aluminium foil and polymer films can also be added, allowing laminates to be constructed that can be formed into liquid packaging cartons. In these applications the paper provides the stiffness, puncture, and abuse resistance, formability on the packaging machine and adds to the ultraviolet (UV) light barrier. The aluminium provides the gas, UV and water vapour barrier. The polymeric film, usually polyethylene, or polypropylene, provides water and product resistance, acts as an adhesive between the aluminium and the paper, and provides an excellent heat seal to ensure the product does not leak and there is no ingress of gas through the sealed surface.

The negative aspects of combining paper with other materials are that it makes the waste more difficult to separate and therefore recycle into its component parts in a commercially viable manner. Water-based barrier coatings (WBBC) have been developed to overcome some of these issues. They provide water barrier, some water vapour barrier, grease barrier and product release and are claimed to be fully repulpable with the paper, negating the need for separation of the coating from the fibres.

10.3 Raw materials

The main raw material used to make paper and paperboard is cellulose fibre sourced from trees and recycled waste. Fibres are mixed with additives to improve performance and control processes and where necessary treated with coatings to further improve performance. The source for and treatment of these raw materials, and how they influence performance will now be discussed in this section.

Fibre length is one of, if not the most important, property with respect to paper and paperboard performance, and the length and shape of the fibre depends on the source. Deciduous trees, i.e. broadleaf trees which lose their leaves in the winter in non-tropical regions, produce short fibres, whereas coniferous trees, i.e. cone bearing, needle leaf trees, produce long fibres. Deciduous trees are also known as hardwoods and coniferous trees as softwoods. Short fibres provide smoothness for printing while long fibres provide strength. Typical ranges of fibre length used in papermaking are:

- short hardwood fibres: 1–1.5 mm in length
- long softwood fibres: 3–4 mm in length

Aspen, eucalyptus and birch trees grown in temperate climates are the common sources for hardwood fibre, and trees such as spruce, larch, fir, hemlock and pine provide the softwood fibres. All of these types of trees are grown in managed forests with a continuous replanting programme and they are used to make virgin pulp. This important raw material is fully sustainable with minimum adverse effect on the environment. Trees are made up of approximately 50% fibre, the rest being lignin and other substances, such as carbohydrates. Lignin is a complex chemical which binds the fibres together, but also causes discolouration.

Trees of a selected species are grown to a specific size, felled, cut into precise lengths, the branches, twigs and some of the bark removed and the logs transported to the mill to be converted into paper and paperboard. Once the debranched trees reach the mill, they are completely debarked and sent to chippers where they are cut to similar sized small pieces before they are sent to the predetermined pulping process.

Recycled fibres vary widely depending on the waste raw material source. In general they are a mixture of hard and softwoods, their fibre length depending on the number of times they have been recycled. Fibres are shortened at each recycling process and, once virgin fibres have been recycled around seven times, the fibre length is considered too small to use further.

Recycled fibres are often contaminated with printing inks, product and 'contraries' such as plastic, baling wire, wax and adhesive. Although the deinking process can remove some of these, others get through, resulting in contamination of the fibrous sheet produced. These can appear as small particulates of colour, text, grease and char spots and produce a reduced whiteness and brightness as well as reduced strength (due to the shortening of fibre length) of the finished substrate compared to that manufactured from virgin fibre. However, the industry continues to develop the quality of recycled fibre, with significant improvement over the past 20 years. Some grades of paperboard made from recycled waste are highly competitive in performance compared to virgin paper and paperboard (see Fig. 10.16). Not all properties are worse than virgin papers and paperboards: for example delamination (IGT) and Scott Bond are at least as good, if not better as can be seen from the figure. There is, however, a concern in certain industries where taint, odour, aesthetics and performance are critical.

10.4 The pulping process

There are two main types of pulping process for the production of virgin fibre and two intermediate methods. These are mechanical (or groundwood), chemical, thermo-mechanical (TMP) and chemical thermo-mechanical (CTMP or semi-chemical). Recycled fibre is prepared in a separate process using a hydrapulper.

10.4.1 Mechanical pulp production

This is the quickest, least costly method of obtaining virgin fibres. The wood chips are washed, to remove any soil, stones or other contaminants, and mechanically ground. The grinding process was traditionally carried out using grindstones similar to those used for grinding flour, to grind the logs directly. This as known as stone groundwood. Nowadays, ridged metal discs called refiner plates are used to process the wood chips. This process separates the fibres individually but in doing so breaks them into shorter lengths. It does not remove the lignin and other impurities, resulting in fibres which discolour with age (see Fig. 10.17).

Mechanical pulp is used for low grade papers such as newsprint and for blending with chemical and semi-chemical produced pulps to reduce costs. It is also used as a sandwich between chemical pulps for the production of folding boxboard (FBB), one of the most popular boards used for carton making in Western Europe.

10.4.2 Chemical pulp production

Here the wood chips are placed in a digester where the cellulose fibres are separated from the lignin and other impurities using heat and chemicals (see Fig. 10.18). If white fibres are required, bleaching is carried out at this stage. This is the most expensive method of producing fibres for paper and paperboard manufacture, due to the lower yield compared to mechanical pulp and the heat energy and chemicals required. However, it produces the strongest and whitest (when bleached) substrates available.

There are two main chemical processes used: the alkaline sulphate process (known as 'Kraft'), which produces the strongest of all the cellulose fibre-based paper and paperboard products, and the acid sulphite process. The sulphate method uses a combination of sodium hydroxide and sodium sulphide in the digestion process. The

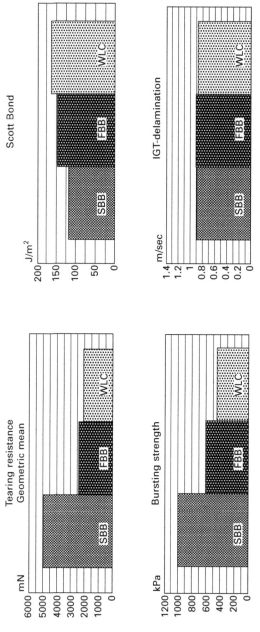

10.16 Properties of some paperboard grades made from recycled waste (courtesy of Iggesund Paperboard; www.iggesund.com).

Paper and paperboard packaging 193

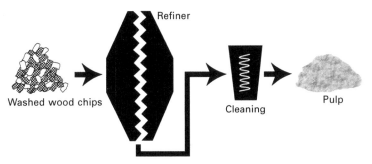

10.17 The production of mechanically separated pulp (adapted from Pro Carton; www.procarton.com).

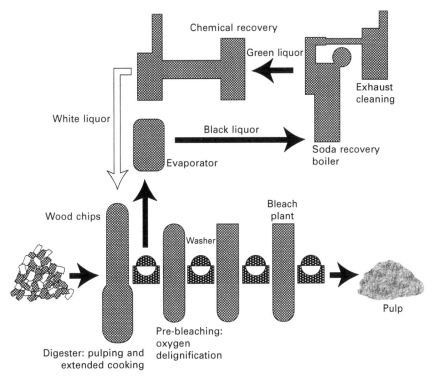

10.18 The production of chemically separated and bleached pulp (adapted from Pro Carton; www.procarton.com).

sulphite method uses metal or ammonium salts of sulphurous acid, producing either sulphites or bisulphites. These chemicals digest the impurities but, in doing so, due to their acidic nature, hydrolyse some of the cellulose resulting in the final paper or paperboard having less strength than that produced by the sulphate process. However, the amount of pulp produced per hour is greater and the quality of bleaching is better, due to its higher purity. Today less than 20% of all chemical pulp is produced by the sulphite process. However, due to its high purity it is used to manufacture Rayon, cellulose acetate and Cellophane™. Cellophane™, though not a plastic, is a clear film suitable for packaging of fresh produce, where breathability is a requirement and for twist wrap films for mechanically wrapping individual sweets, where its excellent deadfold properties are utilised.

In Europe chemical pulp can be approximately twice the cost of mechanical pulp (this is not necessarily the case elsewhere, e.g. North America) and is therefore used where maximum strength is required, such as multi-wall sacks and the liners for corrugated cases, especially in damp, high humidity and wet environments. Chemical pulp is also used where maximum toughness, whiteness and purity are required. Examples are the outside plies of folding boxboard, cartons for high value cosmetics and liquid packaging form fill seal cartons. Paper made from chemical pulp is often referred to as 'wood free'. This does not mean the fibres come from material other than trees; it means there is no groundwood or mechanical pulp included in the paper.

The two intermediate pulping processes of TMP and CTMP referred to earlier are used to either improve the properties or reduce the costs of fibre production. In the TMP process, hot water is used to soften the fibres and render them more supple, resulting in less damage during the mechanical process and consequently higher strength compared with the basic mechanical pulping process. The CTMP process, consisting of some chemical digestion and use of heat, but less than in the chemical process, takes less time and therefore reduces total costs. The process partially digests the wood mass, removing some of the impurities prior to mechanically grinding the softened fibres.

Referring back to Section 10.2.3, and taking into account the fibre length as well as the pulping process, natural (unbleached) Kraft made from softwood fibres produced by the chemical sulphate process is the strongest paper or paperboard available, and mechanical or groundwood pulp made from hardwood fibres produces the weakest paper or paperboard available. Packaging formats are produced from combinations of fibre and pulping processes to meet the performance requirements of the final pack, taking into account the needs of the whole supply chain (see Table 10.2).

Virgin pulp is not always produced at the mill that makes the final paper or paperboard. Many small and some large mills buy in some or all of their pulp and disperse it into water using a hydrapulper. This is a large vessel with an agitating

Table 10.2 Pulp preparation

Pulp preparation	% lignin removed	Yield	Process
Mechanical	nil	95%+	Grinding
Thermo-mechanical	nil	95%+	Steam heating + Grinding
Semi-chemical	10–80%	65–85%	Cooking with chemicals + Grinding
Chemical	80–90%	45–65%	Cooking with chemicals
Chemical + Bleaching	100%	45–55%	Plus bleaching agents
Dissolving	100%	45–50%	Plus sodium hydroxide (cellulose film)

blade. The pulp, often in the form of large sheets, is dropped into the water contained in the vessel and dispersed in it to the required concentration – between 0.3 and 3.0% depending on whether paper or paperboard is the intended product. The higher concentrations are used for paperboard. The dispersion of fibre in water, plus the other additives and process aids, is known as the 'furnish' (see Fig. 10.19). This will be covered in more detail in Section 10.7.

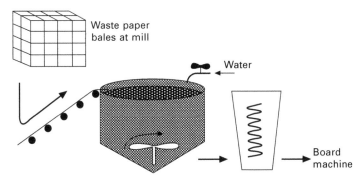

10.19 Pulp production using a hydrapulper (adapted from Pro Carton; www.procarton.com).

10.4.3 Recycled fibre production

Recycled pulp is also produced by using a hydrapulper. Selected, de-inked waste paper and paperboard is dispersed in water in the same way as pre-formed sheets of virgin pulp as described above. For every tonne of waste material, less than 90% is recovered, the loss being due to material being unfit for use and having to be discarded, and to the de-inking and other cleaning processes, which result in a loss of fibre. Recycled pulp is usually much more price competitive than virgin pulp, but produces an inferior product, both from a performance and an aesthetic viewpoint as already discussed. That does not mean it is not fit for the purpose for which it is intended.

Some recycled paper and paperboard materials need to be supplied at up to 20% greater basis weight to provide the same performance characteristics, compared to some virgin paper and paperboards. Paperboards made from recycled materials are referred to as waste-based board (WBB) or coated recycled board (CRB) or white lined chip (WLC). There are other terminologies used in some European countries (see Section 10.2.2).

10.5 Post-pulping treatment of fibres to improve performance

Once the fibres have been extracted by any of the above pulping processes, they are still not in the shape and condition required and further treatments are carried out. The main two post-treatments of fibre are beating and refining (one process) and bleaching.

10.5.1 Beating and refining

The beating and refining of fibres is one of the most critical processes in the production of paper, and to a lesser extent paperboard (see Fig. 10.20). The process bruises/splits the fibres, increasing their flexibility and extends their surface coverage, but in doing so reduces fibre length distribution, weakening the overall paper or paperboard.

Refining is carried out as part of the stock preparation process, by passing the pulp, suspended in water, across rotating surfaces. This causes the fibres to fibrillate and swell. This process is either continuous (where minimal refining is required) or a batch process (where considerable beating is required to produce greaseproof and glassine grades of paper). The latter uses a more sophisticated beating process where the fibres are passed through rotating discs or cones (see Figs 10.21 and 10.22).

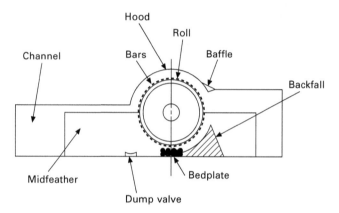

10.20 Beating of fibres (source: Paper Industry Technical Association (PITA); www.pita.co.uk).

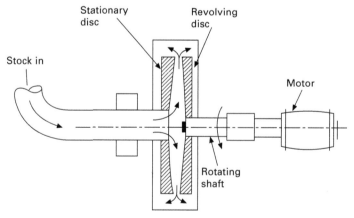

10.21 Rotating disc refiner (source: Paper Industry Technical Association (PITA); www.pita.co.uk).

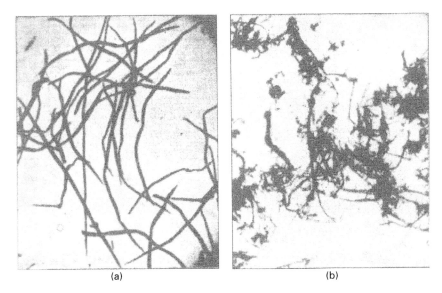

10.22 Fibres before (a) and after (b) beating (source: Paper Industry Technical Association (PITA); www.pita.co.uk).

While refining the fibres improves the paper properties, too much refining will lead to deterioration. As the process is prolonged:

- burst, tensile strength and resistance to the permeation of air through the substrate of air all increase
- tear resistance increases initially, but quickly decreases as fibres reduce in length
- resistance to folding increases initially but then reduces as fibres decrease in size
- uniformity of paper increases, resulting in improved print surface and formation (appearance), all as a result of reduced fibre length induced by the refining process (see Fig. 10.23).

10.5.2 Bleaching

Bleaching is another post-treatment of fibres, carried out once the fibres have been separated from their source. Paper and board produced using bleached pure cellulose made by chemical pulping has a bright, white appearance and shows little or no tendency to fade or yellow when subjected to sunlight. Mechanical pulp, even if bleached tends to yellow over time.

Traditionally bleaching was done using chlorine, which dissolves some of the lignin remaining in pulp. However, due to the environmental disadvantages and potential safety hazards of using chlorine, the amount of pulp bleached in this way has declined. It is currently thought to be around 25% worldwide, the remainder having been replaced by one of two common processes:

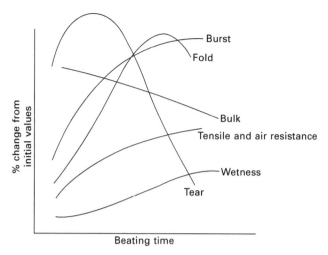

10.23 The effect of beating on paper and paperboard properties (source: Paper Industry Technical Association (PITA); www.pita.co.uk).

- *ECF – elemental chlorine free*. The bleaching sequence uses chlorine dioxide, with no use of chlorine gas or sodium hypochlorite. It is sometimes known as chlorine dioxide bleaching (CDB).
- *TCF – totally chlorine free*. Here the bleaching sequence uses only oxygen-based chemicals such as oxygen, ozone and alkaline or acidic peroxides. It is sometimes known as oxygen chemical bleaching (OCB).

A further category, specific to recycled paper is known as PCF – process chlorine free. This indicates that, while the production of the primary fibre may have involved the use of chlorine and chlorine compounds, these materials are not used during the recycling process; the fibre is either not bleached at all, or if bleaching has been carried out it is done using oxygen-based systems. This is sometimes known as secondary chlorine free (SCF). Both ECF and TCF processes are vast improvements on those which use chlorine. TCF claims to use less toxic starting chemicals than ECF, and to have a reduced environmental impact. However, TCF pulps generally have lower strength properties than ECF pulps, although this can be compensated for by lower brightness, if this is aesthetically acceptable. Another consideration is that the number of mills producing TCF papers is much lower than those producing ECF and is currently thought to be around 7%, mostly in Northern and Central Europe.

10.5.3 Additives

Paper and board are not produced by using fibres alone. Additives and process aids are required to ensure that the important properties required by the converter, packer/filler and end user are controlled consistently.

The types of additives used in paper and board manufacture and their purposes are discussed below:

- Fillers such as kaolin and chalk are added to the furnish to improve printability of the paper, and others such as titanium dioxide are added to improve whiteness. Fillers also improve surface smoothness, control brightness and control opacity. They are also used to reduce the cost of manufacture of the paper and paperboard.
- Pigments may be added to colour the paper, although the range of shades available is limited and most coloured effects are achieved by surface printing.
- Whitening (FWA – fluorescent whitening agents) and brightening (OBA – optical brightening agents) are also added to improve the whiteness, especially under retail lighting.
- Binders such as starch (farina from potatoes, maize, wheat and tapioca) are used to increase strength by linking the fibres together restricting their movement and resulting in a unified mat. Binders are also used to prevent the fillers falling out of suspension in the furnish.
- Size is used to control water and ink penetration and its use is crucial to address problems due to the natural absorbency of cellulose fibres. Totally unsized papers would allow ink to soak in and spread throughout the fibres. Sizing agents used are, for example, AKD (alkyl ketene dimer), aluminium sulphate ('Alum'), modified starch and gelatine for surface sizing and AKD, ASA (alkenyl succinic anhydride), rosin and 'Alum' for internal sizing. Surface size is added to the paper web using a size press and can be applied to one or both sides depending on the final properties required of the substrate. Internal size is added to the furnish before the paper is formed. AKD is an amphipathic lipid, i.e. a molecule which is mostly non-polar (hydrophobic) in structure, but at one end having a region that is polar or ionic (hydrophilic). The hydrophilic region is usually referred to as the head group, and the hydrophobic portion is known as the tail. They work by surrounding the fibre with the hydrophobic tails outermost, pushing the water away from the fibre and thus reducing the amount of water that will penetrate the fibre. Size can vary in pH from acidic through neutral to alkaline and as a result the choice of sizing agent also controls the pH of the final substrate.
- Wet strength resins, based on urea and melamine formaldehyde, can be added to reduce the effect of water on the initial strength of the paper. An example would be paper for multi-wall sacks for use outside, or for carrier boards used for collating packs in wet conditions such as for bottled carbonated beverages.
- Grease resistance can be achieved with additives. Fluorocarbon chemicals can be added, either to the furnish or at the size press. They work by surrounding the cellulose fibres and protecting them from any oil or grease which may penetrate. They have been used in the past for the production of dry petfood bags, sacks and cartons, where the fat content of the product is often over 20 percent. They have also been used in the manufacture of wrapping paper for butter and margarine, replacing the traditional parchment paper. Concerns about taint and odour have greatly reduced their use in recent years.
- Other chemicals such as acrylic resins are added to improve water resistance and wax to improve strength and water resistance, the latter becoming much less common due to issues with recycling of the waste material.
- Process aids are also necessary. These include anti-foaming agents, bactericides

(restrict microbiological activity), flocculating agents (improve dewatering of the furnish as the web is formed) and special chemicals to reduce/prevent the resins from the wood depositing on the paper web, possibly causing web breaks and contamination both during papermaking and in the later printing processes.

The fibre and the additives are now suspended in water and fed into the mill ready for the manufacture of paper and paperboard.

10.6 The manufacture of paper and paperboard

The principles of papermaking are basically the same today as they have always been, with three distinct stages:

1. A dilute suspension of fibre and additives in water is prepared.
2. The dilute suspension is formed into a sheet of intertwined fibres.
3. Water is removed at all stages of the process, via drainage, pressure, vacuum and evaporation, until a suitable substrate is achieved.

Although there are many variations of individual paper and paperboard making machines, they are all made up of a wet end, where the sheet is first formed, and the water is removed by mechanical means, and a dry end, where heat is used to bring the substrate to its required moisture content, and a number of surface treatments are carried out. There are two main types of paper or paperboard mill discussed in this chapter:

- wire or Fourdrinier
- vat or cylinder

10.6.1 The Fourdrinier method

The principles of paper and paperboard manufacture by the Fourdrinier process are very similar (see Figs 10.24 and 10.25). Dilution of fibre in water varies from 0.3 to 3% depending on the weight of paper or paperboard being manufactured. Weights vary from 12 gsm for the lightest tissue paper to 600 gsm for solid bleached paperboard

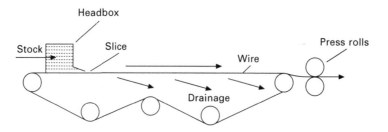

10.24 Fourdrinier papermaking process: basic wire process (adapted from Kirwan, M.S. (ed.), *Paper and Paperboard Packaging Technology*, Blackwell Publishing Limited, Oxford, 2005. Reproduced with the permission of John Wiley and Sons Inc.; www.wiley.com).

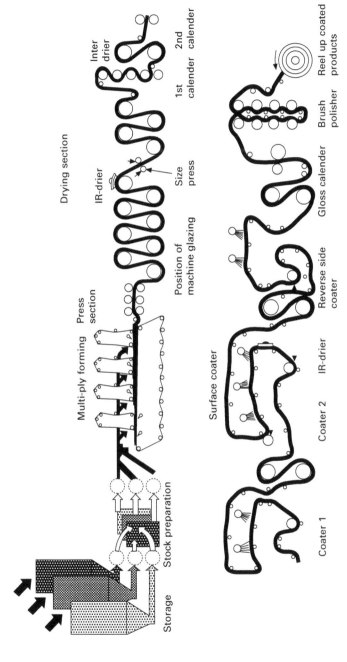

10.25 Fourdrinier papermaking process using multi-formers (courtesy of Iggesund Paperboard; www.iggesund.com).

made from chemical pulp. The heavier the paper/paperboard being manufactured the higher the concentration of fibre in water.

The furnish is held in tanks, each tank holding a specific fibre and additive combination and it is fed into a headbox, which in turn spreads it onto a wire via the slice, which controls the flow (see Fig. 10.26). Today the 'wire' is a plastic mesh. The first layer of fibre, which may be the only layer if making paper and will be the underlayer if making multi-layer paperboard, is delivered onto the wire as consistently as possible. The water drains away through the wire, usually assisted by vacuum suction boxes. The movement of the wire means that the fibres align themselves preferentially in the direction of travel, i.e. the machine direction (MD). This is not ideal as the aim is to produce a sheet where there is no obvious fibre alignment. To try and achieve this aim, the wire is often shaken in the cross direction (CD), to redistribute the fibres.

Depending on the final thickness/weight of the paper/paperboard, other layers are added until the required specification is reached. Each additional layer requires another headbox and slice to deposit the fibres on top of the other layer(s). However, as the layers build up, the rate at which water will flow through the wire decreases significantly and it is necessary to use top wires where the water is sucked upwards using vacuum. Sometimes the last layer (top layer) is formed as an independent sheet and added to the rest of the construction as a final operation. The moisture content of the paper or paperboard is still high at this stage at around 60–70%.

The wet substrate is then sandwiched between felt blankets and passed through the press section where steel rolls remove more water by pressure and vacuum from the fibrous web. The web then passes to the dry end of the machine, firstly through the drying section where more water is removed by evaporation using steam heated steel cylinders. At this stage it can pass over a large polished metal cylinder resulting in a smooth surface finish being produced without compressing the web, which would reduce its thickness and stiffness. This type of finish is known as MG (machine glazed).

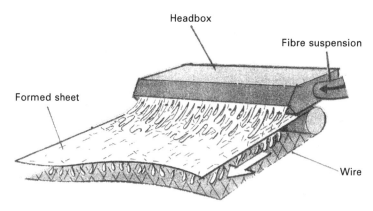

10.26 Fourdrinier forming process: headbox and slice (courtesy of Iggesund Paperboard; www.iggesund.com).

Immediately after the drying section, surface sizing can be carried out using a starch solution which can be pigmented if required. This prevents fibres shedding from uncoated surfaces and improves surface strength and smoothness, hence printability (see Fig. 10.27).

Calendering, which is a type of ironing process giving a uniform thickness and smoothness to the paper or paperboard, takes place once the substrate has been sized. The dry substrate is passed between cylinders, which can be cold or heated and water may be applied to enhance the smoothing effect. The cylinders on paper machines are often a combination of steel rolls and ones made of composite material to provide very smooth glossy finishes. Some papers are produced by 'super calendering' (e.g. glassine), which is carried out on a separate machine having up to 14 rolls, to produce a translucent paper (see Fig. 10.28).

Where required, the web is now coated with a white mineral pigment base (clay or chalk). This gives a hard smooth surface suitable for high quality printing. There are many ways to add the coating and, depending on the colour of the web (brown, grey or white), between one and three separate coats are applied. The amount of coating is also governed by the final smoothness required and the initial smoothness of the web. The smoother and whiter the web initially, the less coating is required (see Fig. 10.29).

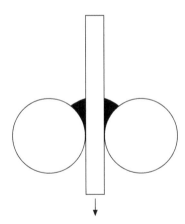

10.27 Size press (courtesy of Iggesund Paperboard; www.iggesund.com).

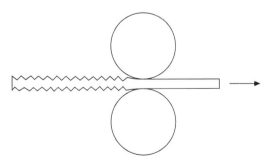

10.28 Calendering (courtesy of Iggesund Paperboard; www.iggesund.com).

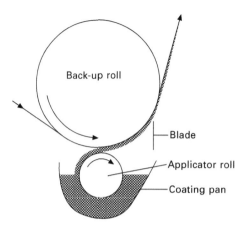

10.29 Blade coating (source: Paper Industry Technical Association (PITA); www.pita.co.uk).

The quality of the coated surface is influenced by the mechanism by which the coating is applied, with the double blade process giving superior results to those using an air knife. Binders are used to achieve good adhesion between the coating and the base web, and between the mineral particles within the coating. Optical brightening agents may also be added to the clay coatings especially for cartons for retail display.

Once the coatings have been applied, the material is ready to be wound up into reels. The web is wound on cores in batches of up to 20 tonnes. These 'parent' or 'mother' reels are wrapped in protective coverings and labelled to ensure product identification and traceability. The mother reels can be slit into the required widths, either *in situ* on the mill or during a separate operation, after the mother reel has been removed from the end of the mill. They are then stored under controlled conditions of temperature and humidity awaiting conversion. If the material is going to be converted into folding cartons and printed via the offset lithographic process, it is often sheeted into various sizes, before being wrapped, labelled and stored.

Not all coating processes are carried out in-line on the paper or paperboard making machine; in some cases manufacturers take the slit reels and apply coatings separately. One example is cast clay coating applied to paper destined for high grade label stock to give a very high quality smooth surface. Other coatings such as wax or synthetic polymers can be applied, either on or off the paper or paperboard making machine. Wax can be applied as an emulsion or as a molten liquid. Where both sides are coated at the same time, a process known as 'cascade coating' is used and where just one surface is coated this is done using 'curtain coating'. Wax coating makes the material difficult or impossible to recycle, but gives it good water and grease resistance, allows it to be heat sealed and adds extra strength and deadfold characteristics. Polymer coating using plastic films has largely superseded wax. This is applied either as an extrusion coating or is laminated to the paper via an adhesion process (see Fig. 10.30). (This will be covered in more detail in Chapter 14.)

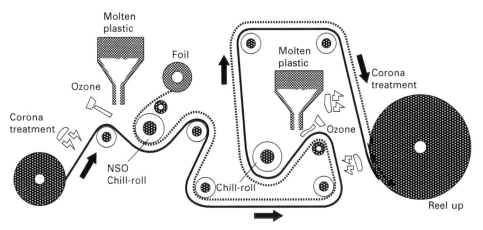

10.30 Extrusion coating and lamination (courtesy of Iggesund Paperboard; www.iggesund.com).

Polymer films hinder recyclability as they have to be removed at or before the hydrapulper. However, correct selection of polymer film can add to the benefits of paper and paperboard, providing barrier to moisture and gas (especially in combination with aluminium foil), grease resistance, water resistance, heat sealability, high gloss finish, and, using reverse print techniques, protection of the decorated surface.

In the past 15 years water-based functional coatings (generally known as WBBC – water-based barrier coatings) have been developed, which when applied to the surface of paper or paperboard can provide similar protection to fluorocarbon, wax and polymer films, but allow the coated substrate to be easily repulped without leaving any undesirable residues. This enables them to be placed in the same waste stream as uncoated paper and paperboard.

Returning to the wet end of the papermaking machine, a variation of the Fourdrinier method is the vertical former ('Vertiformer') which is a twin wire former. The furnish is supplied to the space between the formers, and picked up by two meshes (wires); the water is removed from both sides of the paper by the two wires. This has two advantages:

- the paper can be produced at a faster speed than when using the horizontal wire method as dewatering is quicker, and
- two ply papers can be produced with identical finishes on both faces (see Fig. 10.31).

10.6.2 The cylinder (vat) method

The second major method of manufacturing paper and paperboard uses individual vats containing the furnish required for each layer (Fig. 10.32). A large screen drum revolves within the vat and as it does so it picks up the fibres and the excess water drains away through the screen, leaving fibre on the outside surface of the screen. This fibre is transferred from the screen onto the underside of a continuous moving

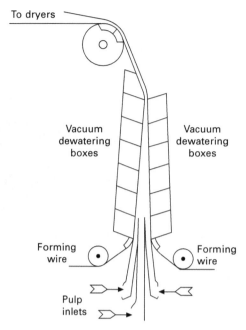

10.31 A vertiformer.

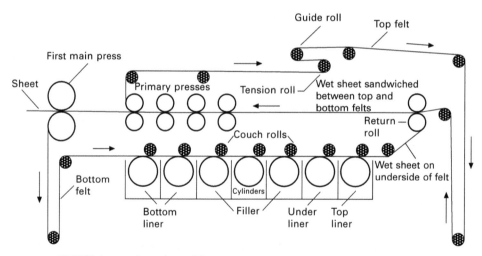

10.32 Vat paperboard machine.

felt. The felt then passes over the next vat where a second layer is added to the first and so on, until the final specification is achieved. Different pulp fibres, e.g. chemical, mechanical, recycled, can be added via the individual vats to build up a multilayer structure.

The cylinder has a differential pressure between the inside and outside which assists in dewatering. The furnish can flow in the same direction as the cylinder rotates ('uniflo'), as shown in Fig 10.33, or it can flow in the opposite direction,

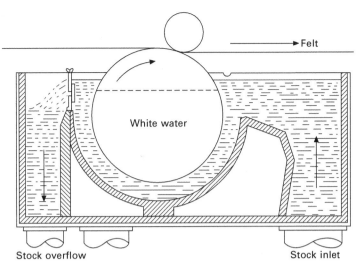

10.33 A 'uniflo' vat cylinder (source: Paper Industry Technical Association (PITA); www.pita.co.uk).

known as 'contraflo'. The uniflo method results in an even, consistent sheet formation, whereas the contraflo method allows a greater amount of fibre to be deposited on the cylinder, resulting in a thicker and heavier board being produced for the same number of vats used. However, the interply bond strength using contraflo is weaker than with uniflo. Also the interply bond strength of cylinder paperboard is generally weaker than paperboard produced by the wire method.

Once all the layers have been applied to the felt, a second felt is placed on the top of the fibrous web (see Fig. 10.32) and the whole is passed through a series of presses which remove sufficient water for the web to be self-supporting and the felts to be removed. From here the dry end section is the same as for the wire method.

10.6.3 Control of quality of paper and paperboard

During paper and paperboard making, controls must be put in place to ensure a consistent substrate is produced which is stable in use. Tests are carried out on and off the machine to ensure conformance with specification and adjustments made to address any unacceptable deviations.

Consistent moisture content is critical and this is monitored on the machine using infrared sensors. These transmit signals to control the activation of aspirated sprays to stabilise the moisture content of the web. Moisture content, thickness, basis weight and stiffness are constantly monitored in the mill laboratory and adjustments made to the machine to ensure the parameters stay within the limits of the specification. It is, however, important to recognise that paper and paperboard have different properties when measured in the cross and machine direction:

- substrates made by the wire method have a MD:CD ratio in the region of 2:1
- substrates made by the vat method have a MD:CD ratio in the region of 4:1.

The relationship between the two directions is a complicated one. It is not just the MD stiffness that needs to be taken into consideration but the CD as well.

Various tests which may be carried out on paper and paperboard to ensure it provides the appearance and performance expected are described in Tables 10.3 and Table 10.4. Most tests are carried out under controlled conditions of temperature and humidity to ensure they are repeatable and completely comparable. The international standard for test conditions is 23°C and 50% RH. When comparing values, the units and test conditions should also be scrutinised to ensure a realistic comparison.

10.7 Conversion processes for paper (<250 gsm)

The most common papers used in packaging and typical end uses are:

- Kraft: bags, sacks, folding cartons, corrugated liners
- sulphite: bags, sacks, folding cartons, dual ovenable trays, liquid packaging, labels, leaflets, corrugated liners
- CTMP: corrugated fluting medium, liquid packaging containers
- test liners (mixture of virgin and recycled fibres): corrugated liners
- recycled Kraft: spiral wound containers
- greaseproof, glassine and parchment: wrappers and bags where grease resistance is important
- tissue: tea bags, industrial packaging, fine jewellery, textiles.

The main packaging formats using paper are:

- corrugated packaging (this is the largest use of paper in packaging and is covered in Chapter 11)
- sacks
- bags
- spiral, linear and convolute wound containers
- liquid packaging cartons.

10.7.1 Paper sacks

A paper sack can be differentiated from a paper bag by its product weight which is usually >5 kg and the fact that it is traditionally made from more than one ply. In the United States paper bags are used as grocery bags referred to as sacks. Paper sacks are traditionally made from between two and six plies of paper, sometimes having one plastic film ply to resist the ingress of water, for example cement sacks for use on open air building sites. Modern developments in sack Kraft papers have led to some sacks being produced from one ply of heavyweight Kraft (circa 120 gsm).

The sack manufacturing procedure is straightforward. The plies of paper, fed from separate reels, are passed through a tubing machine where each one in turn is glued with a water based emulsion adhesive to produce a multi-wall tube in which each ply is free to move independently. This freedom of movement of the plies allows the sack to remain flexible and absorb the bursting forces which would otherwise rupture a more rigid construction. The tube is cut to size and the bottom formed

Table 10.3 Tests carried out for surface appearance

Test	Explanation	Test method
Surface strength	Determination of resistance to picking – for coating and printing it is important for the surface of the substrate to be stable	ISO 3783-2006
Surface tension	This method is used to determine how easily a surface wets out. A good surface energy is required to ensure adhesives and inks do not reticulate on application	ISO 8296
Whiteness	Determination of CIE (International Commission on Illumination) whiteness. A measure of how white a surface is – whiteness affects the final brightness of an ink printed upon it. Not suitable for fluorescent-treated paper and paperboards	ISO 11476-2000
Brightness	Measures the reflectance from fluorescent-treated paper and paperboards – important for products sold under fluorescent light source – for example in supermarkets	ISO 2470.2-2008
Opacity	Measures the amount of light which passes through a paper or paperboard by defuse reflectance	ISO 2471-2008
Surface roughness	Bendtsen method of measuring how smooth the surface of a fibre substrate is by recording the rate at which air leaks between the test piece and the substrate surface	ISO 8791.3-2005
Porosity	The Bendtsen method measures the amount of air which will pass through a substrate in a given time. This is important as porosity determines how much coatings will penetrate and how easily a vacuum sucker will hold the substrate in place	ISO 5636
Gloss	Measurement of reflectance from a standard beam of light shone at an angle of 20° to the substrate	ISO 8254.3-2004
Rub resistance	A measurement of how resistant a printed or coloured surface is to abrasion from a predefined surface, e.g. paper	ISO 105-X12-2001
Surface pH	pH is the measurement of hydrogen ion concentration in water. It is scaled from 1 to 14. 1 is high acid, 14 high alkaline and 7 is neutral. The pH of the surface of a paper or paperboard substrate affects its performance	Tappi T529 om 09
Ink absorption	A measurement of the ability of a coated surface to absorb ink	Tappi T553

and sealed. Sealing can be by stitching, pasting with adhesive or heat sealing, where hot melt, polymer film or other heat seal coating is incorporated onto the sealing surface. There is a wide range of sack designs starting from simple open mouth to block bottom, gusseted valve sacks.

Open mouth sacks can be supplied as sewn flat (a); sewn gusseted (b); pasted

Table 10.4 Tests carried out to control performance properties of paper and paperboard

Test	Explanation	Test method
Thickness	Thickness is important for evaluating the density of the substrate and for printing. Printing requires an even thickness of substrate to ensure an even depth of print	ISO 534-2005
Basis weight	This is a method for accurately determining the number of grams in a square metre of paper or paperboard	ISO 536-1995
Water absorption (Cobb test)	Water penetration (absorption) is critical for many applications. The usual way of measuring this criteria is the Cobb method. This requires a given amount of water at a given temperature to be placed on a known area of substrate for a given period of time. The amount of water absorbed by the substrate over a given time is recorded. This time can vary from 1 minute to 30 minutes, depending on the expected absorbency of the substrate	ISO 535-1991
Moisture content	Paper and paperboard contain moisture. Control of the moisture content across the web is critical as is even drying, if a stable paper or paperboard is to be achieved. Moisture content can be tested in many ways, a quick method where the moisture is driven out of the substrate using a hot plate or iron or a more controlled, but time-consuming method where the test specimen is placed in an oven, set at 105°C until constant weight is achieved. The quick method is used on machine and the more accurate method for quality assurance	ISO 287-1985
Bending resistance/ stiffness	Stiffness and resistant to bending must be measured in both the cross and machine directions. The stiffness of a paper or paperboard can be used to predict the compression strength of a package once the width, depth and height are known, and the ability to maintain the shape of the final package	ISO 2493-1992 ISO 5628
Short span compression strength	This is one criterion which has been developed over the years to provide better guidance for final compression characteristics than stiffness	ISO 9895-2008
Tensile strength (dry)	Paper substrates in particular need a tensile strength high enough to ensure they will run through the converting process without breaking. It is not just the breaking point that is important but the elongation at break (how much it has stretched). This is especially important for 'sack Kraft'. Tensile and burst strength are among the most important quality control properties for paper manufacture. Tests are carried out in the wet condition as well as dry, especially for wet strength papers and sack Kraft	ISO 1994.2-2008
Burst strength	This is a very useful test when determining how paper sacks and bags will resist bursting open if dropped or in normal handling (cement sack for example)	ISO 2758-2003
TEA (Tensile energy absorption)	This is a test used on sack Kraft. High performance sack Kraft paper must be strong, with high tensile energy absorption (TEA). TEA can be defined as the area under the tensile–elongation curve and is therefore a combination of total tensile strength and stretch of the pulp	ISO 1294
Elmendorf tear resistance	Tear resistance helps to evaluate how the substrate will perform in use. Controlled tear resistance in both MD and CD is important to ensure that packages open in use but do not tear unnecessarily when being converted	ISO 1974-1990

Table 10.4 Continued

Test	Explanation	Test method
Interply bond strength (z direction tensile strength)	Interply bond strength is important. During manufacture the plies of fibre interact with each other and bond (hydrogen bonding). The strength of this bond needs to be greater than the rupturing forces applied during processing. Paper and paperboard made by the vat method is generally weaker in the z direction than that made by the wire process. If the interply bond strength is too weak, multi-ply paper and paperboard will delaminate during the conversion process	ISO 15754-2009
Coefficient of friction	Although this test was introduced for plastics, it is very useful to determine the resistance to slip of a paper or paperboard product	ISO 8295-1995
Taint and odour Robinson sensory test	This test has been incorporated into the EN legislative protocol to assess whether paper and paperboard products are fit to be used for direct food contact. It is a sensory subjective test, but extremely important for high fat and bland foodstuffs where taint and odour can be transmitted from the paper and paperboard unless controlled. Tainting and odorous chemicals contained in paper and paperboard substrates can be identified by using chromatography	EN1230.2-2001

flat (c); pinched closed flat (d); pinch closed gusseted (e); pasted double folded flat (f); pasted double folded, gusseted (g) (see Fig. 10.34). Plastic film layers can be included and, if this plastic film is the innermost ply, the sack can be heat sealed. The non-gusseted versions are like a pillow in shape when filled and therefore require care when being palletised due to their instability. They are commonly used for the packing of animal feeds and powdered foods and ingredients.

The other main sack design is the valve sack (Fig. 10.35). Typical designs are sewn flat (a); sewn gusseted (b); pasted, flush cut, flat (c); pasted, stepped end flat (d); pasted and sewn, flat (e). During manufacture, one end of a valve sack is completely sealed while the other has a filling spout or valve built in. Valves can be internal (see c, d and e in Fig. 10.35) and can be constructed of plain paper, polymer coated paper, or a layer of polyethylene film. During the filling operation the valve is located on the filling nozzle and the product dispensed into the sack usually with vibration to speed up the process. Once filling is complete, the valve is closed.

Internal valves without any polymer layer rely solely on the weight of the product to close the pack at the valve end; while this is effective for some purposes, it does not prevent a degree of leakage and possible ingress of contaminants. Also, it is not tamper evident. A more effective closure is achieved by heat sealing, which can, of course, only be carried out when polymer-coated paper or polyethylene film have been used in the valve construction. Sacks with external valves are closed by folding the protruding valve section and tucking it inside the sack, again accompanied by heat sealing if the valve construction allows. The folded section can also be secured by applying a self-adhesive tape or label. Thus external valves provide easier and more secure closing, although they are slightly more difficult to load onto the filling nozzle, especially on high speed filling lines and they use more materials.

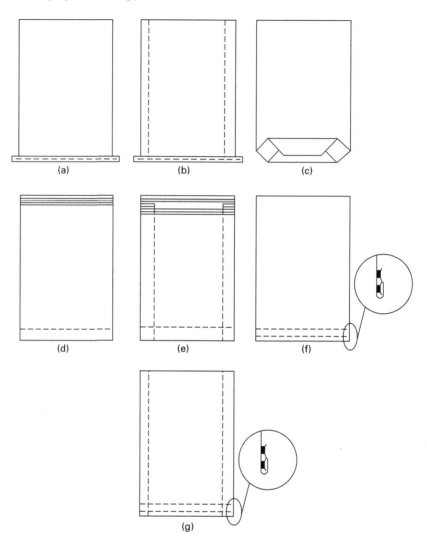

10.34 Types of open mouth sack design (source: Kirwan, M.J. (ed.), *Paper and Paperboard Packaging Technology*, Blackwell Publishing Limited, 2005. Reproduced with the permission of John Wiley and Sons Inc.; www.wiley.com).

Valve sacks are used for the packaging of granulated and powdered products such as sugar, plastic pellets and cement. Compared with open mouthed sacks, valve sacks provide a faster means of filling but must be vented to allow the displaced air to flow out as quickly as the product is entering the sack. Some sacks are perforated to allow for this, but recently sack Krafts have been developed with a porosity which allows the air to dissipate through the single-ply sack wall at an acceptable rate. Filled valve sacks are more regular in shape than open mouth sacks and thus are easier to palletise in stable loads.

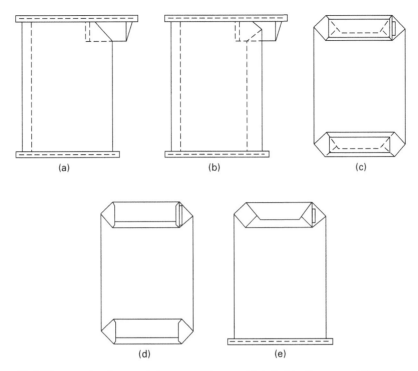

10.35 Types of valve sack (source: Kirwan, M.J. (ed.), *Paper and Paperboard Packaging Technology*, Blackwell Publishing Limited, 2005. Reproduced with the permission of John Wiley and Sons Inc.; www.wiley.com).

Table 10.5 shows the usual tests that would be carried out on the paper and paper sacks to ensure they meet the required specification. The different standards do not necessarily employ identical test methods and therefore care must be taken to ensure the properties of competitive materials are comparable.

10.7.2 Paper bags

Paper bags cover a wide range of designs and uses, from the grocery sack in the United States to the fine art bags used in the gift trade. The main types of paper bag are:

- flat and satchel
- strip window
- SOS (self opening satchel).

Flat bags are the most basic form. They are two-dimensional and confined almost entirely to point-of-sale use (Fig. 10.36). Satchel bags have gussets which allow the bag, once opened, to become three-dimensional making it much easier than flat bags to handle and fill. Like flat bags, their main use is at point of sale (Figs 10.37 and 10.38). Satchel bags can be supplied with strip windows in one side allowing the product to be seen. These bags were developed for the bread and baguette trade where the window film used is a breathable film, often micro-perforated polypropylene; bio-

Table 10.5 Specifications for multi-wall sacks

Property	ISO	Unit	EN	Scan-P	DIN	Tappi
Grammage	536	g/m²			53104	410
Thickness	534	µm	20534		53105	411
Density	534	g/cm³	20534		53105	411
Tensile strength	1924	kN/m		67:95	53112/1	494
Stretch	1924	%		67:95	53112/1	494
TEA (tensile energy absorption)	1924	J/m²		67:95	–	494
Tear strength	1974	mN	21974	11:96	53128	414
Bursting strength	2758	kPa		24:77	53113/141	403
Bending resistance (static bending force)	2493	mN		29:84	53121	543
Bending stiffness	5629	mN*m		64:90		535
Wet tensile strength	3781 (15 min)	kN/m		20:95	53112/2	456
Surface strength Denison						459
IGT, picking velocity	3782, 3783	mm/s, m/s				
Bendtsen porosity	5636/3	ml/min		60:87		
Roughness Bendtsen	8791/2	ml/min		21:67	53108	538
Roughness Bekk	474	ml/min			53107	479
Roughness PPS, H10 (Parker print surface)	8791	µm				
Roughness Sheffield	8791	ml/min				
Friction	15359	–			53375	815
Cobb 60s (water absorption)	535	g/m²	20535		53132	441
WVTR (water vapour transmission rate)	2528	g/(m²*24h)				
Air resistance Gurley	5636/5	s		19:78		460
Moisture	287	%	20287		53103	412
Ash	2144	%		5:63	54371	413
Opacity	2471	%			53146	519
Brightness	2470	%			53145	452
Lightness L	Cie lab 1964	%				425
Gloss		%				480
pH cold water extract				14:65	53124	435
Measurements of empty sacks			26591-1			
Valve position in paper sacks			26591-1			
Paper sack volume	8281/1					
Dimensional tolerances for paper sacks			28367-1			
Butt drop test for paper sacks	7965/1		27965-1			
Flat drop test for paper sacks	7965/1		27965-1			

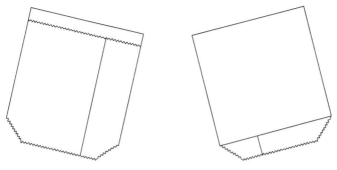

10.36 Flat bag design (source: Kirwan, M.J. (ed.), *Paper and Paperboard Packaging Technology*, Blackwell Publishing Limited, 2005. Reproduced with the permission of John Wiley and Sons Inc.; www.wiley.com).

10.37 Satchel (gusseted) bag design (source: Kirwan, M.J. (ed.), *Paper and Paperboard Packaging Technology*, Blackwell Publishing Limited, 2005. Reproduced with the permission of John Wiley and Sons Inc.; www.wiley.com).

compostable polylactic acid (PLA) film is also starting to be used. The breathability ensures that no moisture builds up on the film, thus the bread remains fresh and crisp. Open mouth potato sacks and bags for clothing also utilise the window concept.

The three stages of storage (a), opening (b) and sealing (c) of SOS bags are shown in Fig 10.39. These bags are often constructed from paper laminated to a plastic film, the film providing product protection as well as protecting the paper from deterioration due to the product, for example rotisserie bags for hot chicken. The laminate construction also allows the bags to be closed by heat sealing. SOS bags are used for pre-packaged dry goods and, when handles are applied, as carrier bags for point-of-sale use. The latter may be printed with high quality graphics, offering the brand owner or retail outlet good advertising opportunities.

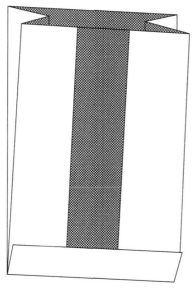

10.38 Satchel bag design with window (source: Kirwan, M.J. (ed.), *Paper and Paperboard Packaging Technology*, Blackwell Publishing Limited, 2005. Reproduced with the permission of John Wiley and Sons Inc.; www.wiley.com).

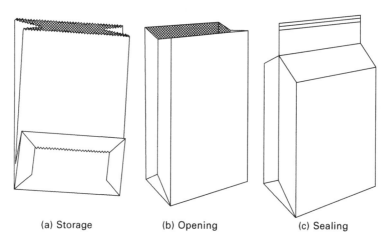

(a) Storage　　(b) Opening　　(c) Sealing

10.39 Storage (a), opening (b) and sealing (c) of self-opening satchel bags (source: Kirwan, M.J. (ed.), *Paper and Paperboard Packaging Technology*, Blackwell Publishing Limited, 2005. Reproduced with the permission of John Wiley and Sons Inc.; www.wiley.com).

10.7.3 Spiral wound containers

Spiral wound composite containers have been in use in packaging for many decades (Fig. 10.40). They consist of three to four plies of paper and paper laminates wound together. The two body plies are composed of recycled Kraft paper, the outer ply is usually a printed paper or paper laminate and the inner ply can be any construction

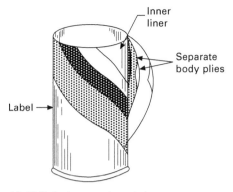

10.40 Spiral wound container.

from plastic coated 40 gsm paper to an aluminium/plastic/paper laminate depending on the end use. If aluminium foil is used in the construction of the inner ply, and properly sealed, then paperboard containers can provide sufficient preservation and protection properties to compete in markets traditionally supplied by metal cans.

The plies are wound around a mandrel, each ply being stepped away from the other to ensure the seams do not lie on top of each other. To prevent a ridge forming when the body plies are overlapped, each is individually skived prior to being overlapped and adhered. Skiving is a process by which the edges of the paper are gradually reduced to minimal thickness, so that when they are overlapped the thickness of the overlap corresponds to the thickness of one ply of paper and as a result no ridge is apparent (Fig. 10.41).

The plies are glued together using emulsion adhesives and the inner ply is skived (if too thick to fold) and hemmed to ensure as near hermetic seal as possible is achieved. Skiving in this instance is a process by which the backing paper is gradually taken away from the edges of the underside of the aluminium/plastic layer, to produce a thinner layer, which is then folded through 180° and heat sealed to itself usually by induction sealing. If the inner liner is thin enough, it can be folded over and seamed without needing to skive its backing paper. The seal so formed is called an 'anaconda' seal. If the seal is left as an overlap seal, moisture and gas will penetrate into the inside of the can and attack its contents.

The body of the container can be sealed to the ends in various ways, just four of these being as follows:

- If a hermetic or near hermetic seal is required, metal end pieces can be seamed onto the body in the same way as on metal cans.
- Paper laminated to plastic and/or aluminium foil can be formed into end pieces and heat sealed onto one end of the body section of the can replacing the traditional metal end. A plastic plug or plastic or paperboard over-cap, with or without a heat sealed diaphragm can be applied to the other end of the can, as used on containers for gravy granules.
- A depressed diaphragm can be sealed to the end of the can body and a plastic plug cap can be inserted into the depression, as used on cans of dry fish food pellets.

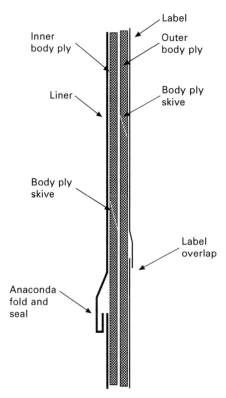

10.41 Skiving, hemming and 'anaconda' fold.

- The can body end can be pre-curled outwards to accept a heat sealable diaphragm and a plastic overcap applied, locating on the outer curl, as used on Pringles snacks.

10.7.4 Linear forming

Spiral wound containers are normally cylindrical in cross section and deviating from this is difficult and usually requires a second operation, thus increasing cost. Linear forming offers more options because, instead of being wound helically around a mandrel with a circular cross section, they are produced by introducing plies along the axis of a shaped mandrel, thus allowing the final body to replicate the shape of the mandrel. Their construction is similar to spiral wound containers in that they have three to four plies and consist of an inner liner, body plies and an outer label. The closures available are also similar to spiral wound containers.

10.7.5 Convolute single wrap containers

Traditionally, convolute can bodies were made by winding a number of plies around a mandrel, to produce a heavy wall container. This has mostly been replaced by a more efficient method in which a single ply of coated or laminated printed paper is wrapped around a conical or parallel shaped mandrel and the body is heat sealed on the overlap. The single ply of paper is made from chemical pulp and has a heat seal layer on at least the underside. Various alternatives are available. The most common are

- paper/plastic
- plastic/paper/plastic
- plastic/paper/aluminium/plastic.

The rim is curled and a disc is placed in the bottom and secured in place by wrapping the base of the body around it and applying pressure. The container is completed by the addition of a push on closure with or without a heat sealed diaphragm underneath, or simply sealed with a printed diaphragm (Fig. 10.42). In use, many of these containers are packed with dairy products such as butter, yoghurt and ice cream as well as cereals, baking ingredients, snacks and biscuits. They are beneficial to the packer filler as they can be made in-house, or supplied as a stack, one inside the other. They are not as robust as the spiral or linear formed containers and therefore often require additional secondary packaging for protection against the hazards of distribution. The body seal is occasionally skived and hemmed to ensure no leakage of product into the body (wicking).

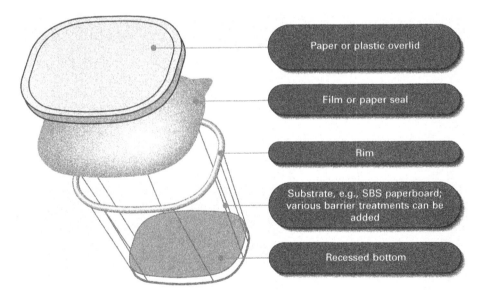

10.42 Convolute can design (source: Paper Machinery Corporation (PMC); www.papermc.com).

10.7.6 Liquid packaging cartons

The type of paper or paperboard (210–250 gsm) used for these packs is the same stock as for single-layer, pre-printed, convolute wound tubes. Multilayer virgin bleached sulphite or CTMP fibre is used due to its excellent strength and/or whiteness. The polymer layer on the outside prevents condensation absorbing into the paper and the polymer on the inside protects the packaging from the product and vice versa (10.43(a)). The whole of the plastic is applied at a coat weight as high as 40 gsm to ensure an integral heat seal. For liquids prone to oxidation, such as long life milk, aluminium foil or EVOH (ethylene vinyl alcohol) are incorporated in the laminate (10.43(b)).

Liquid packaging cartons come in two basic shapes: brick and with a gable top (Fig. 10.44). Openability has been strongly criticised by consumers and features such as plastic pour spouts and more convenient shapes have been added to provide convenience in use (Fig. 10.45). The brick shaped containers are produced by vertical

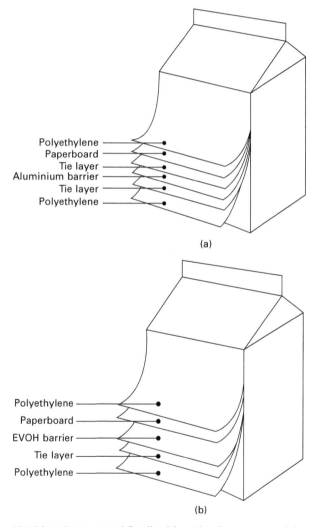

10.43 Laminates used for liquid packaging cartons: (a) standard and (b) incorporating EVOH.

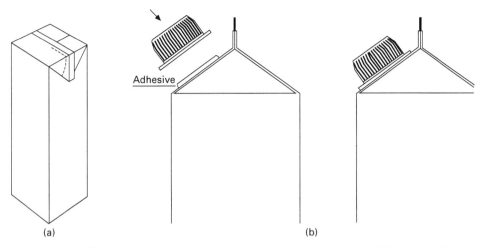

10.44 Standard brick design (a) and standard gable design (b) incorporating screw top.

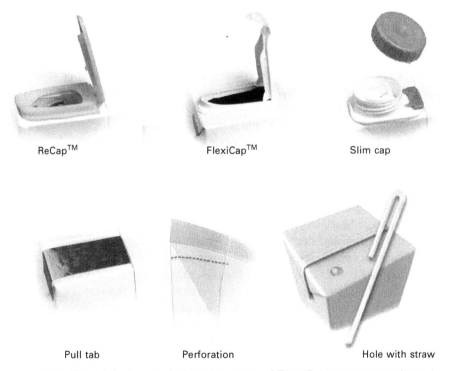

10.45 Tetra Pak closure designs (courtesy of Tetra Pak; www.tetrapak.com).

form seal technology (see Chapter 20) from a reel of printed laminated paper stock. The gable top containers are printed, cut, creased and heat sealed along the glue flap, similar to a folding carton (see later in the chapter).

10.8 Conversion processes for paperboard (>250 gsm)

The packaging formats made from paperboard discussed in this section are:

- folding cartons, multipacks and trays
- rigid boxes
- pressed paperboard trays
- blister cards and Euroslot multipacks.

10.8.1 Folding cartons, multipacks and trays

Cartons can be defined generally as small to medium sized containers made from paperboard or in some instances paper (< 250 gsm) or plastic (toothpaste and some cosmetics cartons, for example), although there is confusion in the packaging industry about the term 'carton'. For example, the so-called egg carton is not a carton at all but a container made from moulded pulp specifically to hold eggs. This section covers folding cartons made from paperboard only, and not 'cartons' made from plastic or corrugated material. Cartons are used for their protective and aesthetic properties, providing a very cost effective means of packing products in a sustainable, recyclable material providing excellent graphics and presentation on shelf. They are, however, restricted in their preservation properties, as they possess poor gas and moisture vapour barrier properties, due mainly to the materials used and the integrity of the seal. Grades of paperboard are selected based on the product that is going to be packed, the machinery requirements to pack the product into the carton, the demands of the supply chain, including the retailer and the consumer, and last but by no means least, the cost and environmental considerations.

The top and reverse side of paperboards may be coated with mineral or artificial white pigments as described in paper and paperboard manufacture. All paperboard grades can also be treated with fluorocarbons to give grease resistance (though this process is declining due to fears of taint) or coated/laminated with wax (many countries have eliminated or greatly reduced the use of wax as it hinders recycling), plastic films, water-based barrier coatings and aluminium foil to provide gas, moisture, grease, water barrier and heat sealabilty, depending on the combination used. The quality of barrier of the finished carton is dependent more on the seal integrity of the carton than on the barrier properties of the materials used. Liquid packaging cartons are either skived and hemmed on the inside vertical seal or the seal is overlain with a plastic membrane to prevent moisture seeping or gas penetrating into the paper (in the same way as for spiral and convolute containers).

Carton styles

European and US folding carton styles are classified by the European Carton Manufacturers Association (ECMA) and its US equivalent ACMA. The basic guidelines are as follows:

- The glue flap should never be showing on the front face of a carton. This is an aesthetic requirement as the edge of the board would be exposed when the carton

is displayed and this bare edge would detract from the quality of the graphic design (Fig. 10.46).
- The glue flap should not be incorporated with the working creases. When cartons are erected from the flat it is important that the carton opens squarely; if the glue flap is incorporated in the working crease the carton will have a higher resistance to opening (Fig. 10.47).
- Glued cartons should be packed in the transit pack (usually a tray, sleeve or case, shrink wrapped or not) on end with the glue flap at the base. This ensures minimal risk of setting of the creases, which would make the cartons difficult to open as the stiffness of the board could be less than the force required to open the creases, resulting in the carton bending rather than opening. This would result in lost time and wastage on the packing line.
- Unglued cartons should be supplied packed flat, directly on the pallet with layer pads between each layer.
- The grain direction of the fibres in the board should always be at a 90° angle to the major creases. This is very important for consistent crease performance and minimal bowing of the carton. It is important to specify grain direction,

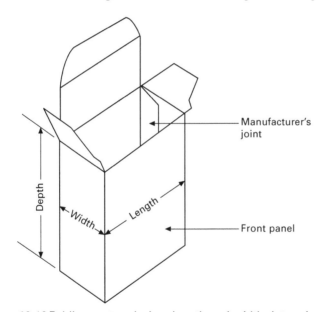

10.46 Folding carton design: length and width determine the location of the opening.

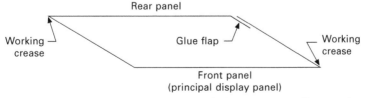

10.47 Folding carton design: relationship between glue flaps and working creases.

especially when using two different suppliers because, depending on the size of printing press used, it may be possible to get an extra carton from a given sheet size by reversing the print in relation to the grain. Although seen initially as a cost saving, this could mean increased costs at the packer/filler and distribution stages (Fig. 10.48).
- Carton dimensions should, follow a common industry pattern (see Fig 10.46).
  - they should be given in the order length, width and depth (these are often termed width, depth, height)
  - length and width should define the carton opening
  - depth should be the perpendicular distance between the openings, which is important as the distance between the top and bottom creases on the front panel of the carton are longer than those between the top and bottom of the side panel, by the approximate thickness of the paperboard. This is important to prevent resistance of the outer flaps to folding.
  - measurement should be from centre of crease to centre of crease.

However, different markets use different nomenclature, and places to measure; so it is always advisable to produce a drawing with the actual dimensions marked on it to prevent any confusion (Fig. 10.49).

Some common carton styles are:

- end load cartons
- top load cartons

End load cartons are designed to be filled horizontally, e.g. bag-in-box cereal carton, or vertically, e.g. direct fill oats carton (Figs 10.50 and 10.51). They consist of four panels, front, rear, left side and right side panels, glued at the side seam. This is a small flap attached either to the rear or side panel and known as the glue flap. The top and bottom flaps can be glued with hot melt or water-based adhesive, or a tuck flap can be incorporated in the design for mechanically closing and opening (see Fig. 10.52). The left diagram shows a reverse tuck carton and the right an aeroplane tuck. The aeroplane tuck style uses more paperboard than the reverse tuck style, but some consider it aesthetically more pleasing (Fig. 10.53). A cut is often made at the

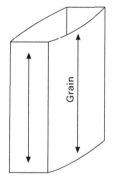 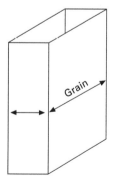

10.48 Folding carton design: carton grain direction should run at a 90° angle to the major creases to ensure maximum carton stiffness.

Paper and paperboard packaging 225

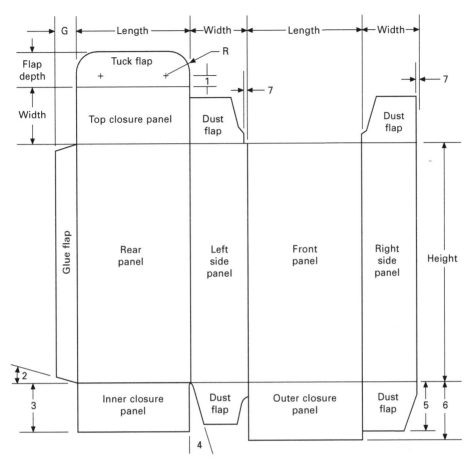

10.49 Folding carton design: diagram showing dimensions (in this case a tube-style carton with a friction tuck top and a full overlap bottom closure).

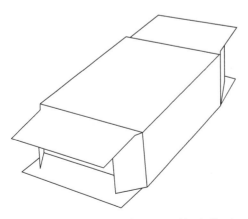

10.50 End load carton (source: Alexir Packaging; www.alexir.co.uk).

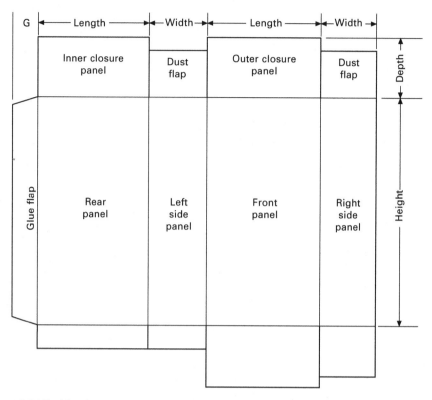

10.51 End load carton: diagram showing basic design.

ends of the flap crease which creates a mechanical lock with the minor flaps helping to prevent the carton opening in transit (Fig. 10.54).

The top load carton is supplied to the packer/filler as a flat blank. It is formed through a die and the side panels are mechanically locked or glued, normally by hot melt adhesive; the product is then filled through the large top aperture and the lid is closed. The lid design can either be a tuck flap as in the diagram, or the flap can be glued to the carton. There is a variety of mechanically locked and glued cartons and trays made in this way (Fig. 10.55). This type of tray (locked or glued) is used to collate cartons prior to shrink wrapping for delivery to the customer. The lock tab version allows for hand assembly on the production line (Fig. 10.56).

The tapered style means that they can be stacked, one inside the other (with the lid up), saving space and ensuring they are ready and open to enable speedy packing of goods at the counter (Fig. 10.57). Plastic coated paperboard can be formed into a heat-sealed web-cornered tray with horizontal flanges which will heat seal to a plastic or plastic coated paperboard lid. This style of tray can be filled with product, the lid sealed on and the whole pack frozen, ready for distribution.

Another style of top load carton, which can be supplied ready glued and folded is shown in Fig. 10.58. A special four or six point gluing procedure with extra diagonal creases allows for these cartons to be laid flat when supplied, negating the need for

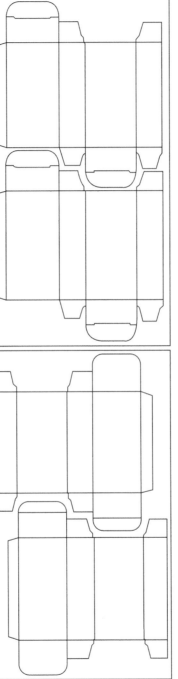

10.52 End load carton: reverse tuck (left) and aeroplane tuck (right).

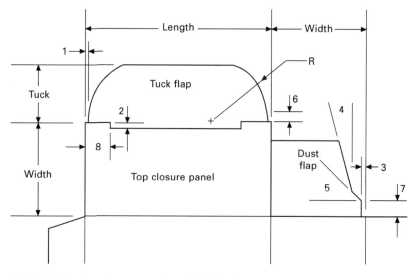

10.53 End load carton: aeroplane tuck design.

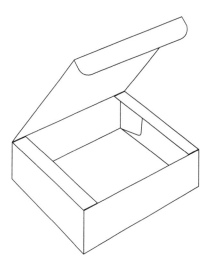

10.54 End load carton: locking mechanism (source: Alexir Packaging; www.alexir.co.uk).

any machine erection and gluing at the customer (e.g. the cake shop). This style of carton often includes a clear window patch.

Cartons are also used to make multi-packs (Fig. 10.59). The board is either a special Kraft board with high water resistance and good wet tear strength or a specially treated recycled board to protect it from moisture and water penetration. These boards are commonly referred to as carrier boards.

Many other designs and styles are available but too numerous to discuss in this chapter. See Section 10.9 for additional sources of information.

Paper and paperboard packaging 229

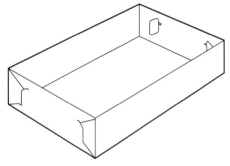

10.55 Top load carton with locked corners (source: Alexir Packaging; www.alexir.co.uk).

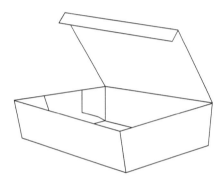

10.56 Top load carton: lock tab design (source: Alexir Packaging; www.alexir.co.uk).

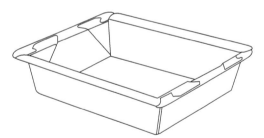

10.57 Top load carton: heat-sealed, web-cornered, tapered tray design (source: Alexir Packaging; www.alexir.co.uk).

Carton making

The carton-making process is varied, but there are some common steps; these are:

- printing
- cutting and creasing
- window patching
- gluing.

Prior to printing, the paperboard as received is conditioned, either in the warehouse or by the side of the printing press, for 72 h. This is to ensure the material is consistent,

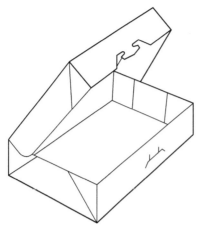

10.58 Top load carton: six-corner, glued, folded design (source: Alexir Packaging; www.alexir.co.uk).

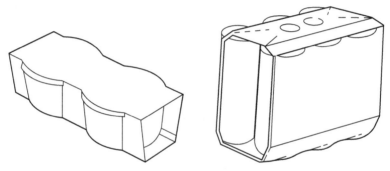

10.59 Top load carton: multi-pack for dairy packs (left) and carbonated drinks (right) (source: Alexir Packaging; www.alexir.co.uk).

especially with respect to moisture content. If it is not, this will affect the print quality. Other areas which may affect print quality relate to the reeled or sheeted board. Checks must be made for cleanliness, especially on the edges, as slitting and cutting dust can transfer to the surface of the substrate creating imperfections on the printed surface. (See Chapter 18 for more detail.)

As the paperboard passes through all the printing presses, it is bent by the tension and feed rolls. This will break some of the fibres, reducing the stiffness and strength of the board and therefore the resulting carton. Mechanical fibre is more susceptible to this than chemical fibre as it is shorter in length and more brittle.

The main three print methods are: offset lithography, which is normally sheet fed, and gravure and flexography, which are normally reel fed. The relationship of the direction of print to fibre orientation within the sheet is very important as discussed earlier. The tolerances and print panels need to be matched to those of the cut and crease die (forme). Failure to do so will result in the carton not being cut and creased in line with the printed design resulting in misregister of print to structural design.

Cutting and creasing is carried out using a flat die for sheet-fed materials or a rotary die for reel-fed materials (Fig. 10.60). The cutting and creasing operation (or box cutting as it is known) is as important as the print operation. If the cartons are not cut and creased correctly, their performance will be impaired during the following stages of conversion, filling and distribution. Cartons are cut and creased using a forme. To enable the cutting and creasing operation to be controlled, a counter plate is placed on the base of the press, exactly in line with the forme.

The forme is made from plywood, with steel cutting knives and creasing rules inserted into grooves which are commonly cut using a laser for high accuracy. Special foam rubber pads are placed at either side of the cutting knives to act as springs which remove the board from the blade after the cutting operation. The cutting knife blades are not continuous but designed in such a way that when the cutting operation is complete, they leave nicks between the individual carton cut-outs. This enables the individual cartons to be held together as if they were one sheet when the waste is stripped away. The nicks are broken at the next stage (Fig. 10.61).

The nick is made by designing notches into the cutting knives. These vary depending on paperboard type and thickness; the stronger the board the narrower the notch (Fig. 10.62). In general, the depth of the notch is made slightly greater than the thickness of the board. The strength requirement of the nick depends on many factors. These include:

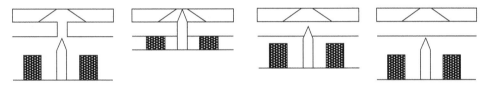

10.60 Cutting and creasing forme with make-ready counter (courtesy of Iggesund Paperboard; www.iggesund.com).

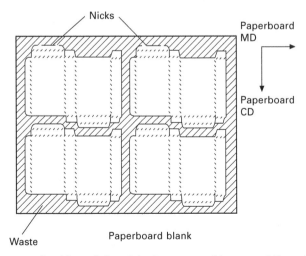

10.61 Position of the nicks (courtesy of Iggesund Paperboard; www.iggesund.com).

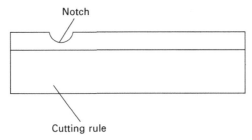

10.62 Die-cutting rule with a notch (courtesy of Iggesund Paperboard; www.iggesund.com).

- grain direction of the paperboard – MD is stronger than CD
- method of making the notches – quality of notch must maintain the integrity of the carton within the sheet without breaking, but must be as small as possible so as not to be seen on the final carton
- strength of the foam rubber to release the board from the knife – the incorrect material could put unnecessary strain on the nick causing it to break prematurely
- dimensions of the nick – normally 0.4–1.0 mm
- how the nicks are arranged is important because if they are not placed evenly across the edge of the carton, undue pressure will be placed on the cut blank resulting in premature breaking of the nicks
- number of nicks per cut edge – more nicks, more strength
- moisture content of the paperboard – as the moisture content increases the strength of the nicks decreases
- quality of fibre used to make the board – the weaker or more inconsistent the fibre, the wider the nick needs to be.

Knife edges blunt over time and need replacing. The speed at which they do so varies with the type of board being cut. Plastic-coated boards often require a specially designed cutting knife. It is uneconomical to use a forme where the knives are blunt or incorrect for the paperboard being cut, as this will result in poor performance on the carton erecting machine and in use.

The other important operation performed by the forme is creasing (Fig. 10.63). Well-formed creases are essential for correctly formed cartons. The quality of crease (and therefore the efficiency of making up cartons on the packaging line) depends on:

- the height and width of the creasing rule with respect to the paperboard being creased – the creasing rule is designed for a narrow range of paperboard thicknesses, and if the length of the rule is too long it will damage the crease area and if too short it will produce an imperfect crease
- the thickness of the make-ready (see below) – if the make-ready is too thin the crease has no room to form properly
- the width of the make-ready groove in relationship to the thickness of the creasing rule – if the make-ready groove is too narrow the crease is likely to form too tightly; if too wide the crease forming operation is not tightly controlled

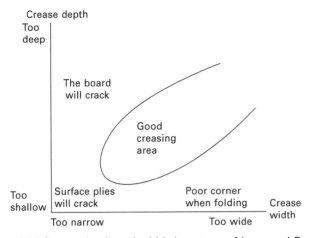

10.63 Crease depth and width (courtesy of Iggesund Paperboard; www.iggesund.com).

- the accuracy and hardness of the make-ready is critical otherwise the crease will be poorly formed, either through distortion of the make-ready or inaccuracy of the crease
- the pressure of the die cutter – if the pressure is too great, undue forces will be put on the paperboard causing it to split; if the pressure is too weak insufficient pressure is available to form a perfect crease.

The 'make-ready' mentioned above refers to the underside of the creasing platen situated beneath the substrate. To ensure an even consistent crease profile, the groove formed as the female form of the creasing rule must have the correct depth and profile and be contained on an even non-deforming metal bed (counter platen). Make-ready matrixes come in three forms:

- self-adhesive metal strips which can be stuck down in place on the counter platen
- plastic (polyester) channel (pre-made) of fixed width and depth
- phenolic resin-impregnated paper/fabric which is pre-etched to match the creasing forme.

The first option is often used for short runs, but relies heavily on the machine operator for accurate placement. The third option is used for complicated designs and long-run work, while the middle option is used for either long or short runs.

Independent of which make-ready is used, they both require the operator to ensure the surface is absolutely even. This is achieved by using specially calibrated self-adhesive tape placed under the low point of the counter to ensure complete overall flatness. This is important because when the forme comes into contact with the substrate and pushes the paperboard into the female form of the counter to make the crease, uneven pressure will be applied if all is not completely flat. This will result in inconsistent creasing which may not show up until late in the converting or packing operation. To form an efficient crease it is necessary to use a multi-ply paperboard.

Once the make-ready and forme are in place, the cutting and creasing operation can take place. The forme is brought down in a rocking motion onto the sheet of paperboard (print side upwards). The creasing rules push the board evenly into the make-ready matrix and a crease is formed. The waste material from the cutting and creasing operation is removed at the next stage, often using pre-set rods to push it out. The waste is then sent to the baler to be compressed ready for recycling. The remaining cartons, held together with nicks, are palletised awaiting the next process.

Where the printed paperboard is in reel form, rather than flat sheets, it is die cut using two sets of profiled metal cylinders; one set for cutting and one for creasing. The cutting cylinders consist of one roll with metal cut away to allow the knife profile to be developed and a second, plain roll to allow for kiss contact of the knives with its surface. Poor setting up of the two rolls will result in the knives blunting, producing inferior cut cartons. The creasing rolls are more complicated as the backup roll mirrors the other profiled in design but the crease area is cut away rather than raised. Due to the high cost of this method of rotary die cutting, simpler and less expensive methods have been developed.

At the cutting and creasing stage, opening features, embossing and cut outs can be included. There are many styles of opening feature, with the two most common being the 'zipper' and the 'concora'. The zipper is produced by cutting a series of tram line perforations through the board, which allow the carton to be zipped open at the consumer's convenience. The concora method is more sophisticated and does not pierce the paperboard, but makes a stepped parallel double half cut (60%) from either side of the paperboard (Fig. 10.64). The outer two cuts are wider than the inner two. When the carton is opened the plies part where cut, resulting in a clean tear. It can be used to form a pouring spout, for example, which can be reclosed using the remaining half cut area to act as a seal for the outer. Concora can only be used on multi-ply paperboard.

Once the cutting and creasing operations have been completed, the cartons can, if required, be window patched. Adhesive is applied to an area around the window, far enough away from the edge so that the adhesive will not spread into the window but close enough to produce a secure seal. The clear window material is then put in place and pressure is applied to the glued area until adhesion is achieved. Water-based emulsions are the preferred choice of adhesives.

Most end load cartons are pre-glued prior to delivering to the packer/filler (Fig. 10.65). This is a high-speed operation, briefly described as follows:

- Flat cartons are accurately fed into the gluer.
- Working creases are pre-broken by folding through 160°.
- For difficult substrates, the surface to be glued is broken to allow for better adhesion.
- Glue is applied to one of the two surfaces to be glued – often coloured pink or blue so it can be seen to have been applied.

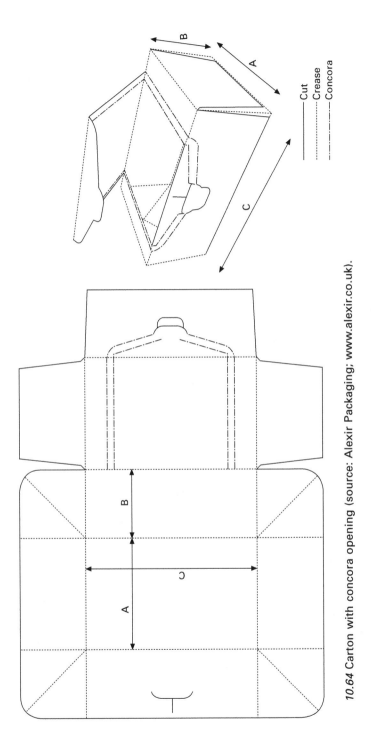

10.64 Carton with concora opening (source: Alexir Packaging; www.alexir.co.uk).

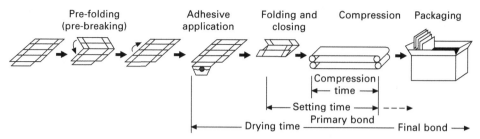

10.65 Glueing operation (courtesy of Iggesund Paperboard; www.iggesund.com).

- The other surface is brought into contact with the glue.
- Pressure is applied and the bond held until the glue is set.

Once set, the cartons are inspected, tested, packed on their glued creases, labelled and put in storage awaiting despatch to the customer.

The normal quality checks carried out on a carton would be:

- graphics and text to agreed specification – see Chapter 18
- coefficient of friction – see Section 10.6.3
- dimensions – dimensional accuracy is very important as without it the graphic and dimensional designs cannot be matched, nor will the final package be dimensionally controlled; dimensions are usually measured with a calibrated steel rule from centre crease to centre crease
- stiffness (MD and CD) – see Section 10.6.3
- moisture content – see Section 10.6.3
- crease bend resistance – see Section 10.6.3
- fibre tear on glue flap – many paper and paperboard packages are glued along the body seam; if this area is not fully adhered the product can force the glued area to fail and spill out of the package
- carton weight – a large variation in the weight of the carton can result in excess product having to be packed into it to enable it to comply with legislation
- carton compression strength – a measure of the force required to permanently deform the package at a given rate of application
- taint and odour – see Section 10.6.3
- direct food contact – all food packaging must comply with current food contact legislation.

The test methods vary depending on country. The tests carried out on the carton are dependent on its end use and not all tests are relevant to every application.

10.8.2 Rigid boxes, carded blister and skin packs

Two other uses of paperboard in packaging are also worth mentioning. They are:

- rigid boxes
- backing cards for blister and skin packs.

Traditionally rigid boxes were the common way of making cartons. Even as late as the 1950s, it was common to find chocolates, hats, shoes and some foodstuffs packed in rigid boxes. Today this style of pack is mainly confined to:

- jewellery and watches
- gifts
- high value perfumes and cosmetics.

Rigid boxes come in many shapes and sizes and are relatively simple, although labour intensive to construct (Fig. 10.66). Blank shapes for the base and lid are cut out of thick, heavyweight, usually recycled board and creased using a scoring wheel. The flaps are then folded along the creases at right angles so they come together to make a tray. The corners are held together (stayed) using wet glued paper tape. A printed cover sheet, usually bleached chemical pulp paper is then wrapped around and glued to the stayed base and lid trays. Other materials can be used, such as leather, fabric, plastic and aluminium foil. Additional sheets can be added to a tray to create a hinged lid. Once dry, the two sections can be inspected, checked for fit and other properties, packed and labelled ready for dispatch to the customer.

Carded blister and skin packs provide a very cost effective and convenient way of displaying small individual components or mini multi-packs and, although most of the traditional paperboard/plastic blisters or skin packs have disappeared, there are still many examples to be seen on retail display. A blister is a rigid, clear

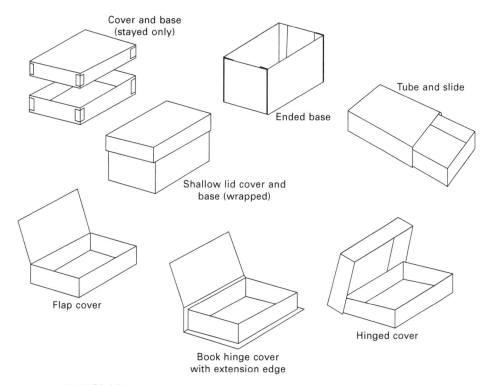

10.66 Rigid box designs.

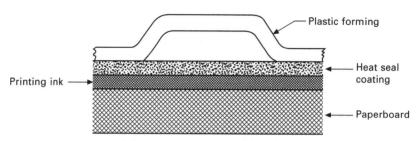

10.67 Blister pack construction.

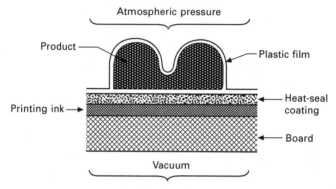

10.68 Carded skin pack construction.

pre-formed thermoformed shape made from a plastic sheet (Fig. 10.67). This blister is often the same shape as the individual packed product, but less defined and larger, to allow for easy placement. The thermoform has a flange around its periphery to enable heat to be applied through the plastic to the low melting point heat seal adhesive coated onto the printed backing card. Environmental considerations have not removed this style of blister as an option just yet; but there is a very powerful lobby driving companies to look for other more environmentally responsible alternatives. The two alternatives that are common in the marketplace are:

- all paperboard pack with special tamper evident features if required
- all plastic pack, with an inserted printed paper for aesthetics and information.

Carded skin packing is similar to blister packing, but in this instance the special grade of transparent plastic skin film is vacuum formed over the product (Fig. 10.68). The paperboard used for this application is printed, coated with a heat seal lacquer and then micro-perforated. This allows for the vacuum to be drawn at the rate required.

10.8.3 Moulded pulp trays and boxes

Moulded pulp items are used to stabilise products as well as to protect them from physical damage. Typical packaging formats in moulded pulp are:

- egg boxes
- punnets for soft fruits

- trays for raw meats
- protective sleeves for glass drinks bottles
- corner protection pieces
- fitments for holding small and large components.

The pulp can be made from recycled or virgin fibres, prepared in the same way as for conventional paperboard. Wet strength performance can be improved by the addition of wax, rosin and polymer resins.

There are two basic ways of producing the items, both involving the deposition of wet pulp on a pre-shaped mould, drying to remove the excess water, and removing the formed item from the mould. The pressure process uses hot air under pressure to remove approximately half of the water from the pulp. The alternative method uses vacuum to remove the water, but only up to around 20%. In both processes the remaining water is removed by using heat. The pressure process is usually semi-automatic and therefore the tooling cost is less than for the more automated vacuum method. It also lends itself better to short runs where a wide variety of shapes and sizes are required. Where long runs of standardised items are needed, the vacuum process is preferable.

The moulded pulp packaging can be bonded to plastic films which provides barrier protection, allowing them to be used for trays for fresh produce where the whole will be flow wrapped. Expanded polystyrene packaging is being replaced by moulded pulp, even for large items such as desk top printers and household electrical products.

10.9 Sources of further information and advice

For further information on carton styles and designs contact:

- ECMA – European Carton Manufacturers Association: www.ecma.org
- Pro Carton – www.procarton.com
- Iggesund – Paperboard reference manual (www.iggesund.com)
- Billerud – Sack Kraft and Sack manufacture (www.billerud.com)
- Korsnäs – Sack Kraft, Carrier Board and Kraft Paperboard for conversion into cartons (www.korsnas.com)
- Kirwan, M. J. (ed.), *Paper and Paperboard Packaging Technology*. Blackwell Publishing, Oxford, 2005.
- PITA – Paper Industry Technical Association (www.pita.co.uk)
- Soroka, W., Emblem, A. and Emblem, H. (eds), *Fundamentals of Packaging Technology*, IOP, Stamford, 1999.
- Sonoco Products (www.sonoco.com)
- Paper Machinery Corporation (www.papermc.com)
- Alexir Packaging (www.alexir.co.uk)
- MMP – Mayr Melnhof Packaging (www.mayr-melnhof.com)
- MMK – Mayr Melnhof Karton (www.mayr-melnhof.com)
- Tetra Pak (www.tetrapak.com)
- Elopak (www.elopak.com)
- Walki (www.walki.com)

11
Corrugated board packaging

T. WATKINS, UK

Abstract: Corrugated fibreboard is manufactured using papers suitable for flat surfaces (liners) and the corrugated inner structure (fluting medium). This chapter covers the papers and adhesives used in board manufacturing processes and the conversion into packaging and display items. It also covers the types of corrugated board produced, their testing and some basic designs.

Key words: liners, fluting and adhesives, corrugator machines, regular slotted and die cut containers together with their decoration and printing, testing methods.

11.1 Introduction

Corrugated fibreboard is the most widely used secondary packaging material. It is typically constructed from two facings or liners bonded to a corrugated (fluted) medium. A corrugating machine forms the medium into a fluted pattern and bonds it to the liners using adhesive. Some basic configurations are shown in Fig. 11.1:

- single-wall (one fluted medium with two liner plies): a standard type for most product types
- double-wall (two fluted medium plies and three liner plies) used for heavier or more bulky products such as machinery, large appliances or furniture, or for display stands, or for products which are stored for extended time periods
- triple-wall (three fluted medium plies and four liner plies) used for particularly

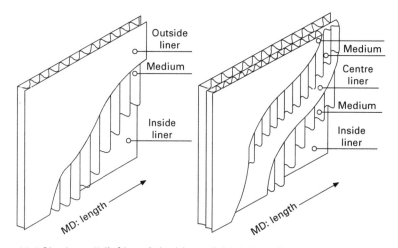

11.1 Single-wall (left) and double-wall (right) configurations.

heavy-duty applications, e.g. as a substitute for wooden containers for pallet boxes or bulk bins.

A single faced material is also available, made up of one fluted layer and one liner; this is a flexible material, suitable for general purpose wrapping.

Corrugated fibreboard has been manufactured since the late nineteenth century, with the first patents being filed in the United States in 1871. Much can be seen today that directly relates back to then, but the major shift in corrugator speeds from 1.8 m/min at 1.8 m. to that of 300 m/min at 2.5 m wide, is due to constant refinement and development. The first big step was the introduction of the Stein Hall process in 1937 which used part of the starch adhesive mix bonding the papers to be gelatinised, whilst the remainder was carried as slurry, enabling a much higher proportion of solids to be used (21–23% starch suspended in water). Drying time was reduced, allowing corrugators to run at higher speeds and produce board with minimal warping. Although this process has in the main changed back to a 'single bag' mix, the starch is modified to enable this carrier and slurry action to take place. It is this type of chemical development combined with automatic reel splicing, direct drive shearing knives, precise computer driven gap and temperature control that enabled this increase in manufacturing speed.

Much the same holds true of the conversion process, with a major milestone being the introduction of case making machines capable of printing, creasing, gluing and folding in a single pass. This equipment has in many cases been augmented with the introduction of rotary die-cutting sections, making it possible to perforate and incorporate locks into designs being manufactured with particular application for retail ready packaging (RRP). The majority of integrated plants, i.e. those with their own corrugators, are fully conveyorised and product is only palletised once ready for transport. Full traceability is maintained, from incoming paper stock to finished cases.

In the United Kingdom some 2.1 million tonnes of paper is used to make 4.1 billion square metres of board, which is converted into cases and trays. Market sector usage is around 17% for processed foods, 17% for fresh foods, 9% for beverages and 19% for industrial products, with the remainder being used for a wide range of durable and non-durable products. A significant influence on the industry in recent years is the move to retail ready packs (RRP), where the secondary packaging provides the retailer with a quick and easy method of shelf loading. Another influence is the widespread use of returnable packaging (mainly plastic crates), which has led to a loss of sales in the industry. Returnable packaging is ideal in closed loop distribution systems such as the transport of fresh produce and many other food products from packer-filler to supermarket, where the loss of relatively expensive crates is minimal. However, the growth of internet and mail order sales has the potential to offset some of this loss of sales, because this sector cannot realistically operate in a closed loop. This direct-to-consumer packaging is almost impossible to recover and reuse and hence has a one-way use. Corrugated board has environmental advantages here, given that in excess of 80% of materials are recycled and some 76% of packs are produced from recycled fibres in the UK.

The demand for RRP has come from the retail sector, with the desire to fill shelves at ever-increasing rates. This challenge has led to significant investment and development by the industry, with die-cutting units being introduced to single pass case-making equipment. These specialised packs can be easily transformed from transit cases into attractive and functional on-shelf display units at the retail store and are commercially attractive due to the improved efficiencies of this type of packaging.

11.2 Materials for corrugated board

There are three basic components required:

- liner or facing materials
- flute materials
- adhesives.

11.2.1 Liner/facing materials

The facings can be made from Kraft paper (named after the chemical pulping process used to manufacture the paper) which has a high proportion of virgin fibres and added recycled papers (known as 'Test' materials). The recycled papers are made from in-process waste and from used cases and trays collected from supermarkets and other outlets. These 'Test' materials owe their name to an obsolete puncture test demanded for railway distribution and are categorised as TL1, TL2 and TL3. The long fibres on the outer ply of TL1 and its highest performance amongst all Test materials mean that it is used as a Kraft substitute, whereas TL2 and TL3 are only distinguished by their short span compression strength (SCT) values and are the most widely used papers. TL3 is the most common with the higher performing TL2 being used in more challenging environments such as frozen food packaging.

Varying amounts of recycled or secondary fibre are used for producing both liners and corrugating medium, a practice which has grown as recycling has increased. Recycled board is made to the same specifications as virgin material, so that stiffness and burst values are similar, although it is slightly thicker to compensate for the weaker recycled fibre (see Chapter 10). Other properties will depend on the source and quality of the fibre. Recycled board typically has a somewhat smoother surface finish (and therefore better printability) and a lower coefficient of friction than virgin Kraft. Recycled board may absorb water significantly faster than Kraft, which could be important when packing wet products such as fish. Whether Test or Kraft, white or brown, corrugated materials can be found performing quite satisfactorily on the wide range of tray, wrap-around and case erecting and packing machines in use.

By using bleached fibre combined with Kraft, a whiter surface providing better graphic presentation can be obtained. If the bleached fibre layer is thin and not uniform, some of the background Kraft shows through, giving it a mottled appearance. Mottled white and Oysterboard are common commercial names used to describe this board. Papers with a thick even layer of bleached fibres have in the main replaced fully bleached white Kraft. The two-ply natural and white Kraft material is used where high-quality graphics are required and is sometimes enhanced with a clay-coated finish. It is typically used in preprint applications, where the outer layer of material is printed before being made up into corrugated board. This avoids the lined effect obtained when printing on a corrugated surface.

A 'balanced' construction is one where the outer and inner liners have identical weights (or grammages). However, corrugated board can be manufactured using liners of different types and grammages, e.g. using a heavier liner to ensure a better outer surface for printing. Unbalanced constructions tend to have more problems with board warping, but can be appropriate on the grounds of cost and/or functionality. For better printing, the heavier liner should be the outer liner, whereas for better compression strength, the heavier liner should be on the inside.

11.2.2 Flute materials

Fluting medium is a one-ply sheet produced from recycled and/or virgin fibres. Where virgin fibres are used, the pulp is prepared using the CTMP pulping process (see Chapter 10), giving a short fibre suitable for corrugating. This is known as semi-chemical fluting and currently represents only a small percentage of fluting used, the majority being made from chemically-enhanced recycled fibres. There is a range of standard flute sizes (Table 11.1):

- A-flute: the largest size
- C-flute: the second largest
- B-flute: the third largest
- E-flute: the fourth largest
- F-flute and other finer flute sizes

Table 11.2 summarises the differences among the main types of flutes.

A-flute is useful for cushion pads, the construction of particularly heavy-duty boxes and the construction of triple-wall board grades where the added thickness is an advantage. However, use of A-flute has been in decline. A-flute's almost 5-mm thickness occupies more space than C-flute and has significantly greater deflection before bearing load when compressed. In theory, A-flute's thicker section should give it the highest top-to-bottom compression strength of the three flutes. This is true under laboratory conditions. However, A-flute has the lowest flat crush resistance (see Table 11.3). This makes A-flute with <127g medium almost impossible to machine

Table 11.1 Standard flute configurations

Flute type	Approx. no. of flutes per metre	Approx. height of flutes (mm)
A	105–125	4.5–4.7
C	120–145	3.5–3.7
B	150–185	2.1–2.9
E	290–320	1.1–1.2
F	400–440	0.7–0.8

Table 11.2 Comparing characteristics of the main flute sizes

Flute height (mm) (ave)	Flute	Flutes per metre (ave)	Stacking strength	Puncture resistance	Cushion	Flat crush	Surface print quality
4.5–4.7	A	110	Best	Good	Best	Poor	Poor
3.5–3.7	C	129	Good	Best	Good	Fair	Fair
2.4–2.6	B	154	Fair	Fair	Fair	Good	Good
1.1–1.2	E	295	Poor	Poor	Poor	Fair	V good
0.7–0.8	F		Poor	Poor	Poor	Poor	Excellent
0.5	N		Poor	Poor	Poor	Poor	Excellent
0.4	G		Poor	Poor	Poor	Poor	Excellent

Table 11.3 Approximate relative flute flat crush values

Medium grammage	Flute		
	A	C	B
127 g	0.70	1.00	1.15
140 g	0.90	1.25	1.45
200 g	1.10	1.50	n.a.

and transport without destroying the flute structure. Engineering studies suggest that A-flute is most efficient when constructed with a 200 g medium.

C-flute has about 10% better stacking strength than the same board weights in B-flute. It is best for applications where the corrugated container must bear some or all of the stacking load. C-flute is sometimes chosen over B-flute for cases that will hold glass bottles, despite C-flute's lower flat crush strength. It is felt that the thicker flute will provide more puncture protection for the glass. Some authorities regard lightweight medium C-flute as having less than minimum acceptable flat crush for shipping applications.

B-flute is used for canned goods or other products where case stacking strength is not required. B-flute's high flat crush strength is an advantage when supporting heavy goods such as bottles or cans. It can also be used to advantage for lighter load applications where high stack strength is not needed, or where the distribution environment is very short. Heavy media cannot be made into the small B-flute. A recent development offering material and space savings over B-flute is R-flute corrugated board. The flutes are smaller and closer together than B-flute, resulting in better printing surface and 20% lower caliper, meaning more board can be shipped per pallet (*Packaging Today*, 3 February 2011).

E-flute and smaller flutes (often known as microflutes) are not generally associated with shipping containers. They are mostly used to replace carton board for heavier or special protective primary packs. E-flute is an excellent choice where the primary container may become an RRP container for some part of its use. There is a range of finer flutes such as F, G, N and T, some specific to certain manufacturers and used when high-quality graphics are required. Gift packs for drinks, etc., is a typical end use, as well as cases for small tools, hardware appliances and home wares.

Double-walled or triple-walled constructions can have a combination of flute types, perhaps E or B on the outer and B or C flute on the inner, offering a flat surface for printing combined with a load bearing flute on the inner. Triple-wall materials are often the exception to this and have load-bearing C and or A flutes throughout to provide the strength needed for industrial applications such as heavy engine components.

Liners and corrugating medium material can be made to virtually any weight or thickness. However, custom and practice has led to the standardisation of certain traditional grades (see Table 11.4). Finished board is described by component grammage: the mass in grams per square metre. Measurements start from the outside. Corrugated board described as 170/127C/170 would have the following components:

- Outside liner = 170 g
- Medium = 127 g formed to a C-flute
- Inside liner = 170 g

Corrugated board specifications sometimes disregard the fact that paper is a natural product. A given paperboard characteristic can easily vary by 8% or more, and tolerance levels must take this into account.

11.2.3 Adhesive selection

Standard corrugated board is made with a starch-based adhesive applied at about 10–14 gsm. Starch loses strength at high moisture levels. However, corrugated board also loses around 50% of its compression strength between 50 and 90% RH, so both need some form of reinforcement or protection in wet conditions. Where higher resistance is needed, starches can be modified by adding various polymers. Water-resistant adhesives are more expensive and are typically used in conjunction with corrugated board that has liners with a membrane or coating to repel water. Cold adhesives, such as polyvinyl acetate (PVA) are normally used to glue several corrugated sheets to form extremely strong materials for heavy duty applications.

11.3 Manufacturing processes for corrugated board

A corrugating machine consists of several steps (see Fig. 11.2). These involve:

- unwinding and conditioning of liners and medium
- corrugation of the medium (to create the flutes)

Table 11.4 The most commonly used grammages for liners and flutings

Kraft/Test	Mottled/White	Fluting
95 g	125 g	90 g
115 g	140 g	100 g
125 g	200 g	127 g
125 g	140 g	
170 g	170 g	
275 g	200 g	
400 g (Kraft only)		

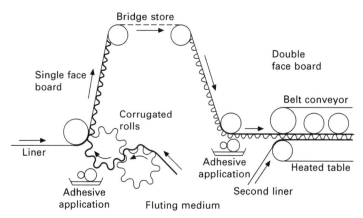

11.2 Corrugating machine.

- bonding the fluted medium to the liners
- drying the finished board
- cutting and scoring the board.

Single-facer stations create the fluted medium and bond it to a single liner or face (usually the inside liner), before passing it on to double backer stations which add the second (outer) liner.

At the beginning of the process, both medium and liner are preconditioned with steam to make them pliable for processing. The medium is fluted by being passed between large rolls with a surface pattern matching the required flute geometry. Older machines used fingers to hold the fluted medium whilst it was being processed, but modern machines use vacuum pressure to hold the medium in place, reducing potential indentations and improving compression strength. Adhesive is then applied to the flute tips. The liner is pressed against the tips, where a combination of heat and pressure bonds the two surfaces together. Typically, greater pressure is used at this stage to ensure a strong bond, resulting in the glue lines potentially showing through. This is why this liner is usually on the inside of the finished case.

The single-faced material, which is still flexible at this point, is sent to a conveyor mechanism known as a 'bridge', where it is draped in an overlapping wave pattern as it travels to the double-backer station. The purpose of the bridge is to separate the single-facer and double-backer stations, allowing for different operating speeds and reel changes. At the double-backer station, adhesive is applied to the flute tips on the other side of the medium and the outer liner is glued on. Corrugators normally have more than one single facer unit, which allows additional plies to be combined at the double backer to create double-or triple-wall boards.

The fully assembled corrugated board with liners on both sides is now stiff. It is passed between two long flat belts for final heat setting and cooling. The edges are then trimmed, and the finished board is slit to widths and cut to lengths as required. At the same time as slitting to the correct width, the machine will also score along the planned flaps for folding. The finished sheets are then stacked ready for assembly.

There is a range of scoring patterns, depending on the container design, starting with scores along the top and bottom flaps (the reverse scores). Three-point or off-set point-to-point scores are used where the case flaps will be folded in one direction to close the container. Where flaps need to be first bent outwards, these bend profiles may produce an uneven break when heavy materials are used. In such instances a five-point score is preferred (see Fig. 11.3).

The resulting joint can be glued, taped, stapled or stitched. Gluing is a relatively simple and fast operation which offers a strong joint. It is the most common joining technique. Double-gluing can be used to reinforce corners. The data in Figs 11.4–11.8 provide typical scored sheet and manufacturer's glue joint tolerances for B and C-flute single-wall cases.

Taping, stapling or stitching are slower, semi-automatic operations. Wire stitching is commonly used on triple-wall constructions and where treatments such as waxing prohibit normal gluing. Taping is used for oversized cases not able to be put through

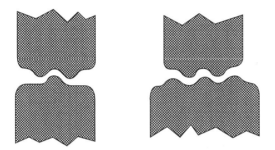

11.3 Profiles of three-point (left) and five-point (right) scoring wheels.

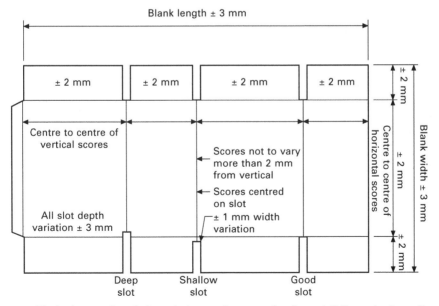

11.4 Typical scored and slotted sheet tolerances for B- and C-flute single-wall cases.

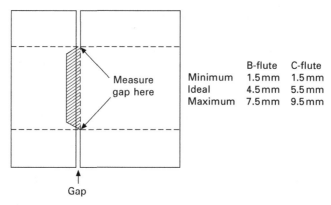

11.5 Gap tolerances for B- and C-flute single-wall cases.

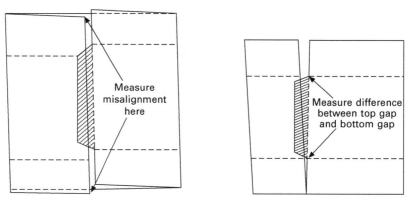

11.6 Gap tolerances for out-of-square and fishtail (U joint) configurations. The maximum allowable difference is 3 mm.

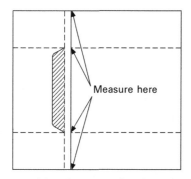

11.7 Identifying overlap (any amount of overlap is unacceptable).

a gluer and for applications where the overlap of the normal glue tab is undesirable. Pressure-sensitive and heat-activated tapes have increasingly replaced gummed tapes since they are easier to apply.

Tapes can also be used in the board manufacture for reinforcing containers. Tapes can be used to provide handhold reinforcement for carrying. They can also include

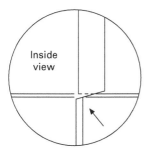 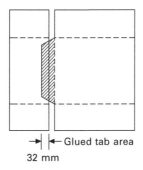

11.8 Length and width of glue tabs. Minimum acceptable width is 27 mm; tabs extending more than 1.5 mm beyond horizontal male scores are unacceptable since they hinder proper step folding and cause out-of-square cases.

Table 11.5 Common case styles

0201	Regular slotted container
0203	Full-overlap slotted container
0204	All flaps meeting slotted container
0312	Half-slotted container
04 Series	Folder

easy-opening tear features and some RRP packs use the tape to turn the container into a tray to make shelf-stacking easier. Application of these tapes can only run in the machine direction of the corrugator.

More complex designs (e.g. those needing angled or curved cuts or creases) require a die cutter which can be a flatbed press (which gives greater accuracy) or a rotary press (which is faster). The crease is pressed into the board using a scoring rule. The die can be specifically designed to incorporate special features. Die-cut designs are correspondingly more expensive than regular slotted containers. One disadvantage is that die-cutting can flatten corrugated flutes, resulting in a loss of compression strength.

11.4 Different types of corrugated board container design

Table 11.5 lists some abbreviations commonly used to describe container types. Figure 11.9 shows examples of common case styles. There are three main basic types of container:

- regular slotted containers (RSC)
- die-cut containers
- multi-component designs.

Regular slotted containers (RSC) are containers in which all scores and cuts are in straight lines only in the machine and cross directions (see Figure 11.10). These containers are easy to manufacture on standard machinery. The manufacturer simply sets the slotters and folder-gluers to the required dimensions. Basic case designs can

250 Packaging technology

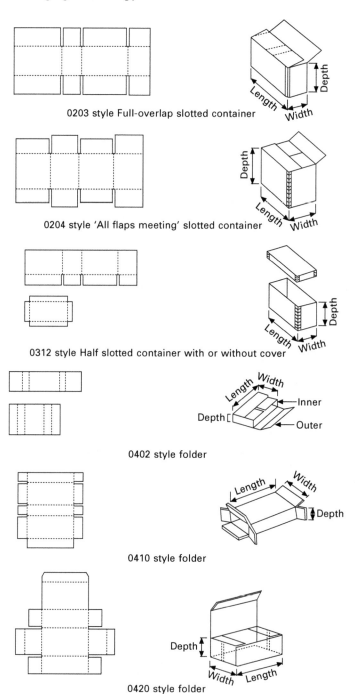

0203 style Full-overlap slotted container

0204 style 'All flaps meeting' slotted container

0312 style Half slotted container with or without cover

0402 style folder

0410 style folder

0420 style folder

11.9 Common case styles.

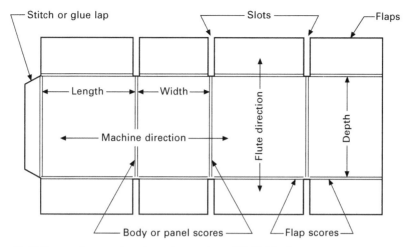

11.10 Parts of a regular slotted container (RSC) blank.

be made with or without top flaps. Variations are primarily differences in how the flaps are cut. A typical RSC has end flaps the same width as the side flaps (see Fig. 11.4). This results in an economical, rectangular master sheet and a case where the end flaps do not meet in the centre. This is known as a 0201 style.

The all-flaps meeting style (0204) is used where it is important to have full, even product support across the bottom or a full, even bearing area across the top. The end flaps are cut to half the case length so that they meet exactly in the centre of the case. This style has a slightly stronger bottom than a standard RSC because of the double board layer, and is more expensive due to additional board usage. Where the container needs to hold a great deal of weight, the side flaps can be cut the full width of the container (0203 style), so that they fully overlap, rather than just meet in the centre.

Trays or open-topped cases can be made simply by eliminating the flaps at one end of an RSC (0312 style). Two trays, one slightly larger than the other, can be telescoped together to form a closed case (known as a 'box and lid'). In the produce business, a similar telescoping container is made by putting the overlapping flaps on the side and end panels, leaving the bottom as a solid sheet. This has the advantage of doubling wall thickness on the case ends, where it will contribute to top-to-bottom compression strength. The construction also uses less material unless the case is tall. Usually the body of such a case is of heavy material while the cover slide is of a much lighter stock. Folders are usually one-piece (0410 style, 0420 style) or two-piece (0402 style) constructions, often used for items such as clothing and home wares sold via mail order and for sending bottles by post.

As has been noted, die-cutting is used for more complex designs (see Fig. 11.11). As well as being more elaborate, die-cut containers are dimensionally more accurate than regular slotted containers. A third design type is shown in Fig. 11.12. These cases are assembled from a number of separate cut pieces, usually a body panel and two end panels, rather than from a single sheet. Using individual cut pieces means

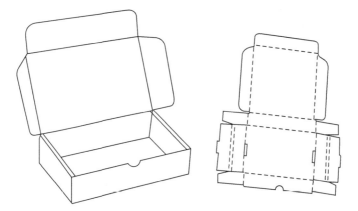

11.11 Typical die-cut case design.

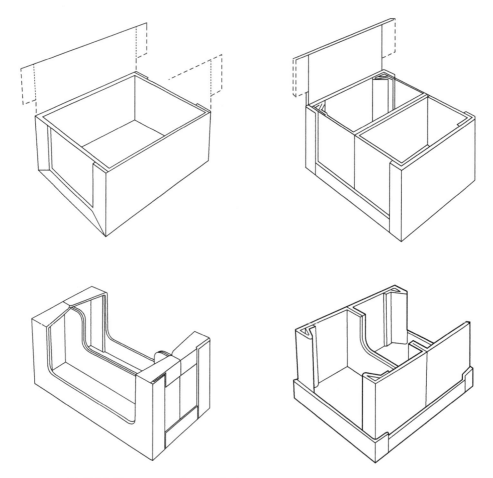

11.12 Multi-component case designs.

maximum utilisation of material and good compression strength. Other design variations include the incorporation of separate H partitions and triangular corner posts for greater strength. Fitments, used to separate items within a case, or to provide added strength, can be made in many styles. Some common ones are shown in Fig. 11.13.

The length of a case is always the greater of the flap opening dimensions (see Fig. 11.14). The depth is the inside dimension between the top and bottom inner flaps. The order in which dimensions are reported records where the opening will be. Top-loading cases have the largest opening for ease of loading but use the greatest board area to enclose a given volume. End-opening cases use the least board for the same volume but have the smallest opening. An end-opening case could be thought of as being the most economical, but this is usually countered by the need to die cut the blank to ensure vertical flute orientation during transit.

When laying out a case, allowance must be made for the material that will form the creases. Corrugated board folds by collapsing in on itself, and the lines drawn on a flat sheet will not be equal to the finished case dimensions (see Fig. 11.15). Scoring allowances are therefore added, based on the case design, flute, and material, and on the type of scoring wheels used. When discussing case sizes with a supplier, it is always best to send the product or a sample case. To size a case accurately:

1. Tear or cut the case open at the manufacturer's joint and lay it face down (see Fig. 11.16).
2. Determine case style, board, and flute.
3. Accurately mark the exact centre of the panel and end scores.
4. Accurately measure between the centre marks of the second and third panels (the W and L dimensions in Fig. 11.16). Do not measure the first or fourth panels.
5. Subtract scoring allowances as dictated by step 2 (see Table 11.6). The result is the case length and width.
6. Repeat steps 4 and 5, measuring between the end scores (the D dimension in Fig. 11.16). The result is the case depth.

11.5 Decoration and printing of corrugated board containers

Kraft board is not an easy medium for high-quality decoration and printing. It is typically a brown colour and has a rough surface made more difficult by the fluting which also creates an uneven surface. Plain corrugated cases are typically printed using flexographic printing. Flexographic printing applies pressure to the corrugated board to effect ink transfer. Compressing the flute structure can lead to a loss in potential compression strength. If top-to-bottom compression strength needs to be maximised, heavy and multiple-colour ink coverage should be avoided. Typical allowable crush per colour is about 0.1 mm. Water-based inks are typically used.

There are various ways of improving print quality. A white Kraft liner provides a better background for printing. Oil-based inks provide a glossy finish and have good rub resistance. However, they need long drying times before further processing which

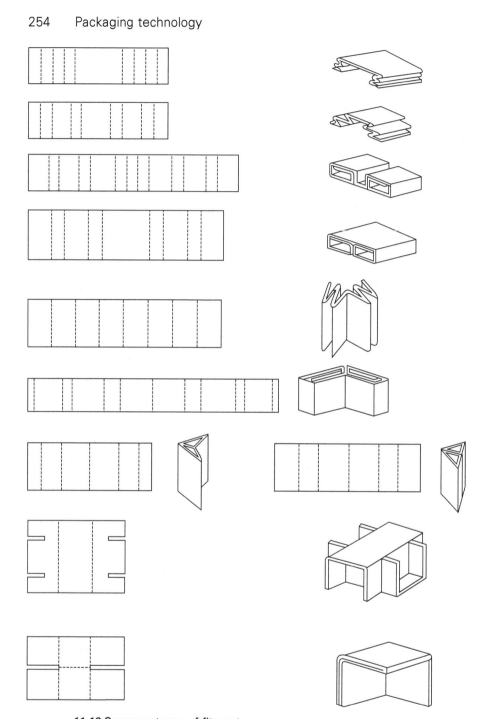

11.13 Common types of fitment.

significantly delays production. Screen printing does not need the contact pressure of flexographic printing which means it can cope better with uneven surfaces. It is also possible to laminate the entire container using lithographic printing. Alternatively,

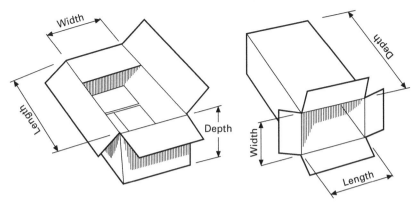

11.14 Case dimensions expressed as the inside dimensions in the following order: length, width and depth.

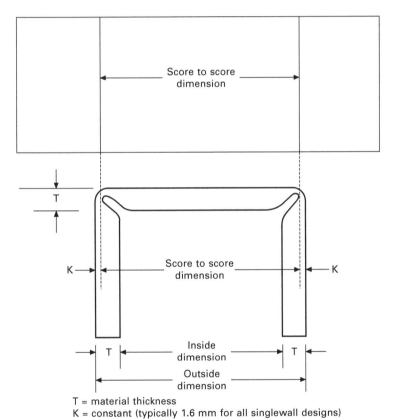

T = material thickness
K = constant (typically 1.6 mm for all singlewall designs)

11.15 A flat sheet laid out with the resulting folded dimension.

labels can be used to decorate a container. White-topped Kraft liner can also be pre-printed before it is converted to corrugated board. Pre-printing eliminates the problems of trying to print an assembled corrugated board. The usual minimum

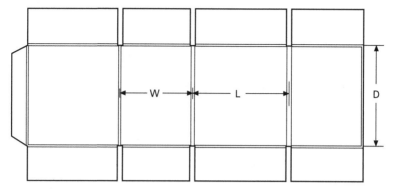

11.16 Sizing a container.

Table 11.6 Scoring allowances for board of various thicknesses

Flute	Typical score allowance
A	5 mm
C	4 mm
B	3 mm
BC	7 mm

amount for pre-printing is 2 mill rolls or several tonnes of paper. This limits the process to fairly large production runs.

There are obviously significant cost implications in the choice of print quality and the material needed to achieve it. However, this has been mitigated by the use of high-quality post-printing technology. This type of equipment is usually installed off-line, but can be linked with a die-cutting unit. As with much of this industry, the basic process has roots well back into the last century, but it is now possible to produce good quality close-registered process work due to the introduction of technical refinements. The print units use thin printing plates with flexible backing, along with vacuum transport systems to reduce undulating surface stripes, finely engraved anilox rolls to closely control ink delivery and individually driven modular print units to maintain colour registration. Such equipment can be used to print upwards of six closely registered colours with a 38 L/cm screen, resulting in good quality graphic reproduction.

11.6 Special board treatments

Most special treatments are concerned with maintaining container performance in humid and wet conditions, particularly tear, puncture and compression strength. Treatments can be added to the paper during milling, applied to the liner or medium as it goes into the corrugator, or added to the finished container. The latter can be advantageous if the surface treatment might make gluing less effective.

Thermoset resins (typically melamine based) can be added at the paper mill to provide greater water resistance. It is important not to exceed a content of 2% if the paper and resulting board is not to be too brittle. Further treatment with acrylic coatings increases wet performance even further. Whilst these coatings provide good water repellency ('run-off'), they are poor in terms of moisture vapour transfer (MVT).

If the container is to be exposed to particularly wet conditions, e.g. because it has to spend some time outside such as cases and trays used at the point of harvesting fresh produce, the liner may need to contain a polymer membrane embedded between two layers of paper. This removes the likelihood of overheating the polymer layer during the corrugating process and avoids potential problems with gluing or printing.

11.7 Testing corrugated board materials and containers

There is a range of tests for materials and containers. Only the most common are noted here. Most board tests are described in methods provided by the British Standards Institution and the Technical Association of the Pulp and Paper Industries (TAPPI). Paper mills concentrate on meeting grammage and burst value requirements. There are no specific liner caliper requirements. However, there is a move towards specifications based on other aspects of performance (e.g. edge crush test (ECT) values).

When conducting testing, it is important to be aware that the mechanical strength of paper depends on the rate at which a load is applied. The faster a load is applied, the greater the ability of paper to withstand it. Dynamic compression strength is thus higher than static compression strength. In addition, under sustained and repeated loading, paper fibres move and distort, further reducing strength. This 'creep' behaviour means that containers tested, for example, at 500 kg load compression strength in the laboratory (a fast loading rate) could be expected to fail in about a year when loaded to 250 kg in the warehouse.

11.7.1 Testing paper properties

One of the most important factors affecting the mechanical properties of paper is moisture content. It is therefore essential that paper testing always be undertaken at the same temperature and humidity. The internationally accepted standard is 50% RH and 23°C. The moisture content of paper depends on humidity and temperature. Paper exhibits hysteresis, i.e. the equilibrium moisture content is slightly different depending on whether the equilibrium is approached from a higher or lower initial moisture content (see Fig. 11.17).

Increasing moisture content leads to reduced compression strength. Increasing moisture content also leads to paper swelling, whilst drying makes it shrink. As an example, between 0 and 90% RH, dimensions can change by 0.8% in the MD and 1.6% in the CD. Poor humidity control can cause distortion in a blank which can compromise subsequent folding into the final container. Uneven moisture content in the outer and inner liners can cause warping as the finished board is dried. Warpage should not exceed 6 mm over a 300 mm span (as measured according to ASTM D 4727).

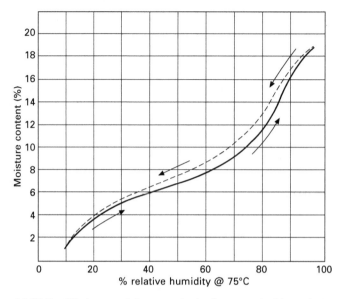

11.17 Equilibrium moisture content of corrugated board at various humidities.

11.7.2 Testing mechanical strength

There is a range of tests of the strength of corrugated board materials and containers. These include:

- Mullen Burst Test (BS 3137/ISO 2759:2001)
- Edgewise Compression Test (ECT) (BS EN ISO 3037:2007)
- Box Compression Test (BCT)
- Short Span Compression (SSC) Test (TAPPI T826 & BS 7325)
- Flat Crush Test (BS EN 23035:1994/ISO 3035:1982); Concora Medium Test (BS EN ISO 7263:2008)
- Thickness of Corrugated Board Test (BS 4817: 1972 (1993)/ISO 3034:1975)
- Puncture Resistance Test (TAPPI T803; ISO 3036 & BS 4816:1972).

The Mullen Burst Test involves slowly applying hydraulic upward pressure to the fibreboard until the liner bursts and the board is ruptured (Fig. 11.18(a)). The burst strength is measured in kilopascals (kPa). If the liners are different, it is important to test both sides. A drawback is that the test is related to paper tensile strength which may not be the most important factor affecting the stiffness and compression strength of the material.

As a result, the Edgewise Compression Test (ECT) is increasingly used as an alternative. The Edgewise test involves placing a specimen of board (typically 50 × 50 mm) vertically in a compression tester and applying a load until failure occurs (see Fig. 11.18(b)). The ECT measures the stiffness of the combined facings and medium and, as a result, is a better indicator of stacking strength. Using ECT measurements, case container strength can be calculated using the simplified McKee formula:

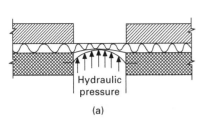
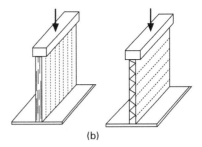
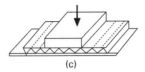

11.18 Testing the strength of corrugated board: (a) Mullen Burst Test; (b) Edgewise Compression Test (ECT); (c) Flat Crush Test.

$$\text{Case compression strength} = 5.87 \times \text{ECT} \sqrt{\text{BP} \times \text{T}}$$

where estimated RSC top-to-bottom case compression strength is expressed in kilo-Newtons (kN)), ECT = edge crush (kN/m), BP = inside case perimeter (m), and T = combined board thickness (mm).

The flat crush test is similar to the edge compression test (ECT) except that the specimen is compressed in the Flat Crush Test (see Fig. 11.18(c)). The test provides a measure of flute rigidity.

The Box Compression Tester is a platen with either four columns (fixed head) or a single central column (floating head) configuration. These columns have lead screws that drive the top plate of the platen downward. An erected box is placed centrally in the tester and the platen is driven down until the point of failure is reached and measured. This test has become particularly relevant when specifying suitable material combinations for multi-piece or perforated retail ready packs (RRPs). It can be used for 'bench marking' and validation where theoretical calculations such as the modified McKee formula are unreliable due to the complexities of such packs.

As there is a strong correlation between the in-plane compression strengths of liners and fluting and ECT values, the Short Span Compression test (SCT) was developed. This allows the ECT and thence the BCT values of a given case to be extrapolated. The paper sample is clamped between the four segments of the apparatus, which has a 0.7 mm gap between the pairs of clamps. When the two pairs are driven together the length of the strip reduces and the stresses increase. As the sample is short relative to its thickness, buckling is prevented and the measured failure is due solely to compression.

Either BCT values or ECT values can be used to specify the boards used to construct a corrugated container. As a general rule, shippers interested in warehouse stacking strength find it advantageous to use BCT. Those shipping in rugged environments or where puncture resistance from product is a concern might also use the Puncture Test (see below).

A test for evaluating laboratory-fluted samples (the Concora Medium Test) also exists. This tests the ability of the corrugated medium to keep the two liners separate. The Thickness of Corrugated Board Test is used to measure whether samples are of the required thickness (known as caliper). Reduced caliper is an excellent indicator that compression strength will be reduced. Caliper can be reduced by improper manufacture, excessive printing pressure, and improper handling and storage.

The Puncture Resistance Test tests strength by measuring the energy required to puncture a board with a triangular pyramidal point affixed to a pendulum arm. The triangular pyramidal point with 25 mm sides is forced through the board under test. The energy absorbed in forcing the tip completely through the board is reported in millijoules per metre (mJ/m).

11.7.3 Testing other properties

There are several other standard tests

- Water absorption: Cobb Test (DIN EN 20535:1994 ISO 535:1991)
- Porosity: Gurley Porosity Test (BS ISO 5636-5:2003); see also Bendtsen Test Method (BS 6538-2:1992)
- Water resistance: Ply Separation Test (TAPPI T812)
- Machinability: Coefficient of Friction Test.

The Cobb Test measures surface water absorption or repellency. It describes the time required to soak a specific amount of water into a corrugated sheet. The standard unit is grams per square meter (g/m^2). The time may vary from 60 seconds to 30 minutes. The Cobb Test is used in various ways, e.g. to investigate potential gluing problems where a reasonable degree of absorption is required, or the degree of water repellency where containers may need a higher degree of water resistance.

The Gurley Porosity Test uses the time it takes for a given volume of air to pass through a paper to measure porosity. The lower the number, the more porous the paper. Porosity can vary from 2 seconds to 200 seconds but averages between 10 and 20 seconds.

The Ply Separation Test evaluates the board's resistance to ply separation when exposed to water. It is used mainly to test the effectiveness of water-resistant adhesives.

It is important to be able to measure how easily corrugated board can be formed and how well it performs in loading. If board slides too easily, it will be hard to process and use in transport and stacking. One useful measure is coefficient of friction (CoF). One method of determining CoF involves placing a weighted piece of test material onto a surface faced with the same material. The angle is gently increased until the material slips, at which point the angle is noted. The tangent of the average slide angle is reported as the static CoF. An alternative, and more accurate test is to use a stress/strain machine to measure the force required to pull a weighted specimen along a flat surface. A stress/strain machine will give both static and dynamic CoF values.

Containers with a CoF between 0.40 and 0.50 would be regarded as acceptable. A CoF below 0.40 is marginal and one below 0.30 is regarded as sub-standard. In this case some form of surface treatment may be needed to increase surface roughness to improve antiskid properties.

11.8 Sources of further information and advice

- For cases (range of standard formats) and FEFCO codes, see standard formats at http://www.fefco.org/fileadmin/fefco_files/fefcocodes/FEFCO_ESBO_code_of_designs.pdf
- For a detailed review of paper types used in corrugated board, see http://www.paper.org.uk/information/technical/5_fibrousmaterials.pdf
- Jönson, G., *Corrugated Board Packaging*, 2nd edn. Leatherhead: Pira, 1999.

12
Basics of polymer chemistry for packaging materials

A. RILEY, Arthur Riley Packaging Consultant International, UK

Abstract: This chapter introduces the basics of polymer chemistry. It explains how the polymers are synthesised, including the various types of polymerisation (addition and condensation polymerisation, copolymerisation and crosslinking polymerisation). It also discusses the factors which influence the way they perform.

Key words: polymer, polymerisation, copolymer, plastics.

12.1 Introduction

'Plastic' describes the ability of a substance to be formed into an object – modelling clay, wax or dough would be good examples. The plastics which are going to be discussed in this chapter are organic and polymeric in nature. That is to say they are made from many (poly) parts (mer). Polymers do occur naturally, for example cellulose, keratin (nails and hair), shellac, rubber and DNA, but the polymers which will be discussed in this chapter are all synthesised, either from oil or other naturally occurring raw materials. All polymers used to make plastics account for less than 4% of oil used worldwide. Polymers and the plastic components made from them can meet nearly every packaging challenge. Choosing the correct polymer for a particular application is difficult without understanding their chemistry and therefore their properties. This chapter explains the essential background whilst Chapter 13 reviews the properties of polymers used for packaging applications.

A polymer is formed when a large number of units known as molecules are joined together to form a larger molecule. The units may be all the same, in which case it is known as a 'homopolymer' or two or more different units may be combined in one polymer, which then becomes a copolymer. For the packaging industry many molecules are based on a common carbon backbone, for example polyethylene (PE); others have a similar mostly carbon backbone, but with additions such as oxygen or nitrogen included in the backbone, for example polyethylene terephthalate and polyamide, respectively. The size of these molecules can be between 100 and 100,000 repeating units.

Polymers can be grouped into two basic classes:

- thermosetting
- thermoplastic, which includes thermoplastic elastomers and biodegradable polymers.

Each polymer has a unique structure and a proper chemical name (see Table 12.1).

Table 12.1 The common packaging plastics

Polymer	Abbreviation	Recycling number	Homopolymer or copolymer
Low density polyethylene	LDPE	4	Homopolymer
High density polyethylene	HDPE	2	Homopolymer
Linear low density polyethylene	LLDPE	7	Copolymer
Polypropylene	PP	5	Homopolymer
Polystyrene	PS	6	Homopolymer
Polyvinyl chloride	PVC	3	Homopolymer
Polyethylene teraphthalate	PET	1	Homopolymer
Polyamide	PA	7	Homopolymer
Acrylonitrile butadiene styrene	ABS	7	Copolymer
Styrene acrylonitrile	SAN	7	Copolymer
Polyethylene Naphthalate	PEN	7	Homopolymer
Ethylene vinyl acetate	EVA	7	Copolymer
Ethylene vinyl alcohol	EVOH	7	Copolymer
Poly vinyl alcohol	PVOH	7	Homopolymer
Polycarbonate	PC	7	Homopolymer
Poly lactic acid	PLA	7	Homopolymer

As these names tend to be lengthy and in some instances difficult to pronounce, the industry has adopted acronyms for all the polymer types. This can, however, be confusing as within a polymer type there are variations which will exhibit different characteristics. Polypropylene (PP), for example, can exist in isotactic, syndiotactic, and atactic forms, determined by the chemical reaction which takes place during the polymerising process; the three forms are quite different in their properties. Copolymerisation can also change the characteristics of polymers, as can chemicals added during and after polymerisation.

The polyethylenes and polypropylene are all classed in one family, known as the polyolefins, which are all hydrocarbons, containing only hydrogen (H) and carbon (C). Substitution of hydrogen can result in a polymer with very different properties compared with the polyolefin. For example, ethylene ($CH_2=CH_2$), the monomer used to make polyethylene $-(CH_2-CH_2)n-$ has a very poor gas barrier, but a good moisture barrier. Replacing a hydrogen atom with a chlorine (Cl) atom results in a polymer with a greatly improved gas barrier but reduced moisture barrier. This resultant polymer is PVC, or polyvinyl chloride.

Traditionally polymers for the production of plastics (the name given to the macromolecule produced by polymerisation) were sourced from oil or coal, although it has always been possible to obtain the basic units to make many of the common plastics from other sources such as plants and trees. Recent developments have resulted in the production of biodegradable plastics such as PLA (polylactic acid) from renewable cereal and vegetable raw materials. This has added to the challenges of recycling as these new polymers are meant to go into special industrial composting and regeneration facilities and must not be mixed with conventional polymers for recycling.

12.2 The basic principles of polymerisation

Packaging polymers are based on the carbon atom. Carbon has the ability to:

- Form four bonds with other atoms. Carbon (C) is an atom; methane CH_4 is formed when carbon bonds with four hydrogen (H) atoms and is called a molecule.
- Join to itself to form long chains, either in a line (-C-C-C-C-C-) or as a branched structure (-C-C-C-C-C-C-):

The structure of the polymer, as will be discussed later, is very dependent on the polymerisation methods and initiators used. In this chapter the chemistry is going to be kept to a minimum. The emphasis will be on the key aspects of the structure and manufacture of polymers and how these affect properties and performance in use.

Monomers are, in the main, obtained from the fractional distillation of crude oil followed by catalytic cracking of the resultant fractions. Fractional distillation involves heating the crude oil in a large fractionating column, where different fractions are separated by virtue of their different boiling points, determined by the number of carbon atoms in the hydrocarbon chain. The higher the number of carbon atoms, the higher the boiling point. The ratio of the different fractions is fixed according to the source of the crude oil, and to obtain more supplies of the most needed fractions, e.g. gasoline, some of the less useful, longer chain structures are 'cracked' or broken down into smaller fractions. Molecules with two or three carbon atoms such as ethene ($CH_2=CH_2$) and propene ($CH_2=CH_2-CH_3$) are the by-products of this cracking process and are, in turn, polymerised to make polyethylene and polypropylene using the addition polymerisation method.

Carbon atoms can easily attach to each other and this can be either via a single bond (CH_3-CH_3), known as ethane which is a very stable structure, or a double bond, forming a less stable structure such as ethene or ethylene ($CH_2=CH_2$). Ethane is stable because it is *saturated*, which means that each carbon atom is attached to the maximum possible number of hydrogen atoms. Ethene, on the other hand, is an example of an *unsaturated* molecule, as it does not contain the maximum number of hydrogen atoms, resulting in the two carbon atoms being joined by a double bond. This double bond can be easily broken down and this is the basis of the polymerisation process to be discussed below.

As stated, monomers can also be obtained from renewable resources such as trees and plants. This includes monomers which can be synthesised to produce common polymers (e.g., polyethylene) as well as the monomers which can be extracted to produce bio-compostable polymers such as polylactic acid.

12.3 Addition (chain growth or coordination) polymerisation of polymers

Many polymers with starting monomers that are similar in structure, such as polyethylene, polypropylene, polystyrene and polyvinyl chloride, are produced by addition polymerisation. There are two different methods described below, that are illustrated by comparing the polymerisation of LDPE with the polymerisation of HDPE.

12.3.1 Polymerisation of low density polyethylene (LDPE)

Low density polyethylene is produced by *free radical polymerisation*. A free radical is a molecular fragment that contains an unpaired electron, often as a result of molecular decomposition or reaction of a molecule with another free radical. Free radicals are obtained from organic peroxides (R–O–O–R′), hydroperoxides (R–O–O–H) or azo compounds (R–N=N–R′). Ultraviolet light and high temperatures are sometimes used to generate the free radicals. Free radicals are unstable in themselves as they need a partner, and they look for stability by attacking an unsaturated monomer, i.e. one with a double bond (C=C). The weak double bond breaks and the free radical attaches itself to the monomer, creating a larger free radical as shown below:

$$R^* + \begin{array}{c} H \ H \\ | \ | \\ C=C \\ | \ | \\ H \ X \end{array} \rightarrow R - \begin{array}{c} H \ H \\ | \ | \\ C-C^* \\ | \ | \\ H \ X \end{array}$$

where C = carbon, H = hydrogen, X = another atom or molecule, for example hydrogen (ethylene), chlorine (vinyl chloride) or a methyl group (CH$_3$), producing propylene. This is the initiation step.

The reaction continues and, each time, a monomer is added to the growing chain, resulting in added repeat units, increasing the size and molecular weight of the polymer to the pre-determined level (n = number of repeat units; * is the free radical):

$$R-\begin{array}{c} H \ H \\ | \ | \\ C-C^* \\ | \ | \\ H \ X \end{array} + \left\{ n \times \begin{array}{c} H \ H \\ | \ | \\ (C=C) \\ | \ | \\ H \ X \end{array} \right\} \rightarrow R-\begin{array}{c} H \ H \\ | \ | \\ (C-C)^*n \\ | \ | \\ H \ H \end{array}$$

This is the propagation stage which continues until the polymer is the size and average molecular weight required.

Next comes the termination stage at the required point. There are many ways of terminating polymerisation. The one illustrated below shows how two free radical polymers are combined. The two free radicals cancel themselves out by reacting with each other, resulting in a stable molecule. There is no longer any propagation and the reaction is stabilised. (NB: m and n are number of repeat units, m is a different number to n):

$$R-(\underset{\underset{H}{|}}{\overset{\overset{H}{|}}{C}}-\underset{\underset{X}{|}}{\overset{\overset{H}{|}}{C}})*n + R-(\underset{\underset{H}{|}}{\overset{\overset{H}{|}}{C}}-\underset{\underset{X}{|}}{\overset{\overset{H}{|}}{C}})*m \rightarrow R-(\underset{\underset{H}{|}}{\overset{\overset{H}{|}}{C}}-\underset{\underset{X}{|}}{\overset{\overset{H}{|}}{C}})n-(\underset{\underset{X}{|}}{\overset{\overset{H}{|}}{C}}-\underset{\underset{H}{|}}{\overset{\overset{H}{|}}{C}})m-R$$

Polyethylene can be produced using either an autoclave or a tubular reactor (Fig. 12.1). Autoclaves are large vessels in which a product is agitated (stirred). These are used in the high pressure (103,000–344,000 kPa) process required for the production of low density polyethylene. Ethylene at this pressure and temperature (125–250°C) acts as both a gas and a liquid and therefore is its own solvent (other monomers use a separate solvent).

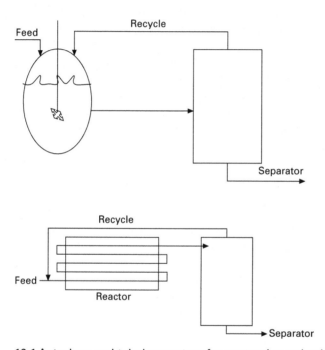

12.1 Autoclave and tubular reactors for processing polyethylene.

The autoclave process produces highly branched random polymers due to the fact that the free radical initiator and the monomer are randomly mixed as the polymer chain is growing. In these circumstances a free radical can react with a monomer as already discussed, or it can extract a proton thus destabilising the atom within the molecular chain. This creates a reaction site within the chain, resulting in a branch being formed.

Depending on where this happens along the polymer chain, short or long chain branching can occur. If the attack occurs near the free radical, as usually happens, a short branch is formed. If the site of attack is a long way from the end of the polymer chain, a long branch is formed. An example of short chain branching is:

$$RCH_2CH_2CH_2CH_2CH_2* \rightarrow RC*HCH_2CH_2CH_2CH_3$$

$$RC*HCH_2CH_2CH_2CH_3 + CH_2=CH_2 \rightarrow RCH(CH_2)_3\,CH_3$$
$$\hspace{4cm}|$$
$$\hspace{4cm}CH_2CH_2$$

An example of long chain branching is:

$$RCH_2CH_2* + RCH_2\,CH_2R \rightarrow RCH_2CH_3 + RC*HCH_2R$$

$$RC*HCH_2R + CH_2=CH_2 \rightarrow RCHCH_2R$$
$$\hspace{4cm}|$$
$$\hspace{4cm}CH_2CH_2*$$

The tubular reactor results in a lesser branched LDPE than the autoclave method, due to the controlled way in which the initiators are added. Tubular reactors are commonly used to make LDPE for film manufacture.

12.3.2 Polymerisation stereochemistry of high density polyethylene (HDPE)

HDPE is produced by *coordination polymerisation* using a catalyst, usually Ziegler-Natta (ZN). This controls the combining of the monomer, polymerisation taking place on the catalyst's surface. Pressure is much lower than in the free radical process, commonly around 2,000 kPa. Polymerisation takes place in the presence of a small quantity of hexene (C_6H_{12}, containing one double bond), which is added to reduce the branching and control the crystallinity to the requirements of its end use. Film grades normally have a density of 0.956 g/cm³ having less than 10 branches per 1,000 carbon atoms). Polymerisation takes place in a solution, slurry or gas phase:

- Solution phase is where the monomer acts as a solvent for the polymer being produced (it is dissolved in its monomer).
- The slurry phase utilises a solvent to dissolve the monomer and suspend the catalyst. As the polymer reaches high molecular mass, it is no longer soluble and therefore drops out of solution, creating a slurry.
- Gas phase polymerisation is carried out in a fluidised bed of the catalyst in a vertical tower. The ethylene percolates through the catalyst, polymerises and precipitates out as a powder. Molecular weight distribution (discussed later) is controlled by the type of catalyst used and the reactor conditions employed.

Tacticity relates to the way non-uniform three-dimensional polymer molecules such as polypropylene are formatted within a polymer chain. In the case of addition polymerisation of polymers other than polyethylene such as PP homopolymer, PVC, PS and PVOH, other interactions can come into effect. One of these is 'head to head' versus 'head to tail' configurations as shown below:

```
      H   X    H   X              H  X  H  X
      |   |    |   |              |  |  |  |
  R — C — C* + C = C  →  R — C — C — C — C*
      |   |    |   |              |  |  |  |
      H   H    H   H              H  H  H  H
```

A head to head configuration is:

$$\begin{array}{cccc} H & X \\ | & | \\ R-C-C^* \\ | & | \\ H & H \end{array} + \begin{array}{cc} X & H \\ | & | \\ C=C \\ | & | \\ H & H \end{array} \rightarrow \begin{array}{cccc} H & X & X & H \\ | & | & | & | \\ R-C-C-C-C^* \\ | & | & | & | \\ H & H & H & H \end{array}$$

The position of the X (e.g. OH in PVOH) when it reacts with the free radical affects its position in the chain. In general, free radical polymerisation produces head to tail, rather than head to head configurations. The CHX group is considered as the head and the CH_2 group the tail. The major reason for this phenomenon is stability of the free radical in the head to tail configuration.

The configuration of a polymer molecule is not just dependent on which atom is attached to which but also tacticity. Polypropylene, for example, can exist in three tactic forms:

- Atactic
$$\begin{array}{ccccccc} H & H & H & CH_3 & H & CH_3 & H \\ | & | & | & | & | & | & | \\ -C-C-C-C-C-C-C- \\ | & | & | & | & | & | & | \\ H & CH_3 & H & H & H & H & H \end{array}$$

All the methyl (CH_3) groups are randomly spread either side of the carbon chain.

- Isotactic
$$\begin{array}{ccccccc} H & H & H & H & H & H & H \\ | & | & | & | & | & | & | \\ -C-C-C-C-C-C-C- \\ | & | & | & | & | & | & | \\ H & CH_3 & H & CH_3 & H & CH_3 & H \end{array}$$

The methyl groups are all on one side of the carbon chain.

- Syndiotactic
$$\begin{array}{ccccccc} CH_3 & H & H & H & CH_3 & H & H \\ | & | & | & | & | & | & | \\ -C-C-C-C-C-C-C- \\ | & | & | & | & | & | & | \\ H & H & CH_3 & H & H & H & CH_3 \end{array}$$

The methyl groups alternate around the carbon chain.

For this tacticity to occur, the polymer chain must contain a 'chiral' carbon, which is one with four different constituents attached to it. For example, in the case of polypropylene:

$$\begin{array}{c} H \\ | \\ R_1-C-R_2 \\ | \\ CH_3 \end{array}$$

R_1 is a polymer chain of length X and R_2 a polymer chain of length Y and therefore there are four separate constituents attached to the carbon. Different catalyst systems can produce different tacticity, resulting in different properties (discussed later and in Chapter 13).

12.4 Condensation (step) polymerisation

Condensation polymerisation results from the reaction between molecules of different functional groups, for example COH (alcohol), CNH_2 (amine) or COOH (acid). The first step is to react, for example, a dialcohol with a diacid to form:

$$HO-R_1-OH + HOOC-R_2-COOH$$
$$\downarrow$$
$$HO-R_1-O-\overset{O}{\underset{\|}{C}}-R_2-COOH + H_2O$$

The alcohol and acid groups react together forming a molecule having two different functional (reactive) groups at opposite ends. (Note that if more complicated monomers were reacted together, having three functional groups, a crosslinked polymer would result.) During the reaction there is the loss of a small molecule, n, in this case water as condensate.

In the second step, the intermediate monomer, having dual functionality, can react with a similar monomer:

$$HO-R_1-O-\overset{O}{\underset{\|}{C}}-R_2-COOH + HO-R_1-O-\overset{O}{\underset{\|}{C}}-R_2-COOH$$
$$\downarrow$$
$$HO-R_1-O-\overset{O}{\underset{\|}{C}}-R_2-\overset{O}{\underset{\|}{C}}-O-R_1-O-\overset{O}{\underset{\|}{C}}-R_2-COOH + H_2O$$

This reaction will continue until the condensation polymer of predetermined molecular weight is achieved.

In the above reaction, if n molecules of terephthalic acid are reacted with n molecules of ethylene glycol, polyethylene terephthalate (PET) is the result (Fig. 12.2). PET can also be synthesised from dimethyl terephthalate. Other condensation polymers can be synthesised. These include polyamides (PA), polyethylene naphthalate (PN) and polycarbonate (PC) as well as polylactic acid.

Polyamide

$$-R'-\overset{O}{\underset{\|}{C}}-\overset{H}{\underset{|}{N}}-R''-\overset{H}{\underset{|}{N}}-\overset{O}{\underset{\|}{C}}-R'-\overset{H}{\underset{|}{N}}R''-\overset{H}{\underset{|}{N}}$$

Polyamides can be formed by reacting a diamine with a diacid. The choice of monomers to form the intermediate monomer in the first step of polymerisation

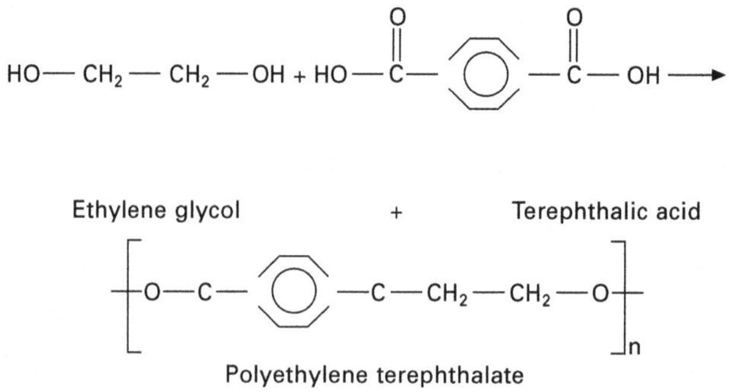

$$HO-CH_2-CH_2-OH + HO-\underset{\|}{\overset{O}{C}}-\bigcirc-\underset{\|}{\overset{O}{C}}-OH \longrightarrow$$

Ethylene glycol   +   Terephthalic acid

$$\left[O-C-\bigcirc-C-CH_2-CH_2-O \right]_n$$

Polyethylene terephthalate

12.2 PET polymerisation.

determines the end molecular structure and therefore the final functional properties. Polyamides can also be produced using amino acids as the starting monomer. In this instance an intermediate is not necessary:

$$n \; x(H_2N-R-\underset{\|}{\overset{O}{C}}-OH) \rightarrow H_2N-(R-\underset{\|}{\overset{O}{C}}-\underset{|}{\overset{H}{N}})_n-H + H_2O$$

Polycarbonate

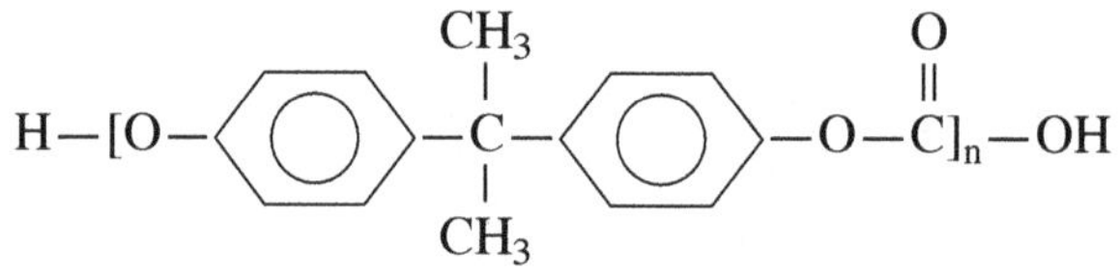

$$H-[O-\bigcirc-\underset{\underset{CH_3}{|}}{\overset{\overset{CH_3}{|}}{C}}-\bigcirc-O-\underset{\|}{\overset{O}{C}}]_n-OH$$

Polycarbonate is produced from phosgene ($O=C-Cl_3$), a poisonous gas which is a derivative of carbonic acid ($HO-\underset{\underset{C}{\|}}{\overset{O}{}}-OH$) and bisphenol A to produce poly(bisphenol A carbonate), known as polycarbonate.

Polylactic acid

Polylactide or polylactic acid (PLA) is a synthetic, aliphatic (containing no benzene ring) polyester produced from the monomer lactic acid. Lactic acid is found in plants as a by-product or intermediate product of their metabolism. Lactic acid can be industrially produced from a number of starch or sugar-containing agricultural products, such as cereals. There are three mechanisms by which lactic acid can be produced. These are:

- ring opening polymerisation (ROP),
- direct polycondensation in high boiling solvents (DP, S), and
- direct polymerisation in bulk followed by chain extension with reactive additives.

ROP is the favoured production route for the majority of polylactic acid produced.

12.5 Copolymerisation and crosslinking polymerisation

A copolymer is a polymer made from more than one monomer type. If the copolymer is made up of say, x units of propylene and z units of ethylene, polymerised together under the same conditions, it will contain the number of units of each in the proportions used. Therefore a large range of ethylene/propylene copolymers is possible depending on the ratio of each monomer used in the reaction. In this way the manufacturer of such a polymer can control some of the properties of the resultant resin. This can be carried out for any monomer suitable for addition polymerisation. For example, if hexene, heptene or octene are copolymerised with ethylene, the result is LLDPE (linear low density polyethylene) (see Fig. 12.3). This copolymer has much more elasticity than LDPE; the larger the carbon chain in the comonomer, the greater the elasticity. The hexene version might be used for carrier bags whilst the octene version, due to its greater elasticity, might be used for stretch wrap.

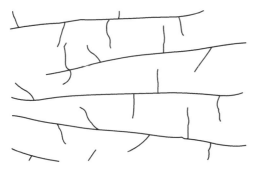

12.3 Linear low density polyethylene.

The molecules of each copolymer can be arranged, with respect to the other, in a random, alternating, block or graft format. This also allows the polymer manufacturer to design the polymer to meet the needs of the customer throughout the supply chain. The main types of copolymers are listed below.

- Random copolymers, as the name suggests, involve an uncontrolled distribution of the two polymers, resulting in a totally disordered polymer chain –OOXOOOXXOOXOOXXXXOOXOOOOXXOOOX–OXOXOXOXXXXXXOO, where X and O are the two monomers. Random copolymers can be formed by adding the second polymer to the first, during random polymerisation of the first, e.g., the addition of ethylene during the random polymerisation of polypropylene.
- Alternating copolymers have a backbone where X and O share equal proportions – OXOXOXOXOXOXOX–OXOXOXOXOXOXOXOX. The polymerisation of an alternating copolymer is more complicated. They can only be produced if the reaction rate of one growing polymer chain (X) is faster than the other (O). If this is achieved, then the majority of the polymer chain is of an alternating structure.
- Block copolymers can vary in the proportion of each monomer used to make the copolymer. Each monomer type forms in a block of its own such as

–XXXXXOOOOOOXXXXXXOOOOOOOO–XXXXXOOOOOO. One way that a block copolymer can be produced is to polymerise each monomer separately to a low degree of polymerisation and then combine these small polymer molecules with one another. Alternatively, if molecules are introduced into the reactor in an alternating sequence, a block copolymer will also result.
- Graft copolymers have a backbone of one type of polymer and branches of another, e.g. high impact polystyrene (HIPS). It has a polystyrene backbone with grafted polybutadiene branches:

```
XXXXXXXXXXXXXXXXXX–XXXXXXXXX–XXXXXXXXXXXXXX
        O                O             O
        O                O             O
        O                O             O
        |                O             O
        O                O             |
        O                              O
```

Graft copolymers are often formed by building reactive sites into the linear backbone; the subsequent branching occurs at these sites.

Combination can occur of copolymer types. For example, acrylonitrile butadiene styrene (ABS) occurs in two stages, combining a random copolymer of styrene acrylonitrile (SAN) with a branch of butadiene grafted onto the random copolymer backbone.

12.5.1 Crosslinking polymerisation

Multifunctional monomers have three or more reactive sites which can bond to one another forming a crosslink, instead of merely forming branches. The polymer chains link together to form a network due to very strong interactive forces to form a two- or three-dimensional structure. Crosslinked polymers are not commonly used in packaging but have important applications, especially where chemical and temperature resistance is required, for example as coatings to protect metal cans from attack during processing and in storage (see Fig. 12.4 and Chapter 8).

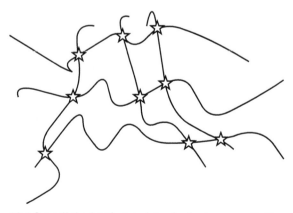

12.4 Crosslinked polymer (star indicates crosslink).

12.6 Factors affecting the characteristics of polymers

There are many factors which influence how a particular polymer will perform and the following will now be discussed:

- the monomer from which the polymer is formed
- tacticity
- the way in which the monomer is polymerised
- the degree of crystallinity of a polymer
- the molecular weight and molecular weight distribution within the polymer
- how many monomers have been used to make the polymer
- the orientation of each monomer in a copolymer
- the type of inter-and intra-molecular forces within the polymer structure
- the additives within a polymer mix, ready for processing into packaging
- the use of special treatments
- mixing polymers in an orderly manner
- the physical orientation of a polymeric packaging plastic.

12.6.1 The monomer from which the polymer is formed

The monomer has a profound effect on the properties of the polymer formed from it. It can be seen from Fig. 12.5 that the four monomers shown are very similar with a hydrogen atom in ethylene being substituted with another atom or functional group. The change may appear to be small but the effect on the properties of the resulting polymer is considerable. For example, polypropylene has the best barrier to moisture, followed by polyethylene, polyvinyl chloride and lastly polyvinylidene chloride.

The size of the replacing atom in comparison to hydrogen is also important. The larger the atom or functional group, the more difficult it is for it to rotate about the central carbon backbone. This is known as steric hindrance. The hydrogen atoms in polyethylene are small and therefore present very little hindrance to rotation. As a result its T_g (glass transition temperature – the temperature at which a plastic changes from a 'glassy' to a 'rubbery' state) is very low at –60°C. Polystyrene has a large benzene ring substituting the hydrogen atom, thus greatly impairing rotation about the carbon backbone, and it has a T_g of +100°C. Polypropylene has a methyl (CH_3) group which hinders rotation less than polystyrene but more than polyethylene and it has a T_g of –15°C. As the steric hindrance increases, stiffness increases. Therefore you can predict that packaging made from polypropylene homopolymer is stiffer than polyethylene but not as stiff as polystyrene homopolymer.

The polarity of monomers and therefore the polymers synthesised from them varies. Non-polar polymers such as polypropylene and polyethylene have very low surface energy; as a result they are difficult to bond and therefore difficult to print. Special treatments have to be applied to the surface of such polymers to enable printing inks to bond to the surface. Such treatments include corona discharge (high voltage, low amperage electric current) and flame treatment of the surface. These treatments oxidise the surface of the polymer, increasing the surface energy to above 40 dyne/cm, which means that there is sufficient energy to accept adhesives and printing

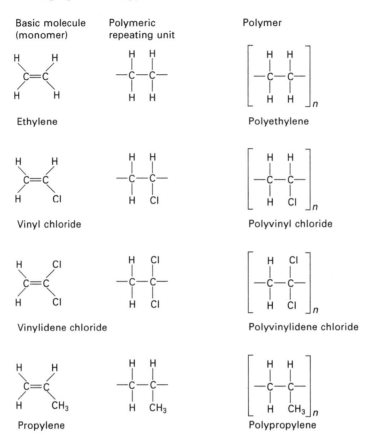

12.5 Four very similar monomers with a hydrogen atom in ethylene being substituted with another atom or functional group. The change may appear to be small but the effect on the properties of the resulting polymer is considerable.

inks. If the surface energy is too low, the adhesive or printing ink will not wet out the surface of the substrate; it will reticulate, forming globules on the surface and hindering bonding.

Another factor affected by the non-polar nature of a polymer is barrier. Polyethylene and polypropylene repel polar water but not non-polar gasses, such as oxygen. Therefore they have good moisture and water barrier but very poor gas barrier. Polyvinyl chloride is much more polar than polyethylene and therefore does not require its surface to be re-energised to accept adhesive or print. It also has a better gas barrier than polyethylene but only a fair moisture barrier. Often polar polymers are mixed with non-polar polymers to enhance their performance. For example, ethylene vinyl acetate (EVA) is often added to polyethylene to give it a cling property, which is useful in stretch wrap film.

The chemical make-up of the polymer influences such properties as density, thermal properties, melting and softening point, solubility, and permeability to gases and moisture vapour. Density is directly related to how many square metres one tonne of polymer will produce. The mass per given volume is dependent on the mass of the atoms in the polymer's make-up. For example, PVC has a higher molecular mass than LDPE due to the higher mass of the chlorine atom. The amount of branching and the strength of the intermolecular forces, which draw the polymer molecules together, also affects density. The closer the molecules are to each other, the less the space between them and therefore the higher the density, for example HDPE compared to LDPE. The atoms contained within a molecule, the forces bonding them to other atoms (intra-molecular), the forces between molecules (inter-molecular) and the randomness of the polymer chains in relation to one another all affect the melting temperature (T_m), the glass transition temperature (T_g) and the decomposition temperature.

Polymers that have no hydroxyl groups, such as the polyolefins (PE, PP, etc.), are hydrophobic (water hating), while others with hydroxyl groups, such as polyvinyl alcohol (PVOH), are hydrophilic (water loving); therefore PVOH will dissolve in water. The polyolefins have a high resistance to water. Polymers which have a percentage of hydroxyl groups such as polyethylene vinyl alcohol (EVOH) and polyethylene terephthalate (PET) will not dissolve in, but will absorb, water.

Polarity of the monomer or monomers which make up the polymer has an effect on whether the polymer will provide a good barrier to gas or moisture: for example, a polar polymer such as PVC has a much better barrier to gases such as oxygen than do the polyolefins which have no polarity. The polyolefins, however, have a much better barrier to moisture than their polar alternatives. The more linear a polymer chain is, the better the barrier properties due to the fact the chains are closer together and therefore there are fewer gaps for the gas and moisture particles to pass through.

Polymer chain length also affects the properties of the polymer. Mechanical strength, melt viscosity and melting temperature (T_m) are just some of the properties which are affected by the molecular weight (MW) of the polymer. The longer the chain length, the greater the number of repeat units, the higher the molecular weight (molecular mass) of the polymer.

As molecular weight increases, the mechanical strength, such as tensile properties and stiffness of a polymer chain increases, due to the greater inter-molecular forces exerted within the polymer. These forces also play their part when a polymer melts and this, combined with the greater entanglement of the polymer molecules with each other, restricts their flow and therefore the resultant melt has a higher melt viscosity. The greater the intra- and intermolecular forces contained within a polymer, the greater the energy required to break down the forces to allow the molecular chains to move freely (melt). The higher the molecular weight, the higher the forces and the higher the temperature required to melt the polymer.

12.6.2 Tacticity

Tacticity of a non-uniform monomer such as propylene in a polymer chain affects the T_g of the polymer. In the last section it was said that polypropylene had a T_g of $-15°C$, but this depends on the tacticity. For example, atactic polypropylene has a T_g of –19°C, whereas isotactic polypropylene has a T_g of –8°C and therefore is much more likely to embrittle at temperatures used in the storage of frozen food (–18 to –40°C). Tacticity can also affect other properties: isotactic polypropylene has higher strength and a higher T_g and T_m than all other common polyolefins, and syndiotactic polypropylene in film form has higher elasticity than other homopolymer polyolefins.

12.6.3 Polymerisation methods, initiators and catalysts

Low density polyethylene is produced by using high temperature and pressure with an organic peroxide providing the free radical initiator. This produces a branched polymer, as shown in Fig. 12.6, with low crystallinity. High density polyethylene is produced at low temperature and pressure and uses an organometallic catalyst. This produces a linear polymer of higher crystallinity with few branches along the polymer chain (Fig. 12.7).

Changing the catalyst (e.g., from a Ziegler-Natta to a metallocene) can change the properties of the polymer, due predominantly to the much narrower molecular weight distribution produced. ZN catalysts have three different sites on the catalytic particles; one site producing high molecular weight (MW), another medium MW and the third lower MW. Therefore the molecular weight distribution (MWD) is wide. Metallocene catalysts have only one site and therefore the MWD is narrower, resulting in lower density polymers with improved physical properties such as clarity in the resultant plastic film or moulding. Effects of molecular weight and molecular weight distribution on the flow properties of a polymer can be determined by measuring the melt flow

12.6 Branched polymer showing short and long chain branching (e.g. LDPE).

12.7 Linear polymer (linear polymers can have very minor amounts of branching. Examples are HDPE, LLDPE, PET, PA, PVC and PS).

index (MFI) of the particular polymer. The wider the molecular weight distribution for a given average molecular weight of a polymer, the less is the resistance to flow (viscosity) and the higher the MFI.

12.6.4 Crystallinity

Crystallinity, as applied to polymers, can be defined as the fraction of a polymer that consists of regions showing three-dimensional order. All polymers have some degree of crystallinity. Those with little crystallinity are known as amorphous polymers, for example low density polyethylene or amorphous polyethylene terephthalate (APET), and those with a high degree of crystallinity are known as crystalline polymers, for example high density polyethylene or crystalline polyethylene terephthalate (CPET).

Figure 12.8 shows polyethylene homopolymer in its two main forms, high density polyethylene (HDPE) on the left, and low density polyethylene (LDPE) on the right. The branching seen in LDPE prevents the molecular chains from packing close together, reducing density and subduing crystallinity. HDPE, on the other hand, has few or no side chains and therefore the molecular chains can pack closely together, resulting in HDPE having a higher density, and because the chains can align themselves to one another, a greater degree of crystallinity.

Table 12.2 shows the progression from LDPE to HDPE and beyond. The conclusions drawn from this are that as density increases so does crystallinity, resulting in increases

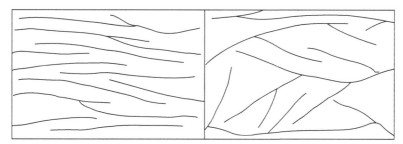

12.8 Polyethylene homopolymer in its two main forms: high density polyethylene (HDPE) on the left, and low density polyethylene (LDPE) on the right.

Table 12.2 Progression from LDPE to HDPE and beyond

0.91	0.92	0.93	0.94	0.95	0.96	0.97
LDPE		MDPE		HDPE		UHDPE
65%			Crystallinity			95%
Lower			Stiffness			Higher
105°C			Softening point			126°C
13.5 mPa			Tensile strength			30 mPa
500%			Elongation			20%
Lower			Barrier			Higher
Lower			Grease resistance			Higher

in stiffness, softening and melting points, tensile strength, barrier to moisture and gases and resistance to oil and grease. At the same time the amount the polymer will stretch and its transparency will decrease. There is, however, an exception to the transparency rule. Crystals are present in all polymers. If the crystals are large enough and of differing sizes, they will reflect and refract (alter the linear path of light by changing the angle of the path it is travelling in) light in all directions, causing the plastic to lose transparency. If the crystal formation sites are kept small by preventing the site from growing and their size is less than the wavelength of visible light, refraction is prevented. Special nucleating agents can be added, providing a wide range of benefits including the deliberate inducement of the formation of small crystals, for example where clarity of polypropylene copolymer is required for bottle manufacture.

12.6.5 Molecular weight and molecular weight distribution

The longer the polymer chain, the higher the molecular weight for any given polymer type. The type of polymerisation and the initiator or catalyst used will affect the molecular weight distribution. As the number of units increases, we move from simple monomers, which are normally gaseous, through to liquids and then waxy materials (oligomers), finally reaching the solid polymeric material we know as plastic. The size of the polymer is normally expressed as molecular weight (MW) or molecular mass.

As the molecular weight (measured as an average molecular weight of a polymer) increases, the melting temperature increases as does the viscosity of the melt, making it more difficult to process, but mechanical properties such as tensile strength and stiffness improve, which is a benefit. Processors therefore have to compromise between ease of manufacture and improvement in mechanical properties, based on the needs of their customers.

However, polymerisation is not that uniform. In every polymer of a particular molecular weight there will be a variation of chain length within the polymer, referred to as the molecular weight distribution. Figure 12.9 is a pictorial representation of the variation in molecular chain length (molecular weight) of an addition polymer such as polypropylene. Chain length can vary from 100 to over 100,000 repeat units.

This variation affects the properties of the polymer. If we can predetermine the molecular weight distribution, for example by using a metallocene catalyst, we can have better control over the final properties of the polymer. The molecular weight

12.9 Pictorial representation of the variation in molecular chain length (molecular weight) of an addition polymer such as polypropylene.

can be the same but if the molecular weight distribution is less, then the mechanical properties such as tensile strength are greater. However, the melting temperature range is shortened as is the heat sealing temperature range and the melt flows less readily; all of which could be detrimental if using a film to heat seal to a rigid container. Polymers with a wide molecular weight distribution will melt over a greater temperature range than polymers with a narrow molecular weight distribution and, if both polymers have the same average molecular weight, the one with the narrower range will have a higher melting point.

12.6.6 Number of different monomers in a polymer

A homopolymer will have the characteristics of its monomer type – ethylene to polyethylene, propylene to polypropylene, for example. However, if we combine the two monomers to form a copolymer we can use one monomer to improve the properties of the other. Polypropylene homopolymer (depending on tacticity) has poor low temperature properties, is difficult, if not impossible, to extrusion blow mould due to its low melt strength and it can be used for blown film only if a special process is used (see Chapter 14). It also has poor clarity. By copolymerising with ethylene, it is possible to overcome most of these difficulties (with the exception of blown film).

Another good example of improving performance by moving from a homopolymer to a copolymer is the copolymerisation of ethylene with one of the higher olefins such as hexene (C_6H_{12}), heptene (C_7H_{14}) or octene (C_8H_{16}), to create a graft copolymer with improved elasticity and elongation vs low density polyethylene, rendering it very suitable for stretch wrap and carrier bags. Selection of the comonomer is very important as the final polymer film can be too easy to elongate or will not elongate enough. As the number of carbon atoms in the comonomer increases, its percentage elongation increases.

Yet another example is in the comparison between general purpose polystyrene homopolymer (GPPS) and high impact polystyrene (HIPS) copolymer. The homopolymer is crystal clear, has poor oxygen and gas barrier and is very brittle. By copolymerising the styrene with butadiene, there is an improvement in impact resistance and barrier properties but a reduction in transparency. An alternative form of copolymerisation to produce styrene butadiene styrene can result in a thermoplastic elastomer (TPE) which is clear and has special properties close to those of a rubber. By copolymerising styrene with acrylonitrile to form styrene acrylonitrile, a copolymer with excellent chemical and scratch resistance and improved gas barrier can be formed. These improvements can be combined by copolymerising all three monomers together to form acrylonitrile butadiene styrene (ABS). The proportion of each constituent can be varied, allowing a range of polymers to be produced with very different properties from the starting materials. Other monomers such as methacrylic acid can be added to replace styrene, and a copolymer of methyl methacrylate, acrylonitrile and butadiene is used in film form for packaging of meat and cheese, chemicals and cosmetics. The methacrylate adds clarity, the acrylonitrile provides a good gas barrier and the butadiene imparts toughness and resistance to puncturing.

Sometimes, where copolymerisation is not possible, polymers are blended together to combine the properties of each individual polymer. An example would be polyethylene vinyl acetate (EVA) added to polyethylene to improve flexibility and adhesion properties, as used in stretch wrap film, for example.

12.6.7 Orientation of monomers in a copolymer

We have already discussed how different comonomers can be orientated in a copolymer. Copolymers can contain the monomer units in a completely random, alternating, block or graft format:

- Random copolymers, such as in the case of ethylene propylene copolymer, produce films and solid packaging forms with greater clarity, lower melt temperature and therefore a better low temperature resistance.
- Alternating copolymers, though rare, have properties associated with those that would be expected if the repeat unit were XO (X = propylene, O = ethylene).
- Block copolymer properties are quite different from those of random or alternating copolymers. For example, styrene butadiene block copolymers have a much higher impact strength than their random form, due to the elastomeric nature of the butadiene block having the ability to absorb much more energy without rupturing than a single unit, especially when paired with a brittle styrene unit. The same phenomenon occurs in ethylene propylene block copolymers which are used where impact strength is required.
- Graft copolymer properties are similar to block copolymer; for example we improve the impact strength of styrene butadiene by graft polymerisation rather than random or alternating. We can graft polymerise onto a chain formed from one monomer as previously described or onto a copolymer backbone as in the case of ABS (acrylonitrile butadiene styrene).

12.6.8 Forces within a polymer

Inter-atomic forces are the forces which bond atoms together to form molecules, for example carbon (C) to hydrogen (H), oxygen (O) or nitrogen (N). These forces make it possible to form very large polymer molecules containing different atoms. There are two types of inter-atomic bonds in polymers, covalent bonds and ionic bonds.

Covalent bonds, which are the more common, consist of two electrons shared by two adjacent atoms. These link atoms together by single, double or triple bonds. Aromatic structures, such as benzene (see Fig. 12.10), contain alternating double and single bonds in a cyclic formation. Polystyrene is a good example of this, where the hydrogen atom of ethylene is replaced by a benzene ring to form styrene which in turn is polymerised to polystyrene.

Sharing of electrons between dissimilar atoms is not equal. One atom having a greater affinity for the shared electron than the other results in a polar covalent bond as in PVC, where the C–Cl bond creates polarity compared to a C–C bond which would be non-polar. Ionic bonds, though rare, do occur where the affinity of two

12.10 Aromatic benzene.

Table 12.3 Inter-molecular forces and their effects on polymer properties

Intermolecular forces	Cohesive energy density	Typical polymer characteristics
Small	Low	Flexible, elastomeric behaviour, high permeability to gases, e.g. polyolefins
Medium	High	Stiffer plastic behaviour, e.g. PET
Large	Very high	High resistance to stress, high strength, good mechanical properties, low permeability to gases e.g. Polyamide (PA), ethylene vinyl alcohol (EVOH)

atoms for their shared electron is so unequal that electrons transfer from the weaker atom to the stronger, creating negative and positive ions. Oppositely charged ions attract each other. This is the mechanism by which ionomers (DuPont Surlyn®, for example, based on polyethylene) are formed. A sodium (Na^+) or zinc (Zn^{2+}) ion is introduced to provide some very different properties than pure polyethylene (see Chapter 13).

As molecules come together, forces are exerted by each on one another. These forces are called inter-molecular forces, i.e. between neighbouring molecules. Forces between different parts of the same molecule are referred to as intra-molecular forces. Inter-molecular forces (often called van der Waals or secondary forces) can be split into three primary categories: dispersion, induction and dipole forces. The primary forces (or bonds), namely covalent and ionic bonds, determine the molecular structure of the material, whereas the secondary forces (or bonds) are responsible for the physical nature of the molecule. Gases have the weakest secondary bonds, solids the strongest with liquids having bonds of a strength in-between the other two.

If the strength of the cohesive energy density (the energy required to distance a molecule from its neighbour so that no significant inter-molecular forces are present) is compared to the size of the inter-molecular forces, we can better understand their effect on polymer characteristics (see Table 12.3). Hydrogen bonding is a subcategory of dipole forces, their high strength and unique characteristics making them worth mentioning separately. Hydrogen bonds are an important consideration in polyamides and ethylene vinyl alcohol as can be seen in Table 12.3. (They are also very important in paper and paperboard – see Chapter 10).

Crosslinking creates bonds which have such a high bond strength that the energy to break the bond is greater than that required to decompose the polymer. Few examples of fully crossedlinked polymers are found in plastic packaging. A few closures and some adhesives, especially for lamination, utilise the performance characteristics of thermosetting plastics. Figure 12.11 shows the structure of a thermoplastic polymer (on the left), either branched as shown, or with little or no branching. Each polymer chain is unconnected with the next, compared to a thermosetting polymer (on the right), where all the polymer chains are connected by a crosslink (see Fig. 12.11).

The structure of a thermosetting polymer provides some advantages and disadvantages when compared to a thermoplastic polymer.

Thermosetting polymers:

- cannot be remelted once crosslinked and therefore cannot be recycled
- have better temperature resistance
- have improved chemical and solvent resistance
- exhibit better resistance to gas and moisture permeation
- do not distort under load, unlike thermoplastics
- have a higher density than thermoplastics.

When thermoplastics are heated, the weak bonds holding the polymer chains together break, which allows the polymeric molecules to move independently of one other, resulting in them melting and being able to be formed. Once cooled in the designated form in which they have been held, for example a film or a three-dimensional moulding, the inter-molecular forces form again, holding the structure together. This can be repeated many times. Thermosetting compounds, such as phenol, melamine and urea, formaldehyde resins, are placed in the mould in a semi-polymerised state. The mould is closed and heat applied. This thermally sets the resin in the mould, due to the cross-linking of the molecules, and it cannot be remelted.

Thermoplastic elastomers (TPE) are polymers which can be processed in the same way as thermoplastics but act similarly to a thermoset elastomer such as rubber, once in the solid state. They are copolymers of rigid and elastomeric monomer units combined in such a way (graft or block) to exhibit special properties. They are rubber-like and quickly recover from deformation; add tactility to a package and, if the correct combination of comonomers is used, provide polymers with near glass-clear clarity. A good example here is styrene butadiene styrene block copolymers, which can be used for soles of shoes but also for the manufacture of clear rigid packaging, depending on the make-up of the copolymer.

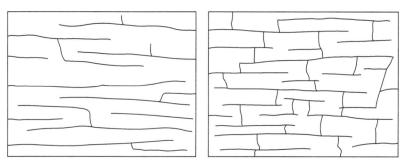

12.11 Thermosetting and thermoplastic polymers.

TPEs consist of one of six categories:

- styrene block copolymers (e.g., styrene butadiene styrene)
- polyolefin blend
- elastomeric alloys
- thermoplastic polyurethanes (TPU)
- thermoplastic copolyesters
- thermoplastic polyamides.

Styrene block copolymer TPEs are the most widely used, an example of which can be seen in Fig 12.12. The large hard, brittle styrene block sits on the flexible butadiene blocks holding them in place as if they were crosslinked. Once heat energy is applied to the monomer, the forces holding everything in place are broken and the molecules move freely. When the polymer is in its solid state, the held elastomeric butadiene chains have very good elastomeric properties providing very good resilience to the clear polymer. Such polymers are being developed for film and mouldings where they are both beneficial and affordable. Cosmetic and toiletry bottles and jars, films (e.g. for cling film), providing glass clarity, tactility and improved barrier and mechanical properties can all be manufactured from thermoplastic elastomers.

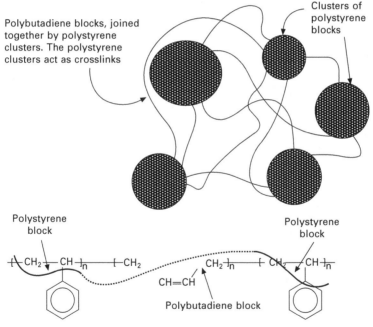

12.12 An example of a styrene block copolymer TPE: poly(styrene-butadiene-styrene), otherwise known as SBS rubber.

12.6.9 Additives

The majority of polymers used to manufacture plastic packaging are thermoplastic. If not compounded with process aids and functional additives, thermoplastics would have a much shorter usage life than they actually do. Additives used in the processing of thermoplastic polymers are:

- Antioxidants, which help to prevent deterioration of the polymer chain as a result of oxidation. They can sometimes cause yellowing of the polymer surface.
- UV stabilisers and inhibitors, which help to slow down the deterioration of the polymer by protecting it from UV light.
- Slip agents, which control the amount of slip (coefficient of friction (CoF)), by blooming to the surface. This can take up to 72 hours before it becomes effective consistently.
- Antistatic agents, which provide a surface which attracts fine droplets of water thus dissipating any build-up of static electricity which would attract dust particles.
- Colours, which are used to decorate and help brand the product. They may also protect against UV and natural light, e.g., carbon black. They are added as a solid or liquid masterbatch (see Chapter 14).
- Fillers, which can be added for special effects, e.g. mica to provide shiny particles, china clay to reduce the cost or glass fibre to improve physical strength. If the filler is added in a very orderly manner and the size of the filler particle is of nano proportions (approx. 200×10^{-9} m), it can increase a polymer's barrier to gasses by diverting the gas around the filler particles, thus increasing the length it has to travel through the polymer (see Fig. 12.13).

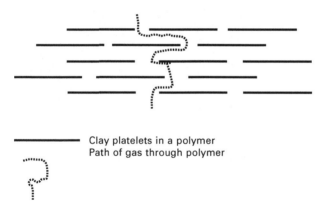

12.13 Nano fillers for improved gas barrier.

12.6.10 Special treatments of polymeric materials

We have already discussed the need to treat the surface of certain polymers (e.g., polyolefins) to ensure there is enough energy to adhere sufficiently to addition of adhesive, coatings and print. There are other special treatments used either for aesthetic purposes or to improve properties. Vacuum deposition of metals and metal oxides on the surface of polymer films increases both gas and moisture vapour barrier as well as reducing the amount of UV light that can pass through the film. The usual materials used for vacuum deposition are aluminium metal, aluminium oxide or silicon oxide. A typical use for a film with an aluminium metal coating is high fat snack products; a bag of potato crisps or chips requires a moisture vapour, gas and UV barrier to prevent the product going rancid and soft (Fig. 12.14). Vacuum deposition can also be carried out on three-dimensional mouldings, for example a clear coating of silicon oxide deposited on plastic bottles for beer and soft drinks, to improve oxygen barrier, and a metallic coating on plastic cosmetic compacts to give the more expensive appearance of metal.

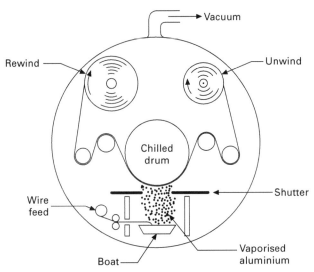

12.14 Addition vacuum deposition chamber.

12.6.11 Mixing polymers

Polymers can be mixed together to form a cohesive mix, as discussed earlier in this chapter. However, if we want to gain the maximum effect, we can combine polymers together in a more orderly fashion. We can coextrude or coinject different polymers together, creating individual layers, which enables us to use the combination of the properties of the individual monomers; for example EVOH has excellent gas barrier but is adversely affected by moisture, PP has excellent moisture barrier but no effective gas barrier. If we combine the two polymers together we get a film or moulding with excellent gas and moisture barrier (see Chapter 14 for details).

Special coextrusions are also effective in improving the heat seal, anti-fogging and cling properties of polymer films. In the modern world of technical advancement, it is unlikely that a polymer film will consist of a single extrusion. Coextrusion is too advantageous to ignore. One important advantage is that the waste material from a conversion process can be incorporated in the middle layers of a film or moulding, thus avoiding costly waste and environmental damage.

12.6.12 Physical orientation of polymers

Mechanical and barrier properties of a polymer can also be improved by stretching the polymer chains of a film or a moulding, in one or both directions. This is carried out at the melting temperature of the polymer. It stretches and aligns the polymer chains, providing improved physical (tear strength, tensile strength, reduced stretch, better clarity, etc.) and chemical (improved gas and moisture barriers) properties but may have an adverse effect on the heat sealing range of a film, as for polypropylene.

Polymer properties of the common packaging polymers are discussed in Chapter 9.

12.7 Sources of further information and advice

Baner, A.L. and Piringer, O. (2008) *Plastic Packaging Materials for Food.* Wiley, New York.

Baner, A.L. and Piringer, O. (2008) *Plastic Packaging: Interactions with Food and Pharmaceuticals.* Wiley, New York.

Giles, G. and Bain, D. (eds) (2000) *Materials and development of plastics packaging for the consumer market.* Wiley, Oxford.

Giles, G. and Bain, D. (eds) (2001) *Technology of plastics packaging for the consumer market.* Wiley, Oxford.

Hernandez, R., Selke, S. and Culter, J. (2000) *Plastics Packaging.* Hanser, Munich.

Nicholson, J. (2011) *The Chemistry of Polymers*, 4th edn. RSC Publishing, Cambridge.

In addition to the above texts, the British Plastic Federation (www.bpf.co.uk) is a valuable source of information on all aspects of plastics, including material properties, industry applications and forming methods.

Other useful sources of information include:

- http://www.britishplastics.co.uk
- http://www.incpen.org
- http://www.packagingdigest.com
- http://www.packagingfedn.co.uk
- http://www.packagingtoday.com
- http://www.pafa.org.uk
- http://www.plasticsinpackaging.com

13
Plastics properties for packaging materials

A. EMBLEM, London College of Fashion, UK

Abstract: This chapter presents a review of the properties and uses of the common packaging plastics, and some of the materials used for specialist applications.

Key words: plastics packaging market, tensile strength, elongation, tear strength, impact strength, surface friction, viscoelasticity, transparency, chemical stability, environmental stress cracking, barrier properties, glass transition temperature, melting temperature, heat sealability, density, polyethylene family, ethene, low density polyethylene, linear low density polyethylene, high density polyethylene, metallocene, collation shrink film, pallet shrink film, pallet stretch film, stretch hooding, ethylene vinyl acetate, plastic sacks, ethylene vinyl alcohol, polyvinyl acetate, polyvinyl alcohol, ethylene acrylic acid, ionomers, polypropylene, isotactic, atactic, syndiotactic, nucleating agents, biaxial orientation, thin wall containers, cast polypropylene, polarity, polyvinyl chloride, polyvinylidene chloride, plasticiser, polystyrene, general purpose styrene, high impact polystyrene, expanded polystyrene, polyethylene terephthalate, polyethylene terephthalate glycol, polyethylene naphthalate, styrene copolymers, styrene acrylonitrile, acrylonitrile butadiene styrene, Barex®, polyamide, nylon, polycarbonate, fluoropolymers, thermoplastic elastomers, cyclic olefin copolymers, liquid crystal polymers, thermosets, phenol formaldehyde, urea formaldehyde, cellulose film, cellulose acetate, bio-based polymers, biodegradation, biomass, oxodegradable, starch-based polymers, cellulose-based polymers, polylactic acid, polyhydroxyalkanoates.

13.1 Introduction

Plastics are an essential part of modern life. Major industries depend on them and products as varied as cars, aeroplanes, electronic equipment, textiles, furniture, household goods, jewellery, shoes and clothing all use increasing quantities of plastics in their construction. Plastics used in packaging are just as widespread, from the commonly used plastic bottles and caps for milk and soft drinks, to the films used to pack a range of goods such as bacon, confectionery, nuts and bolts and textiles, and the rigid boxes/cases used for CDs and computer games. This chapter follows on from the basic principles of polymer chemistry discussed in Chapter 12, and it is essential to refer back to it when working through this chapter. Of particular importance is the section on factors affecting the characteristics of polymers, as this contains many examples relevant to individual material properties discussed here.

13.2 Market overview

Plastic is the most rapidly expanding sector of packaging materials, despite being the newcomer when one reviews historical development. From Table 13.1 it can be

Table 13.1 Overview of the development of plastics

Date	Material
1907	Phenol formaldehyde ('Bakelite') thermosetting resins
1927	Polyvinyl chloride (PVC)
1927	Cellulose acetate
1936	Polyvinyl acetate (PVA)
1938	Polystyrene (PS)
1938	Nylon 66
1939	Polyvinylidene chloride (PVdC)
1941	Polytetrafluoroethylene (PTFE)
1942	Low density polyethylene (LDPE)
1944	Polyethylene terephthalate (PET)
1948	Acrylonitrile butadiene styrene (ABS)
1955	High density polyethylene (HDPE)
1957	Polypropylene (PP)
1957	Polycarbonate (PC)
1960	Linear low density polyethylene (LLDPE)
1964	Ionomers, e.g. Surlyn
1970	Thermoplastic elastomers (TPEs)
1972	Ethylene vinyl alcohol (EVOH)
1994	Polyethylene naphthalate (PEN)

Table 13.2 European polymer demand for packaging vs. other markets, 2007 (000s tonnes)

Material	Packaging	Other markets	Packaging as % of total market
Low density/linear low density polyethylene	5,976	3,110	66
High density polyethylene	3,357	2,609	56
Polypropylene	3,414	5,789	37
Polyvinyl chloride	748	5,707	12
Polystyrene	1,511	2,218	68
Polyethylene terephthalate	3,096	147	95
Other	132	3,084	4
Total	18,204	22,664	45

Source: Applied Market Information Ltd (www.amiplastics.com).

seen that the first plastics were developed in the early years of the twentieth century. However, it was not until the 1950s that their use became widespread, when the relatively low cost polyethylenes brought plastics out of specialist applications and into the mass market. These 'new' materials became increasingly popular due to their ready availability (made from by-products of the oil industry); easy processability (most of those listed in Table 13.1 are thermoplastics, which soften at relatively low temperatures compared with materials such as glass and metal, and are easy to form into a range of different shapes); and the ability to tailor their properties closely to the intended end use, leading to a wide range of market applications.

Packaging is not the only market for plastics, but it is certainly a significant one. Table 13.2 shows the European market for plastics in packaging was just over 18 million tonnes in 2007, which is double the 1987 tonnage. Of this 18 million tonnes, 58% is used in rigid packaging (bottles, tubs, caps, etc.) and 42% in flexible applications, i.e. films and laminates. Not surprisingly, 70% of all plastics packaging is used for food and drink, the remainder being used for household chemicals, pharmaceuticals, cosmetics and toiletries, and a range of other non-food uses, plus industrial products such as shrink and stretch film, and carrier bags.

13.3 Key properties for packaging applications

Compared with other packaging materials, plastics are generally lighter in weight, more easily formed into different shapes, and extremely versatile. This versatility is largely responsible for their growth in usage: the ability to carefully tailor the performance of a plastic container to the needs of the product, the market and the demands of the supply chain means that there is a 'plastic' solution to almost all packaging problems.

When considering the performance of plastic materials, the properties of most interest in packaging applications, and their relevance in use, are summarised as follows.

Tensile strength and elongation are probably of most practical relevance in flexible packaging, especially during form, fill and seal machine operations. Films and laminates must withstand the tension of being pulled through a forming station without stretching or breaking, and bags/sacks must withstand the force of the product weight being dropped into place without distortion. For pallet stretch wrapping, a film with high elongation is required, the most commonly used material being linear low density polyethylene. Tensile strength is a directional property and specifications will state values in machine and cross direction.

Tear strength is also relevant in flexible packaging applications, both in terms of resistance to tearing during machine forming and subsequent handling. Tear properties also influence consumer convenience in ease of opening. In general, plastics have good resistance to tearing, compared, say, with paper, and if the consumer is expected to tear open a pack it is advisable to facilitate this by using a paper/film laminate and/or incorporating an easy-open device in the pack, such as a laser cut.

Impact strength is a measure of a material's resistance to breakage when subject to a sudden impact, such as a falling object. It is probably more meaningful to evaluate the impact strength of a filled plastic component, e.g. a drum or a sack using a series of drop tests, than to rely on data generated from impact tests on the plastic materials.

Surface friction influences the ease with which a packaging material or component will travel through the packaging line, such as a film moving over a forming collar. It also influences the stability of palletised loads during transit and of primary/ secondary packs on display. Surface friction can be modified to the desired level by the use of additives during film manufacture.

Viscoelasticity or 'creep' is a major consideration when deciding which plastic to use for a given application. All thermoplastics exhibit some degree of permanent deformation under load. This typically manifests itself when filled containers are stacked on pallets during storage. Stacking tests under load, using actual packaging components, are essential to evaluate the true extent of a material's suitability.

Optical properties such as transparency and gloss mainly influence pack aesthetics. If the product has an attractive appearance and is stable on exposure to light, then a highly transparent pack may be desirable. Totally amorphous polymers are transparent; as the degree of crystallinity increases, transparency decreases and the materials become hazy and, eventually, opaque.

Chemical stability describes the extent to which a polymer is suitable for direct contact with a range of different product types, such as organic solvents, oils, acids or alkalis. Table 13.3 provides guidance on the general suitability of the common packaging plastics with a selection of such products, although this must be verified in each specific case, preferably using actual packaging components and products.

Environmental stress cracking is the failure of a plastic component under the combined effects of stress and an aggressive environment. Stress in plastic mouldings will bring about mechanical failure over time, but the presence of stress cracking agents such as detergents and oils can dramatically accelerate this failure, sometimes with catastrophic results.

Moisture and gas barrier properties of plastics determine their suitability to provide the required shelf life of the product. The data shown in Table 13.4 may be used as a starting point in selecting which plastics may be suitable, but again, storage tests using actual packs and actual products are essential. Refer back to Chapter 3 for more detailed information on this.

Temperature data of plastics has been referred to in Chapter 12, which differentiates between the glass transition temperature (T_g) and the melting temperature (T_m). T_g is the point at which the material changes from a glassy to a rubbery state and it occurs in the amorphous regions of a polymer. T_m is the temperature at which the crystalline regions of a polymer melt and is therefore always higher than T_g. Totally amorphous polymers (e.g. atactic polystyrene with its large benzene ring on the main carbon chain) do not have a melting temperature and thermosets do not have a glass transition temperature (due to their high degree of crosslinking). The temperature information given in Sections 13.4 and 13.5 refers mainly to T_m (except for polystyrene) as this is of most practical importance when selecting which plastic to use for a given application, such as suitability for sterilisation, microwave or conventional oven cooking, and the material's heat sealability.

Heat sealability is an important property to consider for a range of flexible packaging such as sachets, bags, flow wraps and pharmaceutical blister packs, as well as semi-rigid packs such as thermoformed trays and pots. The temperature at which the sealing medium softens and starts to flow must be compatible with what can be achieved by a combination of sealing head temperature, pressure and dwell time on the packaging machine, so that correctly sealed packs are achieved. Pack seals must survive the rigours of the supply chain, i.e. they must maintain the containment, protection and preservation functions, and be openable by the final user.

Density, as shown in Table 13.5, is an important consideration for cost in use. Plastic resins are traded by weight and for applications such as flexible packaging it is the area of material which is of interest, i.e. the number of square metres of

Table 13.3 Chemical resistance of the common packaging plastics

Material	Dilute acid	Dilute alkali	Oils and greases	Aliphatic hydrocarbons	Aromatic hydrocarbons	Halogenated hydrocarbons	Alcohols
LDPE/LLDPE	****	****	** variable	*	*	*	****
HDPE	****	****	** variable	*	*	*	****
PP	****	****	** variable	*	*	*	****
PVC	****	****	*** variable	****	*	** variable	*** variable
PS	**	****	**	****	*	*	* variable
PET	****	**	****	****	**	**	****

**** Very good, *** Good, ** Moderate, * Poor.
Source: British Plastics Federation (www.bpf.co.uk).

Table 13.4 Moisture and oxygen barrier properties of the common packaging plastics

Material	MVTR	OTR
LDPE/LLDPE	15–25	7000–8000
HDPE	5–12	1500–2000
PP	3–7	1800–2500
PVC (U)	20–60	50–80
PS	70–160	4000–6000
PET	16–20	60–120

MVTR = moisture vapour transmission rate for 25 μ film in $g/m^2/day$; test conditions 38°C, 90% relative humidity.
OTR = oxygen transmission rate for 25 μ film in $cm^3/m^2/day$; test conditions 25°C, 50% relative humidity, 1 atmosphere.

Table 13.5 Density of the common packaging plastics

Material	Density kg/m^3
LDPE/LLDPE	910–930
HDPE	940–965
PP	880–920
PVC (U)	1230
PS	1070
PET	1360

Exact figures will vary with grade.

material available from 1 kg of resin. For any given thickness of film, this area can be calculated from the data in Table 13.5. It can be seen that material densities vary considerably and this must be taken into account, along with the price per tonne of resin, when comparing alternatives for a given end use.

13.4 The common packaging plastics

The ability to tailor material performance to requirements has already been emphasised as one of the most significant contributors to the growth in use of plastics in packaging and it is the case that most solutions are to be found amongst polyethylenes, polypropylene, polyvinyl chloride, polystyrene and polyethylene terephthalate (see Table 13.2). Even allowing for inaccuracies in data collection, it is clear that these materials dominate the packaging market. This section will review the key properties and uses of these common packaging plastics, while Section 13.5 will address the 'other' types of plastics used.

Note that it is not the objective of this chapter to provide fully comprehensive information on every packaging application of every plastic, which would be unrealistic within the space available. Several sources have been consulted in putting together this section (see Section 13.8) and some give quite extensive additional information. This section and Sections 13.5 and 13.6 aim to provide short notes on each material, sufficient to begin to make informed choices for given end uses, but the reader is required to explore other sources for more detailed information.

13.4.1 The polyethylene family of plastics

Ethene ($H_2C=CH_2$) is a gaseous by-product of the process of cracking of long chain hydrocarbons to produce much sought-after products such as petrol and aviation fuels. The development of polyethylene (original trade name Polythene) and its commercialisation in the 1940s/1950s have already been mentioned as an important milestone in the development of plastics for packaging.

For convenience, polyethylenes are usually classified according to density. The more branched chains in the structure, the less these chains can be packed together (i.e. there is more steric hindrance) and thus the lower the density of the material. As chain packing increases, so does tensile strength, barrier to moisture and gasses, heat resistance and opacity.

Low density polyethylene

Low density polyethylene is produced by a high pressure process, and has a mix of long and short branched chains and around 50–65% crystallinity, making it translucent in appearance. It is soft and flexible with good elongation before breakage and good puncture resistance. It has a fair moisture barrier and poor oxygen barrier and softens at around 100°C (lower for some grades), making it an economical polymer to process and readily heat sealable, but of course unsuitable for cook-in packs. In common with all the polyolefins, it is non-polar and must be surface treated prior to printing or laminating.

Linear low density polyethylene

Linear low density polyethylene is a copolymer of ethylene and other alkenes such as butene, hexene or octene. This results in an essentially linear chain arrangement, with the comonomer (i.e. the butene, hexene or octene) forming short, regular chains on the main carbon backbone. It has similar properties to LDPE although it is tougher with slightly better barrier properties. It requires slightly more energy to heat seal and its operating range for sealing is narrower than LDPE, making control of seal temperature on the packaging machine more critical. It is also a little more transparent than LDPE. As already stated, it has very high elongation before break, which accounts for its widespread use in pallet stretch film.

Around 70% of all LDPE and LLDPE is used as film. This includes secondary/tertiary packaging such as pallet stretch film, collation and pallet shrink film, as well as primary packaging such as sacks, retail carrier bags, produce bags and frozen food bags. Film manufacturers are now using blends of LDPE and LLDPE for these applications, the inclusion of the latter offering opportunities for increased strength and decreased gauge, and thus the consumption of LLDPE is increasing at the expense of LDPE.

Metallocene catalysed grades of LLDPE are also now readily available, offering high puncture resistance and good clarity, and these are finding particular application in the high performance end of the pallet stretch film market, especially for pre-stretch machines. Most stretch film is produced by the cast process (see Chapter 14). The choice of comonomer in LLDPE depends on performance required; octene and hexene offer superior strength but are more expensive than butene LLDPE, hence they are used for more demanding applications.

Collation shrink film for secondary packs uses LDPE and LDPE/LLDPE blends, as well as coextrusions to impart specific slip characteristics and thus ensure the stability of a palletised load. If transparency is an important factor, e.g. for collation packs of bottles of mineral water and soft drinks which are sold directly to the consumer, metallocene grades are suitable, offering opportunities for significant down gauging with no loss of strength.

Pallet shrink wrap film uses mainly LDPE, which may be blended with LLDPE and/or HDPE for improved strength and down gauging. In Europe, pallet stretch wrapping is far more common than pallet shrink wrapping, as the latter requires higher capital investment and is more expensive per pallet due to the amount of film used and the need for energy in the form of heat. However, there are some specific applications where it remains the optimum choice, e.g. securing heavy loads such as glass bottles in transit from the producer to the packer/filler, or for providing good protection against dust contamination for palletised sheets of paper/board in transit from the paper mill to the printer/converter.

One further application of LDPE/LLDPE film in secondary/tertiary packaging is that of pallet stretch hoods. Stretch hooding uses highly elastic film to form a hood which is stretched over the palletised load and pulled down to be secured onto the pallet. It requires no heat (unlike pallet shrink wrapping) and is a faster process than stretch wrapping. Films are usually coextrusions including EVA (ethylene vinyl acetate) copolymers and elastomers.

Plastic sacks are used mainly in the horticulture market (for compost, etc.) and for chemicals, fertilisers, building materials and animal feed. LDPE is widely used, often blended with LLDPE for puncture resistance, and MDPE (medium density polyethylene) and HDPE for added stiffness to aid performance on the filling line. Coextrusions are used to impart the required levels of slip and heat-sealing characteristics.

Apart from film uses, both LDPE and LLDPE are used as heat seal coatings, extruded onto paper, board, other plastic films and aluminium foil, as well as for lamination. One significant market is liquid packaging cartons (see Chapter 10). Rigid packaging uses of LDPE include small, squeezable bottles, tubes, push-in and push-on closures.

High density polyethylene

High density polyethylene is made using a low pressure process and Ziegler-Natta or Phillips initiators to control the chain formation, resulting in a highly linear (unbranched) structure. Metallocene catalysed variants are also available. HDPE is more crystalline than LDPE, and hence is more opaque, more rigid and has a higher tensile strength. It has a good moisture barrier and fair oxygen barrier. T_m is around 135°C and thus it withstands boiling water, which makes it suitable as a heat seal coating for steam sterilisable pouches and as a film for boil-in-the-bag foods.

Most of the applications of HDPE are in rigid packaging such as bottles for milk and household chemicals, as well as heavy duty items such as pallets, drums, crates and intermediate bulk containers. It is prone to environmental stress cracking and this should always be checked prior to selection. In the past it has been used widely for bottles for toiletries, although PET tends to be preferred in this sector due to its excellent transparency. It is also used for short-life screw caps, e.g. for milk and soft drinks bottles. Where screw caps are required for multiple uses, e.g. products such as shampoo and other toiletries, HDPE is liable to break in use and PP is a much better option. HDPE is also used for heavy duty sacks and bags, e.g. fertiliser, sand and aggregate.

Ethylene copolymers

The following ethylene copolymers are included here as part of the polyethylene family, although in usage they fall into the more specialised category of materials considered in Section 13.5.

Ethylene vinyl acetate

Ethylene vinyl acetate (EVA) is a random copolymer of ethylene and varying amounts of vinyl acetate (VA). The VA comonomer interferes with chain packing, reducing crystallinity and thus lowering T_m and improving transparency when compared with LDPE. As the VA content increases, crystallinity decreases, until at 50% the EVA is totally amorphous. Flexibility and toughness at low temperatures are also improved making it a good choice for frozen food bags. It has good hot tack and adhesive strength and a wide sealing temperature range, accounting for its uses as a heat seal layer and in hot melt adhesives.

Ethylene vinyl alcohol

Ethylene vinyl alcohol (EVOH) is another ethylene copolymer, this time using the comonomer vinyl alcohol (produced by the hydrolysis of vinyl acetate). It has excellent barrier to oxygen (less than 2 cc/m^2/day) but the –OH groups make it hydrophilic, i.e. it attracts water, which decreases the oxygen barrier. For the oxygen barrier to be effective, EVOH must be 'sandwiched' to protect it from moisture. This is commonly done by coextrusion, examples being PET/EVOH/PET for bottles for sauces and mayonnaise and/or PET/EVOH/EVA for modified atmosphere packs for processed meats.

Polyvinyl acetate and polyvinyl alcohol

Polyvinyl acetate (PVA) and polyvinyl alcohol (PVOH) are further examples of ethylene copolymers. PVA is used as an emulsion adhesive for bag, sack and carton making. PVOH is produced by hydrolysis of PVA and the strong hydrogen bonding imparted by the –OH groups means that pure PVOH is water soluble. The degree of water solubility is controlled by the amount of hydrolysis. Specific packaging uses for PVOH are in unit doses of detergents, where the entire pack is placed in the washing machine, and for agrochemicals, where the pack is placed into a tank of water and mixed. In both of these examples the advantages are that a measured dose of product is used, with no spillage on decanting and, in the latter case, limited exposure of personnel to what may be a toxic product.

Ethylene acrylic acid

Ethylene acrylic acid (EAA) is, as the name suggests, a copolymer of ethylene and acrylic acid (AA). As the AA content increases, crystallinity decreases (due to interference with chain packing) and heat seal temperature decreases, and the increase in polarity means that adhesive strength increases. Its excellent adhesion to a range of substrates makes it a good choice as the adhesive layer in foil laminates for composite containers, toothpaste tubes and sachets.

Ionomers

Ionomers are unusual in that they have ionic as well as covalent bonds in the polymer chains. They are made by reacting metal salts (commonly Na^+ or Zn^{++}) with acidic copolymers such as EAA or ethylene methacrylic acid (EMAA). The ionic bonds act like crosslinks between the polymer chains, resulting in tough, puncture resistant materials with excellent heat-sealing characteristics over a wide temperature range, and the ability to seal through contamination. Bonding to aluminium foil and paperboard is excellent. Ionomers also have very good resistance to oily products, making them useful as heat-sealing layers for processed meats. They are also used in rigid form for closures. There is a large range of options from which to choose, the main suppliers in the packaging field being DuPont, under the Surlyn® brand and Exxon Mobil under the Iotek™ brand.

13.4.2 Polypropylene

Polypropylene has the lowest density of the common packaging plastics (Table 13.5), which gives it an economic advantage over other materials. It is formed by the polymerisation of propene and thus differs from polyethylene in that it has methyl groups on the carbon backbone chain. The arrangement of these methyl groups can vary and, as already mentioned in Chapter 12, PP exists in three different tacticities:

- Isotactic: Methyl groups are all on the same side of the chain.

- Atactic: Methyl groups are arranged randomly on both sides of the chain.
- Syndiotactic: Methyl groups alternate around the chain in a regular manner.

The isotactic form is the one most commonly used in packaging and, unless otherwise stated, this is the form referred to in the literature on PP. It is a tougher and stiffer material than the PE family, with good resistance to creep and to environmental stress cracking, and has a good barrier to moisture (see Table 13.4). It has a higher T_m than HDPE (around 160°C) and can be used where HDPE may otherwise fail, e.g. hot-filled bottles and tubs for soups and sauces and steam-sterilised applications.

In contrast, atactic PP is a soft and rubbery, mainly amorphous polymer. Its chief use is in adhesives, although there is some development of atactic/isotactic block copolymers which form thermoplastic elastomers. The third variant, syndiotactic PP has a T_m of around 130°C. It is softer than isotactic PP but still tough and more transparent. Its resistance to gamma ray sterilisation is being exploited in some film applications for medical packaging.

At very low temperatures PP homopolymer becomes brittle and thus is unsuitable for use in long-term deep freeze conditions. To overcome this problem, propylene is copolymerised with ethylene (typically 1–7%), producing polypropylene random copolymers. As well as improved clarity, other benefits are increased strength and toughness, a broader melting range and a lower melting temperature which improves heat sealability. All of these benefits have led to the widespread use of PP copolymers.

The clarity of PP is also improved by adding nucleating agents (which reduce the incidence of crystalline regions) and this is utilised in making contact-clear ice cream containers, for example. Another way to impede the formation of crystallites is to subject the molten polymer to rapid cooling, as in the cast film process, resulting in highly transparent and sparkling films.

Polypropylene is a versatile material, with important applications in both flexible and rigid packs. As shown in Table 13.6, film is the most significant market, followed by thin wall packaging. PP film is usually biaxially oriented (BOPP) and this has become the material of choice for snacks, biscuits, cakes, confectionery and tobacco products, due to its high strength and adequate barrier properties, even at very low gauges. See Chapter 14 for descriptions of the orientation processes.

BOPP has excellent clarity, and can also be produced in pearlised and opaque

Table 13.6 Packaging applications for polypropylene

Pack type	
Film	32%
Thin wall containers	30%
Bottles	5%
Returnable transit packs (RTPs)	6%
Closures	13%
Other	14%

Source: Applied Market Information Ltd (www.amiplastics.com).

forms, giving a wide range of aesthetic effects. It is also produced in grades with good surface smoothness for subsequent metallising, which gives a shiny metallic appearance without the use of aluminium foil, and at the same time improves the barrier properties of the base film. Technical performance such as barrier, heat-seal characteristics and printability can also be improved by coating and coextrusion, and most BOPP is multi-layer, with suppliers offering a number of tailored options to suit a myriad of uses.

While BOPP is the most commonly used PP film, CPP (cast, non-oriented PP) film has important applications. As already mentioned, rapid cooling after casting results in a highly transparent film, used for textile bags and flower/plant packaging. Food packaging uses include retortable pouches.

Thin wall packaging describes thermoformed and injection moulded pots and tubs, such as those commonly used in the dairy industry for desserts, yoghurts and spreadable fats. PP offers good forming characteristics and good strength at a lower component weight, compared with polystyrene, which is its main rival in this sector (see Section 13.4.4). Thermoformed containers are made from sheet produced by cast extrusion, and as with film, coextrusions are possible to tailor the properties to the performance required.

Polypropylene belongs to the polyolefin family of polymers and therefore, like polyethylene has very low surface polarity and thus must be surface treated before printing.

13.4.3 Polyvinyl chloride (PVC)

PVC is made by the polymerisation of vinyl chloride (chloroethene) which is itself made by the chlorination of ethene. The chlorine atoms on the carbon backbone chain account for the much higher weight per unit volume of PVC compared to the polyolefins (see Table 13.5) and their syndiotactic arrangement means that the polymer is largely amorphous, and hence has good optical clarity. It has a good gas barrier and although the moisture barrier is poor, this is improved by coating with polyvinylidene chloride (PVDC) which has the added advantage of being readily heat sealable, a property utilised in blister packs for pharmaceutical tablets. PVC also has very good grease and oil resistance, and hence was the first alternative to glass bottles for cooking oil, and is still used for bath oils. It has a low T_m (around 90°C) and is thus easy to form, leading to its widespread use for thermoformings such as display trays. It also accepts ink readily. See Table 13.7 for the main uses of PVC in packaging.

Table 13.7 Packaging applications for polyvinyl chloride

Pack type	
Thin wall containers (including trays)	49%
Film	44%
Bottles	7%

Source: Applied Market Information Ltd (www.amiplastics.com).

Unplasticised PVC is naturally brittle, and as the proportion of plasticiser increases, so does flexibility. Plasticisers lower the glass transition temperature, thus improving processability, although they also decrease barrier properties, hence there is a wide range of grades available to suit end uses, from rigid bottles for oily products to highly flexible films such as 'cling' film. There is concern about the possible effects of plasticisers migrating into food products and this has led to a decline in the use of PVC in food packaging. A further contributor to this decline is the considerable environmental opposition to PVC because of the production of hydrochloric acid and dioxins on incineration (although these emissions can be abated, given the right choice of technology). Given these concerns, several companies have taken the decision to move away from PVC for packaging applications, although it continues to be widely used in the construction sector (window frames, pipes and cables) and for flooring.

13.4.4 Polystyrene

Polystyrene is the addition polymerisation product of phenyl ethene (commonly known as styrene). It is a highly transparent, glossy material, with a poor barrier to moisture and gases and limited chemical resistance. It softens at around 75°C and is liquid at around 100°C, and is thus easy to form. It accepts ink readily and thus is easy to print using flexographic or gravure printing for films, and dry offset letterpress or screen printing for three-dimensional components. General purpose styrene (GPS), sometimes known as crystal styrene is very brittle and most polystyrene used in packaging is the high impact variant (HIPS) which is a styrene-butadiene copolymer. The butadiene provides flexibility and lowers the softening temperature, and these two improvements account for the use of HIPS in thermoformed thin-wall containers for the dairy market. It is also thermoformed into sandwich packs and trays for salads, and made into biaxially oriented film for bags for fresh salads and other produce. The poor barrier properties are beneficial for products which respire, e.g. fresh fruit and vegetables, as they help to prevent a build-up in the pack of moist air which is conductive to mould growth.

A further application in packaging is expanded polystyrene (EPS). This is a lightweight material with fairly good compression strength and resistance to moisture. It is a good insulator against both temperature change and shock and is used for fast food trays, boxes for fresh fish and for home delivery of chilled and frozen foods, cell-packs for growing plants and as both loose-fill and pre-formed shapes in packaging fragile items such as china and glassware, domestic appliances and electronic goods.

Specialist copolymers of polystyrene are covered in Section 13.5.

13.4.5 Polyethylene terephthalate (PET)

PET is a condensation polymer made from a diacid (e.g. terephthalic acid) and a dialcohol (e.g. ethane 1,2-diol, commonly known as ethylene glycol). The acid and the alcohol first react together to form ester molecules, which are then polymerised, hence the common term 'polyester'. This term refers to a large family of compounds with many diverse applications; PET is the common thermoplastic polyester used in packaging.

PET is available in both amorphous (APET) and crystalline (CPET) forms, with the former being more commonly used. Moisture barrier is about the same as that of LDPE, but gas barrier is higher than most of the common packaging plastics and this can be further improved by metallising with aluminium or by coating with, for example, PVDC or silicon oxide (SiOx). Almost 90% of the PET used in packaging is for bottles, primarily for soft drinks and mineral water which are now routinely packaged in PET rather than in glass, due to the weight savings and associated transport cost savings available. PET bottles are also being increasingly used for household chemicals such as laundry products, where the transparency of APET is often seen as an advantage over both HDPE and LDPE.

A significant growth market for PET is in bottles, jars and other components such as lipstick cases in the cosmetics sector, where the good oxygen barrier provides protection against degradation of oily products and loss of perfume. Early uses tended to be for lower-price products such as shampoo, but now thick-walled jars in transparent and coloured forms are widely used instead of glass for expensive creams. The materials used in this sector are mainly copolyesters such as polyethylene terephthalate glycol (PETG), made by varying the starting diacids and dialcohols and variants are available for injection and extrusion blow moulding, as well as for thin-wall and thick-wall containers.

A potential growth market for PET is in beer bottles, with the commercial availability of surface coatings and multi-layer technology to improve gas barrier. At the current time the use of PET for beer is mainly confined to outdoor events, where safety is the main driver, but developments such as PET/PEN (polyethylene naphthalate) blends and coextrusions could bring PET into the mainstream beer market as a realistic alternative to glass.

Flexible packaging uses of PET include biaxially oriented film for pouches and lids for ovenable ready meals. The orientation improves stiffness and the high clarity enhances visual impact, especially in laminates where the PET is reverse printed. PET sheet is thermoformed into trays for salads. The crystalline (CPET) variant is opaque and is used for trays for ovenable ready meals, due to its high temperature resistance.

Uses of recycled PET (r-PET) for primary packaging are increasing, both for food and non-food products. As with all developments, compatibility and shelf life studies are essential. A recent study in the Cosmetic Science department at the London College of Fashion showed encouraging results using PET bottles containing 30% post consumer recyclate (PCR) in contact with common cosmetics such as baby oil, shower gel and shampoo (Talarek, 2011). It is thought that higher PCR content would also be compatible with these products, although the decrease in bottle clarity may be aesthetically unacceptable.

13.5 Specialist polymers used in packaging

Table 13.8 lists the materials which fall into this category. Most of these materials are engineering polymers and their main uses are in markets such as automotive, aeronautical, electronic and others, with packaging end uses representing a small percentage of overall consumption. They are invariably much more expensive than the commodity polymers discussed in the previous section, and are selected for packaging applications only when the commodity polymers do not have the required technical performance, and the product value justifies the additional cost. (Note that EVA, EVOH, PVA, PVOH, EAA and ionomers have already been discussed in Section 13.4.)

13.5.1 Styrene copolymers

In this category the main copolymers used in packaging are styrene acrylonitrile (SAN) and acrylonitrile butadiene styrene (ABS), along with some specialist variants.

SAN is a copolymer of styrene and acrylonitrile ($CH_2=CHCN$) in a ratio of around 3:1. It is amorphous, rigid and has good resistance to acids, alkalis, oils and greases. Its high transparency and easy surface printability make it a good choice for cosmetics compacts and jars, and it is available in a range of colours and grades. At this ratio of styrene to acrylonitrile it has a poor gas barrier, but this can be improved by increasing the acrylonitrile content.

ABS is a random styrene acrylonitrile copolymer grafted to butadiene (see Chapter 12). Like SAN, it has good chemical resistance, is readily printable and has good resistance to environmental stress cracking. Many variants of composition are available to suit the required technical performance and aesthetic qualities. ABS is naturally opaque, but the variant methyl methacrylate acrylonitrile butadiene styrene (MABS) is highly transparent. Both materials are used in the cosmetics sector for

Table 13.8 Specialist polymers used in packaging

Ethylene vinyl acetate	EVA
Ethylene vinyl alcohol	EVOH
Polyvinyl acetate	PVA
Polyvinyl alcohol	PVOH
Ethylene acrylic acid	EAA
Ionomers	
Styrene acrylonitrile	SAN
Acrylonitrile butadiene styrene	ABS
Polyvinylidene chloride	PVDC
Polyethylene naphthalate	PEN
Polyamide	PA
Polycarbonate	PC
Fluoropolymers	
Thermoplastic elastomers	TPEs
Cyclic olefin copolymers	COCs
Liquid crystal polymers	LCPs
Thermosetting plastics	
Cellulose materials	

mascara and lipstick cases, compacts and closures. ABS is also suitable for tubs for butter, margarine and desserts such as yoghurt, where it offers improved toughness compared to polystyrene, albeit at a higher purchase price.

As already noted, increasing the acrylonitrile component in these copolymers improves the gas barrier and one example of this is Barex®. This is a copolymer of acrylonitrile and methyl methacrylate (75:25 ratio) and nitrile rubber; it has excellent gas barrier, good chemical resistance and good sealability, making it useful for packaging meat and cheese. It also has excellent impact strength, withstands repeated flexing without cracking and is resistant to sterilisation by gamma radiation and ethylene oxide, making it a good choice for demanding applications in the medical and pharmaceutical sectors.

13.5.2 Polyvinylidene chloride (PVDC) copolymers

PVDC is made by the addition polymerisation of 1,1-dichloroethene (vinylidene chloride). The homopolymer is very difficult to process as it decomposes below its melting temperature. It thus has no commercial importance and all the PVDC used in packaging is copolymerised with vinyl chloride or alkyl acrylates (e.g. methyl acrylate). The most useful properties of these copolymers are excellent resistance to oils and fats, high gas barrier and heat sealability. They can be used as surface coatings, in coextrusions and as films. Coatings can be water-based (e.g. PVDC/methyl copolymers) and solvent-based (e.g. PVDC/acrylonitrile copolymers) and provide a cost effective way of improving the barrier properties of PP, PET, etc., films, and of enhancing the performance of paper and board. Coextrusions of PVDC copolymers and polyolefins are suitable for multilayer films for products such as meat and cheese, and single layer PVDC copolymer films can be used alone as food wrapping and in laminates wherever moisture and gas barrier are important requirements.

As with PVC, there are environmental concerns about the use of PVDC and these have prompted companies to seek out alternatives. EVOH is one such alternative, especially with regard to gas barrier although, as already noted, it cannot be used as a coating.

13.5.3 Polyethylene naphthalate (PEN)

PEN is a condensation polymer of ethylene glycol and naphthalate dicarboxylate. It has been available since the 1980s and PEN film is used in electrical insulation. Packaging uses are limited as yet, almost certainly due to its high price, although it has significant advantages over PET in the drinks bottle market, and refillable PEN bottles for beer have been launched in Denmark and for fruit juices in Germany. The advantages of PEN homopolymer over PET are its higher resistance to heat and 4–5 times better barrier to moisture and gases. It also blocks out UV light. If the end use demands an improvement over PET without the full properties of PEN homopolymer, a cost effective solution may be found in the range of PEN/PET copolymers and blends available.

13.5.4 Polyamide or nylon (PA)

Polyamides are condensation polymers made from diamines and diacids and are named according to the number of carbon atoms in the starting materials, e.g. Nylon 6,6 is produced from hexamethylene diamine $H_2N-(CH_2)_6-NH_2$ and 1,6-hexanedioc acid $HOOC-(CH_2)_4-COOH$ (commonly known as adipic acid). They can also be made from amino acids, in which case they have just one number, representing the number of carbon atoms in the starting substance, e.g. Nylon six is made from caprolactam, a cyclic compound with a total of six carbon atoms and Nylon 11 is made from the straight-chain amino undecanoic acid $H_2N-(CH_2)_{10}-COOH$.

Polyamides are mostly used in engineering applications, due to their strength and toughness, and good heat and low temperature resistance. Gas barrier is very good, but the material is moisture sensitive and as the moisture level increases the oxygen barrier decreases. Some grades have excellent clarity. Packaging applications mainly utilise the high oxygen barrier properties of Nylon 6, for example in packaging of frozen meat joints, where the low temperature resistance and puncture resistance make this an ideal laminate (with LDPE or EVA for heat sealability). If exceptionally high mechanical strength and heat resistance are important, Nylon 6,6 may be a better option.

13.5.5 Polycarbonate (PC)

Polycarbonate is a polyester made by the condensation of carbonic acid $HO-CO-OH$ and bisphenol A $HO-(C_6H_6)-C(CH_3)_2-(C_6H_6)-OH$. It is glass clear and has exceptional impact resistance and good UV resistance, properties which account for its uses in glazing, safety spectacles and in automotive applications such as headlamp lenses. It is also heat resistant and readily sterilisable, which explains its main packaging use for returnable containers for office water dispensers. These large containers are subjected to rough handling and can be repeatedly sterilised without deterioration. Other packaging applications are currently limited, due to the high resin price and the availability of alternatives such as PET.

13.5.6 Fluoropolymers

Fluoropolymers are addition polymers of halogenated alkenes. Perhaps the best-known one is polytetrafluoroethylene (PTFE), better known under its trade name Teflon®, which is highly crystalline, inert and has a very low coefficient of friction and high melt temperature, accounting for its use as a non-stick coating. Polychlorotrifluoroethylene (PCTFE), under the trade name Aclar®, is used in pharmaceutical packaging applications such as thermoformed blisters and novel drug delivery systems. It has exceptionally high moisture barrier and when laminated to Barex® (noted for its high gas barrier – see above) the combined material can provide the long-term shelf stability demanded in such applications, along with justification for its high cost.

13.5.7 Thermoplastic elastomers (TPEs)

TPEs have already been discussed in Chapter 12. They combine the easy processing of thermoplastics with the elastic properties of rubber. The ability to recover from deformation and good chemical resistance make TPEs highly suitable for closure wads and plugs, and a thin layer of TPE on the outer surface of containers and closures provides a soft, rubbery feel which gives good grip, as well as aesthetically pleasing tactile properties. See Chapter 12 for a review of the different types available.

13.5.8 Cyclic olefin copolymers (COCs)

COCs are produced by addition polymerisation of a cyclic olefin (such as cyclopentadiene) and a conventional straight chain olefin such as ethene, using conventional or metallocene initiators. They have been available commercially since the 1990s, although their high cost has limited their use in packaging to specialist pharmaceutical and medical applications. High strength and clarity, combined with excellent moisture barrier make them viable alternatives to glass for items such as pre-filled syringes, vials and ampoules for injectables. There are grades suitable for film extrusion, bottle blowing and injection moulding and they can be used in blends with polyethylene and in individual layers in coextrusions. Resistance to acids, alkalis and polar solvents is good.

13.5.9 Liquid crystal polymers (LCPs)

Liquid crystal polymers are thermoplastic polyesters copolymerised from rigid and flexible monomers. The rod-like rigid segments are connected by flexible segments which allow the materials to flow on heating. During processing, e.g. into film or injection moulded parts, the rigid segments align in the liquid state and retain their crystal-like spatial arrangement on cooling. This results in materials with high tensile, impact and tear strength, high melting point and excellent barrier to gases and moisture. They have been commercially available since the 1980s and used in applications such as surgical devices, audio visual components and business machines. Flow properties are excellent and thin-walled sections can be moulded to close tolerances. Packaging applications are limited as yet, and are likely to be based on the very high barrier properties, e.g. LCPs are available with an oxygen barrier around six times better than EVOH, without the associated deterioration in humid conditions.

13.5.10 Thermosets

As stated in Chapter 12, in thermosets the polymer chains are arranged in a matrix fashion, with strong bonds called 'crosslinks' connecting them. As a result, they are dense, rigid materials with excellent chemical resistance and high heat resistance. The crosslinking takes place during polymerisation and thermosets cannot be reshaped once formed, hence they cannot be recycled. Thermosets such as phenol formaldehyde were amongst the first plastics used (see Table 13.1) but their usage in packaging declined markedly with the commercial availability of thermoplastic resins which offered easier processing at lower price. Both phenol formaldehyde and urea formaldehyde are still used for screw-threaded closures especially where high chemical resistance is important, e.g. laboratory reagents such as strong acids and alkalis. Also, unlike thermoplastics, thermoset closures are not prone to creep and the associated loss of torque over time. Other advantages are the minimal shrinkage on cooling, which means that thick-walled sections do not show sink marks, and the fact that the thermoset forming process allows very high precision and dimensional accuracy, along with high definition. This accounts for its use for closures with an embossed, highly detailed brand logo, as used on some alcoholic spirits.

Other packaging applications for thermosets are crosslinked acrylic, polyurethane and epoxy lacquers and adhesives.

13.5.11 Cellulose materials

As mentioned in Chapter 10, cellulose fibre can be made into cellulose film and this was the first transparent packaging film, widely used for confectionery, snack foods, biscuits, cigarette overwrapping, etc. It was also the material of choice for adhesive tape. The advent of polypropylene film in the 1960s changed this due to its lower cost, and the use of cellulose materials in packaging has now declined to a few specialist applications. This section provides an overview of cellulose film and cellulose esters.

Cellulose materials are derived from cellulose plant matter, usually wood pulp from managed forests and thus they can claim to be made from renewable sources. In technical performance they possess many of the properties of paper, such as excellent folding characteristics and the ability to maintain a fold (deadfold), easy printability and freedom from static. Like paper, uncoated cellulose materials have no heat-sealing capability and they absorb moisture (due to the high number of hydroxyl groups), although all these issues can be addressed by the use of appropriate coatings.

Cellulose film is commonly known by its trade name Cellophane™, and is widely used in packaging for twist wrapping of confectionery. This is due mainly to its deadfold properties, although the excellent clarity, sparkle and availability in a range of attractive colours are also important factors. In addition, the fact that the material is static free (unlike polypropylene, for example) is a key benefit for high speed packaging machines which can be running at 1,000 pieces per minute, where the build-up of static would lead to packs sticking together, causing line stoppages.

As stated, cellulose film can be coated to impart specific properties and Cellophane™ is available with copolymer, nitrocellulose or PVDC coatings, depending on the level of moisture barrier, gas barrier, heat sealability and thermal stability required. Films can be used alone or in laminations and typical end uses in addition to confectionery include wrapping of bakery goods and soft cheeses. Microwaveable films are available with semi-permeable coatings which allow some ventilation during heating, making them ideal for pastry products such as pies and quiches where the release of built-up water vapour maintains the product crispness.

NatureFlex™ is a more recent cellulose film which, as well as being based on renewable resources, has the added benefit of being compostable to EN 13432. For the purposes of this chapter, it is categorised as a bio-based polymer and will be discussed in Section 13.6.

Cellulose esters used in packaging include cellulose acetate (CA) and cellulose acetate butyrate (CAB). CA is used as a clear window in paperboard cartons, especially when moisture permeability is required, such as for cakes where the moisture would otherwise be trapped inside the carton and cause fogging, obscuring the product. It can also be used to make 100% transparent cartons as gift packs for products such as cosmetics. CAB has greater resistance to moisture permeability and is used for clear rigid tubes suitable for a range of products such as confectionery, soap and bath products.

13.6 Bio-based polymers

Most polymers discussed so far in this chapter are derived from crude oil. They are long life materials, destined to provide product protection and preservation, and to withstand degradation. They offer reliability in service and will last for many years, usually far longer than the products they contain. This longevity is desirable in the sense that the consumer can confidently purchase a plastic bottle of milk, shampoo or cooking oil, knowing that the bottle will remain in acceptable condition until the contents are used up. But resistance to degradation becomes a negative property if and when the bottle is taken to landfill at the end of its useful life, where it could remain almost unaltered, probably for hundreds of years. Of course, as already discussed in Chapter 5, there are alternatives to landfill, such as recycling into the same or different end uses and thermoplastics are ideal for this, being readily melted and formed into new structures. Other end-of-life options are reuse (e.g. plastic pallets and drums) and the production of energy from waste plastic.

Returning to the resources used to make most plastic packaging, i.e. crude oil, it should be remembered that plastic packaging is not the major end use of oil-derived products, and should not be held responsible for fossil fuel depletion. Nevertheless, concerns about this have instigated efforts to explore the use of alternative starting materials such as agricultural crops, and it is this category of materials which will now be reviewed.

Bio-based polymers are derived from plant matter, commonly known as biomass (although 'biomass' can also mean animal-derived material). As the crops used, for example maize and sugar cane, can be replanted, these polymers can claim to be made from readily renewable resources, unlike crude oil which takes millions of years to form. Using such renewable resources would appear to be a positive step, although agriculture itself is not free from negative environmental effects. Land has to be prepared for farming, pesticides may have to be used and transport of crops is required, thus generating carbon dioxide emissions. There is also the consideration of using land for the production of polymers, at the expense of food production.

To the general public, the word 'biopolymer' has become synonymous with biodegradable but this is misleading and confuses polymer sourcing (i.e. crop-based) with polymer functionality. This chapter adopts the British Plastics Federation's approach, by using the sub-title 'bio-based polymers' and differentiating between natural bio-based polymers and synthetic bio-based polymers. For example, sugar extracted from crops is used to make bioethanol which can be used as the starting material for making ethylene and polyethylene. Polyethylene produced by this route is a synthetic bio-based polymer, having the same performance properties as the oil-derived polyethylene already discussed, which is not biodegradable. The claim 'derived from renewable resources' may be made, but not the claim of biodegradability.

Clarification must also be made between oxodegradable and biodegradable polymers. Oxodegradable refers to oil-derived thermoplastics such as PE and PP with the addition of initiators which accelerate the breakdown of the backbone carbon chains when the plastic components, e.g. film or bottles are exposed to air, causing them to break down into fragments. This breakdown starts when the components are produced and takes around six months, depending on the exposure conditions.

Biodegradation refers to the process of breakdown of organic matter into its basic raw materials by the action of microorganisms. The compostability standard EN 13432 lays down a set of standard conditions (e.g. time, temperature, relative humidity) under which biodegradation must take place. It also defines the acceptable level of breakdown in terms of particle size and impact on the compost and the environment (e.g. production of heavy metals) for a material to be declared compliant.

A brief overview of the sources and properties of some of the bio-based polymers currently available now follows, although this is not presented as an extensive study and readers are advised to seek out more detailed information for specific applications.

There are two main ways of treating biomass to make bio-based polymers. Either it is processed to extract the natural polymers such as the polysaccharides starch or cellulose which are present in all plant matter, or methods such as hydrolysis, fermentation, or other microorganism attack are used to produce monomers, which in turn are then synthesised into polymers, e.g. lactic acid to polylactic acid (PLA).

Starch-based polymers are commonly derived from maize, although other crops such as potatoes and rice are also used. Just one example of commercially available starch-based materials in Europe is the Mater Bi® range from Novamont. These plant-derived materials conform to EN 13432 with respect to compostability, can be processed using traditional plastics processing techniques such as extrusion and injection moulding, and are readily printable without the need for surface treatment. A wide range of grades is available for food and non-food uses and includes flexible packaging and thermoformed trays, as well as coatings for paper and board, and foams for cushioning against shock. Laminates are available where moisture and gas barrier are important requirements, e.g. for biscuits and snacks.

Reference has already been made to the compostable range of cellulose-based films under the NatureFlex™ brand, from Innovia. They are made from renewable resources (wood pulp from managed forests) and in their uncoated form have excellent transparency, deadfold, anti-static and easy-tear properties. Coated grades give enhanced performance such as moisture barrier, heat sealability and resistance to grease and oil. Metallised grades are also available, making the range serious competitors to oil-derived polypropylene in the bakery, overwrapping, fresh produce and confectionery sectors, for companies wishing to make both renewable resources and compostable claims.

PLA is produced by fermentation of biomass, using microorganisms to convert the starch to lactic acid monomer. Maize is commonly used as the starting material, although sugar cane/beet and wheat can be used, indeed any crop with a sufficiently high sugar content in the plant starch. Lactic acid exists as two optical isomers, the D and L forms, depending on the spatial arrangement of the groups around the central (chiral) carbon atom. PLA made from 100% L-lactic acid is highly crystalline (known as C-PLA) while copolymers of the L and D forms are more amorphous and thus more transparent. PLA resins are available for extrusion into film, thermoforming into pots and injection moulding, although processing equipment designed for conventional polymers such as PET may need modification. Generally, PLA has similar barrier properties to polystyrene and current uses include bags and thermoformed tubs and pots for fresh produce. Non-packaging uses include disposable cutlery and other catering items.

With regard to product compatibility, initial studies conducted in the Cosmetic Science department at the London College of Fashion showed some softening of PLA bottles and film in contact with shampoo, although contact with baby oil appeared to have no detrimental effects (Talarek, 2011).

At the end of life, PLA is compostable in industrial facilities, but not in home composting due to the high temperature (above 50°C) and humidity required.

A further category of bio-based polymers is the generic class of copolyesters known as polyhydroxyalkanoates (PHAs). These are commonly made up of the simplest PHA, polyhydroxybutyrate (PHB), copolymerised with other hydroxyalkanoates such as hydroxyvalerate (PHBV). PHAs can be made directly by fermentation, without going through the polymerisation stage as used in the production of PLA. They are generally tough materials, with high temperature resistance, and grades are available for film and sheet extrusion, and injection moulding.

13.7   Conclusion

Compared with oil-based polymers, bio-based polymers are in their infancy and are currently used in niche rather than mass packaging markets. Global production is projected to increase from 700,000 tonnes in 2010 to 1.7 million tonnes by 2015, up from 262,000 tonnes in 2007. However, even the 2015 projection represents <1% of annual plastics consumption (tacooper@post.com; www.pcn.org/cooper.htm). Whilst there are sophisticated manufacturing facilities and high levels of capital investment, these are just a fraction of what has been expended over the past 60+ years of oil-derived polymer production; hence resin prices are relatively high. However, as technologies develop and demand increases, prices may become more comparable. Increased demand could be brought about due to pressure on mainstream brand owners by the 'green' consumer and government policies.

13.8   Sources of further information and advice

Baner, A.L. and Piringer, O. (2008) *Plastic Packaging Materials for Food*. Wiley, New York.

Baner, A.L. and Piringer, O. (2008) *Plastic Packaging: Interactions with Food and Pharmaceuticals.* Wiley, New York.

Giles, G. and Bain, D. (eds) (2000) *Materials and development of plastics packaging for the consumer market.* Wiley, Oxford.

Giles, G. and Bain, D. (eds) (2001) *Technology of plastics packaging for the consumer market.* Wiley, Oxford.

Hernandez, R., Selke, S. and Culter, J. (2000) *Plastics Packaging.* Hanser, Munich.

Nicholson, J. (2011) *The Chemistry of Polymers*, 4th edn. RSC Publishing, Cambridge.

Talarek, K. (2011) Sustainability in Cosmetic Packaging, undergraduate thesis, London College of Fashion.

In addition to the above texts, the British Plastic Federation (www.bpf.co.uk) is a valuable source of information on all aspects of plastics, including material properties, industry applications and forming methods.

Other useful sources of information include:

- http://www.britishplastics.co.uk
- http://www.incpen.org
- http://www.packagingdigest.com
- http://www.packagingfedn.co.uk
- http://www.packagingtoday.com
- http://www.pafa.org.uk
- http://www.plasticsinpackaging.com

Applied Market Information Ltd (www.amiplastics.com) is a recommended source of market data for plastics used in packaging, specifically the following reports:

- *Plastics Packaging Producers – A Review of Europe's Largest Players 2011*
- *European Plastic Industry Report 2011*

Note: Sources used in the preparation of this chapter also include teaching and learning materials written and used by the author in the delivery of courses to a number of organisations, such as the Packaging Society, Loughborough University, University of Warwick, University of Bath and London College of Fashion (University of the Arts London).

14
Plastics manufacturing processes for packaging materials

A. RILEY, Arthur Riley Packaging Consultant International, UK

Abstract: Polymers are finding new opportunities in packaging every day. The inventiveness of the polymer producer and the converting machinery manufacturer has seen plastics become the most used packaging material on a value basis. Plastic packaging is replacing metal retort cans, glass bottles, paperboard cartons and even corrugated cases. It provides barrier, ensuring food is kept safer for longer, reduces the weight of packaging, provides convenience, transparency where required, and still only uses less than 4% of oil resources. This chapter will provide the understanding of how polymers can be used on their own, or combined together to form packaging for all needs.

Key words: moulding, extrusion, thermoforming, coextrusion, laminating.

14.1 Introduction

This chapter will concentrate on the forming of thermoplastic packaging components from oil-based polymers. Polymers, as covered in the last two chapters, are highly versatile. A major advantage of plastics is that they can be combined together to provide packaging which is lightweight yet provides the barriers required, is tough and strong yet can be made easy to open. Most plastics used today are copolymers rather than homopolymers and are increasingly supplied in multiple layers rather than as monolayer. Using special techniques, plastic components can be produced to closely rival the absolute barriers available from metal and glass. This chapter will examine the processes required to make flexible, semi-rigid and rigid packaging formats for a variety of applications and the controls that are required to ensure consistency of supply is achieved.

To convert polymers into useful packaging requires specialised equipment and an understanding of their chemistry and properties. Polymers are converted into films, coatings, trays, bottles, jars, cans, closures and blister packs. They are combined together through coextrusion, coinjection and lamination processes; combined with paper and aluminium foils; coated with other polymers and undergo many chemical and physical treatment processes; all with the aim of changing their properties to suit the needs of the marketplace.

14.1.1 Selection of materials

When developing any new pack, there are some basic considerations with respect to the product and the pack which influence the choice of packaging materials, in this instance the polymer(s) that may be suitable. For example:

What aspects of the product are important?

- What are the barrier requirements – gas, moisture, alcohol, UV, etc.?
- Is the product likely to be aggressive to the packaging – acidic, alkaline, greasy, etc.?
- Is the product hazardous or dangerous – poisonous, high/low pH, etc.?
- What is the expected shelf life of the packed product?
- What are its physical properties – solid, liquid, gas, powder gel, flake, etc.?
- Does it need to be visible through the packaging?
- At what temperature will the product be filled – hot, warm, cold?
- At what temperature will the product be stored – ambient, chilled, frozen?

What aspects of the packaging are important throughout the expected life of the product?

- How many size variants are there going to be?
- How many of each variant do I require per annum?
- How soon do I require the packaging?
- Is it going to be a flexible or rigid container?
- What shape is it going to be?
- Is absolute clarity important?
- What quality of decoration is required?
- What temperatures does it have to withstand?
- What type of closure/closing mechanism will be used?
- What surface finish is required?
- Does it require tamper evidence?
- Does it have to meet child safety requirements?
- Will it meet product contact regulations?
- Is it environmentally responsible?
- What can I afford to pay?

The reason for asking these questions is that polymers, though excellent packaging materials, are not as definitive as paperboard, metal or glass in their absolute properties. Paperboard has little or no barrier to gas and moisture; glass, and metal (over 20 μm in gauge for aluminium) are total barriers and metals provide a total barrier to UV light. Polymers have a wide range of properties, depending on their chemical structure, the materials and the coatings added and how they are converted into packaging materials, and thus there are far more variables to consider. See Chapter 13 for more detail about the range of materials available.

14.2 The plasticating extruder

Excluding regenerated cellulose film (e.g., Cellophane™), rotational moulding and the thermosetting materials, all thermoplastic materials are converted using one or more plasticating extruders (Fig. 14.1). The polymer resin, which must be pre-dried if necessary (e.g., PET, PLA and PA) is fed into the hopper of the plasticating extruder, together with any additives such as colour or process and performance

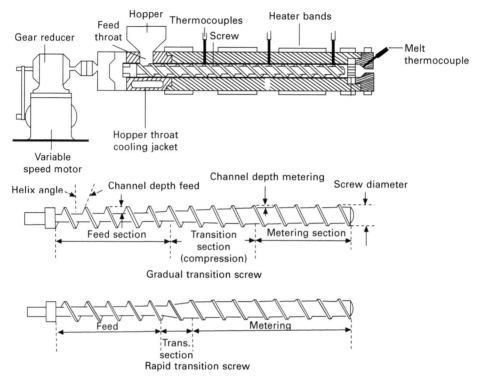

14.1 Plasticating extruder.

aids. See Chapter 12, Section 12.8.9 for a list of typical additives and their functions.

Most resins are in the form of small pellets and are delivered to the converting company either in bulk, in which case they are blown into storage silos, or, for smaller quantities of specialised materials, in 25 kg sacks or larger intermediate bulk containers (IBC). Whatever the delivery format, there must be an incoming quality assurance process to ensure the resins, and the additives, are correct to the relevant specification.

New technologies are being developed to control the particle size of functional fillers, such as carbon black (anti-static) and clay (improved barrier). These particles are nano in size, the filler being reduced to this size within the process, thus reducing any potential health hazards. They are exfoliated onto the polymer surface prior to the polymer being added to the extruder (Fig. 14.2).

The polymer and additives travel along the heated barrel of the extruder. Different temperatures are applied to separate zones down the length of the outside of the extruder. These heater bands ensure that sufficient heat energy is applied to the resin, melting it prior to it reaching the breaker plates and filters. The breaker plates and filters are situated between the end of the screw and the melt thermocouples. The filters are there to ensure unmolten polymer (often high molecular weight polymer, known as gels) and debris are held back and do not contaminate the component.

14.2 The use of carbon black nanofillers to produce anti-static polyethylene carbon nanocomposites (reproduced courtesy of Polyfect Solutions Limited; www.polyfectsolutions.com).

The resin enters the throat of the extruder and is immediately transferred down the barrel by the screw. The screw is designed so that the core diameter increases along its length. This is to ensure that as the resin melts the decrease in occupied volume is accounted for so that the polymer melt continues to be worked, providing most of the energy to change the polymer from solid to liquid. To keep the feed hopper cool and to ensure the polymer does not overheat and therefore degrade, cooling systems are placed around the barrel alongside the heater bands.

Temperature control of the polymer melt is very important. There are normally three to eight temperature zones, depending on the size of the extruder. The heating is controlled by electric heating bands and the cooling by forced air or chilled water contained in pipes. The die has separate heating zones (no cooling required) to control the temperature of that area independently (Table 14.1).

Polymers melt at different rates. The more crystalline the polymer, the shorter the temperature range from start to completion of melting and therefore the quicker the volume loss. Extrusion screw profiles are designed for specific polymers (Fig. 14.1 shows the difference between a gradual and rapid transition screw). Some polymers, for example PVC (polyvinyl chloride) and PVA (polyvinyl acetate), give off acidic fumes when processed through the extruder, therefore extruders and extruder screw and parts need to have special coatings (e.g., chromium plate) to ensure the steel is not subjected to excessive wear.

Pressure control is also critical. The required pressure depends very much on

Table 14.1 Plastic conversion process: process temperatures for some common thermoplastics

Polymer	Acronym	Processing temperature (°C)
Low density polyethylene	LDPE	150–315
High density polyethylene	HDPE	200–280
Polypropylene	PP	205–300
Linear low density polyethylene	LLDPE	190–250
Polyethylene vinyl acetate	EVA	150–205
Polyethylene vinyl alcohol	EVOH	200–220
Ionomer	None	180–230
Polystyrene	PS	180–260
Polyvinyl chloride	PVC	160–210
Polyethylene terephthalate	PET	260–280
Polycarbonate	PC	245–310
Polyamide	PA	240–290

the melt viscosity of the polymer and the forming process. For example, injection moulding requires high pressure to force the molten material through a small orifice whereas the relatively wide die used in extrusion moulding means less pressure is needed. Pressure can build up inside the extruder and in certain circumstances, if not controlled, be high enough to cause an explosion. Extruders are equipped with safety devices which rupture if the pressure builds to a dangerous level. Pressure fluctuation also leads to inconsistent output leading to inconsistency in the formed components.

The extruder feeds a die, which defines the shape and/or quantity of polymer which is fed from the extruder. The die can be a simple slot or annular die form, used for cast sheet and film manufacture, or more complicated to manufacture a wide variety of solid and hollow profiles. For coextrusion processes there are multiple extruders feeding into one die. Extruders for injection moulding are of a different design and are covered later in this chapter.

14.3 Sheet and film extrusion

Sheet is thicker than film, but there is no numerical value to define this. Sheet plastic as used for thermoforming is mainly, but not exclusively, made by the cast extrusion method. Films can be made by two main methods:

- cast extrusion, using a slot die
- blown extrusion, using an annular (ring) die.

There is little difference in the properties of the film produced by each method, although mono orientation is difficult in blown extrusion and the optical properties of blown film can be less than cast. However, the MFI (melt flow index) and melt strength of a polymer affect its suitability for each process. Melt flow index is a measure of the resistance to flow (viscosity) of the polymer melt at a given temperature under a given force for a predetermined period of time.

The polymer's molecular weight and molecular weight distribution have a direct influence on the MFI: the higher the molecular weight of the polymer the higher the viscosity; the less the polymer flows over a given time, the lower the MFI for any individual polymer. The wider the molecular weight distribution the less resistance to flow; the more material flows over a given time, the higher the MFI. PET homopolymer and PP homopolymer both have low melt strength which makes it difficult to produce either of them using the conventional blown film process, where the bubble is blown upwards. Thus PET is more commonly made using the cast process, and while PP is made using the blown process, it requires a special adaptation to be successful (see later).

14.3.1 Cast sheet and film extrusion

The cast extrusion process can be used to produce film or sheet as a monolayer or a multilayer coextrusion. Multilayer coextrusion requires one extruder for each different type of polymer used. The number of polymers required is often greater than the number of functional polymers used, as tie layers (adhesive) are required where two incompatible polymers (e.g. EVOH and PP) are adjacent to one another in the multilayer construction. The molten polymer is transported by the extruder(s) into a slot die. It is here where the polymer layers combine as shown in Fig. 14.3.

The slot die in Fig. 14.4 has a narrow opening, which is adjustable and controls the flow rate as well as the initial thickness of the emerging film. There is also a

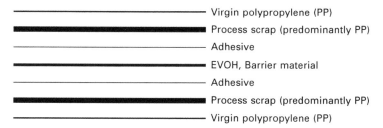

14.3 Typical PP/EVOH/PP barrier layer coextrusion for a thermoform sheet (including process scrap).

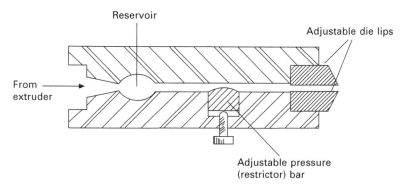

14.4 Cross-section of a slot-orifice die.

reservoir to help prevent polymer surge through the die, which would result in uneven film thickness and thus uneven performance in use. The polymer is extruded through the slot die which is often a series of small slots adjacent to one another rather than one wide slot, and falls onto large chilled metal rolls to form either sheet or film. Providing the film or sheet are not stretched, there is no orientation of the material (see below). It remains the same width and thickness as controlled by the slot die and has no stiffness, stress or other change to its physical and barrier properties (see Fig. 14.5).

For sheet applications the plastic melt is often extruded onto a temperature controlled three-tier calender stack instead of a chill roll. This smoothes out the surface of the sheet and adds a textured finish if required. The calendered sheet is then cooled by passing through a number of chill rolls or a quench tank, before being wound up ready for despatch for further conversion such as thermoforming into pots and tubs. In some cases, especially when using polypropylene homopolymer which has a low melt strength, it is better to extrude the sheet in line immediately before the thermoforming process. This overcomes the difficulties in controlling the re-heated web on a conventional thermoformer, fed with unmolten preformed sheet requiring reheating.

Film properties can be improved by physically orientating the film in one or two directions (as mentioned in chapter 12) (see Fig. 14.6). Mono-orientation can be achieved by pulling the film in the machine direction (MD) at a faster rate than it is being extruded, i.e. the take-off speed is increased. This realigns the polymer molecules in the direction of stretch, rather than leaving them in their 'natural' random state. Stretching can occur immediately the melt comes into contact with the chill roll or, more commonly, after the first chill roll, often requiring reheating before it is stretched. Mono-orientated film can also be produced by using the stentering method, which stretches and orientates the film in the cross direction (CD). Mono-oriented film is used for shrink sleeve and roll on shrink on labels and pallet strapping. If both orientation mechanisms (MD and CD stretch) are performed in one process, the film becomes biaxially oriented. This method is commonly used to manufacture polypropylene (BOPP) and polyester films. The film is stretched in the machine direction first; grips then take hold of the edges of the film and gradually stretch it in the cross direction. Films can also be stretched in both directions simultaneously.

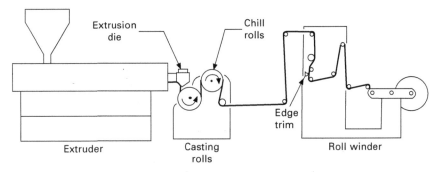

14.5 Cast-film extrusion line.

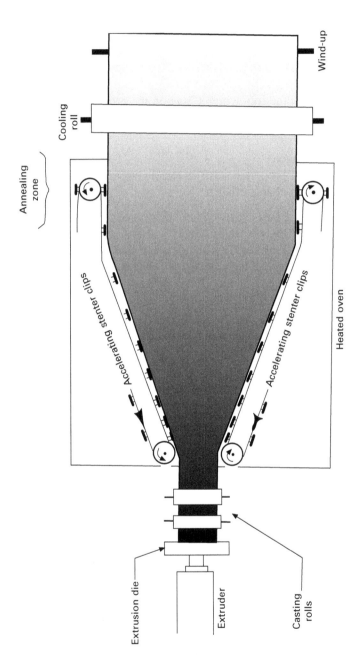

14.6 Orienting cast film: clips grasp the film along each edge and stretch it in the cross direction, while acceleration in the machine direction orientates the film in the machine direction.

If the polymer being orientated is crystalline it must be below its melt temperature (T_m) but higher than its glass transition temperature (T_g) to maximise orientation. If the temperature is too high the less the orientation achieved; if the temperature is too low uneven stretching causes thin spots and even rupture. Where the film must remain thermally stable in use, it is annealed at controlled temperature and then cooled to 'freeze' the orientation before the tension is released. This helps to overcome the tendency of the polymer to return to its natural, more random molecular arrangement when heat is applied.

Biaxial orientation has the following effects on the properties of a film:

- improved moisture barrier
- improved gas barrier
- improved tear resistance
- improved mechanical properties such as tensile strength
- impaired heat-sealing characteristics.

Most films produced using cast extrusion are coextruded or coated films, especially where coefficient of friction, barrier and heat seal are important. Using a coating or polymer layer with a lower heat seal temperature than the main film means that the resultant multilayer film can be heat sealed without melting the main substrate (Fig. 14.7).

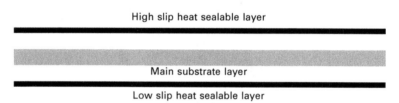

14.7 Coated overwrap film for tobacco packaging (shrink properties can be added to ensure a tight overwrap of the carton).

14.3.2 Blown film

Blown film, for all but polypropylene homopolymer (which is easier to blow downwards – see later), is routinely carried out by forming a bubble vertically upwards, as shown in Fig. 14.8. From the plasticating extruder, the molten polymer enters the annular die and is formed into a tube of material. This tube is taken up to the nip rollers where it is sealed, then air is introduced inside the tube to inflate it, creating a bubble. The inflation of the bubble increases its diameter thus orientating the film in the cross or transverse direction. The greater the ratio of the diameter of the bubble to the diameter of the annular die, the greater the orientation. This is known as the 'blow up ratio' or 'blow ratio', which is determined by the melt strength of the polymer; the greater the strength the higher the blow ratio that can be used.

Chilled air is blown on the outside of the film to cool the polymer bubble below its melting temperature T_m. The frost line is the point at which crystallisation occurs as the melt solidifies; as a result some transparency is lost. It is important therefore

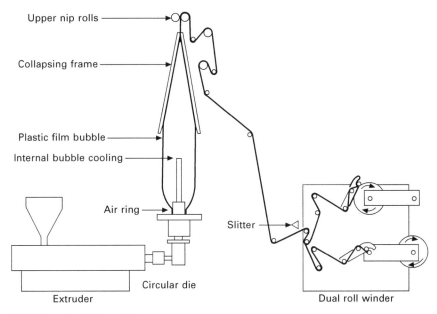

14.8 Standard blown-film process.

that the cooling speed is controlled carefully, to manage the change from liquid to solid state. The slower the film is cooled, the larger the crystals formed, resulting in less transparency and gloss of the resulting film. Some extruders employ internal cooling to increase production rates. The external and internal cooling air is normally refrigerated to allow for better control of the final properties of the film.

To achieve machine direction (MD) orientation, the film is stretched in the longitudinal direction by drawing it through the nip rolls at a faster rate than it is coming out of the die. The final thickness is controlled by the die gap and the amount of orientation imparted to the film. Orientation in both directions takes place while the polymer is still molten (Fig. 14.9).

It is important to control the symmetry of the bubble, i.e. the area of film on both sides of the centre line must be equal. If the film is uneven in either thickness or solidification, the bubble will be asymmetrical and the thicker side of the bubble will not expand as much as the thinner side, thus the gauge of the film will be uneven across the web. Also, as the thinner side will expand to a greater extent than the thicker side, it is possible that this will lead to excessive thinning and the bubble will burst. Certain parts of the film have high spots caused by imperfections in the process. These are often in the same place across the web of the film and when it is wound up these high spots can multiply, resulting in a ridge in the finished roll of film. To prevent this happening either the die, or more commonly with mono films the bubble frame and nip, are rotated to and fro (oscillated) to evenly distribute the high ridge, greatly reducing any adverse effects.

Once cooled the film approaches the nip roll and the bubble is gradually flattened into what is known as lay flat tubing. The nip rolls, one metal and one rubber transport the film to the in-line slitters and roll winders at the base of the line (see Fig. 14.8).

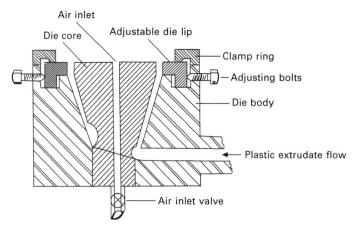

14.9 Cross-section of an annular blown-film die.

The slitting of the film can be carried out off-line if a high level of dimensional accuracy in slit width variation is required. When very wide films are required (e.g. agricultural film) the lay flat tube is slit on one side only, to allow the film to be opened out to its full circumference in use.

The bubble film method is used mainly to produce monolayer or co-extruded polyolefin films, used for stretch and shrink wrap materials. Stretch wrap is used to hold pallet loads of transit packaging in place. It works because when stretched, the polymer molecules want to return to their original formation, providing the elastic limit has not been exceeded. Most stretch films are made from modified LLDPE. EVA blends and plasticised PVC are also used. Coextrusions are produced to add cling features to one or both sides of the film. Ideally stretch film should have:

- a smooth surface to improve cling
- good elasticity (elongation and recovery)
- transparency (some special films are opaque)
- high tensile strength in the machine direction (MD)
- high fatigue resistance (relaxation with time)
- low creep properties
- good puncture resistance
- good tear resistance
- low neck-in properties (reduction in film width when stretched).

Some stretch wrap used for wrapping pallets (especially in the brewing industry) is perforated to allow the free flow of air through the pallet, reducing the effects of condensation.

Shrink wrap also relies on a film's tendency to want to return to its unstressed state. Shrink wrap relies on the in-built stresses in the film being stable until heat energy is applied. This allows for the shrink wrap to be loosely applied to the pack and heat sealed into a loop. It is only when heat is applied to the film that it shrinks tightly around the pack, holding the contents in place.

Shrink film is made by orientating the bubble in both directions and then freezing the stressed molecules by freezing the film as quickly as possible. When heated the in-built stresses are released and the film shrinks. Lightly crosslinked materials with elastomeric properties are often used to increase the shrink properties of the film. Shrink wrap is made from a variety of polymers, including LDPE, LLDPE, PP copolymer and PVC. The properties required from shrink film are similar to those required for stretch film:

- transparent (some special films are opaque or all over printed)
- controlled shrink ratio is both directions
- good puncture resistance
- good heat-seal characteristics
- appropriate slip characteristics
- low creep properties
- high fatigue resistance
- good tensile strength.

While polypropylene copolymer films can be made using the blown process described so far, as already mentioned polypropylene homopolymer has low melt strength and a better approach is to blow vertically downwards as shown in Fig. 14.10. Molten polymer travels to the annular die as already described, but instead of being immediately blown, it is formed into a cast tube which runs vertically downwards. This tube is then reheated to its softening point and inflated to form a large transparent bubble, which effectively orientates the material equally in both directions. The bubble is then collapsed and two knives are used to slit the film into two webs. These are annealed to reduce their tendency to shrink and may also be surface treated to improve print adhesion (as discussed in Chapter 12). BOPP film produced in this way is very similar in properties to cast BOPP already discussed and competes in the same markets. BOPP PP film has many uses, from overwrap for cigarette packets and chocolate boxes to FFS packaging for fresh produce and in its coated and laminated form, bags for potato chips (crisps).

14.3.3 Cellulose film

There is one other transparent film type used in packaging which was the forerunner of BOPP. That is cast regenerated cellulose (Cellophane™ and much more recently the biocompostable regenerated cellulose film NatureFlex™). As it is not made from oil, but from cellulose fibre, it is neither thermoplastic nor thermosetting and is not a true plastic. However, it is a useful packaging film. Wood pulp produced by the chemical process (see Chapter 10) is chemically converted into a thick liquid form called viscose, which is extruded through a flat die into a regeneration bath (Fig. 14.11). At this point the viscose converts into a solid thin film form. Many processes incorporate two dies on the same machine, allowing the manufacturer to double output on the same casting machine. The web is carried down the casting machine on rollers, through a series of baths which wash and soften the film in order to produce a kind of 'transparent paper'. At this stage the film is transparent and glossy but has no heat seal and moisture barrier as one would expect from cellulose. In most cases,

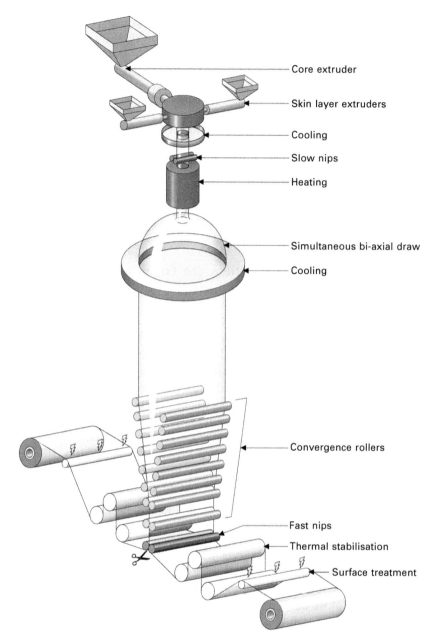

14.10 The 'bubble' process for producing biaxially-oriented polypropylene (BOPP) (courtesy of Innovia Films, www.innoviafilms.com).

an anchor resin is applied in the final bath, prior to drying, to prime the surface to make it receptive to secondary coatings applied off-line. The secondary coatings are tailored to provide the heat seal and barrier properties for the intended use, (e.g., PVdC (polyvinylidene chloride) to provide a heat seal, gas and moisture barrier).

Plastics manufacturing processes for packaging materials

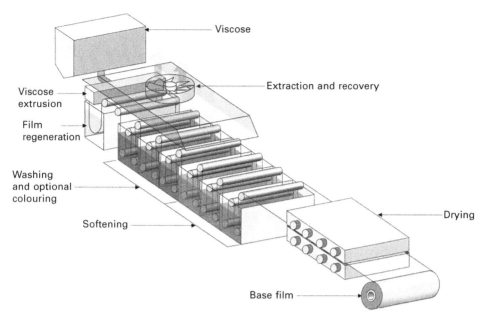

14.11 The casting process for producing regenerated cellulose film (courtesy of Innovia Films, www.innoviafilms.com).

14.3.4 Coextrusion of cast and blown films

Coextrusion is the process used to combine two or more different polymers during the extrusion process. The use of three layers is common but more than nine is possible to achieve a variety of functional benefits by careful choice of each layer. Aesthetic effects such as coloured layers and layers with coloured stripes can also be achieved. The combining takes place while the polymers are in the molten state, just before the extrusion die in the cast process and just after the extrusion die in the blown film process. This allows the different polymer layers to bond together, without mixing, to form a laminar structure. The purpose of using this technique is to maximise the properties of polymers at optimum cost (see Fig. 14.12). Coextrusion can also be used to produce cast sheet, as explained earlier, with coloured stripes or layers as well as sheet for thermoforming containing barrier layers such as EVOH.

Each polymer type requires its own extruder. All the extruders feed into an adapter, known as the feed box, before entering the die or directly into the multi-manifold coextrusion die (Fig. 14.13). In the cast process the individual extruders connect with the feed box, where the polymers combine in layers. This permits a simpler die design. However, multi-manifold dies (as used for blown film) are used where the flow properties of the polymers are widely different. The multi-manifold system provides a shorter flow path before the polymers solidify and therefore less chance for distortion at the interlayer interface.

When some materials are combined together (e.g., HDPE and PA or PP copolymer and EVOH), their adhesion to each other is very weak, which would result in delamination during subsequent conversion processes such as printing or bag making.

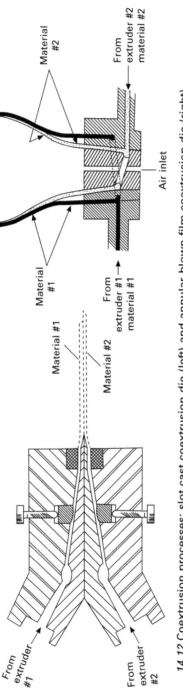

14.12 Coextrusion processes: slot cast coextrusion die (left) and annular blown film coextrusion die (right).

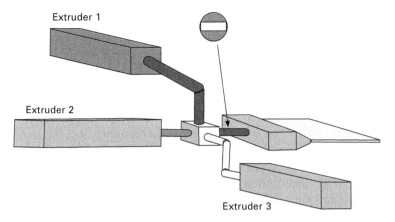

14.13 Cast film coextruders feeding into the feed box and die.

To overcome this, a third component has to be added, to act as a tie or adhesive layer. As mentioned previously, this means the use of an additional extruder for every tie layer. Processing of the coextruded film is the same as for the monofilm, described earlier. Once films are reeled, they often have to be left for 48 h for the molecular structure and slip additives to stabilise before proceeding to the conversion stage.

14.3.5 Performance parameters of polymeric films which require strict control

Packaging films require control of many performance parameters to ensure that there is consistency throughout conversion and use of the film. The main areas which must be specified are listed below:

- *Thickness*: Control of thickness is very important for ensuring consistency of heat sealing, printing and mechanical strength. Thickness of film is difficult to measure by hand as individual points will vary considerably. It is therefore important to take a number of measurements over a small area using a micrometer with a 25 mm head, and to average the results.
- *Basis weight*: Basis weight, also known as grammage is an alternative to thickness as a way of controlling film.
- *Density*: Once thickness and basis weight are known, the yield can be determined, i.e. the number of square metres that can be obtained for 1 kg of polymer.
- *Moisture barrier*: Where appropriate, moisture vapour transmission rate (MVTR, also known as water vapour transmission rate, WVTR) needs to be accurately measured at a predetermined temperature and humidity.
- *Gas barrier*: Where appropriate, gas barrier (e.g., oxygen, carbon dioxide, nitrogen) needs to be measured at a known temperature and humidity. Separate measurements need to be taken for each individual gas.
- *Grease barrier*: Not all polymers have a grease barrier that is required for high fat content products, such as butter and dry pet foods. The type of fat is also

important. There are several test methods which usually reflect the actual filling and storage conditions using the product provided. Where this is not possible, tests are carried out with chemicals selected to provide a guide to the barrier properties of the particular film.

- *Coefficient of friction (CoF)*: Coefficient of friction is the reciprocal of slip. The more slippery a surface is, the lower is, the coefficient. There are two types of test. One uses an inclined plane where the angle is altered until the sample slides. The second and far more accurate method uses a moving sled on a flat bed. This method is used to determine both the static (CoF at the point the sled starts to move) and the dynamic (CoF as the sled is moved) at a constant rate under controlled conditions of temperature and pressure. It is very important to have a constant CoF within a film, otherwise it will not move over packaging machinery smoothly, having a negative effect on the line efficiency.
- *Heat-seal temperature*: Polymer films are complicated, ranging from films made up of one or many monomers having varying properties to films that are coextruded, laminated or coated with other polymers. The ideal film for heat sealing is one where the outer part of the film has a significantly higher sealing temperature than the inner, and the inner part of the film has a very wide sealing range. To ensure the correct film is selected and that the heat-sealing characteristics are uniform throughout the web, tests are carried out where the temperature, dwell time and pressure are varied to determine the ideal sealing conditions.
- *Cold-seal strength*: Sometimes, for example when packaging a chocolate bar, a cold-seal adhesive is used. The same type of test is carried out as for heat seal to ensure consistency of performance.
- *Tensile strength and elongation at break*: Tensile strength determines how much force is required to break the film of a given thickness and the elongation determines how much it will stretch before it breaks. Both these parameters are important especially for a printed film. The film has to pass through the tensioning rollers of the packaging machine without breaking or distorting the print.
- *Stiffness*: Stiffness of films is very important when making bags, sachets and pouches which need to stand up without sagging. It is also important when placing a bag into a carton. The thicker the material the greater the stiffness but the higher the cost; therefore for many applications stiffness at lowest thickness is a very important attribute.
- *Puncture resistance*: Plastic films are used to pack many items which have sharp edges. If the puncture resistance is too low then the product will place a hole in the film thus significantly reducing any barrier which has been engineered into it.
- *Surface energy*: Polyolefins have a poor surface energy which results in print and adhesive not bonding to the surface of the film. To overcome this, the surface is activated by corona discharge, flame treatment or application of a special coating. This increases the surface energy to about 42 dynes/cm which is sufficient to ensure the surface print or adhesive keys to it.

Plastics manufacturing processes for packaging materials

Other characteristics of the film may also be tested. These would include gloss, haze, optical density, anti-fog, anti-blocking and direct food contact. There are other market specific tests which need to be carried out, such as:

- reel geometry – accuracy of slitting and reeling – edges parallel
- which way wound? – side A or side B on the outside
- how palletised? – care of edges and flattening of the reel – reels of film should be stored and transported suspended between A frames or for less critical films (e.g., stretch and shrink) on their ends with or without edge protection.

14.4 Film treatments after forming

Coextruded, laminated, coated or single material flexible packaging can be found in thousands of different specifications, specially developed to suit the needs of the product, machinery, distribution chain, aesthetic, convenience and environmental considerations. Examples are shown in Figs 14.14, 14.15 and 14.16. Films can, for example, be given special treatments or coated to improve their properties. Examples include corona discharge to improve surface adhesion and vacuum deposition of mineral oxide, metal oxide or metal particles (Fig. 14.17). Films are treated with a vacuum deposition to improve aesthetic and barrier properties. Aluminium is the

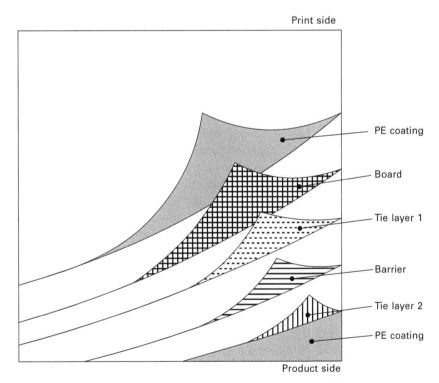

14.14 Example of a typical packaging laminate (courtesy of Elopak; www.elopak.com).

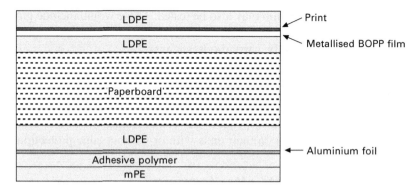

14.15 Example of a laminate for a carton containing liquid (courtesy of Tetra Pak; www.tetrapak.com).

most common vacuum deposition material used in packaging, providing a very bright silvery surface improving the UV, gas and moisture barrier of the film. Aluminium oxide and silicon oxide are also used where clarity is important, but improvement of gas and moisture barrier is still necessary. The two main films metallised are PET and PP copolymer. The new biocompostable films (e.g., PLA) are now being treated by vacuum deposition in an attempt to improve moisture barrier.

14.4.1 Lamination and coating of flexible materials

As already discussed, it is possible to combine polymers using coextrusion, but this is not always practical, for example for short-run lengths of specialised materials. Also, it is not possible to include layers such as aluminium, paper and paperboard in coextrusions. Lamination processes may be required, in which webs of individual materials are combined using adhesives (Fig. 14.18).

The simplest laminate is paper or paperboard extrusion coated with a polymer on one or both sides. Polymer granules are placed in the extruder where they melt and pass through a slot die. The extruded film width and thickness are controlled by the die and the speed of application of the plastic to the substrate. The melt is at such a temperature that it will adhere to the substrate after which time it is passed over a chill roll which cools the melt prior to it being wound up into a reel. If both sides of the substrate are to be coated, the procedure is repeated or another extruder is situated in line with the first. This process addresses some of the inadequacies of the base material and typical end uses include ream wrap, sandwich cartons, frozen food, pet food and soap powder bags, sacks and cartons. The polymers normally used are PE, PP and PET. Hot melt adhesives are also applied in this way or via a roller application (Fig. 14.19).

More complicated extrusion coated/laminated structures are used for liquid packaging cartons. Here two or more extrusion heads are used to produce a laminate of plastic/paper/plastic/aluminium/plastic. The paper gives rigidity and a good printing surface, the aluminium provides barrier to UV, oxygen and moisture vapour

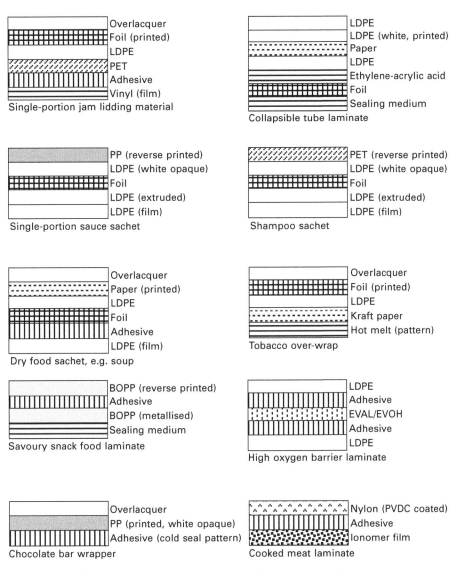

14.16 Examples of laminates for specific product applications.

and the plastic seals the surfaces together. The plastic also provides an external barrier to condensation and an internal barrier to product penetration into the paper.

In addition to extrusion lamination, other laminating techniques which can be used, depending on the combination required and the end use, are dry bond lamination and wet bond lamination. Dry lamination can be achieved by a variety of means. The oldest is probably wax bond lamination (Fig. 14.20). This will bond substrates together but as it is a non-polar material, it relies mainly on mechanical bonds, for example to paper, but can achieve moderate chemical bonds to aluminium and plastic

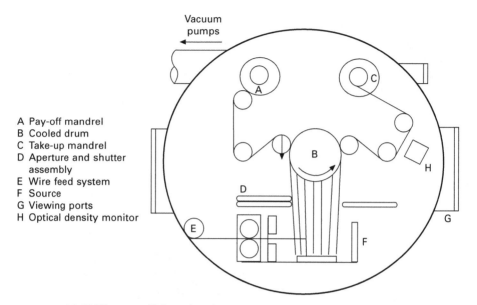

A Pay-off mandrel
B Cooled drum
C Take-up mandrel
D Aperture and shutter assembly
E Wire feed system
F Source
G Viewing ports
H Optical density monitor

14.17 Film metallising chamber.

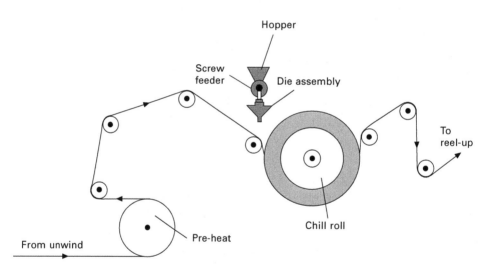

14.18 Extrusion coating process (courtesy of Walki Group; www.walki.com).

substrates. The wax is applied molten to the substrate via a wheel applicator; whilst the wax is still molten a second substrate is applied to the wax coating on the first and the whole is bonded by passing through a nip roller and cooled via a chill roller immediately prior to being reeled up. One of its most common uses is in lamination of foil to paper in the wadded closure used for jars of coffee. The laminated wad in the closure is induction sealed to the top of the glass jar. The heat generated melts the wax which is absorbed into the paper. When the jar closure is unscrewed the wax

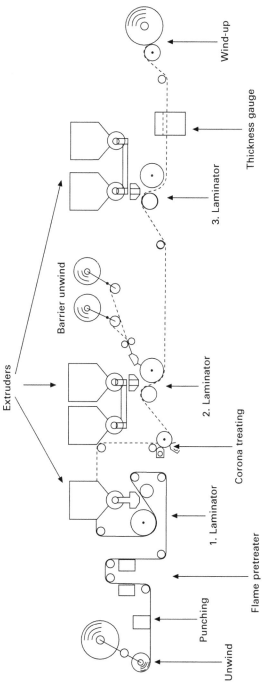

14.19 Process for coating both sides of a substrate (courtesy of Elopak; www.elopak.com).

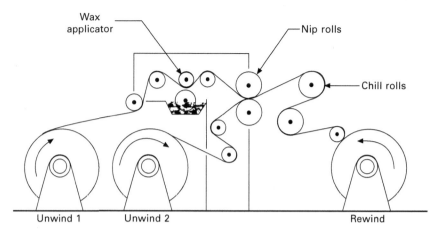

14.20 Wax bond laminator.

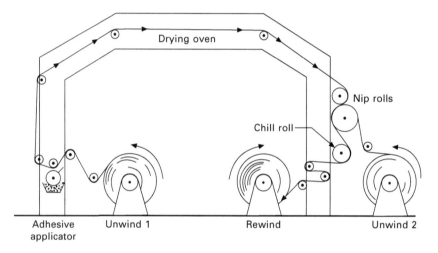

14.21 Dry bond laminator.

bond breaks leaving the paper wad in the closure and the aluminium foil diaphragm sealed to the jar.

Dry bond adhesion is very useful when two non-porous materials need to be laminated (Fig. 14.21). Water- or organic solvent-based adhesives are applied to one surface, and the solvent is driven off in a heated oven. The hot tacky adhesive-coated substrate is then laminated to a second substrate, using a nip roller which is often heated to ensure the dried coating is still active. It is important that all solvent is driven off prior to bringing the two substrates together, otherwise it is trapped between the substrates, which can have a negative effect on adhesion and also can contaminate the packed product. It is also necessary to consider the management of the volatile solvents given off in the drying process, to comply with relevant legislation.

The final type of dry bond adhesion to be discussed is the use of two-part polyurethane adhesives (Fig. 14.22). These are 100% solid and therefore, as with extrusion lamination, do not require drying. They are applied to the web at ambient conditions and rely on a chemical crosslinking reaction for the bond to be completed. This can take up to 48 h before the laminate can be further converted, e.g. slit into smaller reels, otherwise delamination may occur. Once cured these adhesives have much higher temperature resistance than their thermoplastic counterparts and therefore can be found in the construction of retort pouch laminates.

Wet bond lamination is used where one of the substrates is porous (e.g., paper). The two substrates can be brought together before the oven, where the heat dries the adhesive and the substrates. Care needs to be taken during drying to ensure the moisture profile along the length and across the web of the finished laminate is stable.

Coatings can be added to substrates during the lamination process (as with wet bond lamination), as part of the printing process (e.g., cold-seal adhesives for sealing of confectionary bars) or as a separate operation (e.g., water-based functional coatings to replace polymer films and fluorocarbons, for pet food and fresh produce). They can be added via a gravure or anilox roller, or knife or rod metering system (Fig. 14.23).

Often the board is preprinted before it is coated with polyethylene. The tie layers ensure good adhesion between incompatible substrates. The barrier layer is usually aluminium but could be ethylene vinyl alcohol (EVOH) coextrusion. In Fig. 14.15 you will notice a difference from the specifications shown in Fig. 14.14. Liquid packaging serves a vast market with products from acidic orange juice to high fat dairy goods, all requiring very different packaging performance.

From lids for single portion jam pots (replacing traditional aluminium) to collapsible tube laminates for toothpaste, film for tobacco overwrap, savoury snacks, shampoo sachets and many, many more end uses, flexible packaging can be specified to replace

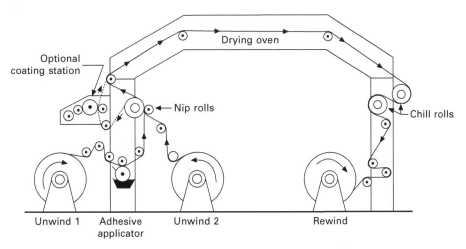

14.22 Wet bond laminator.

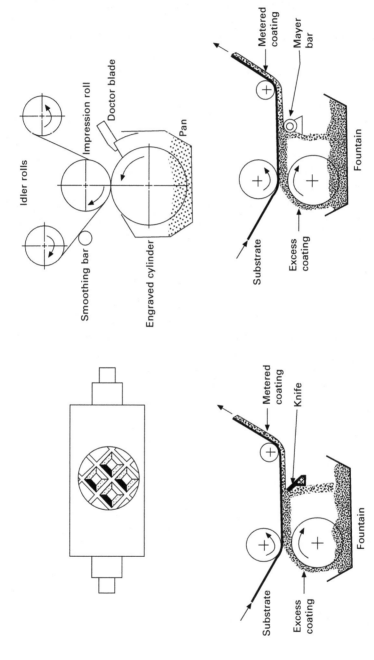

14.23 Gravure coating process (engraved cell shown enlarged on left).

other more traditional forms of packaging. It uses minimal materials and can be formed into sachet, bag, carton, can and bottle shapes. For example, retort pouches are currently replacing metal cans, sachets and pouches with reclosable pour spouts are replacing traditional bottles, cartons are being replaced with block bottom bags with reclosable zip systems. These flexible packs rely on judicious choice of materials to meet the end use and the requirements of the packaging machinery on which they will be formed. These requirements will be covered in Chapter 20, but at this point it is important to note that all of these processes rely on the sealing properties of the packaging material being compatible with the sealing characteristics of the relevant forming machine.

14.5 Thermoforming process for making plastic packaging

Thermoformed packaging produces a less dimensionally accurate moulding and less complicated shapes are achievable (e.g., undercuts) but is often quicker and less expensive to produce than an injected moulded container. In its simplest form the thermoforming process involves heating a sheet (which can be mono material, a coextrusion or a laminate) of even thickness and drawing it over, or into a mould to form a rigid or semi-rigid shape. The excess material is trimmed off usually, leaving a rim around the finished article. The greater the depth of the object to be formed, the more likely it is that the material will thin, even to the point of breakage, and this is one of the most serious disadvantages of this relatively simple process. This unwanted thinning can be reduced by various means as will be mentioned below.

The thermoforming process is used to make many different packaging articles, for example, tubs, pots, display trays and blister packs, and can be broken down into seven basic steps:

1. Making the sheet – normally by the cast extrusion method.
2. Heating the sheet.
3. Forming the pack by stretching the sheet either into or over a mould.
4. Cooling the formed packaging.
5. Cutting and trimming the multi-unit moulded sheet into individual units. This can be carried out at the same time as the forming process.
6. Printing or decorating as required.
7. Stacking the individual units before packing and labelling for despatch to the customer.

As with all processes, quality control throughout is crucial.

Sheet extrusion is described earlier in the chapter. The sheet can be foamed (e.g. EPS expanded polystyrene or cellular polypropylene or polyethylene) if required. As already noted, if polypropylene is being cast, it is usual to extrude the sheet in line with the thermoforming. Coextruded multilayer sheet is the most popular for thermoforming as it provides the performance and barrier characteristics required. Many packaging suppliers make the sheet and thermoform it into the required pack shape, while some packer-fillers thermoform trays from a pre-supplied reel of sheet material, place the product into it, and heat seal a flexible lid onto the tray, all on one machine.

Once the sheet is heated, the container can be formed using different methods. The traditional method was to drape the heated sheet over a cavity or plug mould, draw a vacuum and form the sheet to the shape of the mould (Fig. 14.24). This method is adequate for shallow thermoformed packaging components of uncomplicated design. If the depth is greater than the diameter then plug assist vacuum forming is the better choice (Fig. 14.25). In conventional vacuum thermoforming, the sheet is formed, the wall thins and there is a risk that the sheet will not conform well to the contour of the mould, especially in the bottom edges. Plug assist overcomes some of these inadequacies by acting as a heat sink and displacing the material in a more even manner, reducing the thinning of the wall section. This is especially useful for deep-drawn items.

'Vacuum snap back' can also be used where overall wall thickness becomes important (Fig. 14.26). This is used for items where the depth of the article is up to 2.5 times greater than its width. The initial stretching of the sheet is free of any contact points and therefore more even than in the direct forming processes already described. As a result, the wall thickness throughout the item being formed is much more uniform. Where the depth of the article is greater than 2.5 times its width, 'billow' forming can be used (Fig. 14.27). Heat and/or pressure cause the sheet to billow upwards, a heated plug is introduced and the vacuum is turned on, forming the sheet over the mould.

Solid phase pressure forming (SPPF) has been developed to produce thermoformed articles with better definition, especially at higher pressures. The process has improved the quality of thermoformed articles to the extent that it can now compete with injection moulding for some moulding designs. It is also possible to mould two sheets, one over the other, at once. In solid phase pressure forming the sheet is reheated inside the machine until it becomes plastic and easy to form (Fig. 14.28). Using plugs and then compressed air, the sheet is pushed into the shape of the mould. It is here that by using high pressures moulding definition is greatly improved, the plug reducing the variation in wall thickness which occurs with conventional thermoforming. Once frozen, the shape is cut out of the web – this is completed while the article is still held to ensure the accuracy of the cut. The articles are ejected through the front of the machine.

Before the moulding is removed from the mould, it must be cooled so that deformation is avoided. This is very important when thermoforming polypropylene as it has a very high shrinkage rate and, if not cooled sufficiently, it can continue shrinking for several days after it is removed from the mould. Cooling is carried out using temperature controlled water directed to cooling channels designed into the tooling. Once the article has been formed and cooled, it needs to be cut from the sheet. The cutting process leaves a rim around the article, which can be used for heat sealing and as a lip for an overcap. The forme used for cutting is similar to that used for cutting paperboard cartons. Modern developments have made it possible to minimise the lip on the article.

Plastics manufacturing processes for packaging materials 337

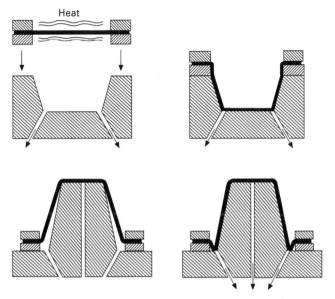
14.24 Vacuum forming over cavity and plug moulds.

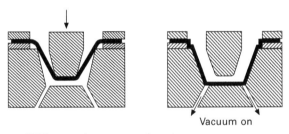
14.25 Plug assist vacuum forming.

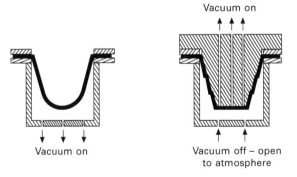
14.26 Vacuum snap back forming.

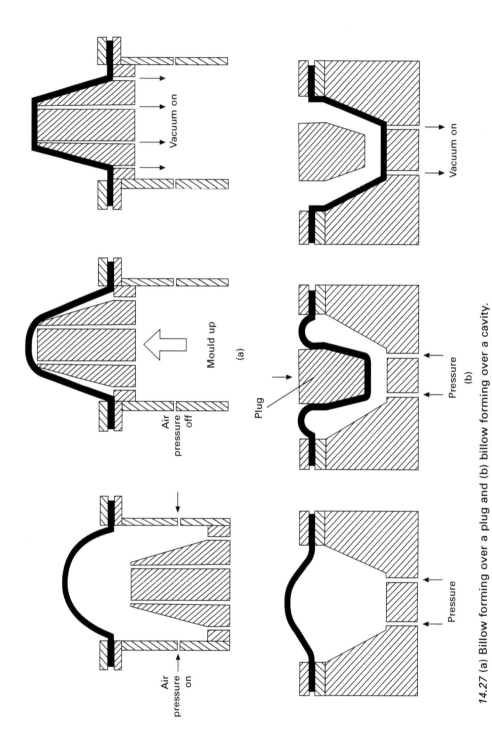

14.27 (a) Billow forming over a plug and (b) billow forming over a cavity.

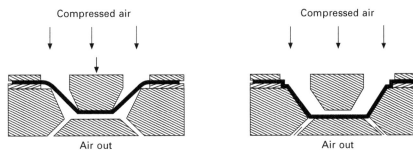

14.28 Solid phase pressure forming.

Thermoformed items can be decorated using the following methods:

- In-mould labelling – this provides the best overall graphics of all the decoration methods; however, the adhesion of the label can be adversely affected if the item is too cold.
- Dry offset (offset letterpress) and screen printing – graphics are not as good as when using in-mould labelling, although quality is improving due to modern developments.
- Colour – can be added to the extruded sheet. Multicolour stripes can be added and placed in such a way to allow one section of the item to be decorated differently from the others.
- Surface finishes – can be designed into the final thermoformed article, but the quality of finish is not as good as with injection moulded items.

14.6 Injection moulding

The injection moulding process is suitable for all materials, only the tool design having to be changed depending on the shrink characteristics of the polymer/polymer combinations. Injection moulding is a process which converts polymer granules into one of the most dimensionally accurate moulded thermoplastic parts possible. It does so using a reciprocating or ram, plasticating extruder (Fig. 14.29). Dry plastic granules are added to the plasticating extruder in the same way as described earlier in the chapter for film and sheet extrusion. Most importantly, the drying time of the polymer and its masterbatch significantly affect the process time, with PET, PA and PLA needing between 6 and 12 hours for complete drying. To overcome this problem, pre-hoppers are used to pre-dry the polymer and liquid colorant is used to negate the need to dry the masterbatch.

Once heated in the extruder, the homogeneous, molten mass is then injected into a mould through a gate, known as the injection point. A predetermined mass of polymer, designed to completely fill the mould, is metered out within the injection moulding machine, by controlling the stroke of the reciprocating screw or the ram piston. The polymer (PP has a much higher shrinkage than HDPE, for example) and the colour used significantly affect the shrinkage of the moulding. This can result in a separate die being required or changes made to the cycle time and cooling conditions used, to ensure the moulding once cooled meets its dimensional specification.

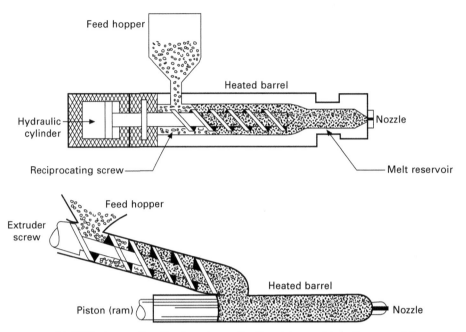

14.29 Reciprocating (top) and ram (bottom) type of injection moulding machine.

The mould consists of two or more steel parts, one with a cavity accurately cut away to form the female section of the moulding, the other with a corresponding profiled section (Fig. 14.30). When the two halves are clamped together, the gap between the male and female sections of the mould corresponds to the shape, finish and thickness of the moulding required after taking shrinkage into consideration. The parts are locked together with a clamping force sufficient to ensure that both single- and multi-cavity moulds stay closed, until the moulding is cool enough to be ejected. Multi-cavity moulds are used to increase the number of units produced over a given time and therefore reduce the unit cost. The number of cavities possible is dependent on the surface area of the cavity and the locking force of the injection moulding machine, the larger the locking force the greater the number of cavities possible.

The mould is cooled with temperature controlled liquid to ensure the mouldings cool as evenly as possible. Injection points are positioned so that the flow of material into the mould is as even and thus stress free as possible, and the whole moulding is free of weak areas where the polymer has flowed together in the mould. Injection points can be visible on the mouldings, therefore consideration must be given to this at the mould design stage. With care, they can be hidden from view (Fig. 14.31).

The injection moulding machine must have sufficient clamping force to prevent any of the injected plastic from escaping at the interface of the two halves of the mould. This would cause an unsightly part-line or excess material, known as 'flash' which has to be removed or, in severe cases, the moulding scrapped. Excessive shot size or injection pressure can also cause flash to occur, as can the use of an old, worn mould.

Plastics manufacturing processes for packaging materials 341

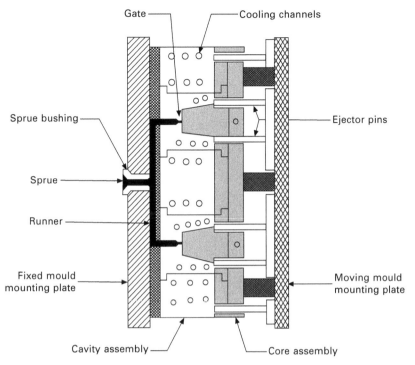

14.30 A multi-cavity mould.

The size of the injection moulding machine is specified by its clamping force and injection capacity and the mould must be compatible with the machine (Fig. 14.32). Moulds are clamped together and opened using mechanical (toggle lock), hydraulic (direct lock), or a combination of both (lock and block or hydro-mechanical) methods. Mechanically operated machines use integrated hydraulic systems for the motive power. However, during more recent years, to improve cleanliness and reduce energy, servo-motors are used, hence the 'all-electric' machines. As stated, clamping forces have to withstand the internal pressure within the mould cavity, and injection pressures of around 2,000 bar (29,000 psi) are common.

Normally the female cavity side of the mould is attached to the stationary end of the injection moulding machine (fixed or nozzle platen) and the male cavity side is attached to the moving end (moving or ejector platen). The molten polymer is injected into the mould through the fixed or nozzle platen. To ensure the two halves of the mould are precisely in line (necessary to achieve accurate mouldings), large guide pins are situated on the four corners of one side of the mould with matching locating holes on the other. If moulds are not vented effectively, the injected plastic cannot displace the air inside and therefore an imperfect moulding will result. The consequences can vary from an imperfect finish on the surface of the moulding to an incomplete moulding, due to the pressure build-up in the mould preventing free flow of polymer.

Moulded parts are ejected from the mould using ejector pins, air pressure or stripper

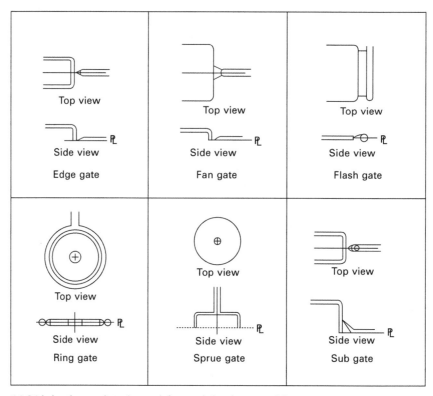

14.31 Injection points (gates) for an injection mould.

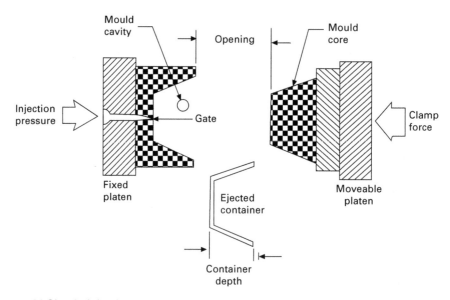

14.32 Simple injection mould.

rings. The ejector pins leave a tell tale ring in the moulding, stripper rings and air pressure are much kinder in this respect. Some polymers are not tough enough to withstand the high pressures employed when using air to eject, for example polystyrene is usually too brittle. Air pressure ejection lends itself to thin section parts rather than thick.

14.6.1 Mould designs

Mould design is an important part of the development process when considering the production of a new moulding. Some considerations are given below.

- The shape of the moulded part has a direct bearing on the time involved in making the mould, the cycle time of the moulding. For example, undercuts and deep screw threads often require a separate moving part which increases the cycle time of each moulding operation.
- Angles on the side walls are necessary to ensure the moulding can be removed from the mould.
- The surface finish of the moulding needs to be decided before the mould is completed. Embossing, etching and other finishes can be added to the moulding to enhance the aesthetic qualities of the moulding. Any imperfection on the finish of the moulding will transfer to every mould made.
- Weight, surface area and thickness of the moulding directly affect cycle time. Where mouldings have significantly varying thickness, cooling has to be controlled very carefully otherwise depressions form (sink marks) on the outer surface of the moulding caused by excessive shrinkage of the thick section of the moulding.
- The number of cavities in a mould is governed by the number of units per annum required. The number of cavities directly affects the size of machine required for the multi-cavity mould; this in turn affects the cost of the mould and the unit cost of the moulding. Where there are many cavities, each one must be uniform compared to the others and the cooling profile needs to ensure that the outer and inner cavities cool at an equal rate, otherwise moulds of differing dimension will result. Hot runner systems are usually used for multi-cavity systems. This reduces the cycle time and amount of waste material formed but increases the overall cost of moulds (see Section 14.6.2).
- The need for an insert or label to be inserted during the moulding sequence has to be considered at the mould design stage.
- All injection moulds require a point or points at which the molten plastic is introduced into them. If not considered at the design stage, this can leave unsightly surface blemishes on the finished moulding which require a further stage to remove them, incurring extra costs.

Successful removal of the moulding from the mould is another critical consideration in mould design. Some mouldings, such as closures, have undercuts, which complicates the mould design. Sometimes the moulding is flexible enough, and the undercut small enough to pop or blow it off the tooling without damaging the moulding. However, this is not always the case and a more complex mould design may be required (see Fig. 14.33). Other options are the use of a collapsible core or, in the case of threaded mouldings, the incorporation of an unscrewing device.

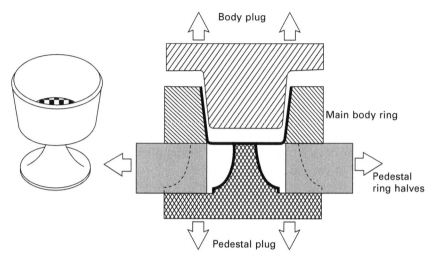

14.33 Mould design for parts with undercuts (in this case a cup with a pedestal).

14.6.2 Hot runner moulds

Figure 14.34 shows a hot runner mould (the figure shows a co-injection design which is discussed in more detail in Section 14.7.2). Feeding of the plastic material to the mould can be carried out using hot or cold runner systems. In a cold runner system, the sprue feeds polymer to the runner, which in turn supplies polymer to each individual mould within a multi-cavity tool. The polymer is then fed through the gate into the individual mould cavities. This feed system cools and is ejected with the mouldings which then have to be removed from the cold runner as a separate operation. The removed plastic is then sent to waste or reground and fed back into the hopper along with virgin material at a controlled percentage. Hot runner injection moulding tools, though more expensive to design and make, overcome these issues. The runner is either insulated within the mould preventing the polymer solidifying, or it is heated to ensure the plastic is held at the most efficient temperature. These 'hot runner' systems reduce waste and increase cycle time. Hot runner mouldings are now commonplace in plastic packaging applications.

14.6.3 Decoration of injection moulded parts

Just like paperboard and plastic films, injection moulded parts can be decorated by vacuum deposition of metals such as aluminium. Other types of decoration are possible and are covered in Chapter 18. Mould surface treatments such as spark erosion is used to give special effects, and mouldings can also be laser etched or coloured in

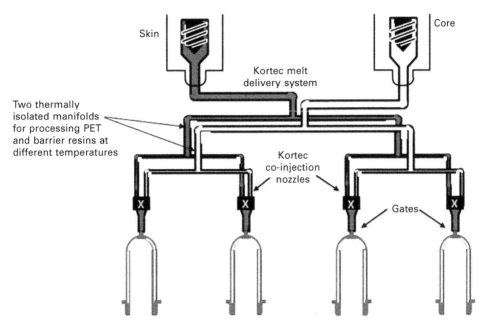

14.34 Hot runner mould (courtesy of Kortec Inc.; www.kortec.com).

a variety of ways. Colours can also be added together, to give unique multicolour effects, by using masterbatches of dissimilar MFI.

14.7 Multi-injection moulding

Multi-injection systems are used where more than one polymer or colour is required. This can be achieved in two ways:

- Two-stage injection moulding and overmoulding
- Single stage coinjection.

14.7.1 Two-stage injection moulding and overmoulding

The two-stage injection moulding process can be used to produce a component with different coloured areas or with a core and an outer skin (Fig. 14.35). For example, the core could be foamed product of one polymer, the skin could be a completely different polymer. Overmoulding of another object is also possible, e.g. overmoulding of a glass bottle with a clear resilient plastic such as an ionomer or a thermoplastic elastomer (TPE). The first polymer is injected into the smaller mould, the item is removed from the mould and inserted into a second mould adjacent to the first. This second mould is closed and the second polymer injected around the first moulding. The first stage can be injected at the same time as the second stage, once the process has started, therefore reducing cycle time. Overmoulding of glass and metal can be carried out in the second mould.

Step 1: Injection material No. 1 and injection material No. 2 simultaneously

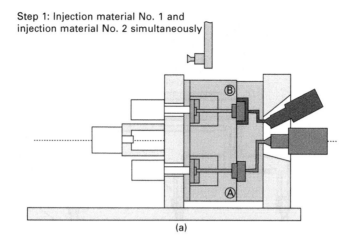

(a)

Step 2: At end of mould opening
- robot descends
- upper and lower ejection forward
- take finished part and substrate

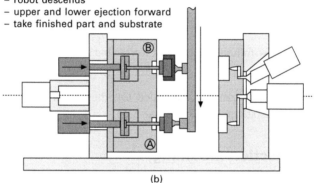

(b)

Step 3: After upper and lower ejection return
1 – robot moves vertically up to an interim position
2 – substrate is inserted to overmoulding cavity
3 – robot moves away

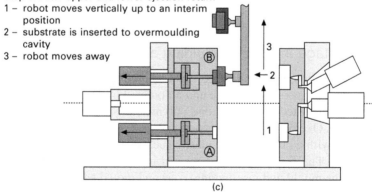

(c)

14.35 Two-stage overmoulding process: (a) step 1; (b) step 2; (c) step 3; (d) step 4, (e) step 5 (courtesy of Husky Injection Moulding Systems; www.husky.ca).

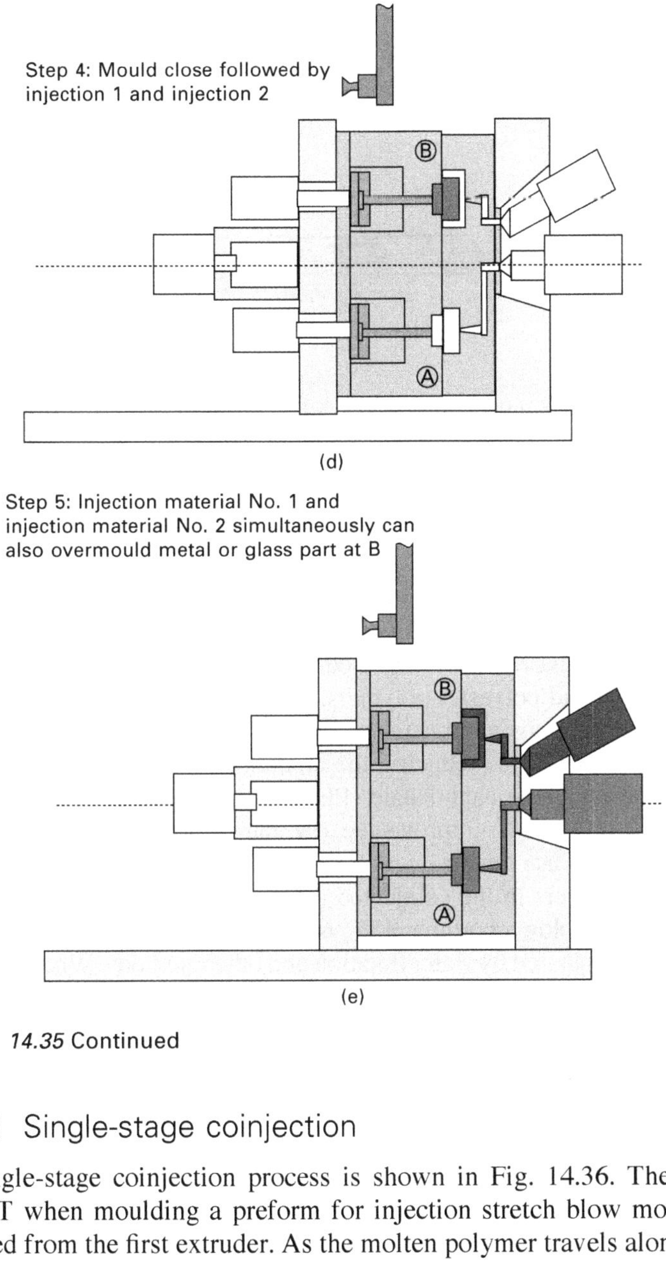

Step 4: Mould close followed by injection 1 and injection 2

(d)

Step 5: Injection material No. 1 and injection material No. 2 simultaneously can also overmould metal or glass part at B

(e)

14.35 Continued

14.7.2 Single-stage coinjection

The single-stage coinjection process is shown in Fig. 14.36. The main polymer, e.g. PET when moulding a preform for injection stretch blow moulded bottles, is delivered from the first extruder. As the molten polymer travels along the hot runner system, a second extruder injects another polymer, for example PA. The more viscous PA travels through the core of the less viscous PET. The melting temperature of the materials should be similar. The main material (PET) is first injected into the neck finish of the mould. The core material is injected into the body portion of the preform. The base, like the neck finish of the preform, contains little or none of the core material to ensure maximum strength is achieved when the bottle is blown. The injection system is carefully sequenced to ensure the core material is delivered to the

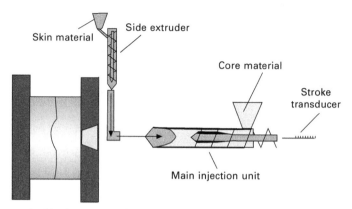

14.36 Single-stage coinjection process (adapted from Giles, G. and Bain, D. (eds), *Technology of Plastics Packaging for the Consumer Market*, Wiley, 2001).

exact areas required, during the coinjection process, using a hot runner coinjection system.

One of the main uses for this technology is PET bottles and jars requiring an improved oxygen barrier and, in the case of carbonated beverages, a carbon dioxide barrier. The two types of barrier materials used are ethylene alcohol-based polymers and special polyamide and polyester polymers. The polyamide and polyester-based polymers actually act as oxygen scavengers, preferentially absorbing the oxygen as it passes through the PET, thus reducing the amount of oxidation of the contents of the container. Polyethylene naphthalate (PEN) is sometimes used as a minority percentage blend with PET, to improve the UV barrier and processability. It is, however, restricted in its use due to its high cost. To allow for easier recyclability, the different material layers in the coinjected preforms are not tied together. Once the neck is cut from the blown bottle and the bottle ground into small particles, the two polymers can be separated by water flotation and other methods. Where preforms are made combining PP and PET, as shown in Fig 14.37, the core barrier layer in the PP preform has to be incorporated all round the preform, including the neck and base areas. This is because PP has less barrier to oxygen than PET.

14.8 Comparing injection moulding and thermoforming

Table 14.2 compares injection moulding and thermoforming. Key issues to bear in mind in choosing between the two include the following:

- The amount of material used to make thermoformed items is always greater than for injection moulded items. This is because the injection moulded items only require the exact amount of material to fill the mould (thick and thin sections are predetermined at the design stage so as to reduce the amount of material). The thermoforming process relies on a sheet of material of even thickness.
- Thermoforming thins the walls as they stretch resulting in more material being required to compensate for the thinning. Once formed, the items have to be cut

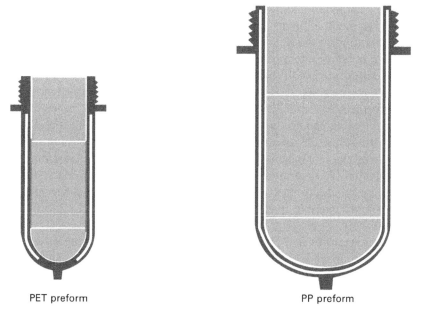

14.37 Coinjected PP and PET preforms (courtesy of Kortec Inc.; www.kortec.com).

Table 14.2 Thermoforming versus injection moulding

Thermoforming	Injection moulding
Not suitable for all materials	Suitable for all materials
Expensive material	Less expensive material
Less expensive tooling	More expensive tooling
Less time to make tooling	More time to make tooling
Multi-barrier less difficult	Restriction on barriers
Less accurate dimension	Excellent accuracy
Undercuts restricted	Undercuts possible
Poor distribution of material	Excellent distribution of material
In-mould labelling difficult	In-mould labelling possible
Good for short runs	Less suitable for short runs
Surface effects limited	Many surface effects possible
Lip or rib on packs	No lip or rib on packs

from the multimould, creating waste. Hot runner injection moulding creates no waste.

- The tooling costs for injection moulding are in general more expensive than for thermoforming, due in part to the extra accuracy required and the pressures involved in the process. However, the costs as well as the time required to make the tooling for both processes is becoming very competitive.
- Multilayer sheet for the thermoforming process is straightforward to produce, but multilayer injection moulding is still being developed, therefore predetermined barrier requirements are less difficult to produce with thermoforming than with injection moulding.

- Thermoformed mouldings do not have the high level of dimensional accuracy of injection moulded items. The thermoforming process relies on pressure and/or vacuum to form the semi-molten sheet into a mould. Injection moulding forces molten polymer into a precisely formed tool, under high pressure, ensuring the moulding conforms to the design of the tool.
- Undercuts are restricted to very shallow ones, on thermoformed mouldings. This is governed by the ability to remove the moulding from the mould. Injection mould tooling can accommodate any undercut by designing special tooling.
- Accurate screw threads are not possible with the thermoforming process.
- The distribution of material when in sheet form is consistent, therefore the material distribution after thermoforming the article is far less controllable than with the injection moulding process. As a result, mechanical and barrier properties are much more variable in thermoformed articles compared to injection moulded items.
- The temperatures and pressures involved in the manufacture of thermoformed articles are much less than in the injection moulding process. This results in the adhesion between in-mould labels and the moulded item being far superior in injection moulding.
- Injection moulding is comparable in cost to thermoforming on long runs (millions of items), especially when thin wall injection processes are used for such items as tubs for yellow fats.
- Thermoforming leaves a lip on packs, where the moulding has been cut away from the remaining sheet. There is also a tendency for a rib to be formed just under the orifice of the thermoforming where undue pressure has been applied to the moulding while still soft. Neither of these phenomena occur with injection moulded items, though in a moulding where there are thin and thick sections, sink marks can occur due to the extra shrinkage of the thicker section.

14.9 Blow moulding

Blow moulding can be achieved via an injection, extrusion or combination process. It is used where the orifice of the moulded item is smaller than the overall cross section of its body, for example bottles and jars.

14.9.1 Injection blow moulding

Injection blow moulding is a combination process. First, we have to injection mould a preform and then blow it into the shape required, thus two moulds are required: one for the preform and one for the final blown form. The following methods can be used for producing blown mouldings from an injection moulded perform.

The two-stage blow moulding process is used for standard and stretch blow moulded items (Fig. 14.38). The preform is injection moulded as a separate stage, in a separate machine. The preform is designed to have a profile and variable wall thickness to provide the correct mechanical and barrier properties in the final blown moulding. Once moulded, the preform is reheated (different zones of the preform are

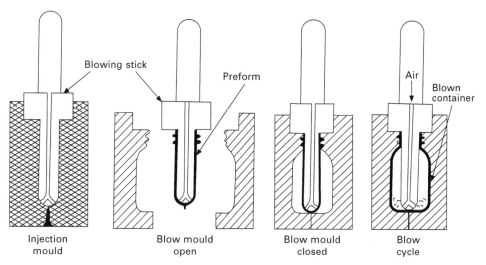

14.38 Injection blow moulding process.

heated to different temperatures to best suit the final blown form requirement) and placed in the blow mould. Air is introduced via the preform neck and the preform is blown into the shape of the mould. Venting is necessary so that all trapped air is removed from between the moulding and the mould. The mould has cooling ducts incorporated into its design. The coolant is usually water maintained at a constant temperature. To produce bottles suitable for carbonated beverages, the gas barrier and mechanical properties (tensile and burst strength) of the final bottle have to be greater than for a non-carbonated product not requiring a gas barrier.

Stretch blow moulding can be carried out on both injection and extrusion processes (Fig. 14.39). In injection (the most common use) stretch blow moulding, the preheated preform held on the stretch rod is placed in the blow mould. Both the rod and the preform are heated to a controlled constant temperature, usually just above the T_g of the polymer, the bottle finish area being kept cool so that it does not distort. Once in the blow mould, the stretch rod pushes the preform to the bottom of the blow mould, air is introduced through the rod which expands the preform to the shape of the blow mould. In this way the material is orientated in both directions; this improves clarity, mechanical and barrier properties. Very lightweight bottles with good moisture and carbon dioxide barrier and pressure resistance properties can be produced using this method. Improved oxygen barrier can be obtained by using barrier materials such as polyamides and polyvinyl alcohol, the former can be added as a monolayer, but usually the barrier (oxygen scavenger) is added as a separate layer in the centre of the preform.

Injection moulds for the bottle industry are multi-cavity (over 100 per tool) to ensure the cost is kept to a minimum (Fig. 14.40). The preform is so designed that once the bottle is blown, the neck finish and base are five times thicker than the wall of the final bottle. The incorporation of a barrier layer to the preform structure extends the shelf life of the packed product (e.g., beer) to an acceptable level. Barriers are also used in wine bottle manufacture but often added in the monolayer, rather

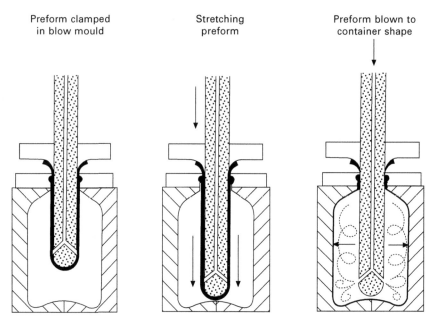

14.39 Stretch blow moulding process.

than coinjected as for beer. As well as providing an excellent oxygen barrier, it is claimed to offer improved clarity compared to coinjected bottles.

Single-stage injection blow moulding requires only one piece of moulding machinery, but still requires two sets of tooling (one for the preform injection mould, one for the blow mould). The preform is injection moulded, transferred from the injection mould to a reheating station and then transferred to the blow moulding station where the item is formed, cooled and ejected. The benefit of this process is that, for small quantities of items (e.g., cosmetics and toiletries) all mouldings can be carried out by the packer-filler, in one process.

14.9.2 Extrusion blow moulding

Extrusion blow moulding is a less expensive process than injection blow moulding and can provide a wide variety of barriers and features such as handles, but the dimensional accuracy is not as well controlled (Fig. 14.41). Some materials such as PET and standard PP homopolymer are difficult to impossible to extrusion blow mould on a commercial basis due to their low melt viscosities. However, polypropylene and polyester copolymers are available which can be extrusion blow moulded acceptably. The parison (hot hollow plastic tube) is extruded through an annular die. The thickness of the parison is controlled in the die by varying the wall thickness of the parison, whilst leaving the outside diameter the same. This is accomplished by having a conical inner sleeve, which can be moved up and down in the die, resulting in a controlled variation in wall thickness of the parison (Fig. 14.42).

This control is important, especially when producing bottles of complex shape with widely different cross-sectional areas from base to neck, for example a cylindrical

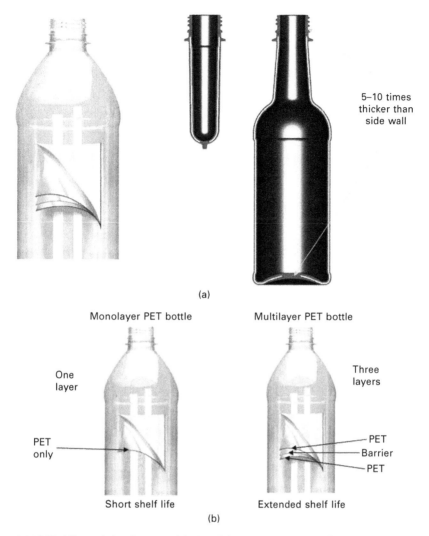

14.40 Multilayer injection moulded PET bottles (courtesy of Kortec Inc., www.kortec.com).

bottle with a heavily waisted section for ease of holding, or figurine-shaped bottles used for children's products. It is equally important for more regular shapes such as square or rectangular cross sections, where the plastic has to be stretched a long way from the centre at varying distances. The control of wall thickness of the parison is also important where, due to its weight, it flows downwards before it is taken into the blow mould (Fig. 14.43).

Parison thickness control is also important to ensure the correct weight of polymer is applied to the correct areas, based on the design of the final item, for example integral handles. To produce a handle, material needs to be 'stolen' from the parison. Sufficient material needs to be placed in the area of the parison where the handle is to be formed to allow this. It is also important to control the handle area so that

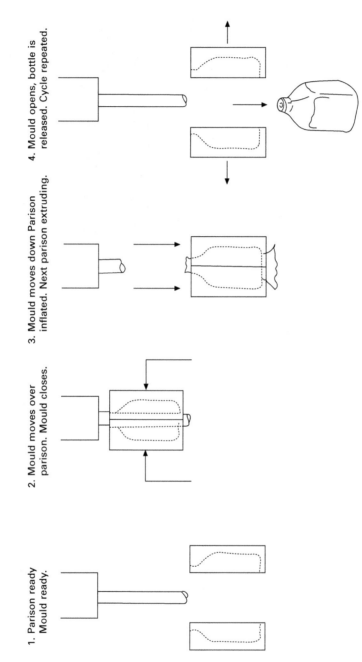

14.41 Extrusion blow moulding process.

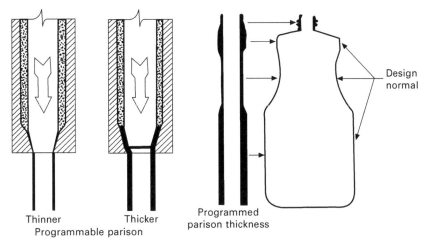

14.42 Control of parison thickness in the die.

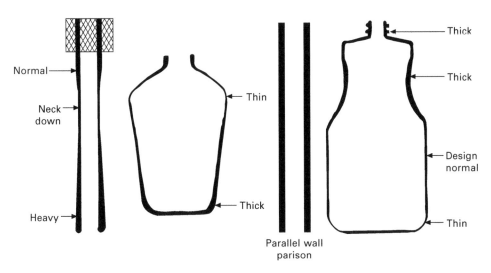

14.43 Problems in controlling parison wall thickness: (left) parison 'neck down' due to weight and (right) inflation of parallel-walled parisons in containers with variable diameters.

a balanced moulding results. Sometimes the neck is designed to be away from the centre of the moulding. This too requires redistribution of the polymer to ensure good performance of the final moulding. The control of the parison wall thickness ensures a minimum amount of material is used to produce the final moulding without compromising its performance. The parison is extruded to a given length (the length of the blow mould) and the blow mould is closed around it. Once closed, air is introduced to the parison, blowing it to the desired shape. The moulding is ejected once cooled sufficiently so that it will not deform as it comes out of the mould (Fig. 14.44).

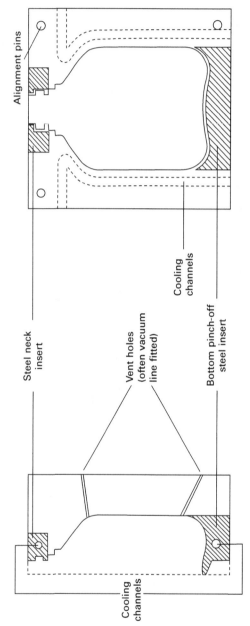

14.44 Cooling of moulding ready for ejection.

The moulding process requires the blow mould to seal the parison by pinching it together. This results in excess material being squeezed out when the mould comes together, and this excess has to be removed. This is carried out by transferring the moulding, contained in the mould to a stripping area, where the excess material is cut off with sharp knives. This action leaves a tell-tale scar on the base of the moulding. This is an easy way to differentiate between an item produced by injection blow moulding, which will show an injection point nipple at the bottom, from an extrusion blow moulding. The waste material is, where acceptable, reground and coextruded, as a sandwiched layer within the parison, thus reducing waste and keeping costs to a minimum.

Multilayer extrusion blow mouldings are now the norm. The combination of different polymers provides a vehicle to increase the performance characteristics of the final moulding, whilst at the same time keeping costs competitive (Fig. 14.45). Multilayer preforms can be so designed as to leave the outside layer free of colour at a predetermined point, allowing for a translucent stripe to be incorporated, so that the product level can be seen. This is very convenient where multi-dose bottles are used.

The normal machine design for most extrusion blow moulding manufacture uses alternating moulds (either single- or multi-head). However, for very high volume on one design, rotary blow moulding machines are commonly used (Fig. 14.46) . Here the air can be introduced via a needle through the side of the parison rather than the more common method of through the centre hole. As the needle is introduced and extracted before the moulding cools the needle hole self seals. This type of rotary system is often used for moulding HDPE milk bottles.

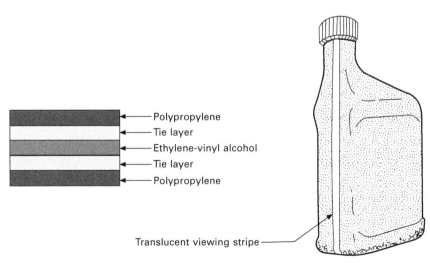

14.45 Multilayer extrusion blow mouldings: layers for a high oxygen-barrier bottle (left) and a bottle with a coextruded transparent viewing stripe to view fill level (right).

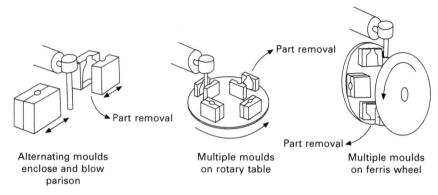

14.46 Alternating and rotary blow moulds.

14.9.3 Design specifications for blow moulded bottles

The standard terminology when describing bottles (and other blow mouldings) should be understood, especially when including in specifications (see Fig. 14.47 and Table 14.3).

Table 14.3 Typical specification for a carbonated soft drinks bottle

---

Resin intrinsic viscosity 0.78–0.82 g/cm^3
Volume
    2,000 ml ± 20 ml (individual)
    2,000 ml ± 10 ml (average)
Fill level drop
    40 mm under 4 bar (function of the bottle shape)
Fill point variation (from bottle to bottle)
    max 2.5 mm (function of the bottle shape)
Bottle creep (24 h filled with 4 vol. CO_2)
    diameter increase max 3%
    height increase max 3.5%
Deviation from perpendicularity
    max 9 mm
Top load (empty and peak values)
    >200 N (function of the design and weight)
Drop test
    no leakage for bottle drop from 2 m height
Thickness
    minimum 0.25 mm
    in the heel 0.20 mm
CO_2 loss, shelf life
    14 weeks with 15% loss CO_2
Stress cracking
    15 min in 0.2% NaOH (weight) at 4 vol.
Burst pressure
    >8 bar

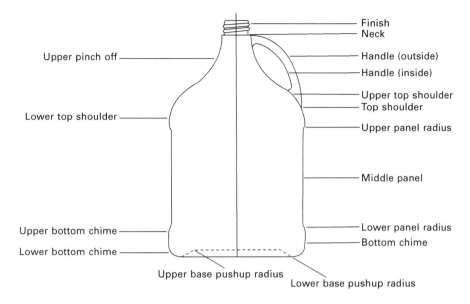
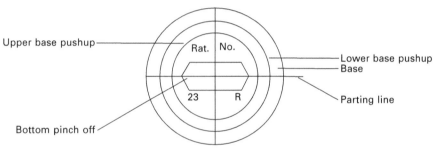

14.47 Standard bottle terminology.

14.10 Environmental considerations in plastic packaging

As covered in Chapter 5, the environment is a serious consideration when choosing the type of packaging for a product. Plastic materials are often viewed by consumers and ill-informed bodies as environmentally irresponsible. However, the consumer only sees them lying in the street, and the complexities involved when determining if a particular packaging material or design is good or bad for the environment compared to the alternatives are not well understood. Whether polymers are used for single- or multilayer items, the properties required for the end use, cost and environmental issues are the main considerations which determine the type and thickness used. Plastic items are all recoverable in some way or another (re-use, recycle, energy recover, biocompost); their main advantage is in reducing materials used and preserving the products inside, at the minimum total cost. One of the difficulties of film with respect to re-use and recycle is the cost of collection and sorting, as not all polymers are compatible. Incineration for energy recovery is possible and this simplifies the sorting process, although it requires special incinerators to ensure there is no pollution from the gases produced during the burning process.

14.11 Sources of further information and advice

Baner, A.L. and Piringer, O. (2008) *Plastic Packaging Materials for Food.* Wiley, New York.

Baner, A.L. and Piringer, O. (2008) *Plastic Packaging: Interactions with Food and Pharmaceuticals.* Wiley, New York.

Giles, G. and Bain, D. (eds) (2000) *Materials and Development of Plastics Packaging for the Consumer Market.* Wiley, Oxford.

Giles, G. and Bain, D. (eds) (2001) *Technology of Plastics Packaging for the Consumer Market.* Wiley, Oxford.

Hernandez, R., Selke, S. and Culter, J. (2000) *Plastics Packaging.* Hanser, Munich.

Nicholson, J. (2011) *The Chemistry of Polymers*, 4th edn. RSC Publishing, Cambridge.

In addition to the above texts, the British Plastic Federation (www.bpf.co.uk) is a valuable source of information on all aspects of plastics, including material properties, industry applications and forming methods.

Other useful sources of information include:

- http://www.britishplastics.co.uk
- http://www.incpen.org
- http://www.packagingdigest.com
- http://www.packagingfedn.co.uk
- http://www.packagingtoday.com
- http://www.pafa.org.uk
- http://www.plasticsinpackaging.com

15
Packaging closures

A. EMBLEM, London College of Fashion, UK

Abstract: This chapter considers the important functions of closures, and reviews a range of different closure types. Closure material properties are not repeated here, and it is necessary to refer to the relevant chapters throughout the text. Specialist closure functions such as child resistance, tamper evidence and dispensing features are covered.

Key words: screw thread, push-fit, tamper evidence, child resistance, crown cork, lever lid, wad, liner, linerless closure, lug closure, peelable lid, ROPP, shrink band, flip top closure.

15.1 Introduction: the role of packaging closures

A closure can be defined broadly as any method for closing a pack so that the product is properly contained and protected. A more specific definition is a device that seals a product within a pack but which can be removed to allow the product to be accessed. There is a huge variety of closure types, including the following, defined by their method of application/achievement of a seal:

- push-fit closures
- screw-threaded closures
- heat-sealed closures (e.g. sachets, film lids and tablet blister packs)
- folded closures, often glued (e.g. cartons and cases, paper bags for flour, sugar)
- mechanical seaming (e.g. on metal cans)
- stitched or stapled closures (e.g. on paper sacks).

This chapter concentrates on closures which are usually added separately to a container, after filling, and used by the consumer to gain access to the product, i.e. mainly the first two categories above. The requirements of the remaining categories are addressed in the relevant chapters of the text, and are not discussed in detail here, e.g. the requirements of heat-sealed closures are discussed in more depth in the chapters on plastics properties and the packaging line.

Regardless of type, the closing method of a pack product containment is the interface between the product and the environment. It is also the critical interface between the product and the consumer and must meet a number of requirements:

- It must not contaminate the contents of the container or be degraded by them.
- It must be compatible with the container and its materials of construction.
- It must be able to withstand processing conditions such as sterilisation, and the

application of significant force (e.g. in tightening a screw cap or crimping a crown cork).
- It must be capable of withstanding vibration and temperature fluctuations during transport and storage as well as potentially rough handling by consumers in use.
- It must provide an adequate seal until the contents are ready for use. This may include protecting the product from air, light, moisture or foreign particles (e.g. dirt), providing an airtight seal or retaining a vacuum or internal pressure in the container.
- It must be convenient and safe to remove by consumers (who may have a wide range of abilities, e.g. the elderly with reduced manual strength and coordination).
- It may need to include child-resistant features which restrict access to potentially harmful products.
- It may need to include tamper-evident features to indicate whether a container has been interfered with in some way prior to purchase.
- It must often be resealable, capable of retaining an adequate seal once resealed and able to withstand repeated opening and closing.
- It must contribute to the overall aesthetic design of the pack.
- The type of closure and the method of application must be compatible with the overall volume and speed of production, particularly the requirements of the product filling line.
- It must meet the cost and operational requirements of the business, including acceptable design and material costs of closures, and acceptable set-up and operational costs of the packaging line.
- It must meet increasingly stringent environmental requirements, including minimising the amount of material used and facilitating recycling of the closure and container.

15.2 Types of packaging closure

This section starts by looking at the basic types of closure, such as push-fit, screw-threaded and lug closures, plus closures made by crimping. It then goes on to explore more specialised types of closure, offering features such as tamper evidence, child resistance and product dispensing or metering. Due to the wide variety of closures in each of these categories, it is not possible to cover all available options and discussion will concentrate on the principles of how an effective seal is achieved, rather than describing specific designs. A list of useful sources of further information is given at the end of the chapter.

15.3 Push-fit closures

Push-fit closures fall into two types:

- Push-in closures: those in which the closure is pushed *in* the open neck of the container, e.g. a wine bottle cork, or the lever lid used on paint cans.

- Push-on closures: those in which the closure is pushed *on* or over the outside of the top of the container, e.g. the metal lid on a tin of biscuits or the plastic overcap sometimes supplied as a resealable feature on containers for dry products such as ground coffee.

15.3.1 Push-in styles: corks, lever lids and plastic push-in closures

For push-in wine bottle closures, a seal is achieved as a result of the compression of the closure material against a rigid container surface which prevents excessive expansion. Cork and low density polyethylene are therefore ideal closure materials, and glass is the ideal container material to maintain a good seal (see Fig. 15.1). The quality of the seal can be enhanced by incorporating fins or serrations, which improve the flexibility and allow a tighter fit, with each fin creating a mini-seal as it presses against the container neck. Seal quality is enhanced by container design features such as a slightly rough surface and reverse tapering lower down the neck.

The push-in, pry-off lever lid design relies on a tight fit between the lid and container. To achieve an effective seal, both closure and container materials need to be sufficiently rigid to withstand the pressure imposed by pressing the lid into the neck of the container, container and closure tolerances need to be very closely controlled and the angles of the contact surfaces carefully calculated. Both the lid and neck typically include a rim and the neck is tapered to allow for some variation in dimensions as well as to increase the effectiveness of the seal (see Fig. 15.2). Metal clips are sometimes used for added security. This type of closure is easy to apply and provides a good seal but is vulnerable to damage from the levering action in opening and the pressure needed to reseal, often distorting the shape of the lid and aperture. Also, the seal area is easily contaminated, which compromises reseal quality. Lever lids (and containers) are traditionally made of metal but plastics are now widely used, e.g. for emulsion paints.

15.1 Push-in cork in glass bottle.

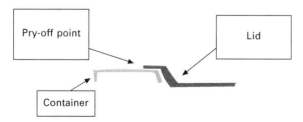

15.2 Push-in pry-off lever lid design.

Push-in plastic closures are common for composite containers for dry products such as cocoa or custard powder. The lid is pushed into the top of the container and the compressibility of low density polyethylene makes it ideally suited to this application. However, the relative flexibility of the container (compared with metal) means that this is not an airtight seal and an aluminium foil diaphragm is often used to provide the required level of preservation, as well as tamper evidence. Once the diaphragm is removed, the seal provided by the push-in closure must be sufficient for the remaining life of the product.

15.3.2 Push-on styles: metal lids, plastic push-on closures and push-on twist-off closures

Push-on closures (also known as press-on closures) are easy to manufacture and apply, allowing high filling and sealing speeds. The simplest type is a metal rimmed lid (known as a slip lid) which fits over a metal container, e.g. as in a tin of biscuits. Since the fit cannot be so tight as to make opening difficult, the seal is relatively poor and, if the product is particularly prone to moisture gain, it will need additional wrapping inside the tin (e.g. a plastic film) to provide adequate preservation. To avoid the lid falling off in transit, and to provide a measure of tamper evidence, adhesive tape or labels can be used across the lid/container junction.

Push-on plastic closures must have good extensibility so that the closure can be stretched across the neck of the container to provide a good seal. Low density polyethylene is particularly suitable because of its good elongation properties. The container usually has a rounded lip to make it easy to remove and refit the closure, and there may be a locating ring to hold it in place (see Fig. 15.3). As with the push-in plastic closure already mentioned, additional features such as the use of an aluminium foil diaphragm may be required to provide appropriate product shelf life and tamper evidence.

There is widespread use of push-on plastic closures in the toiletries and household chemicals sectors, often incorporating flip-top or spray devices (see later). Design possibilities are numerous, with a wide variety of shapes and colours of both closures and containers. Locating features on the container are common, in some cases preventing the easy removal of the closure. Due to the relatively long usage period of such products, which means that the closure is subjected to repeated opening and closing, polypropylene is likely to be the material of choice, owing to its excellent resistance to flexing.

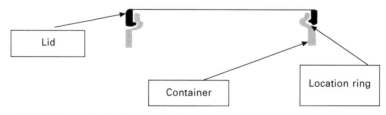

15.3 Push-on plastic closure design.

Push-on, twist-off closures (see Fig. 15.4), as commonly used for baby foods, are made from a tinplate shell, injected with a soft thermoplastic sealing compound. The product is hot-filled into the container (usually glass) which has an interrupted thread, and the closure applied. As the product cools, a partial vacuum is developed, pulling the metal closure down onto the finish of the container. The heat involved causes the soft flowed-in liner to conform to the shape of the threads. The final seal is effected by applying an appropriate torque in the capping machine, and the consumer opens the pack by unscrewing the closure. This type of closure is often given additional tamper-evident features such as a vacuum button (see Section 15.5) and/or an extended plastic skirt which has to be removed prior to unscrewing the metal part.

Categorising closure types can never be an exact science. Some screw-threaded closures, for example, are applied by push-on actions and these will be discussed in the next section.

15.4 Screw-threaded closures

Screw-threaded closures have a thread following a helical (spiral) pattern on the inside, which must be matched up with the thread on the neck of the container. Ideally, the minimum thread rotation is 360°, giving one complete turn of thread engagement between container and closure. This is known as a 400 or R3 neck finish. Increasing the thread engagement results in a tighter seal, but requires more turns by the capping head on the packaging line. Standards exist for one and a half turns of thread engagement (410/R6 neck finish) and two complete turns (415/R4 neck finish).

As well as the number of turns of thread, there are also options for the thread profile (see Fig. 15.5). The original profile, developed for glass containers, was rounded. However, using this same profile for plastic containers and closures did not allow the closure to be easily and accurately seated on the container, and severely limited the capping torque. Two alternatives were developed, the symmetrical buttress or L-style, in which the threads are angled at 30°, and the modified buttress or M-style in which the thread angles are 10° (lower) and 45° (upper). The latter has now become the most popular thread design for plastic containers and closures, as it provides both structural strength and ease of movement, allowing smooth closing and opening whilst maintaining reasonable seal quality when closed. It is also more resistant to the danger of over-tightening. Matching up the thread profile on container

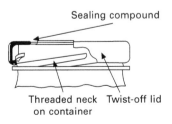

15.4 Push-on twist-off closure.

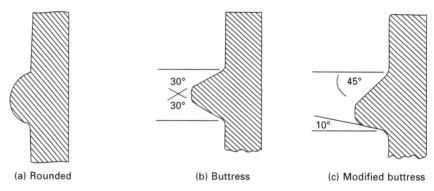

15.5 Thread profiles: (a) rounded; (b) symmetrical buttress L-style; (c) modified buttress M-style.

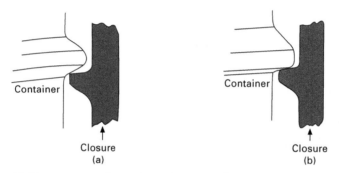

15.6 Importance of matching thread profiles: (a) symmetrical buttress container thread and modified buttress closure thread; (b) modified buttress thread profile used on container and closure.

and closure is essential for an effective seal. Figure 15.6(a) demonstrates that using a symmetrical L-style container thread and a modified buttress M-style closure thread provides minimum engagement compared to using the modified buttress style for both container and closure (see Fig. 15.6(b)).

Although many closure designs are specific to a particular product, closure manufacturers invariably offer a range of standard types. Dimensions are dictated by standard sizes of container necks and typically range from diameters of 12 mm to around 100 mm. These limits are restricted by the ease with which consumers can grasp and twist the closure to open it. There is a standard nomenclature for container dimensions, as shown in Fig. 15.7 and these must correspond on the closure thread, i.e. the container's 'T' dimension (diameter across the outside of the thread) must match up with the closure's 'T' dimension (diameter across the root of the thread). The same applies to the 'E' dimension, and the tolerances on the 'T' and 'E' dimensions on both container and closure must be such that:

- when the closure dimensions are at the minimum of the specification, and the container dimensions are at the maximum, the closure can still be easily applied to the container

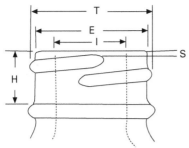

I Diameter at smallest opening inside finish.
T Thread diameter measured across the threads.
E Thread root diameter.
H Top of finish to top of bead or to intersection with bottle shoulder on beadless designs.
S Distance from the top of the finish to the top of the start of the thread

15.7 Standard nomenclature for container neck dimensions.

- when the closure dimensions are at the maximum of the specification, and the container dimensions at the minimum, the closure still sits correctly and delivers an effective seal.

This emphasises the importance of ensuring that tolerances are agreed prior to finalising specifications and that they are maintained throughout manufacture of the components. Should the need arise to investigate causes of leakage, a good starting point is to check the agreed component drawings to ensure that the specifications are correct, followed by measurement checks of both components, taking at least one sample from each cavity of container and closure, where multi-cavity tools have been used.

The 'H' dimensions of container and closure are also relevant. When the 'H' dimension of the container is at its maximum and that of the closure at its minimum, the capped container must still be aesthetically acceptable, i.e. the closure must not look too short for the neck of the container, with an unsightly gap between it and the neck ring or shoulder of the container (see Fig. 15.8(b)). More critically, when the 'H' dimension of the closure is at its maximum and that of the container at its minimum, there must still be engagement between the two components at the 'land' or sealing surface of the container. If not, there will be no containment, as there will be a clear path for product leakage (see Fig. 15.8(a)). This latter point raises the subject of how a seal is achieved in screw-threaded closures and how this differs in wadded and linerless closures.

Wadded closures are delivered to the packer-filler with a wad inserted by the closure manufacturer. The purpose of this wad is to act as a gasket between the top surface of the container and the inside top surface of the closure. When the threads of the two components engage and a tightening torque is applied, the wad is compressed, forming a seal (see Fig. 15.9). The effectiveness of the seal relies on correct choice of wad (see below) as well as correct closure placement and tightening. The wad must also be firmly secured in the closure, and this is achieved either by adhesive,

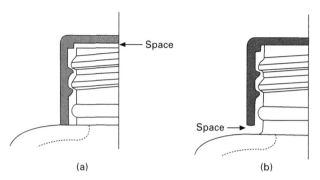

15.8 Accuracy of 'H' dimensions: (a) container 'H' dimension is too short, resulting in product leakage; (b) container 'H' dimension is too long, resulting in an unsightly gap.

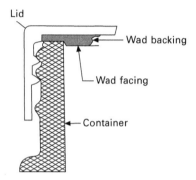

15.9 Wadded closure.

or by using an oversized wad which is forced into the space available, often held there by a retaining ring. If the wad falls out when the consumer opens the pack, and it is not reinserted into the closure, there will be little or no seal on reclosing the pack, resulting in potential contamination and/or leakage.

It is the wad which is in direct contact with the product, and therefore the wadding material must be compatible with the product and it must provide appropriate barrier properties to spoilage, e.g. due to gain/loss of moisture or gases. The wad must also be compressible, and exhibit some elasticity to recover from the compression exerted in use, especially for closures designed for multiple reopening and reclosing. Paperboard and foamed plastics both provide compression, with the latter providing better recovery. Paperboard alone will not provide barrier properties and thus it is usually laminated to a facing material such as aluminium foil, or films such as PET. Foamed plastics may suffice as barrier materials, but are also laminated to films (e.g. PET) for additional preservation.

Linerless closures, as the description suggests, have no wad and rely on direct contact between closure and container neck to effect a seal. They were developed as a lower cost alternative to wadded closures. Most commonly, they rely on a ring inside the closure fitting into the inside of the container neck (see Fig. 15.10). This type of seal is often referred to as a 'bore' seal and it relies for its effectiveness on very close control of the dimensions of both the closure and the container neck. In a wadded system, the internal bore diameter of the container is of minimal concern, as the seal is achieved on the top surface, but now that the bore has become the sealing area, this dimension becomes critical. This may dictate the container manufacturing method, making it necessary to use the two-stage injection blow moulding process rather than the simpler extrusion blow moulding process (see Chapter 14). There are many design variants available, with different ring depths (i.e. the extent to which the ring goes down into the container neck) and shape of ring, for ease of application via a simple push-fit, followed by tightening. Because of the pressure on the cap, the seal tends to weaken over time due to creep, and while high density polyethylene may be adequate for single or minimal use packs (e.g. mineral water, milk) polypropylene is a better option for multiple use over extended periods. Alternative designs involve creating the seal at the top of the closure, for example as shown in Fig. 15.11; as the cap is screwed on, the ridges bend under pressure to create the seal.

Screw-threaded closures can be designed to be linerless during consumer use, but are provided with a wad at the packer-filler stage, to ensure the pack withstands the rigours of distribution. A common example is the shallow closure used for milk bottles. This is supplied to the packaging line with a membrane inserted and it is applied to the filled bottle in a push-on manner and then the capped container is passed between induction heating coils. The membrane is usually a three-layer structure, composed of paper/aluminium foil/low density polyethylene. A metal layer is essential. The aluminium foil absorbs heat, which softens the polyethylene layer, which in turn bonds to the neck of the container, thus providing secure containment and tamper evidence. On removing the screw closure, the consumer peels off the membrane and recloses the pack using the plastic closure. As this is unlikely to be leakproof, it should be stored upright.

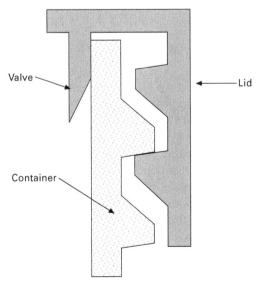

15.10 Linerless closure with bore seal.

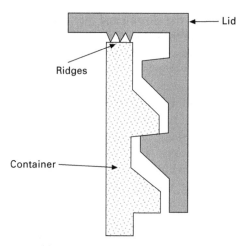

15.11 Linerless closure with top seal.

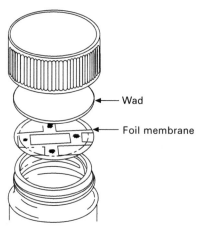

15.12 Induction sealed combined wad and membrane seal.

Another variant is the use of a combined wad and membrane as shown in Fig. 15.12. The foil membrane is loosely bonded to the paperboard or foamed plastic wad, usually using a wax adhesive. During the induction heating process, the foil membrane adheres to the container via its polyethylene coating, as described above. The wad to membrane bond is weakened by the melting of the wax adhesive, so that when the consumer unscrews the closure, this bond breaks, leaving the membrane as a tamper-evident seal, and the wad retained in the closure for an effective reseal. Thus, unlike the membrane-only example above, the consumer can reseal the pack, and still maintain adequate sealing for the life of the product. Uses of this type of seal include jars of coffee and pots of vitamin tablets.

A further consideration is the material used to manufacture closures. Metal-threaded closures are made from malleable materials such as tinplate and aluminium, coated with various protective and decorative lacquers and inks, all of which must be suitable for the product and withstand all application and processing methods. Thermosetting plastics are used in some closures (see Chapter 13) due to their rigidity, resistance to creep and their ability to be moulded in high definition. However, the most commonly used plastics for closures are thermoplastics, especially the polyethylenes and polypropylene, which are easily formed and coloured. PP has significantly better resistance to creep than PE and is the better choice for multiple use closures.

Finally, for all screw-threaded closures, the correct application torque is essential, so that the final fit is not too strong (which could lead to distortion, and/or the consumer being unable to safely open the pack) or too loose, which could result in leakage and contamination. This makes setting the right parameters on the machinery for applying screw caps to containers particularly important (see Chapter 20 on packaging operations). As well as application torque, which measures how tightly the closure is applied, other measures include removal torque, i.e. the amount of force necessary to loosen and unscrew the closure, and stripping torque, which measures the force required to override the closure so that the threads no longer hold it in place.

15.5 Lug closures

Twist-off metal lug closures, which require less than one turn to apply and remove (typically less than 90°), were developed by the White Brothers in the United States in the early 1950s. For this reason, they are often referred to as 'White' caps. Lug closures include lugs or protrusions on the inner edge, designed to engage with an interrupted thread on the container neck (see Fig. 15.13). They are secured by placing on the container, exerting a downward pressure and twisting so that the lugs slide under the thread to hold the cap in place. A seal is achieved between the closure and the container via a liner, which is usually a soft polymeric compound which is 'flowed in' during the closure manufacturing process, often applied only in the rim area.

Given the stress exerted on both closure and container, this type of closure is usually manufactured from heavy gauge steel and is restricted to rigid containers such as glass. They are available in 27 mm to 110 mm sizes with either smooth or fluted side walls, the latter allowing better grip for removal. Because less rotation is required, they can be applied at a higher speed than screw-thread closures. A further advantage is that the lug finish is easier to keep clean and free from sticky product than a helical thread, making this type of seal an excellent choice for jams and other preserves.

This type of closure is often fitted with a pop-up 'vacuum button' which indicates to the consumer, prior to opening, whether there is still vacuum inside the headspace. This provides both a tamper-evident and safety feature, warning a consumer if the product has been previously opened or if the container is defective. The vacuum button is typically created by embossing a ring in the top of the closure. The internal vacuum

15.13 Lug neck finish and closure.

produced during the packing operation is sufficient to pull down the vacuum button (see Fig. 15.14). The button pops up once the cap is unscrewed and the pressure is released.

15.6 Crimped crown cork closures

These are widely used on bottles of beer and are typically made from heavy gauge metal, e.g. tin-free steel, with suitable coatings on both the inside and outside. They have a sealing wad inside the closure which may be cork (hence the name 'crown cork') or a compressible plastic, or, more commonly now, a flowed-in liner of soft plastic material around the inner circumference of the cap (Fig. 15.15). The closure is preformed and then placed over the neck of the container, often using magnets, and the outer circumference is compressed around the lip of the bottle and the edges tucked under a retaining ring on the lip to secure the closure in place, leaving a characteristically ribbed effect.

Crown corks are typically used for sealing glass bottles containing carbonated beverages, and are ideally suited to withstand the internal pressure of such products.

15.14 Tamper-evident pop-up button on metal lug closure.

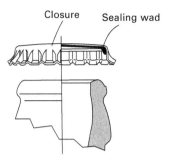

15.15 Crimped crown cork closure.

The high seal integrity prevents air ingress or loss of carbonation, maintaining product quality. Unlike most closures, a bottle opener is required to remove the closure from the bottle (although a twist-off version is available) and once removed, the closure is deformed and cannot be reapplied. The simple design and method of application makes this type of closure relatively cheap and suitable for high-volume, high-speed production.

15.7 Peelable seal lids

In some cases sealing materials can be used as the lid, e.g. on yoghurt pots. Typical lidding materials include aluminium foil, paper, metallised PET or PP, often used in combination. A polymer layer on the inside is essential to effect the required heat seal. These lids are removed by peeling back, and the seal must be sufficiently strong to protect the product over its intended shelf life yet not so strong it cannot be opened by consumers. Many packs include a tab as part of the seal which the consumer can use to peel back the lid. Some designs include a sticky hot melt adhesive on the rim of the container protected by a polyethylene layer. Peeling back the lid fractures the polyethylene layer and exposes the adhesive. The lid can then be resealed. Materials and processes issues relating to this type of closure are discussed in Chapters 9 on aluminium foil, 13 on plastics properties and 20 on packaging machinery.

15.8 Tamper evidence

As already mentioned, a closure is the interface between the consumer and the product, and it must be convenient and safe to remove to gain access to the product. This means that it may be vulnerable to being opened before purchase, and, depending on the product, the brand owner may wish to provide some means of identifying whether or not such 'unauthorised' opening has taken place. Note that a 'tamper proof' or 'pilfer proof' pack is an unrealistic expectation and this section is concerned with the subject of tamper evidence, this being defined as 'having an indicator or barrier to entry which, if breached or missing, can reasonably be expected to provide visible evidence to consumers that tampering has occurred' (US FDA, www.fda.gov).

The key point in this definition is that it is important that the consumer can see if a container may have been tampered with before or at the point of purchase, so tamper-evident features must be obvious, rather than subtle. Where necessary, clear instructions should be given to the consumer (as shown, for example, in Fig. 15.14). It should also be noted that whilst most attention is paid to the pack closure, i.e. the main means of gaining access to the product, a determined tamperer may take a less conventional approach and mount an attack on another seal area of the pack. Thus, when assessing the tamper evidence of their products, brand owners need to consider the whole pack and not just its closure.

Tamper-evident closures abound in the fast-moving consumer goods (FMCG) sector. Some examples are:

- roll-on pilfer-proof (ROPP) metal closures and screw-threaded plastic closures, both of which result in a ring breaking away once the closure is unscrewed
- plastic closures with a tear-off section which has to be removed before opening the pack
- metal closures with pop-up buttons (already discussed)
- shrink seals or bands round the container/closure interface
- adhesive labels and tapes.

15.8.1 Roll-on pilfer-proof (ROPP) metal closures

Although inaccurately named, this is a firmly-established type of tamper-evident closure, made of soft-temper aluminium with a partially perforated ring (or skirt) at its lower edge (see Fig. 15.16). It is delivered to the packer-filler as a printed shell, without threads, and with a wad in place. It is placed on the filled container and the threads are rolled into the outside surface of the metal shell, conforming to the threads on the container. See Chapter 20 for details of its application on the packaging line. The pressure of the thread-forming rollers limits this type of closure to use with highly rigid containers such as glass bottles, and it is widely used for wines and spirits. When the closure is unscrewed, the perforated ring should fall away, leaving clear evidence of opening. The closure has the inherent design weakness that, since the metal is soft, the thread may be lost if it is over-tightened by the consumer when reclosing.

One variant of the ROPP design has the aluminium body of a roll-on closure, with a plastic moulded insert which is pressed in to give a tight fit between the two components. The insert (usually low density polyethylene) has internal threads moulded

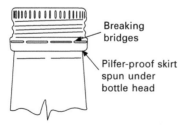

15.16 ROPP closure.

into the side walls and is fitted with a wad. It provides two additional features to the standard ROPP closure:

- As the thread is internal, and provided by the insert, there are no unsightly threads on the outside, making it more aesthetically pleasing.
- As the internal thread of the closure is made from moulded plastic, it is particularly suitable for use with products which have high sugar content, such as liqueurs, where the traditional ROPP aluminium closure tends to become sticky over time, and difficult to remove.

15.8.2 Plastic tamper-evident closures

Plastic tamper-evident closures with a perforated ring or skirt added at the bottom of the cap are applied as conventional screw-threaded closures. Extra pressure is required to push the perforated ring at the bottom over a ridge or bead around the neck so that it snaps into a specially designed groove which holds it in place. The ring is held in place either by the thickness of the groove which keeps the ring stationary as the cap itself is unscrewed, or by teeth which engage with ratchets moulded as part of the groove. These prevent the capture ring moving when the cap is unscrewed. Opening the container breaks the perforations and allows the cap to be removed, leaving the ring in place.

Push-on closures can also be designed to include tamper-evident features, by using a deeper than usual skirt, the bottom of which snaps over a bead on the container neck. Above the bead, the closure is perforated so that the lower part can be torn off using a tab, thus allowing it to be removed completely. One disadvantage with this design is that, once removed, it may not be obvious to a consumer that a container had been interfered with, since the closure itself will still be in place. Sometimes the perforations are not carried all the way around the skirt, leaving a small connecting piece between the cap and the retained ring. This provides a 'captive' cap that cannot be easily misplaced.

15.8.3 Shrink seals or bands

As well as incorporating tamper-evident features in the design of the closure itself, secondary tamper-evident features can also be included, such as printed shrink sleeves. These are cylindrical sleeves of plastic film, placed loosely over a filled and sealed pack and then shrunk in place using heat (see Chapter 20 for packaging line application). Printed shrink sleeves are often used as pack labelling (e.g. soft drinks) and by extending the sleeve over the top of the closure they can also provide tamper evidence. In such cases, the sleeve should be perforated vertically to allow the consumer to gain access to the closure, and then horizontally so that the top part can be torn off without removing the entire label (and thus losing the branding). Alternatively, short shrink bands can be applied just to the closure area, again preferably perforated so that they can be easily removed by the consumer. As it is important that the consumer realises that there should be a shrink band in place, they should be printed so that they stand out as part of the product image.

15.8.4 Adhesive labels and tapes

Label seals can be self-adhesive substrates or ungummed substrates applied using adhesive applicators. They can be applied across the junction of a bottle/jar and its closure, or across the flaps of cartons. The label substrate must tear easily when the pack is opened, and must remain in place, so that it is clear that the label has been torn and the container tampered with. It must not be removable or repositionable. Some labels are constructed so that they disintegrate and others have a series of cuts that extend partially across the label to make it tear more easily and make it difficult to realign. Label seals should always be printed for maximum effectiveness.

There are various tapes used as tamper-evident devices, from simple printed self-adhesive tapes applied to corrugated cases to highly sophisticated double-layer materials which open to reveal warning statements such as 'void' or 'tampered with'. The latter have uses in the pharmaceutical sector, where the need to maintain pack integrity from packer-filler to consumer is crucial.

In summary, the extent to which packs are given tamper-evident features, and the types of features used, is determined by evaluating the risk of tampering and its possible consequences. Ultimately, brand owners are responsible for placing safe products on the market and it is up to individual companies to assess the risk to their products and take appropriate action.

15.9 Child-resistance

Packaging which is difficult for young children to open has been around since the early 1970s and there are international standards which such packs must meet. It is important to understand that these standards apply to packs, and not just to closures, and that alongside being difficult for children to open, the packs must not be difficult for adults to use. The standards also define 'young' children as those less than 52 months. As well as some pharmaceutical products, child-resistant packs are widely used for household product such as bleach and should be considered for any product which could be harmful to young children. Of course, this does not take away the responsibility of parents and others to keep harmful products out of the sight and reach of young children.

There are two relevant standards: BS EN ISO 28317 which applies to reclosable packs and BS EN 14375:2003 which applies to non-reclosable packs. Compliance with the standards is demonstrated by the results of two tests, a child test and an adult test. In the child test a panel of 30–200 children aged 42–51 months is given five minutes to open the pack, followed by a non-verbal demonstration of how it is opened and then another five minutes. Eighty per cent must fail to open it after demonstration. In the adult test, the panel is made up of 100 adults aged 50–70; 90% must be able to open, and (if relevant) reclose the pack in one minute. For non-reclosable packs, e.g. blister packs, 90% of the adults must be able to remove one tablet within one minute.

There are many designs of reclosable child-resistant packs, mostly applying the principle of having to carry out two different actions at the same time, which is generally not intuitive for young children, e.g. the need to both press down and turn a closure, or to squeeze and turn. In some of these designs a significant force is required to carry out these actions, adding a further barrier to a child gaining access to the container. The combination of coordination and force can be a problem for some groups of adults with reduced motor skills (e.g. the elderly). The requirement for cognitive skill (e.g. to understand and follow instructions for a set of coordinated actions) is preferable as a child-resistant feature to the requirement for significant manual strength which may be beyond some adults.

Single-use pharmaceutical products, whether tablets or edible capsules containing a liquid, are often contained in blister packs sealed with foil or a foil laminate. Access is gained by pushing the tablet or capsule through the foil. A key requirement is that both blister and foil must be opaque so that it is not possible to see the product inside and they must be durable enough not to tear easily.

15.10 Dispensing and metering closures

Designing a closure to aid the dispensing and/or metering of the product adds significantly to enhancing the convenience function of packaging, and the use of such features can make a difference to the selling function. The possibilities are many, from the simple use of the closure as a measuring device, e.g. laundry liquids, to the more complex systems used on pharmaceutical products such as asthma relievers.

Bearing in mind that most containers are filled through the open neck, one of the key requirements for packaging line efficiency and high speed is that the neck diameter is as large as possible. However, when it comes to consumer use of the product, this is a disadvantage as it could lead to too much of the product being poured out in one go. One easy way of dealing with this is to use some device to reduce the size of the neck orifice, after filling. This can be via a separate plug inserted in the neck, or a closure which includes a dispensing feature, such as a flip-top.

15.10.1 Flip-top closures

Flip-top closures have become particularly popular for such applications as controlled dispensing of a container's contents. Because of its particular requirements, this type of closure is typically made of polypropylene which can create a flexible hinge that does not break with repeated use. Flip-top closures can be either screw-threaded or push-on and are usually injection moulded as a single piece. The top of the inner part has an aperture of an appropriate size for the product, over which a hinged lid fits (see Fig. 15.17). A plug in the hinged lid section seals the aperture when the lid is in the closed position. The aperture may contain a one-way valve which holds the contents in even when the lid is open and the container inverted, e.g. in containers of shower gel designed to be suspended upside down. When closed, the lid is kept in place in the same way as standard push-on lids. It is usually designed to be opened by flicking it upwards, with one hand.

There are many variants of this basic design. The aperture, for example, may contain a nozzle, which may be fixed, hinged or may be extended out from the

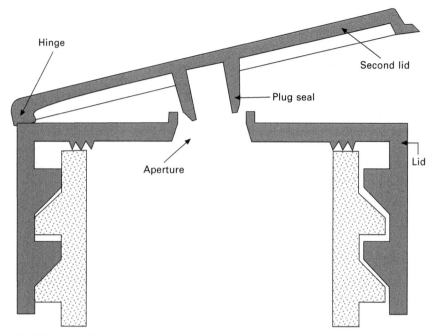

15.17 Flip top closure.

aperture to allow more accurate dispensing of product. One variant used for bottled water and sports drinks has a screw cap, an extendable nozzle and a protective overcap, incorporating tamper evidence as well as ease of use when 'on the run'.

As for other closures, a key requirement here is the careful specification of dimensions for all closure parts (as well as the container, of course) and tight control of the manufacturing processes to ensure that specified tolerances are met. Injection moulding is invariably used for closures and, depending on choice of material and level of accuracy demanded, injection blow moulding may be essential for container manufacturing (see Chapter 14).

15.10.2 Dispensing pumps

Trigger-operated spray pumps are ideal for dispensing liquids such as household cleaning products. They can be fitted with either screw-threaded or press-on closures, the latter being easier to apply on the packaging line. Because the action of squeezing the trigger can loosen a screw closure, bottle necks for spray pumps frequently incorporate an 'anti-back off' feature that locks the closure once it is fully engaged. This prevents accidental loosening during use but still allows the pump to be removed either to allow refilling or to separate for recycling. Lined or linerless valve seals are used to create an effective seal. Again, closure and bottle neck dimensions must be closely controlled.

Dispensing pumps are used for a range of products, including high viscosity liquids such as hand wash and emulsions such as hand lotion. Here, the usage action is to press down the dispensing nozzle, which can usually be done with one hand, releasing an appropriate amount of product. They are sealed and applied in the same way as trigger spray pumps. To prevent accidental spillage before their intended use, the pump is often locked in the down position and needs to be rotated so that it can be released.

15.11 Testing closure performance

Regardless of closure type, testing its performance is essential, throughout design and development, manufacture, application and consumer use, as well as during storage and distribution. As already mentioned, close attention to dimensional tolerances is paramount and this must start at the design stage and include the container to be used.

During development, samples must be checked for conformance to specification, including all key dimensions, presence of the correct liner (if applicable) and liner retention. Fit to container for correct seal efficiency should be checked at this stage, before full-scale component manufacture is sanctioned. Options available include using a coloured dye around the sealing surface of the container, to check that closure and container are properly aligned, with maximum contact at the seal area. Other types of tests are designed to check if the closure has adequately fulfilled its functions of containing and preserving the product, e.g. weight or volume tests after storage for a predetermined period (possibly using accelerated conditions of temperature and humidity) or microbiological testing.

Usage tests should be carried out, to confirm that the closure fulfils its intended features of openability, dispensing the right amount of product, and ease of reclosing. For maximum benefit, such testing should be done under actual conditions of use, so testing a closure for a shower gel in the office environment will provide limited information. Carrying out consumer usage trials is a better approach, although it adds to the cost and time of development.

During closure manufacture, samples must be checked for conformance to specification as above, as well as for visual aspects such as surface defects, e.g. flash at the mould part line, prominent injection point, or unfilled moulds. Colour should also be checked against the agreed standard.

On the packaging line, seal efficiency can be checked by a variety of methods, both on-line and off-line. Typical on-line tests include simply inverting each filled pack, or subjecting each to a squeeze test, by narrowing the gap through which each passes. Channelling packs through a height detector may also be used to check and reject any which are above the standard, possibly due to the closure not being seated correctly and fully screwed/pushed into position. Statistical off-line tests (i.e. taking an agreed number of samples off the filling line at agreed time intervals) include measuring take-off torque, as well as carrying out visual checks.

The degree of testing to be undertaken will vary from product to product. A new design for a pharmaceutical dispenser closure, or one for a hazardous chemical will require a more rigorous approach than, say, a push-on lid for a tub of safety pins.

Regardless of this, agreed test protocols are essential and testing must be carried out by trained personnel using properly calibrated and maintained equipment.

15.12 Bibliography and sources of further information

Several general texts on packaging materials and technology will contain sections on closures. The following texts are specific to closures:

Emblem, A. and Emblem, H. (2000) *Packaging Prototypes 2: Closures*. RotoVision, Crans-près-Céligny.

Theobald, N. and Winder, B. (2006) *Packaging Closures and Sealing Systems*, Sheffield Academic Press, Sheffield.

The following internet reference may also be useful:

http://www.packaging-gateway.com/features/featuresafety-first-child-resistant-packaging/

16
Adhesives for packaging

A. EMBLEM, London College of Fashion, UK and
M. HARDWIDGE, MHA Marketing Communications, UK

Abstract: Adhesives are ubiquitous in packaging, whether applied to a packaging component by the converter or the packer-filler. This chapter explores the theories of adhesion, i.e. what makes materials stick together, and then reviews the properties of the main classes of adhesives used in packaging. A brief overview of adhesives application methods is given.

Key words: starch, dextrin, polyurethane, acrylic, casein, latex, polyvinyl acetate, hot melt, cold seal, borax, peel strength, coating weight, tack, green bond, wetting out, viscosity, fibre tear.

16.1 Introduction

Adhesives are critical to the structure of most packaging, whether applied during the conversion process or on the packaging line. Selection of the appropriate adhesive is vital to make sure that the pack will meet its performance criteria throughout the production, distribution and retail chain, and consumer use. From a production perspective, adhesive choice can significantly affect line efficiency and production performance. Adhesives are central to pack performance and it is important to understand how the various types work, the broad principles underlying their operation, how to get the best performance in use and how to troubleshoot adhesion problems.

Adhesives come in many forms and types and choice will be determined by the substrates being bonded (the 'adherends'), the machinery in use in the process and other factors, for example the potential requirement for food safe materials. Most adhesives are applied via specially designed machinery, adding another layer of complexity to the selection process and requiring adhesives with specific properties to match the operating parameters of the equipment. In some cases, adhesives are designed specifically for a particular machine type or model. Some of the terminology used throughout this chapter may not be familiar. Refer to Section 16.9 for explanations of commonly used terms in the adhesives industry.

16.2 Adhesives in packaging

An adhesive will either be supplied directly by an adhesives manufacturer for the packaging converter or packer/filler to apply on the production/packaging line, for example hot melt adhesives for carton closing or casein adhesive for bottle labelling; or it may be pre-applied to a substrate, for example as a self-adhesive label or a heat-seal coating for blister packing or lidding. Probably the majority of adhesives are supplied 'unseen' to the packer/filler and then to the consumer, as part of the structure of the packaging, for example laminating adhesives in flexible packaging, starches used in corrugated board making and cardboard tube winding, or PVA adhesives used in the side seams of folding cartons.

Examples of adhesive use by the packaging converter include:

- case and carton manufacture
- paper bag and sack making
- tube winding
- flexible packaging lamination
- remoistenable gummed tapes and labels.

Examples of adhesive use by the packer/filler include:

- sealing of cases and cartons
- labelling of cans, bottles and other containers
- closing, and in some cases forming of paper bags, e.g. for sugar and flour.

16.3 Theories of adhesion

In practical terms, an adhesive exists to bridge two surfaces, creating an intimate connection between them. In order to do this the adhesive has to completely fill the gap between them, and has to be sufficiently strong in itself that its structure will not break down under the adhesive joint's normal working conditions. That is, it must be not only adhesively strong but must also be cohesively strong. In production line operations, adhesives are often blamed for pack failures when the failure is in the adherends themselves, for example a weak laminate structure. In order for an adhesive bond to be made, a number of key criteria have to be satisfied. Firstly, the adhesive has to be able to make intimate contact with both surfaces to be bonded. That is, it has to have a sufficient affinity for both adherends so that it is able to 'wet out' the surfaces, filling all irregularities in the materials and flowing over the entire sealing surface without leaving voids. The adherend surfaces must be clean and free from contaminants and the amount of adhesive applied must be just sufficient to coat the surfaces.

As implied above, adhesives are generally applied as liquids which then become solid as they lose their carrier or solvent, or as they cool. The major exception to this is the pressure sensitive adhesive most commonly encountered as case sealing tape or as a pressure sensitive label. This behaves differently in that the adhesive layer must remain 'liquid' until the tape or label is applied to the required surface. Pressure sensitive (or self-adhesive) labels are covered in Chapter 17. The different theories of adhesion will now be discussed in terms of the mechanism of bonding. It is important to note that more than one mechanism may be present within any one bond.

16.3.1 Mechanical adhesion

In mechanical bonding the adhesive flows into the surfaces of the adherends and anchors them together once it has solidified, using an interlocking effect. The adhesive sits at the interface between the two adherends. The bonding of paper and board (e.g. carton side seams) is an obvious example of mechanical adhesion. The greater the penetration of adhesive into the two surfaces, the better the bond, thus the properties of surface roughness and absorbency are important. Surface lacquers can seriously impair adhesive penetration, and this is why any adhesive areas must be left clear of lacquer during printing of paper and board. Similarly, heavy clay coatings which are likely to be both smooth and non-absorbent will reduce adhesive penetration; one way of addressing this is for the carton converter to incorporate small cuts in the clay coating during the cutting and creasing stage of manufacture. As already mentioned, there is rarely only one mechanism of adhesion in place, and in the example above it is likely that there is also a component of the bond that is due to a specific or chemical attraction between the adhesive and the adherend. As long as the *cohesive* (internal) strength of the adhesive film is sufficiently high, failure of a mechanical bond is almost always due to insufficient strength of the substrates.

16.3.2 Specific or chemical adhesion

The bonding of rough and absorbent surfaces has been explained by the mechanical adhesion theory but this does not explain the bonding of smooth surfaces such as metal or glass, or relatively impenetrable materials such as polymer films. Specific adhesion theory proposes that there is an attraction between surfaces in intimate contact via short range molecular attractions called van der Waals forces. These are very weak electrostatic attractions which exist between all atoms and molecules, at all times, due to their positive and negative charges. The larger the atomic or molecular mass of a substance, the stronger the van der Waals forces. There must be sufficient adhesive at the interface between the adherends to make maximum use of the available bonding area, and the substrate surfaces must be receptive and not repel the adhesive. This is especially important for polyolefins, which have low surface energy. Refer to Chapter 12 for further explanation, including surface treatment options.

Any smooth surfaces brought into contact will weakly attract each other. A common example is the bonding between a microscope slide and a cover slip; the addition of a very tiny amount of water makes it very difficult to lift the cover slip from the slide, because the polar water molecules are able to form bonds with the two glass surfaces. The small surface irregularities in the glass increase the surface area covered by the water so increasing the tensile strength of the bond.

16.3.3 Diffusion

The diffusion theory of adhesion relies on the solubility of the materials being bonded, and is of particular relevance in packaging for bonding polymeric substances. Put simply, if two materials which are soluble in one another are brought together in intimate contact, they will form a solution at the point of contact. Instead of a separate interface between the two surfaces, which forms a third layer with its own properties, here there is an *interphase* made up of the two materials, diffused into one another. Thus there is no mismatch of properties in the bond area, no inherent stress and a very strong bond is formed. For maximum bond strength the materials being bonded (adherends and adhesive) should have the same solubility parameter.

Examples of diffusion bonding include solvent welding of plastics, where the presence of the solvent allows the polymer molecules to diffuse into one another, and when the solvent evaporates it leaves an interphase of the two adherends. Using heat (e.g. by ultrasonic welding) has the same effect, allowing the polymer molecules to diffuse into one another (www.specialchem4adhesives.com). Another example of using heat to bring about diffusion is in the coextrusion or coinjection of molten polymers to form multilayer constructions, which are virtually impossible to separate.

16.4 Adhesive types

There are various ways in which adhesives may be classified, none of which is perfect; the following types are discussed in this section:

- water-based adhesives, both natural and synthetic: these include starch and its derivatives, casein, latex (for cold seal adhesives) and synthetic emulsion systems such as polyvinyl acetate (PVA), acrylics and polyurethanes
- solvent-based adhesives, in which the carrier is an organic solvent rather than water; these include polyurethanes and acrylics (although both of these are also available in water-based formulations)
- 100% solids adhesives which have no carrier solvent; these include reactive solventless liquids and hot melts.

16.4.1 Water-based adhesives

Starch and derivatives

Like cellulose, starch is a complex polysaccharide which occurs naturally in plant matter. Commercially, the plants used are corn, wheat and potato, with some use of rice, tapioca and sago. Starch composition (e.g. amylose:amylopectin ratio) and particle size vary with source and adhesive manufacturers adjust their processes to suit. Starch granules suspended in cold water have no adhesive properties and the granules must be broken down, usually by heating in water with the addition of metal salts or caustic soda. The granules swell and at the gelatinisation temperature, which can be 50–70°C depending on starch source, burst open to give a thick colloidal suspension which now functions as an adhesive.

Unmodified starches produced in this way have low solids content and high viscosity and have to be further treated to give the stable rheological properties needed for commercial applications. Treatment with alkali lowers the gelatinisation temperature and can be used to produce jelly gums, developed in the 1950s for applying paper labels to glass bottles. Acid treatment lowers viscosity but maintains solids content; and chemical oxidation using hypochlorite improves tack. It also results in low colour which makes these starches useful in paper making (www.specialchem4adhesives.com).

One of the most common uses of starch adhesive is in the production of corrugated board. Due to the large quantity required, it is common for the corrugator to prepare the starch adhesive on site, the starch being delivered in sacks, intermediate bulk containers or tankers. Different formulations and preparation methods are used, including two-phase and single-phase, but essentially the unmodified cooked starch produced as described above is mixed with water, caustic soda and borax, and pumped to the points of application on the corrugating machine, i.e. the fluting rollers at the single facer and the point of bonding the second liner at the double backer (see Chapter 11). Typical solids content of the prepared adhesive is 20–30%. Borax is added to give the required rheological properties and it improves initial tack by causing chain branching in the starch polymer. The heat of the fluting rollers causes the starch to gelatinise immediately, forming a high initial tack which firmly secures fluting to liner. The speed of bond forming and the initial bond strength are crucial for modern corrugating machines running at speeds of 350–400 m/min. The 'green' bond continues to develop to its full strength as water evaporates from the adhesive. Cold corrugating processes are currently being developed, requiring different starch formulations (www.specialchem4adhesives.com).

Starch adhesives are ideal for bonding paper-based materials and are also used for corrugated board making and for tube winding. The raw material is relatively low cost, readily available in good quality grades and biodegradable. Also, being a thermosetting polymer, starch has good heat resistance. Various additives may be used to impart specific performance properties, such as urea formaldehyde to give good moisture resistance and polyvinyl alcohol/polyvinyl acetate to improve resistance to cold water. Biocides are also used to inhibit fungal growth.

Dextrin adhesives are derivatives of starch. The starch is depolymerised by acid and/or heat treatment and the molecules are then repolymerised to produce highly branched structures which are soluble in water, the extent of their solubility being determined by the acid/heat treatment. A wide range of dextrin adhesives is available, with different viscosities and applications, and modifications are possible using the additives mentioned above. Borated dextrins contain borax to increase tack. Dextrins generally have higher solids than starch adhesives, i.e. less water, which means they dry faster and thus support higher line speeds. As well as being suitable for bonding paper-based materials, e.g. bags/sacks and tube winding, dextrins can also be used in high speed labelling of cans and bottles.

Casein

Casein is the protein present in milk, rendered soluble by the addition of an alkali in water. Casein adhesives have an aggressive tack and are not 'stringy' in application. They can also absorb significant amounts of water without significant change in viscosity, making them suitable for high speed labelling of glass bottles and jars in cold or wet conditions, for example in beer bottle labelling. They have excellent resistance to ice water, which means that labels will not fall off the bottles in chilled conditions, but they can be removed when required, for example on returnable glass bottles, by soaking in an alkaline solution. However, casein is a high cost raw material which is becoming more expensive, and alternative casein-free options are now available for the beverage sector. These include water based starch formulations modified with various polymers and resins, as well as wholly synthetic alternatives (www.specialchem4adhesives.com).

Animal glues

Animal glues are also protein adhesives, derived from bones, skin or blood and many ancient artefacts in museums testify to their long time usage. They are mostly solid at room temperature and are dispersed in water to give the required solids content and applied at a slightly elevated temperature by a roller/doctor blade, though they may also be jetted. Initial tack occurs as the temperature drops below the gelling point, typically 25°C, and then a final bond forms as water evaporates. Although largely replaced by starch or synthetic alternatives, some animal glue is used in rigid box making and fancy box covering.

Latex

Natural rubber latex is commonly used in the production of cold-seal adhesives, which are widely used in confectionery packaging. Early pressure sensitive adhesives were also based on natural rubber latex but have largely been replaced by acrylates. Latex has the interesting property that it adheres relatively poorly to any material other than itself. When a water-based latex adhesive is applied to a polymer film, and dried to remove the water, the resulting coating is non-tacky and thus the film can be reeled up. On the packaging line the film is unwound and wrapped around the product (typically on a horizontal flow wrapping machine) and the latex-coated surfaces will seal to each other simply by the application of pressure. No heat is required, which allows for high speed wrapping and there is no deterioration of heat sensitive products such as chocolate. The cold-seal coating is usually applied in a pattern by gravure printing, to match the sealing areas, although this can lead to winding problems due to the uneven layers. Offsetting onto the printed side of the web is another potential issue.

As well as natural rubber latex, formulations may also include plasticisers to improve flexibility, resins to improve tack and bond strength, and antioxidants to limit oxidation of the rubber, as well as other ingredients. Blends of latex and copolymers such as ethylene vinyl acetate are also being used as cold-seal adhesives. Concern about allergies associated with natural latex has led to the development of synthetic alternatives.

Polyvinyl acetate (PVA) and ethylene vinyl acetate (EVA)

These synthetic water-based emulsion adhesives are the well-known 'white glues' used throughout the packaging industry for applications such as carton side seaming, carton closing, bag making and many other uses. Strictly speaking, they are dispersions rather than emulsions, in which the insoluble polymers (the dispersed phase) are uniformly distributed in water (the continuous phase). This uniform distribution is brought about by using a protective colloid which effectively surrounds the solid polymer chains and holds them in a stable state in the water carrier. Suitable colloids include hydroxyethyl cellulose and polyvinyl alcohol, which also has surfactant properties, providing stable formulations with good wettability. Solids content is 50–70% and thus these adhesives are faster drying than starch and dextrin. After application, once sufficient water is removed, the emulsion 'breaks' and the polymer rapidly forms an adhesive bond. This 'green' bond should be sufficient to hold the substrates in place and it will continue to develop as more water is evaporated.

PVAs and EVAs are ideal for use with cellulose-based substrates which readily absorb the water carrier. They are widely available in formulations for a range of applications, are easy to apply using conventional roller or jet applicators, and application equipment is easy to clean. They must not be allowed to freeze in storage, as this will cause separation of the solid and liquid components.

Acrylic

Acrylic adhesives are available as water-based emulsions and can be formulated to meet a diverse range of applications such as: wet bond lamination of polymer film to paper and board; dry bond lamination of polymer films; and pressure sensitive labels and tapes. They are relatively low cost adhesives, with low odour, based on acrylates such as 2-ethylhexyl acrylate or iso-octyl acrylate and can be crosslinked for improved heat or chemical resistance. They tend to be especially useful for bonding low surface energy materials such as the polyolefins. When used for cellulose-based substrates, their low water content avoids excessive wetting and hence swelling of the fibres compared with PVAs. Crosslinking formulations used in lamination develop a good initial bond on application and heating, and then the crosslinking reaction continues in the reel during storage at ambient temperature.

In the pressure sensitive sector, acrylic adhesives can be formulated to provide removable or permanent properties. UV curable urethane acrylate solvent-free systems are also available. Vinyl acetate acrylic copolymers are being used in adhesive formulations and are claimed to have excellent strength. Again, they allow a high degree of formulation latitude and can be made into stable emulsion systems suitable for film lamination or pressure sensitive coatings for labels and tapes (www.specialchem4adhesives.com).

Polyurethane

Water-based polyurethane adhesives are available for film laminating and have excellent adhesion properties with a range of different substrates. They are based on polyesters and isocyanates and, like acrylics, can also be crosslinked, with the majority of the crosslinking taking place in the reel during storage. They are more expensive than acrylates but may be more appropriate if heat resistance is important, such as hot-fill applications.

16.4.2 Solvent-based adhesives

Adhesive systems using organic solvents as the main carrier were much more widely used in the past than they are now, due to increasing environmental and health and safety concerns. Over the past 10–20 years these concerns, plus legislation limiting the emission of volatile organic compounds (VOCs) into the atmosphere have encouraged user companies and adhesive formulators to seek alternatives. These alternatives fall into two main classes: water-based formulations (e.g. acrylics and polyurethanes as already discussed) and 100% solids solventless systems, which will be discussed below.

When comparing solvent- and water-based systems, the latter will almost always take longer to dry than their solvent-based counterparts, thus requiring more energy and possibly resulting in reduced line speed. These factors should be taken into account when comparing environmental impact and operating costs. It is also worth noting that water-based formulations are not necessarily organic solvent free, and thus VOC emissions may still present a concern. With regard to performance (specifically bond strength), water-based adhesives are generally regarded as less effective, although this is a very broad generalisation and performance must be evaluated for each end use; it is quite feasible that while a water-based adhesive may provide a weaker bond than a solvent-based version, that bond may be sufficient for the job required.

Solvent-based adhesives still have a role to play, for example polyurethanes as dry bonding adhesives for retortable pouch laminates, which have to withstand high temperatures for extended time periods, and pressure sensitive acrylics and high solids content styrene block copolymers for some specialist tape and label applications. Converters using solvent-based adhesives are likely to have to install solvent emissions abatement systems, at high capital investment, to comply with legislation.

16.4.3 100% solids adhesives

Solventless adhesives

Solventless adhesive systems come into this category, many developed in response to the requirement for reduced solvent usage and emissions, referred to above. Many are based on polyurethane chemistry using isocyanates and can be one-part or two-part systems. For film lamination, one-part systems rely on moisture to promote cross linking after the adhesive has been applied to one substrate and nipped to the second substrate, which means storing the reels until the adhesive is cured before slitting and packing to the customer's requirements. The extent of crosslinking depends on the moisture available and is not always consistent. Two-part systems allow better control, provided metering of the two components is accurate, and mixing is efficient. However, once mixed, the adhesive will start to set and this can give operational issues on the laminating equipment, reducing the appeal of these systems (www.specialchem4adhesives.com).

Developments in formulation now allow more rapid crosslinking and thus bond integrity, reducing the waiting time before slitting. Urethane acrylate systems cured by ultraviolet (UV) or electron beam (EB) are also available, both for laminating and as pressure sensitives for labels and tapes, allowing high production speeds without damaging heat sensitive substrates. Isocyanates are toxic and must be handled with care in the workplace, and the development of non-isocyanate solventless polyurethanes may have some applications in flexible packaging and pressure sensitives. A general property of polyurethane adhesive systems, either solvent-based or solventless, is that they are thermosetting and thus have good heat resistance in use.

Hot melts

Hot melt adhesives account for the major part of the 100% solids solventless sector, and are used for a wide range of applications both by converters, for example pressure sensitive coatings for labels, heat-seal coatings for lidding and film lamination; and packer-fillers, for example closing of paper bags and sacks, cartons and cases. Formulation options for hot melts are wide and depend on end use, for example styrene block polymers can be used for pressure sensitives, and polyolefins such as polyethylene, ethylene vinyl acetate and polypropylene can be used for coating and laminating, as well as for the packer-filler end uses noted above. Being based on thermoplastics, obviously hot melts do not offer heat resistance in use. As well as the base polymer, hot melts also include tackifying resins, waxes to lower viscosity, antioxidants to inhibit polymer chain breakdown and stabilisers to reduce deterioration due to heating.

In the packer-filler sector, hot melts are usually delivered cut or extruded into small pieces to allow for easy melting in a holding tank. The liquid adhesive is delivered via heated carrier hose to the applicator, which can be a dauber, wheel or jet. Application temperatures vary from around 170°C to around 100°C for 'warm melts'. The main advantages of hot melts are that they are generally fast setting, allowing high production speeds, and they have good gap filling properties, useful for rough surfaces. However, they require continuous energy to keep them in the molten state during application, which is a potential health and safety hazard as well as an economic consideration. If overheated they may be prone to degradation which can result in unacceptable levels of odour and taint or even contamination with charred fragments. Also, as already stated, they are not suitable for use in high temperatures.

16.5 Selecting the right adhesive

As can be seen from the previous section, there are several adhesive options available for different end uses and thus selecting the most appropriate adhesive is not always straightforward. The best approach to this task is to regard the adhesive in the same way as any other packaging component, and consider the key functions of the finished pack (refer to Chapter 3) and what part any adhesive(s) will have to play in meeting those functions. The pack design and development process (Chapter 18) starts with a brief and involves the consideration of options for meeting that brief; these options should always include the adhesives, i.e. adhesives should be considered at the start of the process and not left to the stage of final specifications. It should be clear by now that knowledge of substrate properties is essential when selecting adhesives, as well as a thorough understanding of the conditions to which any adhesives bonds will be exposed throughout the life of the packaging components and the finished pack.

16.6 Adhesive application methods

Adhesives are applied by a variety of methods, which can be divided into those in which the adhesive covers the entire surface of the substrate, and those in which only certain areas are coated with adhesive. Broadly, the former are more likely to be part of the converter operation and the latter part of the packer-filler activities, although these are not exclusive categories and there are exceptions.

Laminating adhesives for film and paper structures are generally required to be applied over the entire substrate surface, as are pressure sensitive adhesives applied to label stock, and heat-seal coating applied to lidding stock. Coating methods include roller coating, blade/knife coating, flexo roller and gravure cylinder. Critical factors are the deposition of an even coating, at the correct weight, and adequate drying/curing of the adhesive, especially when laminating impermeable substrates because of the potential risk of trapping 'wet' adhesive between the substrates, which would lead to poor adhesion and odour problems.

Cold-seal coatings are applied as part of the converter operation, but are most often applied in a pattern to coincide with the pack sealing points, rather than being applied all over. Gravure cylinders are normally used for this, although current developments include formulations which can be applied by flexo plates.

Many packaging line operations involve the application of adhesive to the closure part of a filled pack, e.g. end flaps on cartons and cases, and folded sections on paper bags and sacks. Such operations involve applying the adhesive to a designated part of the pack and this is achieved by means of daubers or jet systems. Again, a critical aspect is the application of just the right amount of adhesive: too little and the bond will be weak, too much and the line speed will have to be reduced to allow time for drying, as well as being a waste of money and resources. Another problem associated with using too much adhesive is that the excess can be squeezed out away from the seal area, causing packs to stick together later in the process. Accuracy of application is also critical, so that the adhesive is in the correct place in relation to the alignment of the two substrates.

For all adhesive operations, cleanliness is important. This applies to substrates (dirt can impair wetting out) and to adhesive reservoirs, delivery systems and applicators such as jets. All possible steps must be taken to prevent contamination, whether by dirt or any other foreign bodies, and also by other adhesives. Adhesive manufacturers generally formulate their products to work 'as is' and it is not advisable to interfere with them by adding other materials. Recommendations on application temperature should be followed closely; the temperature of an adhesive will affect its viscosity, which in turn determines how much is applied. The temperature and humidity of the operating environment are also important.

16.7 Evaluating adhesive performance

Probably the most relevant parameter to check on any adhesive bond, both during pack development and as an on-going quality check, is the bond strength, and there are two aspects to this: firstly, the resistance of the bonded substrates to being peeled apart, and secondly, their resistance to the application of a shear force. The former is more likely to be of use in packaging applications, but both are mentioned for completeness.

Peel strength can be measured using a tensile tester or similar device; essential requirements are a means of clamping the two substrates of the test sample and a means of applying a consistent force to peel the adhesive bond apart. Sample dimensions, speed of peel and environmental conditions of temperature and humidity will all affect the result obtained and thus for accuracy must be standardised. International standards exist, such as the American Society for Testing and Materials (ASTM) International which specify test conditions. Another factor which affects the measured peel strength is the angle of peel and the choice of test should reflect the most likely in-use conditions.

Test options are:

1. T-peel, in which each substrate is clamped vertically in the jaws of the tensile tester, e.g. ASTM D1876-08.
2. 90° peel, in which one substrate is pulled away from the second substrate at an angle of 90°, e.g. ASTM D6252-98(2001) (note that this is a quality assurance test method specially for pressure sensitive labels).
3. 180° peel in which one substrate is pulled back on itself away from the second substrate, e.g. ASTM D903-98(2010).

A practical approach may be to use ASTM or other test methods as the basis of a test programme, introducing modifications to mimic actual use, provided the conditions used are consistent from test to test. This could apply when evaluating alternative adhesives for a given application, and when carrying out regular quality checks on the packaging line. One of the difficulties associated with the latter is that the bond is unlikely to have reached its full strength during the packaging operation and this is especially the case with water-based adhesives. However, it is still possible to set a standard for on-machine bond strength. In addition to the adhesive bond strength, other properties of bonded substrates which may be evaluated at the development stage include temperature resistance, odour and colour.

16.8 Troubleshooting adhesive problems

As has already been noted, adhesives generally work in harmony with specialist equipment, and the equipment will have certain requirements with respect to viscosity, running characteristics, tack and open time. For example, on a high speed carton closing line, a hot melt adhesive may be applied to the flap of a carton, the flap is folded to close the carton and the adhesive bond is held under light compression for a period of time. Closed cartons are then automatically packed into outer cases which are then palletised. There are several demands which the adhesive must meet, in addition to the obvious one of being compatible with the carton and its intended end use. It must remain open or active until the flap of the carton is closed; it must then develop a sufficiently strong bond so that when the compression is released it does not spring open; and the bond at that stage must be strong enough to withstand the rigours of the casing and palletising processes.

If any of the operating parameters moves substantially away from the optimum conditions, the bond strength will be impaired and ultimately the pack can fail. For example, a failure in the upstream process (such as a delay in filling the cartons) which necessitates reducing the line speed, albeit for a short time period, may mean that, by the time the carton is closed and under compression, the adhesive has already lost its tack, i.e. it will no longer form a bond. If the speed of the packaging line is increased for some reason, the bond may not be developed sufficiently when compression is released, resulting in the carton springing open. Even worse, the pack may retain a certain percentage of the bond strength, so that the failure only becomes apparent once it is in the storage and distribution environment, including on the retailer's shelf.

All automatically applied adhesives will have the same issues. They are typically designed to work within certain parameters and, particularly when troubleshooting, attention should always be paid to any changes that may have been made to the operation of the packaging line. Any planned changes to the operation of the packaging line should be made with reference to the adhesives manufacturer, who will be able to advise on any required modifications to the grade of adhesive in use.

Faced with what appears to be an adhesive failure, the first step is always to examine the adhesive film. Adhesives fail in one of two ways, cohesively and adhesively. A cohesive failure is a failure of the adhesive film itself, while an adhesive failure is the failure of the adhesive to stick to one or other of the adherends. An immediate indication of which of these the packaging technologist is faced with is to examine the adhesive film. If it is 'shiny', with no visible fibre tear from the adherend, then it is likely that the problem is one of adhesion. If there is adhesive on both sides of the joint, then it is most likely that the adhesive film itself has failed, which is a cohesive failure. There is a further failure mode, evidenced by the surface of the adhesive film being covered by fibres from the adherend. This is a failure of one or other of the substrates, not the adhesive.

It is important to note that an adhesive 'failure' may just be a symptom of a change in the packaging process, and therefore the first question to ask is 'what has changed?'. There should be a record of operating parameters on the production line, and these should be checked to identify whether any modifications have been made. This does, of course, assume that the packaging line has previously been running trouble-free.

16.9 Common adhesive terminology

Many of the following terms are to be found on adhesive date sheets:

- *Closing time or setting time*: This is the time during which the adherends and adhesive must be in intimate contact such that the bond is sufficiently secure

that it will withstand subsequent handling. In most packaging operations, this is determined by the layout and speed of the packaging line.
- *Coating weight*: This is the amount of adhesive applied, usually quoted as grams per square metre. Adhesive manufacturers often recommend typical coating weights for given applications. Applying too little means inadequate bond strength and applying too much is costly and may lead to an open time that is too long for the machine and bond failure.
- *Cohesive strength*: This is the strength of the adhesive film itself. Once the adhesive film has formed, the bond can still fail because the polymer is not strong enough. A self-seal envelope and a cold-seal bond on a confectionery wrapper are both examples of bonds where the adhesive strength is higher than the cohesive strength of the adhesive film.
- *Fibre tear*: This applies only to bonds involving paper and board. A good bond is one which, when an attempt is made to peel it apart, results in tearing of the cellulose fibres, i.e. the adhesive is stronger that the inherent strength of the paper.
- *Green bond*: This is the bond achieved prior to drying or curing of the adhesive.
- *Open time*: The maximum amount of time between when an adhesive is applied and when the substrates can be brought together and still make an effective bond. This varies from a few seconds in the case of many hot melt adhesives to several minutes with some PVA or EVA types.
- *Solids (non-volatile) content*: The percentage of the adhesive which will remain once the carrier has dried off. For example, a hot melt adhesive is described as 100% solids, because it moves from a liquid to a solid state without losing any of its content. A PVA water-based emulsion is typically around 50% solids, so the volume of the final adhesive is half of the volume of the adhesive as applied. When calculating adhesive costs, it is important to take solids content into account.
- *Tack or grab*: The property of an adhesive which enables it to form a bond of measurable strength (the 'green' bond) immediately after the adherend and adhesive are brought together under low pressure (ASTM). If it is desirable to reposition one substrate after forming the bond, an adhesive with a lower tack will be required.
- *Viscosity*: Viscosity is a measure of resistance to flow. It must be appropriate for the method of application being used so that the correct coating weight is applied consistently.
- *Wetting out*: The ability of an adhesive to spread across the surface of an adherend such that it forms a bond with it; failure to wet out will result in uneven bonding.

16.10 Sources of further information and advice

As adhesives are so widely used in many industries, finding sources of information specific to packaging applications can be difficult. One highly recommended web-based source used to inform the writing of this chapter is www.specialchem4adhesives.com. This is free to access after registration and has an easy-to-use search facility which identifies articles of both technical and market relevance, as well as patents and data sheets.

Market information is often available from trade associations, such as:

- The British Adhesives and Sealants Association: www.basaonline.org
- The Adhesive and Sealant Council (North America): www.ascouncil.org
- FEICA, The Association of European Adhesives and Sealants Manufacturers: www.feica.com

Useful textbooks are:

- Petrie, E. 2007. *Handbook of Adhesives and Sealants*, 2nd edn, McGraw-Hill, London.
- Pocius, A.V. 2002. *Adhesion and Adhesives Technology*, 2nd edn, Hanser, Munich.

17
Labels for packaging

A. R. WHITE, AWA Consulting, UK

Abstract: This chapter discusses the importance of the label in the packaging process. The different types of labels are outlined along with trends in types of labels and the substrates used. Basic specification requirements are listed. The market for labels and the opportunities for the future are mentioned.

Key words: self-adhesive, markets, trends in labels, uses of labels, sleeves.

17.1 Introduction

Labels are an integral part of the packaging industry and have been for well over 100 years. In the early days they were simple, handwritten and used for identifying bales of cloth by batch and colour. Their use spread quickly and nowadays they are used in a variety of ways and applications.

The label is a versatile tool with the basic function of providing information, be it brand recognition, legislative information, as a selling tool or for many security applications. The most important information which a label can contain includes weight, usage instructions, hazardous warnings, etc. Labels are produced in many shapes and sizes often with focused, defined characteristics as we will see later in this chapter.

Wherever you look, labels are in evidence. A trip to the local supermarket will reinforce how important the label is to product recognition and thereby to the selling function. Look at the many bottles, jars or cartons which use the humble label. The end use markets (consumers) for labels include food, pharmaceutical, cosmetic, industrial, wine and spirits etc. (see Fig. 17.1), each requiring specific characteristics, e.g. resistance against chemicals, abrasion, freezing, water, etc. In addition to normal labels, derivatives include in-mould, sleeves and pouches. 'Smart' labels are able to indicate the freshness of a food product, some can provide oxygen scavenging properties, while others offer anti-counterfeiting measures and tamper evidence. They can even be applied directly onto fresh food products like apples, oranges and bananas.

An expanding and developing market is the use of RFID (radio frequency identification) tags which are becoming more widely used as they are becoming cheaper to produce. These tags are used in a variety of tracking and security applications, e.g. libraries, retails shops and for tracking products through a supply chain. They will gradually replace bar codes for pricing and promotional campaigns.

When all the information required cannot be contained on a single label, multi-page labels can be produced which may be used to supply information in several

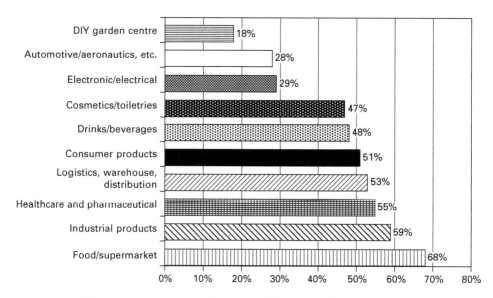

17.1 Main end user sectors for labels (M. Fairley, Tarsus Group Ltd).

languages or the many contra-indications required in the pharmaceutical industry. A very recent application is the provision of Braille characters which has become a mandatory requirement on all retail medicines.

Although label purists would say that shrink sleeves are not part of the label world, they do play an important role in decorating bottles which are an unusual shape and they are, in general, manufactured using a narrow web press.

17.2 Trends in label types

For the first half of the twentieth century glue applied labels were the only method of application, until the 1950s when Stanton Avery 'invented' the self-adhesive (pressure sensitive) label. Figure 17.2 shows the current usage of self-adhesive labels compared to other types of labelling techniques. In-mould labels and sleeves now feature prominently in the label industry.

17.3 Self-adhesive (pressure sensitive) labels

This type of label is now by far the most popular and has taken a significant market share from wet glue applications during the past 10 years. They can be used in a wide range of applications as they can be applied to the product more quickly and can be printed on many different substrates. With modern printing presses and finishing equipment, very fast changeovers from one design to another can be achieved quickly and easily.

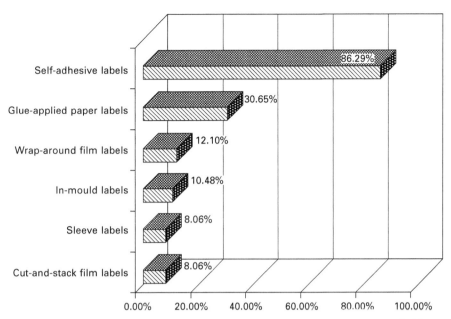

17.2 Types of labels produced by European label converters (M. Fairley, Tarsus Group Ltd).

17.3.1 Label production

In general, self-adhesive labels are produced on a roll-to-roll press for efficiency and supplied finished, ready for application to a product using a fast label application machine.

The label face material, which can be paper or a film composite (there are more than 1,200 different face materials available that can be used in label production), is provided in roll form with an adhesive on the back. It can be printed using several printing techniques although there are still labels which are provided with no printing on initially. For visual effect, the finished label can be laminated or varnished.

In order to stop the material sticking to itself, a backing sheet or liner is used which is discarded when the label is applied to the product. The composition of the liner is very important as it must have enough tack to keep the label on the liner yet release it easily when it is applied to the pack or container. A silicone coating is the normal means of preventing the label sticking to the backing sheet.

When the label is printed, it is usually cut to shape using hardened metal dies which can be in a flat bed or rotary configuration. Laser cutting technology is gradually replacing metal dies as it is faster and can be adjusted to the desired shape of the label much quicker and it is easier to produce complicated designs. An important factor in die cutting is that the die must only cut through the face stock and not the liner otherwise, when it comes to the application stage, the liner would come away with the actual label.

When the label has been printed and die cut, the finished label is separated from the selvedge or waste (the skeleton which is the area not required for the actual label) and wound up ready for use on the labelling machine, the waste being collected for disposal or recycling if possible.

17.4 How a label manufacturer approaches a job

Once the label manufacturer has determined the end use for the label, he will purchase the label stock to the correct specification, already coated with the appropriate adhesive on the back and the liner in place. It is then printed on the face material with the required design in exact register, not only with the multiple colour images but also the die cutting position. During the production run the label is cut to shape in line by means of a rotary (usually) cutting die which, as already mentioned, must be carefully set so that it only cuts through the face material and the adhesive, and not the silicone layer or the carrier web. If this happens the web can break during label application, causing downtime and loss of product. Labels may be butt-cut, with no space between each one on the carrier web or, more frequently, die cut to a specific shape with a defined space between each label.

Labels for semi-automatic and automatic application are invariably required to be spaced on the carrier web, as determined by the requirements of the application machine. Labels are normally supplied on reels.

17.4.1 Label substrates

Paper was the first material used for self-adhesive labels and is still widely used. Factors governing choice of paper will be covered in Section 17.5.

Plastic films are being used to great effect as they offer a wide range of extra benefits over paper. Some of the reasons are outlined below.

- The wide range of aesthetic effects available: gloss, matte, opaque, transparent, pearlescent.
- More effective recycling opportunities by using one type of plastic for the whole product enclosure, e.g. a polypropylene label on a polypropylene bottle also fitted with a polypropylene closure.
- Only plastic films can offer a 'no label' look on transparent containers such as glass and PET bottles. This type of label is becoming increasingly popular especially for spirits and water bottles. The total cost is much less than decorating the bottle (usually glass but not always) by any other means. It also means that one type of bottle can be utilised for a wide range of products simply by changing the label. This reduces storage space requirements for incoming packaging materials and enables a more cost effective solution to be offered for the filling company.
- Plastic films are more durable and versatile than paper in moist conditions such as bathrooms and kitchens.
- Plastic films offer improved product resistance where there is likely to be product spillage down the sides of the container and/or the product is likely to be handled in dirty or damp conditions such as garages and gardens.

(The performance of paper in the last two cases can, however, be improved by surface lacquering or film lamination.)

17.4.2 Adhesive selection

As well as selecting the most desirable face material for a given end use, it is also necessary to select the correct adhesive from a range of standard options. Selection criteria are based on the requirements of the substrates and the performance expected from the final product, including a consideration of whether or not the label is meant to be a permanent feature of the pack. A label applied to a pharmaceutical product is almost certainly required to be permanent, whereas an information label on a decorative tin of biscuits may be designed to be peelable so that it can be removed when the initial use is complete. Self-adhesive labels are available in permanent and peelable options, as well as water removable for reusable containers or recycling purposes. Depending on the storage conditions of the product and the substrate to which the label is applied, even peelable labels can be difficult to remove over time and so-called permanent labels can peel off.

17.4.3 Label application

Self-adhesive labels are normally applied by semi or fully automatic labelling equipment, depending on the speed required. In fully automatic systems, the label applicator is an integral part of the packaging line and speeds in excess of 600 packs per minute and more are possible. A sensor detects the presence of the item to be labelled, causing the label to be automatically dispensed from the backing paper. The label is picked up by direct contact with the item, or blown on (common in the application of top 'saddle' labels). The label is then wiped firmly onto the container to ensure full adhesion with the substrate.

Fully automatic systems require special tooling known as a 'beak' which is specific to the label dimensions and shape. The purpose of the beak is to fold the carrier web back on itself, allowing the adhesive-coated edge of the label to be exposed, ready to be picked up by the container. The carrier web is wound up separately and disposed of using specialist disposal contractors.

17.4.4 Storage requirements

Self-adhesive labels should be used (once printed) within a reasonable period of time after their manufacture, preferably not more than six months depending on storage conditions. There are two reasons for this: firstly, the adhesive tends to bleed slightly around the die-cut edge, which means that adhesive can offset onto the back of the release paper, which will cause unwinding problems on the application machine. This is made worse by storing the reels of labels in warm conditions. Reels must always be stored horizontally on the flat, cut edge and never on the face. This applies to part reels removed from production as well as to new material from the supplier, which should be left in its individual reel wrapping and stored in cool, dry conditions. A second potential problem associated with long-term storage is that, like any reeled material, the labels can take on a permanent curvature, which becomes more pronounced towards the core. This can lead to the label not standing out sufficiently from the beak to be picked up by the container at the point of application.

17.5 Wet glue (gummed labels)

This type of label can be produced on paper or film and, in general, requires the application of an adhesive to allow it to be attached to a product container. This technique has enjoyed widespread use (and still does in some areas) in fixing labels to bottles, primarily in the beer and spirits markets. They used to be used almost exclusively for wine bottles, but in recent years this market has given way to self-adhesive labels. They are still used widely for wraparound labels on canned goods and some soft drinks applications.

17.5.1 The application of labels to the product

The labels are normally pre-printed and delivered to the filler machinery in stacks, pre-cut to the right shape and dimension. A stack of labels is placed into a magazine on the filling machine and a label picked from the stack using glued pickers or vacuum transfer over a glued roller, and then applied to the container. The container must be firmly gripped to prevent excessive movement which would result in mis-applied labels. The label is wiped onto the container with rollers or brushes. For wraparound labels on cylindrical containers (e.g. cans), a line of adhesive is applied to the can and this acts as the pick-up mechanism for the label. The can rotates to wrap the label around the surface and a second line of adhesive is applied to the overlap. The usage of adhesive is thus kept to a minimum which also makes the label easier to remove for recycling.

17.5.2 The choice of substrate

Printed paper is the most economical choice for labels for canned goods, while beers and spirits are more likely to be labelled using film laminate structures, aluminium foil/paper laminates or metallised paper. Embossing and foil-blocking are good ways to add visual effects to a label and give a wide range of special effects.

The selection of suitable papers for ungummed (and other) labels is dependent not only on the aesthetic effects required but also on the environment in which it will be used. For example, if the label is to be used in freezing conditions then the label must stay on the container until it is defrosted and the contents used. Some might require resistance to scuffing, others to being re-wet constantly as in shampoo bottles. The economics of producing and placing a label on a product can be an important feature of the whole production chain.

Papers can be clay coated for surface smoothness and opacity giving high quality finishes suitable for high quality printing processes. Where this is not important a lower, cheaper grade of paper can be chosen. As with all other packaging materials, the packaging technologist should seek out the most cost effective solution, commensurate with meeting all the other requirements of the product pack.

Paper selection is also governed by the label application method to be used. For instance, where a vacuum pick-up is to be used, the paper's porosity to air is important as is the degree of stiffness with regard to application to tightly-radiused surfaces. Where the curvature of a container is tight, a lightweight, flexible label is easier to apply – and more likely to remain in place – than a heavier weight, stiffer material. Moisture absorbency (measured by the Cobb value) will affect the speed of wetting out of the adhesive and thus the speed of application.

An important property of paper in its use for labels is its degree of curl. When paper is wetted during the application of ungummed labels, the fibres swell due to the absorption of water, the degree of swelling being greater in the cross direction than the machine direction which causes the paper to curl parallel to the machine direction, away from the wetted surface. The effects of paper curl can be minimised by the choice of paper and by keeping the amount of water used in the adhesive application process to a minimum. The grain direction required should always be specified to the supplier. Usually, but not always, the grain direction is parallel to the base of the label.

Wraparound labelling of PET bottles for soft drinks uses paper or plastic labels, the latter providing better flexibility on the flexible bottle, less wrinkling due to moisture absorbency and good aesthetic effects. Other benefits of plastic versus paper are improved scuff resistance of the print (by reverse-printing the label) and resistance to tearing and damage during use. Plastic film labels on bottles destined for refilling are more easily removed intact than paper labels, which are likely to disintegrate in the cleaning process leaving a slurry which is difficult to clean out.

A wide range of materials is used in the label market. Paper substrates are by far the most popular followed by plastic films as shown in Fig. 17.3. Film labels are usually supplied on the reel, which eliminates the cutting stage during label production and are easier to handle and are less prone to damage than stacks of cut single labels. A cutter on the label applicator cuts the label to size as it is applied.

The major uses of ungummed labels are in the high-speed bottling of drinks and canning of food. Machines with application capabilities up to 100,000 containers per hour are available.

17.6 In-mould labels

In-mould labels are applied to a container during the manufacturing process. A pre-printed label is placed in the mould and the label becomes an integral part of the finished item, with no requirement for label application equipment on the filling line.

In-mould labelling (IML) is carried out on blow mouldings such as polyethylene and polypropylene containers or bottles. The label substrate can be paper, in which case it is coated with a heat sensitive adhesive, or films such as polypropylene which fuse directly to the blow moulded container.

In-mould labelling of injection mouldings most commonly uses film labels,

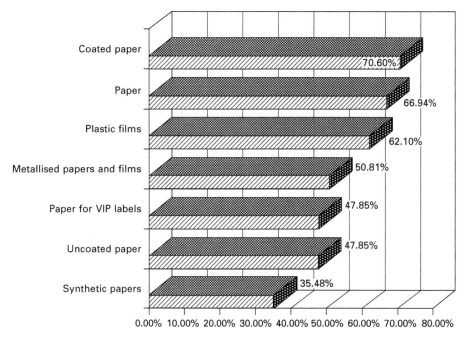

17.3 Main types of materials printed or converted (M. Fairley, Tarsus Group Ltd).

which are placed into the injection mould. The molten plastic is injected and the label fuses to the component surface. Good, all-round decoration can be obtained using this method, which is currently used for tubs for butter and margarine, and for large containers for biscuits. A similar process exists for in-mould labelling during thermoforming.

The pre-printed and die-cut labels are delivered to the moulder in stacks and are usually picked up and placed automatically into the mould during the opening cycle. Accuracy of label placement in the mould is critical for good finished effects.

IML offers high quality printing (film labels are printed by gravure or flexographic processes), over a large surface area, with excellent adhesion and resistance to scuffing.

17.7 Sleeves

Shrink sleeve labelling is a relatively low volume but fast-growing sector of the market, offering all-round decoration of a container, scuff resistance (by printing on the reverse side of the film) and the option of tamper evidence, by extending the sleeve over the closure of a container. Shrink sleeves are also used to combine two packs together as one sales unit, often as a promotional offer. Where tamper evidence is required, it is usual to incorporate a tear-strip (either as a separate tape, or by the use of two lines of vertical perforations) to gain access to the closure, and horizontal perforations around the sleeve to avoid removing the entire label when opening the product.

Shrink sleeves are made from a flat web of plastic material, usually PET, OPP or PS. PVC has been used in the past and has good shrink characteristics, with minimal distortion at low temperatures but has some environmental issues. Therefore PET and OPP are gaining market share. PET shrinks very quickly and requires tight control of the shrink tunnel temperatures to avoid distortion. OPP shrinks more slowly, requiring a longer dwell time.

Shrink sleeves are normally utilised in the following way:

1. The film is printed and then formed into a sleeve and welded.
2. Sleeves are delivered to the user either pre-cut for manual or semi-automatic placement over the container, or in reel form for automatic cutting and application on the filling line.
3. After placement the sleeve must be located in position on the container which is done either manually or by rotating flails or bristles.
4. The loosely sleeved container is then transferred to a shrink tunnel which may use hot air, radiant heat, or steam to apply heat to the sleeve to shrink the label onto the container in the correct position. The direction of the heat onto specific areas is important to provide an even and correct level of shrinkage without excessive distortion.

17.8 The choice of printing process

The choice of the printing process to produce a pre-printed label will be largely determined by:

- quantity of labels required
- finished quality required
- limitations of the label type
- whether the copy is fixed or variable (e.g. price/weight information).

Labels may be printed by rotary letterpress, flexo, gravure, screen or litho, depending on quantity and quality requirements, and may be foil blocked and/or embossed for special effects (Fig. 17.4). The use of digital printing for labels is expanding rapidly, as it enables the brand owner to enter a market with the minimum of waiting time and minimises development costs. 'Spur of the moment' promotional campaigns can be mounted quickly in response to changing market conditions. Shrink sleeves are printed in the reel prior to tubing, using flexography or gravure.

Ink jet printing is an 'on demand' process which is more commonly used for on-line printing of date codes, 'best before' data, batch identification, etc., directly onto packaging components such as tins and bottles. This application is not considered to be high quality, as the main consideration is that the information is legible. It is a non-contact process, with special ink being dispersed into very fine droplets through a nozzle, giving the familiar dot matrix appearance.

Printed labels can be dated, coded and given variable information by ink jet printing, in which case, if the label is lacquered, an area must be left unlacquered for satisfactory adhesion of the ink droplets. On high-speed lines where the capital

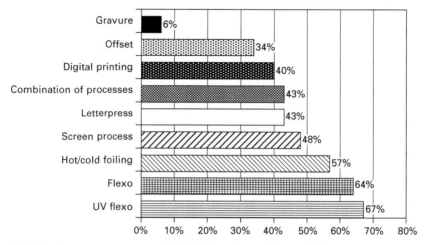

17.4 Printing processes used (M. Fairley, Tarsus Group Ltd).

cost can be justified, laser printing of additional information can be carried out. The laser effectively etches the printed surface, leaving the label substrate exposed.

17.9 Label specifications

Printed labels must meet all the requirements of any printed surface for a particular application, such as resistance to fading, scuffing, specific products, etc. They must be easy to read under the normal usage conditions within which the product is used. This must be a prime consideration at the design stage. Bar codes must be guaranteed to be readable at the point of sale and at any other relevant points in the distribution process, and should be verified at some stage in the production cycle.

Label dimensions, reel dimensions, core dimensions and unwinding direction must be specified before the label is printed to allow the maximum compatibility with the downstream operations. Reels should be carefully wrapped to protect them from damage during transit to the filling machine. Storage conditions should be carefully monitored to ensure that the labels arrive in good condition ready to be used and to minimise excessive waste.

Labels must meet all the service requirements of the final product and have good adhesion to the relevant surface on which it will be mounted. Any particular requirement must be met during the useful life of the product, e.g. a shampoo bottle (continual wet and damp conditions), tinned vegetables (long shelf storage), etc., and finally disposal/recycling. This means that during the packaging development process, the label must be regarded as an integral part of the product/pack mix and be fully tested to ensure that it conforms to all requirements.

The current (2009) *FINAT Technical Handbook* outlines 27 test methods which can be used to test many aspects of self-adhesive label materials for specification and production. Although these are not 'official' standards, they are widely used by suppliers and printers throughout the world. They are perceived as a universal standard for determining certain characteristics. This publication, together with the

FINAT Educational Handbook, provide a very useful source of information on label production for the student.

17.10 What can go wrong?

It is important to get the production of the label right first time as the further down the production chain a problem arises, the more expensive it is to put right and the longer it takes to produce. The opportunity cost of re-printing can be very expensive. Apart from technical problems during the printing and finishing stages, there are many areas where disputes can occur between the parties involved in the production process. It is essential that proofs are signed off at each stage of the production process; in this way everyone knows what is expected of them and corrections or modifications can be made early in the process.

Starting well before printing and pre-press artwork is produced, it is imperative that the specification of the label is accurate and accepted, and is understood and signed off by all parties. This will include the brand owner, the designer, the printer and the filling company. This stage is essential and time should be taken to ensure that everyone in the supply chain understands what is expected at each stage in production. This is the time when the specification of the substrate and adhesive is agreed. This is also when the end use of the label is specified, the type of container, the printing process, the method of filling, and the expected life of the label on the product, to name just a few criteria.

One of the most important factors for the brand owner is colour consistency not only from pack to pack but from batch to batch over a prolonged period of production. Meeting the colour specification is probably one of the greatest areas of contention between brand owner and printer, which is why exact colours must be specified if possible using a 'standard' colour system such as Pantone as a reference.

Another problem area is the actual printing. Although the label is a transitory element in the life cycle of a product, it must be presented to the highest quality commensurate with the type of end product it is to be used on. The biggest single fault is mis-register between one colour and the next. This of course has to be within accepted tolerances. For example, this may not matter on a tin of peas, but if this occurs on a pharmaceutical label, this could prevent the information from being easily read. This also applies to 'heavy' printing where overimpression or overinking can make some typefaces (especially 4pt and below) illegible.

The golden rule is to specify and obtain agreement at the earliest stage in the label production chain.

17.11 The label market

What is the current status of the label market and what are the trends for the future. What is affecting the growth of the label industry?

There is increasing globalisation of suppliers to the industry and their customers, and major brand owners are also becoming more global. This means that these brand owners are looking for global converters/printers who can supply labels in the locality of the product manufacture. There are relatively few global printers which indicates that there are potential business opportunities available.

Self-adhesive labels dominate the label converting sector which is also increasingly producing other types of labels. Sleeves are seen as a potential expansion area as are wraparound film labels. As mentioned elsewhere, wet glue applied labels are steadily losing market share even though their market volume is increasing. In-mould labels are still perceived as a relatively small niche market. The boundary between flexible packaging and label converting is becoming less distinct.

During the past few years flexography has become the dominant process worldwide with UV flexo growing rapidly. In addition digital printing has become a mainstream process in most developed markets. Letterpress continues to decline in volume yet offset litho is enjoying a revival. Foiling is becoming more popular due to the introduction of cold foil techniques.

17.12 The digital revolution

Almost every aspect of label production has a digital input at some stage: design and artwork, scanning and cameras, proofing and page make-up, plate-making, printing, finishing, quality inspection and die cutting. More than 12% of new label presses are digital or have a digital unit. Stand alone digital presses are being introduced which operate at commercially acceptable speeds to make them viable alternatives to conventional printing techniques. The ultimate would be control of all the operations digitally through management information systems (MIS).

17.13 Conclusion and future trends

The use of digital printing of labels and tags will become more widespread. RFID and other smart technologies will grow rapidly. The use of label printing technology will become widely used in the new and evolving 'printing for electronics' market; this is seen as an additional segment not as a replacement technology. Anti-counterfeit and product authentication applications are growing to respond to the enormous current counterfeiting activities. There are major new advances in nano-materials which will influence the way labels are employed on high value and pharmaceutical products along with developments in anti-microbial and anti-bacterial label products. Techniques are in the development stage to use labels to detect MRSA, E-coli, BSE, Asian bird flu and many other viruses. There will be nano sensors to track food from the farm to the plate. As these techniques are introduced so the importance of security inks and materials will become more important. Smart, active and intelligent labels will offer new opportunities.

The opportunities for the label industry in the future are enormous and will see the label industry grow and diversify considerably over the next 5–10 years.

17.14 Sources of further information and advice

Fairley M., *European Label Survey 2007*, Tarsus Group, London.

Fairley M, *Label Market Trends 2008*, Tarsus Group, London.
Fairley M, *World and European Trends 2009*, Tarsus Group, London.
Labels and Labelling International magazine (monthly). Visit www.labelsandlabelling.com
NarroWebTech magazine (quarterly). Visit www.flexo.de
Spring, R. (ed.) (1996) *FINAT Educational Handbook: Self Adhesive Labelling*. FINAT, The Hague. See also: *FINAT Technical Handbook (Test Methods) 2009*. FINAT, The Hague.

Part III
Packaging processes

18
Packaging design and development

B. STEWART, Sheffield Hallam University, UK

Abstract: This chapter firstly considers the packaging design process, describing the interaction of elements and activities that make up a typical design study. It then examines each of these in greater detail, beginning with the brief and progressing through research of both technical and market issues. Description of the design phase follows, looking at stimulating creativity, generating concepts, working with both structural and graphic elements through to analysing design candidates and recommending solutions. Finally, the process is illustrated through a case study.

Key words: packaging, design, graphic, standout, branding, brief, lifestyle.

18.1 Introduction

The fundamental needs for packaging are echoed throughout this book but we can be sure that packaging must be effective in containing, protecting, identifying and promoting products and it should do so with the least impact on the environment and at minimum cost. Clearly, we have here a mixture of technical, environmental, financial and communication issues to resolve even at the most basic levels of packaging design. Frequently the emphasis on each of these issues varies. Few of us are likely to question the level of packaging required to ensure, for example, that hospital supplies, swabs, needles and other medical items are sterile and unambiguously identified for the benefit of the medical team using them and, ultimately, for us, the patients, wherever in the world we are being treated. Here, the emphasis is to deliver packaged products that guarantee unwavering performance at minimum cost. Considerations of environmental impact and branding, while not ignored, are of less importance. In fact, even here, good packaging design can provide benefits for medical staff by, for example, being easy to open and use under medical conditions.

By contrast, the packaging of fast moving consumer goods, particularly within self-service retail environments, is a direct consumer interaction. It demands that the function of identification be greatly augmented to include promoting the brand and, importantly, allowing the packaged product to stand out from competitors' products. The pack in this scenario is often the only channel of communication between potential purchaser and product and, while pack cost and environmental performance are still vital components, sales performance dominates.

When we move to luxury goods, the relationship between functions becomes even more distorted. Here, in the case of packaging for fragrances, for example, the pack cost may exceed the product cost. We are buying the image, the dream, the brand and the envy of others. Packaging in this scenario becomes primarily the purveyor of status with cost and environmental considerations relegated to the background.

Less dramatically perhaps, creating packaging that reinforces the emotive bond between product and user is often a key component of a brief across a range of products. Costs and environmental performance are still key issues in all design studies.

In this chapter, we shall consider how the design process works, how it incorporates these issues and the parts played by each of the component parts of the process itself. Using a case study, we shall gain a practical insight into how packaging designers tackle the task, how creativity might be stimulated and what tools they might use to help express their ideas.

18.1.1 The design process

Figure 18.1 shows the design process represented by a logical progression of events, beginning with a brief, followed by a research phase, conceptual designs, through to developing design candidates worthy of progression, testing, refinement and final recommendations. The chart also indicates how the elements of structural and graphic design fit within the process and how the requirements of consumers, technical aspects and sales are fed into the study. While all these functions need to be considered even for the most modest packaging project, it should be understood that the linear nature of such charts tends to conceal a cyclical process also taking place within the process itself where different design concepts are being originated, developed and assessed. Frequently designs will be tried, refined, tried again and so on, ultimately being progressed as potential design solutions or rejected along the way.

There is seldom just one solution to any design problem. It is more likely that a number of design solutions emerge, some perhaps more costly but offering consumer benefits, others perhaps providing better distribution efficiencies or improved environmental performance. In many instances, the overall design is likely to be a compromise but one that must have a strong rationale for supporting it.

Figure 18.1 purposely simplifies the design process for clarity. Most packaging design companies break down a study into a series of stages. Figure 18.1 shows a typical three-stage packaging project where stage 1 involves research and concept creation. This is usually the most important and longest stage where creativity is being challenged. At the end of this stage, it would be usual to present conceptual work to a client, together with recommendations for further development of preferred design candidates. The project then progresses to a second stage where concept development and product/pack testing and evaluations take place, ending as before with a client presentation. Finally, in stage three, with specifications, drawings and artwork completed, the project ends.

This represents a typical packaging design project but a major project can extend over many months, involving multiple suppliers within different countries using different machinery. Consumer testing, again often in different countries, can yield results requiring slight modifications or even a radical design rethink. We can now consider the component parts of the design process in greater detail.

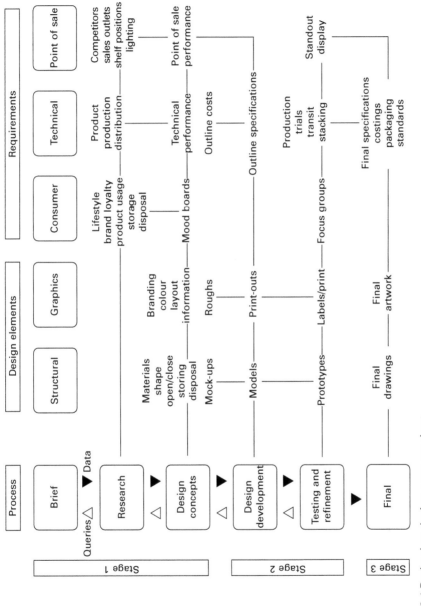

18.1 Packaging design process chart.

18.1.2 The brief

The project brief is probably the most critical part of the design process. It not only begins the process but becomes a reference throughout the lifespan of the project and, ultimately, is the yardstick by which success or failure will be measured. It is not something to be read before design work begins and then discarded. It accompanies work at every stage so that the design team can question whether or not their concepts meet the brief. No matter how clever or ingenious we might consider our individual design ideas to be, if they do not meet the brief we are simply wasting time.

It is true that briefs can vary and some packaging designers might claim that they worked from a brief written on the back of a paper serviette. The majority, however, will be more familiar with receiving a verbal briefing by the client accompanied by a comprehensive written briefing document. Most of the larger corporations and major brands work in this way, where the brief is provided by marketing staff or brand managers, often to consultancies on their 'roster' of design companies. As might be expected, frequently briefs from these sources are comprehensive, citing brand values, amongst many other attributes, that must be incorporated into the design work. The brief might be accompanied by market research results, competitor analysis, an advertising strategy or other data that helps the design team really understand the exact task ahead.

Smaller companies may not have the resources to employ such specialists and may encourage some assistance from designers to help establish a valid brief. For designers, this too can be rewarding as packaging designs at this level may be of greater significance in shaping company strategy. The design team members have the opportunity of demonstrating their expertise in design but might also augment that by contributing their experience of market sectors, human interaction or of other areas useful to the project and the client company.

The structure and organisation of client companies clearly show wide variations. Some employ their own in-house packaging specialists and design team, where technical packaging experts, purchasing managers, graphic and structural designers respond to marketing strategies. In this scenario, familiarity with product, production, suppliers and distribution capabilities provide a clear advantage over bringing in outside designers. A disadvantage sometimes results from such familiarity however, by a reluctance to challenge established company conventions or disturb departmental budgets. Understandably, a production manager, for example, may not welcome a design proposal that requires installation of new, unfamiliar plant and machinery that uses up his entire budget. Conversely, other production managers may be delighted at the prospect and have been pro-active in pressing for it. Many companies also respond to supplier led innovations, where packaging suppliers and client company work together developing new concepts.

Clearly there are many routes to packaging design that may involve designers working alongside specialists from marketing, production, advertising, distribution, suppliers, merchandising, sales, legal and environmental/sustainability departments, and, of course, packaging technologists. In this chapter, in order to highlight the packaging design process itself, we have elected to consider design as a separate entity. So, while reference is made to design consultancies, the design process remains broadly the same for in-house design teams and other organisational arrangements.

As with client companies, the structure of design companies also varies in size, structure, speciality and track record. On one hand, there are small studios with just two or three designers; on the other, major companies with offices throughout the world. Somewhere in between lie many well-established consultancies employing 20 or 30 staff.

Whatever the nature of both client-company and designer, design projects begin with a brief. Typically, a packaging brief will contain the information summarised in Table 18.1. While the emphasis is likely to vary from project to project, all the information categories listed here should, at least, be considered on all packaging design studies.

Many design consultancies view the brief as being of such significance that they take the time to review it with their team and feedback their understanding of the brief to the client. In this way both parties can be sure that all details are confirmed and no misunderstandings are likely to surface as the project progresses. This is also the point that a design consultancy will normally respond to a potential client with an indication of fee costs and timings, often broken down into stages.

18.2 Research

While the brief establishes the aims and parameters of the design project, further research is inevitably required before any design work begins. This may be, for example, to help the designer or design team fully understand a market that is

Table 18.1 Checklist for a packaging brief

Market information	
Product	Sizes/weights/volumes – purchase motivation – brand loyalty – brand values
Market	Size – value – trends – brands, brand share – seasonality
Target audience	Age – gender – lifestyle – purchaser/end-user – decision maker
Consumer benefits	Carrying – opening/closing – dispensing – storing – after use – disposal
Competitors	Brands – products – categories – cannibalism
Technical information	
Product protection	Moisture – light – temperature – gases – mechanical damage
Product compatibility	Materials – shelf life – deterioration
Production	Filling – closing/sealing – printing/labelling – secondary pack
Distribution	Warehousing – transport – tagging/tracking
Merchandising/pos	Outlets – fixture types/sizes – shelf position – lighting
Environmental and legal information	
Material sources	Renewable resource – recycled – sustainability – energy – impact
Pack construction	Monomaterials – separation of components – weights/bulk – energy
Disposal	Reuse – recyclable – compostable – toxicity
Marking	Symbols – weights and measures – copy sizes – hazards

unfamiliar to them or, perhaps, see for themselves what point-of-sale conditions are like. There are a number of research areas designers need to consider at the outset of a design project, some of which may be very specific to the nature of the product or area. While, for example, the packaging of medical products for hospital use may place a different emphasis on the direction of research, nevertheless, all the research shown in the following sections should at least be considered in all packaging studies. Broadly, research can be considered in two sections: marketing considerations and technical considerations.

18.2.1 Researching marketing considerations

Referring to Table 18.1, we can see that, even before any design work can begin, there is a substantial amount of background information to be gathered. Without an understanding of the market and a knowledge about the consumer/purchaser/end-user, designs would be created in a vacuum and lack any clear focus. The main areas for further investigation are outlined below.

Product positioning

Client marketing groups will often spend time explaining how they anticipate the product being positioned in the market. We can consider an example to illustrate how this can be done. A confectionery manufacturer briefing the design team explained how the product, a chocolate truffle bar, was to be positioned as an 'indulgent treat'. They had identified the market as being almost exclusively women who wished to reward themselves with a small treat for accomplishing a task. The rationale was that, having taken the children to school, completed the ironing or taking a break in the office, a little time could be set aside for a treat and the indulgence justified in terms of calories expended finishing the task. We also see this approach in advertising, the *'just because you're worth it'* campaign by L'Oreal, for example or, more directly, Nestlé's Kit Kat, *'have a break, have a Kit Kat'*. It is important for designers to investigate the purchasing motivation of the purchaser or consumer as it will have a direct outcome on the design work. In the above example, rich dark reds and golds were used to underpin the positioning of the product in the indulgence market sector.

In some instances products may be bought as 'distress purchases'. These are items that are usually bought quickly, reacting to a domestic crisis or forgotten birthday for example. Product category identification and branding are particularly important here as purchasing decisions are being made almost instantly. A 500 ml bottle of mineral water with a distinctive shape associated with an established brand is likely to be selected as a safe bet, preferable in circumstances of limited time, to examining the alternatives before making a decision. If visual standout against competitors is critical in such circumstances, it is also important in almost every other situation. Even when purchasing behaviour is more considered and less pressured by time, rapid brand and product identification is one of the most important attributes of packaging.

Brand values

Brands are often a company's most valuable asset, often representing, in monetary terms, many times more than the value of company land, plant and machinery combined. Company take-overs are frequently about the acquisition of brands rather than just physical company assets. A successful brand, therefore, has a monetary value based on its ability to sell products, together with an emotional appeal to its target market, carefully nurtured by marketing and advertising strategies usually over a considerable period of time (Table 18.2).

If the project concerns packaging a branded product, it is vital that the values of the brand are accurately reflected, reinforced and promoted by the packaging design. To achieve this requires that the designers understand and become familiar with the brand's values. We, as consumers, might assume that we buy a particular brand because our experience of it has been positive, for example a cleaning product that we have found to be effective in the kitchen or a savoury sauce that we really like. Both are examples of products that physically satisfy our needs. In addition to this, however, we form emotional bonds with such brands, often based on trust. Probably all baked beans are much the same but we might select Heinz, for example, because it is our 'old friend', the one we trust not to let us down. The Heinz brand has changed little in over 100 years and it is of little surprise, therefore, to find that the packaging of Heinz baked beans is distinctive, standing out from the competition and making in-store selection an easy task. For many, choosing 'our' brand for the price of a small premium makes product selection quick, easy and risk free.

As with 'trust', most brand values are expressed in human terms, reflecting their emotional value. Descriptors like 'serious' or 'fun' are often used. Brand values associated with Apple, for example, or Apple's brand personality to put it another way, could include 'imaginative', 'rebellious', 'passionate' and 'different'. An iPhone or iPad becomes an object of desire rather than simply functional electronics. Apple has been successful in promoting its brand values and creating a loyal following of consumers eager to purchase and, importantly, to be seen by others, to purchase Apple products. Whatever brand the packaging designer might be working with, it is vital that brand values become intertwined with all stages of the design process.

The market

Packaging designers need to gain an insight into the market that the project is addressing in order to respond to its demands. In particular, they need to get a feel of how the market is developing, what trends are driving it, which brands are succeeding and why this is the case and which brands are losing market share. It may be important to the study if there are seasonal factors that affect the market and, if so, how these can be incorporated within the design task. Later in this chapter the case study reveals how market information drives the direction of design work.

Table 18.2 Top-branded products

2009 rank	Brand	Country of origin
1	Coca-Cola	US
2	IBM	US
3	Microsoft	US
4	General Electric	US
5	Nokia	Finland
6	McDonalds	US
7	Google	US
8	Toyota	Japan
9	Intel	US
10	Disney	US
11	Hewlett Packard	US
12	Mercedes Benz	Germany
13	Gillette	US
14	Cisco	US
15	BMW	Germany
16	Louis Vutton	France
17	Malboro	US
18	Honda	Japan
19	Samsung	Korea
20	Apple	US
21	H&M	Sweden
22	American Express	US
23	Pepsi	US
24	Oracle	US
25	Nescafe	Switzerland
26	Nike	US
27	SAP	Germany
28	IKEA	Sweden
29	Sony	Japan
30	Budweiser	US
31	UPS	US
32	HSBC	UK
33	Cannon	Japan
34	Kelloggs	US
35	Dell	US
36	Citi	US
37	JP Morgan	US
38	Goldman Sachs	US
39	Nintendo	Japan
40	Thomson Reuters	Canada
41	Gucci	Italy
42	Phillips	Netherlands
43	Amazon	US
44	L'Oreal	France
45	Accenture	US
46	ebay	US
47	Siemens	Germany
48	Heinz	US
49	Ford	US
50	Zara	Spain

Source: interbrand, Best Global Brands, 2009

Target audience

All packaging design work needs to provide a communication between the product, brand and the target audience. In many situations including all self-service transactions involving packaged goods, the pack plays a critical role in communication. In most situations the purchaser may also be the consumer but often, say in the case of a mother shopping for her family, the purchaser may be buying for someone else. If mum is accompanied by family members, she may be influenced in her choice by them. We must establish, not only who the principle targets are that we are designing for but begin to understand their motivations. The target audience will often have been identified by the client in the brief but now, at the start of a project, our task is to identify and understand what motivates them, identifying their wants, needs and desires. This can be done in a number of ways.

Demographic data provides statistical and numeric information about populations that can provide a useful input to a design study. It is helpful to know, for example, in a study concerning speciality teas, that the fastest growing market sector is amongst the 25–35 year olds, although the majority of tea drinkers are currently in the 55+ age group. Failure to understand such details can drive design work in the wrong direction, in this case aiming at an older market when, in fact, younger consumers are more promising targets.

Psychographic data, by contrast, is concerned with people's beliefs, opinions and lifestyles and seeks to identify the motivations of groups of like-minded people. Often these groups are given names, reflecting their shared lifestyles. For example, we might describe 'urban adventurers' as city dwellers, driving black 4×4s, brand aware, 'cool', health club members, living in a flat within a gated development, chrome and glass décor, skiing in France, enjoying dining out or elaborate meals at home with friends, and so on. Descriptors like this are immediate and vivid, helping designers understand the target audience and providing clues about what might motivate them and drive their purchasing decisions.

Both the above types of data can be sourced from published reports and surveys. There are many lifestyle magazines that reveal, through their articles and the type of advertising they include, details about particular groups of people. Additionally, field research conducted by designers themselves can supplement this information and contribute towards creating a market profile. This might involve direct observation of a target group's behaviour, organising focus groups to discuss lifestyle choices, brand selection and motivations for purchase or, as we will see in the case study, simply talking to friends and family if they happen to be in the target group concerned.

By following these research techniques, a consumer profile begins to emerge. It can be further developed by probing deeper into the lifestyles of the target group. We may, for example, want to question which brands they might buy, what car they might drive, where they are likely to go on holiday, their choice of music, and so on. Designers often try to encapsulate this information by creating mood boards. A mood board is simply a collection of images, tear sheets from magazines, sketches, materials, colours or any other items that represent the target market's lifestyle. It helps encapsulate target market research into a visual reference, meaningful to designers or a design team. By considering the target audience in this level of detail, it becomes easier to design products and packaging that will appeal directly to them, creating an emotive response that is likely to encourage and maintain product and brand loyalty.

Consumer requirements and benefits

How packs perform from purchase, through storage, in-use and disposal is an important consideration for consumers and, therefore, important also for designers. Some products might benefit by allowing inspection prior to sale. Abrasive papers in the DIY market are an example here. Consumers might also expect packs to comply with any market sector protocols and be confused if they do not. A green coloured chicken stock cube breaks the 'yellow for chicken' that has become an established convention.

Good packaging design can recognise consumer needs and build in consumer benefits through an understanding of these areas; for example, simplifying the opening and re-closing of containers, the ease of dispensing or pouring product and providing containers that are stable in the environment where they are used (Fig. 18.2). A shower gel that can be dispensed one-handed and does not topple over when placed on a shelf would clearly provide a consumer benefit compared to other brands that might have adopted a pack format that has not considered end use.

For some specific target groups, building in these types of features should be part of the brief. With an increasing market of people aged over 60, for example, it is becoming more important to consider issues of manual dexterity, when joints become stiff and painful and where the ability to grip is weakened. The over 60s, however, are not one coherent market. The category fragments into sectors from fit and healthy through elderly and infirm to those in care. Any packaging design that aids opening a tin, for example, must be inclusive, suitable for all ages and not stigmatise one sector. Currently, research is ongoing to make ring-pull tops easier to use. The study, taking place at Sheffield Hallam University, has identified that many people struggle to lift the ring-pull into a position where leverage can be exerted. One simple solution, amongst several being trialled, is to incorporate a recess under the ring making the operation much easier – for all, not just the elderly. Clearly, where brands and products are perceived as being designed with end-users in mind, in terms of ease of use, storage and disposal, they are likely to encourage repeat purchase and brand loyalty.

Competitor products

Products, brands and packs compete against each other on-shelf. Clearly, for designers, it is important to recognise the strengths and weaknesses of competitors, particularly within a supermarket environment, where the retailers follow their own merchandising strategy, largely outside of the control of brands. This becomes increasingly important where brands introduce product variants. Five different flavours of crisps from the same brand can begin to compete for market share with each other without actually

18.2 Using packaging to benefit the consumer. Consumers require a pack that is stable and easy to use within its environment. Here, the pack is designed to be always on display, reinforcing the brand at every use.

increasing overall sales. This is referred to as cannibalism, where, for example, smokey bacon flavoured crisps might erode the sales of cheddar cheese flavoured crisps, while, overall, total crisp sales remain static. Designers should also be aware that the product or brand they are working on may compete against a different category of product adjacent to it in the aisles. Loose tea and tea bags often, in this way, compete with roast and ground coffees. Category, product and brand standout are even more important in circumstances such as that. Standout is probably the most significant challenge for packaging designers within retail markets.

18.2.2 Researching technical considerations

To ensure that all information is in place, the checklist (Table 18.1) provides a detailed list of considerations. Not all apply to every design study but, by checking them all, you may be assured that nothing has been overlooked. The sections below indicate some of the principal areas that designers have to address.

Containment, protection, preservation and compatibility

As covered in Chapter 2, the technical functions of packaging are fundamentally concerned with containing, protecting and preserving products. Containment seems obvious, but in many instances the pack design must be effective in not permitting any unwanted product loss during the total product lifecycle, including such times when the product is partially used and then stored. Packs must protect their contents against degradation or spoilage, most frequently caused by moisture, oxidation, UV light, microbiological contamination, temperature extremes and odours. Additionally, any known compatibility problems between product and packaging material should be recognised as, clearly, this will impact on the choice of packaging materials, often eliminating some at the outset. There are often shelf life requirements that will also eliminate some packaging materials from subsequent design considerations. Although, ultimately, packaging design culminates in producing packaging specifications, in the early stages of a study, a broader approach is acceptable. If, for example, we know that flexible laminate films work for similar products, we do not need to discuss detailed specifications until later in the study.

Production, distribution and point-of-sale

In the research phase, before design begins, designers should, wherever possible, gain an insight into how products are produced, packed, distributed and sold. Seeing how jam is made, for example, how jars are filled, sealed, labelled, collated and packed into secondary cartons, provides a useful backdrop to any packaging design study. Designers begin to get a feel about the nature of the product at different stages and the speed of production lines. They can see, at first hand, the importance of retaining contact points on jars, for example, helping to eliminate conceptual work on new shapes that will not provide sufficient pack stability on the lines.

Similar practical details will be revealed when warehousing, handling and transport operations are visited. For packaged products, particularly those produced in large volumes, packaging design impacts directly on the 'bottom line'. Efficiencies gained in palletisation, warehousing, transport utilisation and weight reduction, all contribute to profitability. Packaging design proposals that challenge any of these areas have to be justified, usually through their potential to increase sales.

There are also efficiencies at point-of-sale where pack size and configuration may influence both technical and graphic design. Technically, shelf optimisation is frequently an issue. Graphically, the main sales panel of the pack should present itself to potential purchasers. It is unrealistic to think that retail staff will have time to spend rearranging packs. In addition, by seeing the point-of-sale conditions, the designers can see what actual conditions are like. Ice cream tubs, for example, might be displayed in a chest type freezer where the lid is the most important panel and where the job of product description and branding needs to be strongest.

By experiencing the above areas through visits, a valuable body of knowledge is created concerning the practical implication of design decisions. There is a danger, however, that feeding this information too early into a study might restrict creativity in the subsequent design work, stifling imagination in favour of practicality. Feed it too late, however, and time might be lost working on concepts that have no practical application. Most design teams deal with this dilemma by having a technical specialist who will allow some creative work to get underway and guide the direction it is taking. For individual designers, the task may be harder, working creatively and then analysing results. Fortunately, we are built with brains that allow us to be creative on one side and analytical on the other. Packaging design challenges both.

Environmental considerations

Although it is convenient to separate out a section on environmental considerations, in practice all design work should be underpinned by environmental considerations at all stages of the design process. We should begin by questioning whether the product needs packaging at all. In some instances, it might not. Certainly, there are instances where point-of-sale (POS) units could remove the need for retail packaging. DIY hardware, drills, sanding discs, etc., are robust enough to survive without packaging. They can be identified by POS material and protected against theft by microchip. If packaging has to be used, then it should be at minimum levels. The packaging industry has quietly but effectively worked in the background, reducing packaging levels, for example, by lightweighting containers. Even so, designers should seek to minimise the impact of packaging by removing it where possible and minimising packaging levels where necessary. By using mono-materials, or, at the very least, enabling different materials to be easily separated, recycling and composting by consumers is eased. In many instances, recycled materials can be used to create new packaging. For some products, a closed loop system of packaging might be a solution, where containers are returned to a central location for refilling.

Although designers should always strive to minimise the impact of packaging on the environment, it should be remembered that the often high value of a product is being protected by the relatively low cost of packaging. In other words, the energy invested in the product is protected by a very much lower level of energy invested in the pack. Should the product become damaged, not only is there an energy loss but additional energy wasted in obtaining a replacement. Additionally, food packaging, in particular, helps to extend product shelf life, reduce product spoilage and increase consumer choice. These are difficult factors to resolve within the complex situation of balancing feeding the world's population while also preserving the planet.

18.3 Conceptual design

With a thorough knowledge of the product, market, consumer profile, production, distribution and point of sale conditions, design work can begin. Of course, while all the above activities have been taking place, some ideas will already have been forming. What we need now is a free ranging supply of ideas, even those that we might discard later. This stage of packaging design is the creative phase where thinking should be lateral as well as logical. It is the most critical part of any design study and often the most extensive in terms of time and cost.

18.3.1 Sources of inspiration

Many inexperienced designers put off starting design work by prolonging the research phase, collating more and more material. Now, faced with a blank pad of paper and empty screen, it can be difficult to get going. In commercial practice, designers often work in teams. This has the benefit of combining the differing skills of individuals but is also valuable in that ideas can emerge during discussion and debate within the group. It is also useful to create a stimulating environment based around the project by bringing in samples of the product, competitor products, mood boards and other materials that relate to the product or brand.

When working on the packaging design for a natural range of products, one highly acclaimed design company created a 'natural' environment within the studio, including an astro-turf floor. This might be going too far but for such a brief involving natural products, why not get the design team into a field, farm or botanical garden? Another agency gave a sum of money to each member of the design team and gave them a time limit of one hour to go out and buy packaged products that they most admired. If the project involves, for example, snacking on-the-go, get the team or individual designer to watch what people actually do in stations, parks, around offices and, if possible, take photographs. The point here is that creativity is stimulated by external factors so that it is worth replacing the situation of blank paper in a sterile office by something a little livelier.

Another frequently used technique is brainstorming. Here, a group of designers is encouraged to suggest concepts, even implausible ones with all ideas being noted for subsequent discussion. It can be surprising how some, initially crazy, ideas can be modified and adapted to provide interesting design solutions. To get the best out of brainstorming, sessions should be properly structured and there are sources of information recommended at the end of this chapter.

18.3.2 Generating concepts

The process for creating conceptual designs will vary depending upon the nature of the project but where both structural and graphic design is required, where do we begin? In this section we show that while the two elements cannot be considered entirely separately, the usual progression of events looks at structural design first.

Ways of working

It is perhaps worthwhile at this point considering some of the ways that packaging designers work. Many work directly with sketches for both structural and graphic concepts, quickly generating ideas and exploring both technical and graphical features as they go (Fig. 18.3). Others, particularly on a structural project, prefer to create three-dimensional rough mock-ups, using simple materials such as paper, board, solid foam, clay, plasticene, wood or by modifying found objects. Whatever medium is used, it needs to allow concepts to be generated quickly without hindering creativity. Precision is not required, just a standard capable of communicating a concept. Computer generated work is far too slow and limiting in this initial stage.

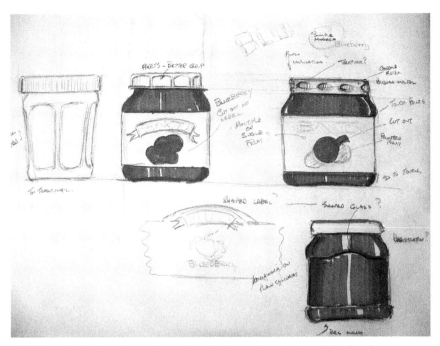

18.3 Sketch of glass jar project showing how the structural design is complemented by working on the label design at the same time. Here cut-out parts of the label reveal the colour of the jam.

If the project is concerned with both structural and graphic design work, each structural concept needs to be accompanied by a rough indication about how the pack form could be decorated. (Decoration is the term used to include all forms of graphics and includes direct print, labels, sleeves, embossing and debossing.) In this way, we might, for example, consider the shape of a new household detergent bottle while simultaneously evaluating how it might be decorated. In practice, however, where structural design is required as part of a packaging design study, it is usual to begin by considering this first, realising that it should not prevent evaluation of the graphic potential of structural design candidates.

Structural design

Structural packaging design concerns working with shape and materials. Any one category of materials imposes design constraints through the nature of the material and its ability to be converted into packaging. It would be usual, therefore, to begin a study by considering packaging options broadly, in terms of packaging types. Could the product be packed into rigid tubes, jars, bottles, cartons, flexible tubes and pouches, thermoformed trays, tins, tubs, sleeves, etc.? Sometimes, importing a pack form from another area provides a distinctive 'new' design, as in the example shown in Fig. 18.4. Here a board container associated with liquids now provides a convenient and easy-to-use dispenser pack for sugar.

18.4 This French pack for sugar demonstrates how a pack from one market sector can emerge as a new and interesting design in a totally different sector. It is also very easy to use and store.

Technology is also providing new materials and creating new opportunities. Working at sub-molecular levels, nanotechnology has already provided self-cleaning surfaces, plastics with the strength of steel and transparent waterproof paper. Electronic technology has established radio frequency identification (RFID) systems that we are all familiar with, particularly in the form of security tags. They are, unsurprisingly, becoming increasingly sophisticated, smaller and capable of storing data and programmes. Already they are used on-pack to monitor transit conditions and product deterioration. When this technology is combined with miniature paper batteries and electroluminescent inks, packs can provide information, display animated graphics and interact with other devices. Microwave packs that self-set the timer, packs that send text instructions to your mobile phone or update your computer shopping list when they are empty – all are possible now. When selecting materials, designers need to be aware of these new and fast developing ranges of possibilities.

At this point, probably designers will be working with sketches as a fast and efficient way of exploring initial ideas. It is a good idea, however, to quickly move to sketching packs at their approximately correct sizes. Working at 1:1 scale often reveals practical issues that can be masked by smaller sketches. Mock-ups are a great way to bring pack concepts to life. They need not be elaborate but good enough to provide a visual reference to the pack form they represent. Here we also need to be able to understand what materials are being represented by mock-ups and sketches and how they might be decorated.

Graphic design

Conceptual graphic design work may be running in parallel with structural design or following it or, more usually, a bit of both. Until the structural design is established and the materials decided, graphic design can only remain conceptual. Nevertheless, it can suggest directions from an early stage. It might explore photography versus illustration, branding and sub-branding, colour, use of imagery, investigate typography, or consider corporate requirements.

Importantly, any graphic design must be effective on the panel most seen by consumers at point-of-sale. Deciding which panel will become the main panel is normally a first step. It is not always the largest panel. Earlier research should indicate how the product/pack will be displayed. Sometimes it might be the end panel of a carton that is seen and, if so, this is where graphic design will be most critical. Having selected the main panel, the field of vision needs to be checked. A cylindrical container offers a restricted width for graphic visibility, for example, often reducing the available area to carry branding and product descriptor. It is better to establish such restrictions early in a design study to avoid spending time developing graphics that do not work when applied to the pack.

Analysis

It would be usual for the design team to hold an interim meeting at this point, putting all their individual ideas on the table and discussing the relative merits of each, relegating some design candidates while promoting others for further development. The brief is always used as a reference for judging the success of any design candidate. In some instances, designs might be found not to meet the brief but most often design concepts will meet some aspects of the brief. It becomes a question of degree. There are also the practical aspects of production and implications for transport, distribution, warehousing and point of sale to be considered, together with costs. These now also begin to become criteria for judging designs.

It is rare for one individual design to succeed at this stage. Design A might provide terrific standout but present filling problems, while design B provides efficiencies in storage and transport but lacks shelf impact and differentiation from the competition. Some design groups use assessment grids to rank design candidates; others do so more informally. Table 18.3 shows some assessment criteria that might be used to rank design candidates. All, however, must now decide which designs need further development and, almost certainly, there will be more than one.

Often, too, some concepts might be identified as benefiting from cross-fertilisation,

Table 18.3 Assessment chart for design concepts

Standout – which design provides greatest standout			
against other design candidates	against competitor products within the same product sector	against surrounding products in adjacent sectors	
Imagery and tone – which design provides			
right 'voice' (serious, fun, healthy, etc.)		resonance with target audience	
Branding – which design			
promotes the brand		represents brand values	
Believability – which design is most believable			
Fits with product sector		Looks the part	
Graphic layout – which design candidate follows design 'rules' in terms of			
typography	legibility	balance	colour	images
Aesthetics – which design looks the most			
elegant	integrated	exciting	effective
Practical and technical issues – which designs are			
cost effective	transferable	allow promotions	not recessive at POS
meet legal needs	environmental	production friendly	transport efficient

combining positive points between designs and creating hybrid design variants. In a commercial packaging design project, at this point the work conducted up until now, the first project stage, would be presented to the client. All work would be shown, recommendations made and a rationale provided for developing favoured designs.

Design development

In many packaging design studies, the development process might extend into many months. On the marketing side, there may, for example, be consumer testing to take place. Technically, there may be filling trials, transit trials and other tests that need to be conducted before any new pack form can be introduced. In order to ensure that all activities are coordinated, it is usual to produce a chart indicating activity, timing and staff resources required. Once agreed by all parties, this document schedules all further stages in the development process through to product launch, imposing deadlines for completion of all interim activities. There are many software programs that can create project charts and provide a critical path analysis, particularly useful for complex operations. Figure 18.5 provides an example of the level of information required and how the development process is structured. It allows for further concept development and refinement before scheduling a series of tasks to be completed leading up to product launch.

As indicated in the previous section, it is frequently the case at this stage that there may be more than one design concept to be developed. In each case, concepts begin to be refined, sketches giving way to accurate drawings and provisional specifications established. Discussions with packaging suppliers are an essential

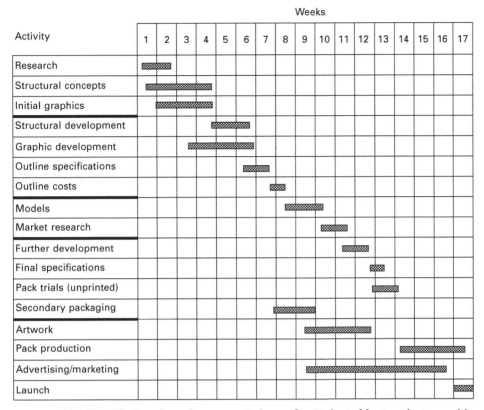

18.5 Simplified project plan presented as a Gantt chart. Most projects would require greater detail and show agreed start and completion dates.

part of the development process where suppliers often identify modifications to design concepts that contribute to production efficiencies. Inevitably, there will be negotiations at this point between designers, perhaps wishing to retain a feature and suppliers who see the same feature as a production restriction or cost factor. For example, incorporating a built-in handle on a blow moulded HD polyethylene container might be championed by designers who see it as a consumer benefit, but a supplier might view it as increasing the weight of the pack. This has penalties in terms of unit cost, transport costs and environmental performance. Here a central neck section and cylindrical configuration provide the best opportunity for reducing weight but might not meet marketing criteria or find favour with consumers. Such contradictory issues might require consumer tests to help resolve them.

It is likely that models now supersede mock-ups produced earlier. These will be to a much higher finish and will include both structural and graphic design elements. The example shown in Fig. 18.6 has been fabricated from vacuum formed plastic components, hand finished to create a realistic pack. Rapid prototyping techniques are also frequently used and specialist model makers often employed by design companies to create one-off models or, in some instances, a series of models for consumer testing. Many companies seek consumer approval of any new packaging

18.6 Highly finished model package. The concept was for a dim-sum steamer. This detailed model provides a realistic representation of how it might look. Mock-ups are not required to meet this standard of production.

design before making a final commitment to the project. Focus groups are often used where members of the public, selected to represent the target audience, are given realistic and often working models of new designs to evaluate. Depending upon the nature of the product, this may involve handling, pouring, dispensing, closing or other tasks. Feedback from consumers might result in design modifications or, more often, helping a company decide between design options. If consumer tests are required, it is important to include realistic timings in the project plan for producing models/ prototypes and to coordinate this with the market research company organising focus groups.

This stage also sees the development of secondary packaging and the evaluation of packaging performance on line and in transit. Often undecorated containers can be used for trials, although designers should be aware that, in some instances, pack performance could be affected by print. Corrugated fibreboard, for example, can suffer some slight crushing of the flutes during printing that might adversely affect performance. Production departments will normally run trials to establish filling, labelling, coding, collation, and stability on line. Transit trials can be organised using trial packs sent on representative warehousing and distribution systems. These should be designed to replicate typical conditions of pallet use, stack heights, transport methods and handling. Alternatively, packs can be evaluated using simulated package testing carried out by specialist companies. This is often quicker than real-time tests and has the advantage of being able to incorporate climatic testing, where packs can be humidity and temperature conditioned as part of the test sequence. Vibration testing can also simulate transport methods allowing the different vibrations from road, rail or air transport to be incorporated. (Pira International, a long-established company offering specialist package testing services has more information on its website, http://www.pira-international.com/Homepage.aspx.) The project plan must identify and incorporate appropriate timings for the required level of pre-production testing. Graphic development would include the extension of initial graphics from main panels onto other surfaces and incorporation of mandatory labelling requirements. The stage would see outline costs being established and would conclude with a presentation of work to the client team.

18.4 Case study: yoghurt for children

To help put the points discussed in the previous sections into context and illustrate how they operate in practice, it is useful to consider a case study. Although, in this instance, the brand is fictional, the case study is based on a real-life project. Here we only have space to show snapshots of the work at an early conceptual stage of the design process. As always, the project begins with the brief.

18.4.1 Outline brief

The company is a well-established manufacturer of dairy products and currently a brand leader in the butter market. It now wishes to establish a greater presence within the healthy food sector by increasing its portfolio of organic products. A key strategy is entry into the yoghurt market with a new range of organic yoghurts. The company has experience in this market through manufacture and supply of own label yoghurt products. This brief is particularly aimed at providing mums with a choice of healthy snacks for their children's lunchbox. The product range, under development, will initially include apple, strawberry, peach and raspberry variants in a creamy organic yoghurt base, using real fruit. Portion size will be between 80 and 100 ml and, unusually, multipacks will contain five portions (one per day). The target audience is 25–35-year-old women with young children aged 4–9. Products will be sold within the chill cabinet yoghurt section of the major multiples. The pack design must work on two levels, appealing to caring mums and engendering brand loyalty from young boys and girls.

Brand values

The butter brand is well loved in the UK and is a tried and trusted friend in many households. It is not yet readily associated with organic products but brand values are, 'friendly', 'trustworthy', 'adult', 'countrified', 'traditional'. For this project, however, a new sub-brand will be used, endorsed by the parent brand. (For reasons of confidentiality, the parent brand will be omitted from any design work represented in this chapter.)

Advertising

Previous butter advertising featured lively animated cows with some of the fun element in the ad, rather puzzlingly for an adult product, carried over to the pack design. An advertising strategy for the yogurt includes TV and magazine coverage combined with a promotion tied to the company's butter products.

Competitors

There are many competitors including those featuring franchised characters from Disney, Bob the Builder and other children's favourites.

18.4.2 The design study

At the outset of a packaging study, depending upon how broad the brief may be, the designer may have a palette of materials and pack forms to choose from. Of course, in many instances this will not be the case and the packaging project may be more evolutionary in nature, directing the designer to pack forms that, for example, can run down existing production lines without major modification. To help illustrate how the packaging design process works in practice, we can consider the approach to a case study where, in this instance, the design team is presented with a broad choice of pack forms and materials.

In this particular study, the advertising agency, a high profile multinational company, provided a two-person copywriter/designer team to work alongside the packaging designers and to contribute to the conceptual thinking. The client company provided a resource of technical and marketing assistance on demand and also presented a full range of competitor products to the creative design team. As the product had not yet been fully formulated, it was confirmed by the client that we could use natural yoghurt for trials. The in-house food technology department provided samples of the range being developed.

Research

Research begins by considering the market, expanding on information supplied by the brief as described in previous sections of this chapter. The market for children's yoghurt should be viewed within the overall UK yoghurt market as purchase is overwhelmingly carried out by adults, although children may influence the purchasing decision. Mintel International provides a readily accessible source of market data and analysis. The following information is typical of the detail provided and is extracted from the Mintel report, Yogurt – UK – May 2009 illustrated in Table 18.4. The total market for yoghurt and fromage frais has grown by 27% between 2004 and 2009 and is estimated to be valued at £1,590 million. Of this, products specifically for children represent around 18%, £275 million. In addition the report highlights some significant points, relevant to this study.

Overall, products are being positioned and repositioned in different ways to reflect market changes:

- Light/diet products are being repositioned as consumers take it for granted that all yoghurts are low fat.

Table 18.4 Leading companies and brands in the UK yoghurt market

Company	Base	Brand	Positioning
Alpro	Belgium	Alpro	Organic, Healthy diet
Arla	Sweden	Lactofree	Lactose free
		Bob the Builder	**Children**
		Mr Men	**Children**
		Scooby Doo	**Children**
Benecol	Finland	Benecol	Lowering Cholesterol
Dale Farm	UK	Spelga	Broad market
		Loseley	Extra premium
		Rowan Glen	Scottish provenance
		Intune	Probiotic
Fage	Greece	Total	Greek provenance
Danone	France	Actimel	Probiotic
		Activia	Creamy probiotic
		Shape	Hunger management
Müller	Germany	Amore	Indulgence
		Corner	Split pot
		Little Stars	**Children, natural**
		Mullerlight	Treat
		Vitality	Healthy digestion
Nestle	Switzerland	Ski	
		Munch Bunch	**Children, calcium source**
		Munch Bunch Squashams	**Children, 5–12**
		Disney fromage frais	**Children, young**
Onken		Onken Natural	Breakfast
		Onken Fruit	Treat
		Onken Wholegrain	Healthy
Rachel's	UK (Wales)	Rachel's Luscious Low Fat	Dietry
		Rachel's Greek Style	Traditional
		Rachel's Forbidden Fruits	Indulgent
		My First Yogurts	Babies
		Taste Explorers Squeezies	**Children**
		Taste Explorers snackpots	**Children**
		Natural	Breakfast
		Divine Desserts	Luxury
St Helen's Farm	UK	St Helen's farm	Goat's milk
Unilever	Flora pro-active	Lowering cholesterol	
Yakult	Japan	Yakult	Scientific
Yeo Valley	UK	Yeo Valley Smooth & Creamy	Organic, fruity
		Yeo Valley Natural	Organic, probiotic
		Little Yeo's (tubs)	**Children, organic**
		Yeo's (tubes)	**Children, organic**
Yoplait	France	Yop (bottles)	Children 10–16
		Petits Filous	Babies
		Petits Filous Frubes (tubes)	**Children**
		Also Co-branded promotions Dr Who, High School Musical, In the Night Garden	
		Frubes Pouches	**Children, 9–16**

Source: Mintel, Yogurt – UK – May 2009. Brands and products in bold, indicate particular relevance to case study target market. Note that other products, fromage frais and fruit compots also compete in this market.

- Active health products (probiotics) have encountered some consumer scepticism about unsubstantiated claims.
- Organic is less important than price and is not providing benefits to justify price differentials.
- Tubes and pouches for children's yoghurts gained ground on the basis of being freezable and easily packed into lunchboxes.

Additionally, the design team also considered other market factors appearing in this and other market reports, surveys and publications, including those investigating:

- market value, seasonality, trends, brand shares
- brand values
- target audience profile, purchaser/consumer/end-user/decision-maker
- consumer requirements, in-use, storage, disposal, opening/closing
- competitor products.

The target audience had been clearly identified by the client company as young women aged 25–35 with young children (Fig. 18.7), so desk research was augmented by unstructured and informal interviews with parents of young children in this category. In this instance, the team used family, friends and parents at a local school.

Clearly, the healthy lunchbox was seen to be a key factor currently driving the market. Although Mintel reports that tubes were gaining market share, most mothers, however, were not in favour due to the mess tubes could cause on opening. They favoured conventional rigid pots where the contents can be eaten with a spoon. Any floppy pack form was less favoured, especially if it could not be resealed. Experience of children snacking in the car convinced parents that the child/tube interface was not controllable. School staff echoed this view through their experiences during school lunch breaks. Some parents, however, disagreed and were of the opinion that if packs could be resealed some children would not finish their yoghurts.

18.7 The target audience in the case study represented by a mum and two children of school age.

It was recognised that 'pester power' influenced brand choice in store. Characters featured on-pack were both a strength and a weakness as children quickly switched character allegiance with age and peer pressure. Most parents, in any case, were becoming fatigued by characters and pointed out that many girl heroes were pink and delicate whereas boy heroes were action figures. They did not want gender issues causing squabbles amongst children.

Amongst the team, there was a feeling that the new brand would be wise to feature natural values, represented in a simple way that could span a wide age group of children rather than adopt more specific and intense characters. Parents were also keen to try and restrain children from messy eating, some making comparisons with Tetrapak-type mini-drinks, where the straw often proved to be more of a device for spraying juice than for drinking it. A small sample of the competitor products are shown in Fig. 18.8, illustrating tubes and pouches, the most radical of pack forms currently in the UK market.

Design concepts

As is common practice, the design team now began by working on unit containers, considering a wide range of pack concepts. It was felt that although research was

18.8 A selection of yoghurt for children already on the market, concentrating here on pouch and tube formats.

already beginning to influence some directions, the team should attempt to be open minded and receptive to all ideas and not introduce too many constraints. The concept range included pouches and tubes but also other forms of squeezable packs (Fig. 18.9). While there were advantages in the tube format, freezability, cost, novelty, and ease of collating five per pack, there were also disadvantages. Secondary packaging would have to be robust as the unit packs could not contribute to stacking strength. Also, the parental reaction indicated a resistance to purchasing this format.

As the conceptual phase continued, two major design routes were emerging. Both favoured rigid containers where yoghurt would be eaten with a spoon. Thermoformed pots were seen as more conventional but in line with brand values. In addition, the company already had experience of this format, supplying own-label yoghurts. It would also be possible to shape the 'pots' to allow five per outer carton or sleeve, meeting the one-per-day requirement of the brief.

The second route explored traditional paper-based 'pots', associated sometimes with high quality ice creams. It was felt that this format could provide a more natural and better quality image and also differentiate the product from competitors. In many ways, this is the opposite of tubes. While it was thought that it would appeal to parents, it perhaps lacks the fun (and hazards) that children might enjoy by squeezing tubes.

In each case, the team considered secondary uses for the packs. Thermoformed packs were considered with simple insect shapes moulded into the base. The team acknowledged that seeing an outline of an insect at the base of a yogurt pot is more fun for a child than an adult but the pack could later be reused as a mould for producing

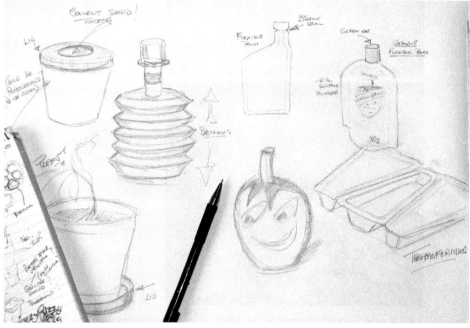

18.9 Initial sketches of possible concepts in yoghurt packaging design.

clay or plaster casts. The paper pot could find secondary use as a small plant pot, using the wooden spoon as a label. The team could see how encouraging the link with a natural product and nature itself could be the basis of promotions. Free seeds, supplied through a web-based 'children's' site in exchange for codes on the packs themselves, could form the basis of a marketing dialogue and user database.

Experimental graphics concentrated on natural elements but began to develop illustrations of animals to feature on-pack and in subsequent advertising (Fig. 18.10). Mock-ups were made of a range of design ideas, the paper pot mock-up carrying experimental graphics shown in Fig. 18.11. The paper pot, however distinctive, was difficult to collate efficiently into fives, a major disadvantage and one that would

18.10 Some rough graphic concepts for yoghurt packaging design. It is often quicker to work with sketches but, in this case, the designer chose to work directly with Adobe Illustrator.

18.11 Mock-up of paperboard yoghurt pot. The rough graphics from the previous figure were converted to an arc shape, printed and simply applied to an ice cream carton – a quick and effective way of creating a mock-up.

challenge the brief. Thermoforming, on the other hand, allowed for production efficiencies, providing web-fed forming, in-line filling, sealing and sleeving.

Design analysis

The environmental performance of packs was important per se but, in the context of a natural organic product, of significance to an environmentally aware target group. The tubs and lids could be made from rigid paperboard; these frequently have a thin plastic coating that interferes with recycling. Newer coatings are claimed to overcome the problem being both recyclable and biodegradable. At this stage there was some doubt about coating performance in contact with acidic yoghurt and a cost implication to be investigated. As mentioned earlier, the thermoformed tray concept provided efficiencies in production and in cost. The plastics were not immediately recognised as recyclable and overall appearance more difficult to market as an organic natural product.

All work was presented to the client company and design recommendations made during a formal presentation. The rationale was explained for making decisions and a programme of further work proposed. Here, unfortunately, we are forced to leave the case study at this point to maintain client confidentiality. The case study, however, provides a flavour of how the design process works in practice and, in particular, how introducing mock-ups begins to bring the project to life. The study shows that understanding consumer groups is key to developing packaging that will meet their needs and reflect their lifestyle values and behaviour. Ultimately, the commercial reality is about designing packs that help sell products.

18.5 Conclusion

Packaging design involves creative, technical and analytical disciplines and, as we have seen, follows a process that seems linear and logical but is often cyclical and occasionally tangential. Nevertheless, the design process as outlined here has been proved to work in countless design studies and with different types of organisations.

18.6 Sources of further information and advice

18.6.1 Packaging design books

The following books provide good sources of design information.

David Dabner (ed.), *Graphic Design School*, London, Thames and Hudson, 2004. Good coverage of the design process with a chapter on packaging design by Bill Stewart.

John Grant, *The New Marketing Manifesto: The 12 Rules for Building Brands in the 21st Century*, London, Texere, 2000. Another slant on how brands are established.

Adrian Shaugnessy, *How to be a Graphic Designer Without Losing your Soul*, London, Laurence King, 2005. Mainly graphic design but useful, showing how designers work.

Bill Stewart, *Packaging Design,* London, Laurence King, 2007. Complete guide to packaging design.

18.6.2 Brainstorming and creativity

There are numerous websites covering brainstorming techniques, but the following source is recommended to help stimulate creative thinking.

Alan Fletcher, *The Art of Looking Sideways*, London, Phaidon, 2001. A selection of thoughts from one of the most interesting, talented and outstanding designers of modern times.

18.6.3 Useful websites

www.monbiot.com – UK environmental activist
www.europa.eu.int – Source for European packaging legislation
www.euromonitor.com – European marketing reports
www.landor.com – Useful packaging design case studies
www.pearlfisher.com/ – Branding, structural and graphic packaging design at its best

18.7 References

www.interbrand.com/best_global_brands.aspx, Best Global Brands, 2009, accessed 12/01/2010.
www.mintel.com, Yogurt – UK – May 2009 – Companies and Products, accessed 12/01/2010.

19
Printing for packaging

R. MUMBY, Chesapeake Pharmaceutical and Healthcare Packaging, UK

Abstract: This chapter reviews key principles and techniques in the colour printing of packaging. It discusses colour description and measurement before going on to review colour mixing and printing techniques. It also outlines key issues in graphic design, reprographics and pre-press processes, including proofing and other quality control techniques.

Key words: packaging, colour printing, reprographics, pres-press.

19.1 Introduction

Colour is one attribute used to describe the appearance of an object. In fact we use a whole host of these when we examine an object, e.g. texture, gloss, transparency or opacity. The sensation of colour is our brain's interpretation of signals received by the eye. Therefore, in order to see colour, three things are required: a light source, the object and an observer (Fig. 19.1). In most cases an object appears coloured as it is reflective; these are termed 'surface colours' although there are other types, namely self-luminous objects such as a television which creates colour by converting electrical energy into a form of light. It is surface colours that are critical in the appraisal of colour for packaging since the surface colour produced in packaging tends to be formed by a pigmented or dye-based surface coating or plastic.

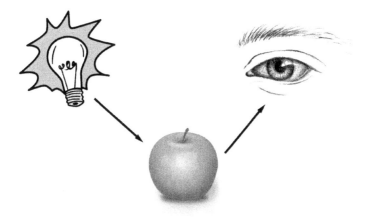

19.1 Requirements for viewing colour.

19.2 Light and colour

Visible light such as sunlight or white light can be described as electromagnetic radiation or waves with wavelengths ranging between 380 nm and 730 nm. Our understanding of this comes from Newton who showed that white light could be split into all of the basic colours via a prism. Figure 19.2 shows the scale of electromagnetic radiation and the visible part of spectrum with approximate wavelengths for each basic colour. It is an understanding of light and its interactions with objects that gives us an understanding of surface colours. Surface colours appear as a result of selective absorption of specific wavelengths from white light. For example, a white light shone onto a plain paper surface is reflected evenly across the spectrum to cause a balanced set of visual signals to be passed from the eye to the brain. The brain interprets this as white (Nobbs, 1998).

When white light is directed at the pigmented/dyed surface of a coating, plastic or fabric it reacts in a different way. The pigments or dyes present within the surface absorb or scatter the light so that the visual signal received by the eye is no longer balanced. When this happens certain wavelengths from the visible spectrum are now missing or reduced. This causes the brain to interpret the signal differently, i.e. coloured. This is a relatively simplistic view of the interaction between light and materials, and for a more detailed understanding of this topic, a study of reflection (matt/gloss effects), refraction (rainbows), defraction (shadow effects) and scattering (opacity) should be conducted.

19.3 The description of colour

A number of terms are used to describe colour:

- visual descriptions such as green or blue, light or dark
- emotive descriptions such as vibrant or mellow
- colourist descriptions such as dirty or clean, strong or weak.

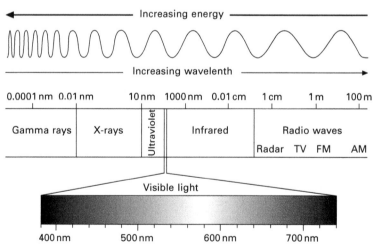

19.2 Electromagnetic radiation spectrum (from Antonine Education, www.antonine-education.co.uk).

There is a standard set of terms which are widely used in the coloration industries. All descriptive terms fit simply into one or more of the three categories used.

19.3.1 Hue

The hue describes the colour or its family and these can be seen in the split of white light from the visible spectrum. There are six colours or hues: red, orange, yellow, green, blue and violet; the colours of the rainbow (indigo, the seventh hue is missing and in fact was added by Newton to include a seventh colour. He added this to create symmetry with the seven notes in the musical scale) (Nobbs, 2002). If one attempts to describe colour, it soon becomes apparent that there are specific hues which are more important than the others. These are known as unique hues or psychological primaries. These psychological primaries are: red and green, blue and yellow. All of the other hues can be described by a combination of these four primaries, e.g. violet can be described as a blue-red. The primaries are also separated into opponent pairs since one can never use the two opponent pairs to describe a colour, e.g. the object is never a yellow-blue or a green-red colour.

19.3.2 Lightness

We can use the term lightness to describe neutral shade objects, e.g. white, black or grey. Lightness can also be used when describing a colour by adding it to the hue for example light blue or dark blue. It is important not to confuse the term lightness with that of brightness since brightness is dependent on the viewing conditions of the object being described, e.g. the more intense the light source the brighter the colour viewed. This is a result of lightness being a relative term; as an example a mid-grey object will always look mid-grey regardless of the viewing conditions.

19.3.3 Intensity of colour

Intensity of colour is best described as the intensity of the sensation when viewing a colour. It helps us to distinguish between strong saturated colours and weak pastel shades. The term used to describe the intensity of colour is chroma. Chroma is a relative term in that it describes the intensity of sensation of an object viewed under a light source compared to white viewed under the same conditions. If the intensity of the light source is altered, the intensity of both the white and the object being viewed alter by the same amount and therefore the intensity of sensation between the two remains the same.

19.4 Colour vision

Now that we understand what is required to see colour and the basics of colour description, it is useful to understand how the eye can visually appraise hue, lightness and intensity.

19.4.1 The human eye

The structure of the human eye is shown in Fig. 19.3. The light sensitive area of the eye where an image is formed is called the retina. Light is focused onto the retina by the cornea and lens. The amount of light focused in this area is controlled by the iris; it is the aperture at the centre of the iris, the pupil that allows the light to pass into the eye depending on the viewing conditions. For example, in relatively low levels of light the iris expands the pupil to allow more light to pass into the eye, in bright conditions the pupil is smaller to restrict the passage of light into the eye. The retina itself comprises two types of photosensitive cells: rods and cones. Activity from these cells is transmitted to the brain via the optic nerve for the brain to process the information into an image or colour sensation. The rods present within the retina provide a monochromatic signal to the brain. They are adapted to provide information at low light levels (night vision). The cones provide signals to the brain at normal light levels and the information is provided in colour.

There are three types of cone cells present within the retina; each type provides information from a specific band of wavelengths from the visible spectrum. They can be characterised as follows:

- short wavelength receptors; these are sensitive to blue light
- medium wavelength receptors; these are sensitive to green light
- long wavelength receptors; these are sensitive to red light.

The intensity of signal from each type of sensor is passed to the brain which interprets the information as colour vision.

19.5 Additive and subtractive colour mixing

The function of the retina and its three wavelength specific photoreceptive cells supports the trichromatic theory developed throughout the 1800s. The theory suggests

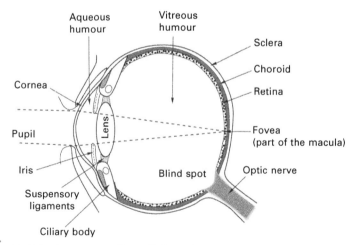

19.3 The human eye (from Editure Education Technology, www.schools.net.au).

19.5.1 Additive colour mixing

Additive colour mixing describes the creation of colour by the mixing of two or more coloured light sources. The most familiar example of colour reproduction by additive colour mixing is colour television in which all of the colours visible on screen are produced by a combination of light emitted by red, green and blue light sources (Nobbs, 2002). Figure 19.4 (and Plate I between pages 460 and 461) demonstrates the principles of additive colour mixing, which is often termed RGB from the red, green and blue colours used to form multicoloured images. It is important to note that each mix of two additive primaries forms each subtractive primary.

19.5.2 Subtractive colour mixing

This type of colour creation has already been described briefly in Section 19.2. Colours are created by selective absorption of specific wavelengths of white light. This is of huge importance when considering packaging as it is the method of image reproduction for printing techniques. The subtractive primary colours: cyan, yellow and magenta are used to control the wavelengths of light absorbed or reflected.

19.6 Other factors affecting colour

Now that we have an understanding of how we perceive colour, it is important to understand some of the things which may affect our perception of a given colour.

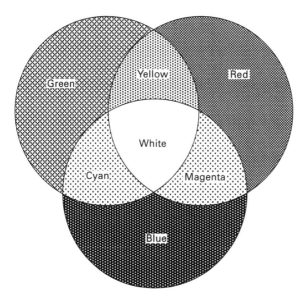

19.4 Additive colour mixing.

19.6.1 Illuminants

The colour of a given object is a product of the light source under which it is viewed. Different light sources have a different distribution of wavelengths of light from the visible spectrum. For example, a colour viewed in average daylight will appear different from the same colour viewed under a domestic light bulb; this is due to the domestic light bulb emitting a greater amount of light in the orange-red part of the visible spectrum compared to the more even distribution of wavelengths from average daylight.

The Commission of Illuminants (CIE) established a set of standard illuminants for viewing colour. These are: standard daylight (D65), incandescent light (A) similar to a standard light bulb and fluorescent light (F) of which there are 12 different spectral distributions. The eleventh F light source TL84 is important for packaging since it is used by a number of high street stores particularly in the UK and in Europe and it has also been adopted in the Far East.

19.6.2 Simultaneous contrast

Simultaneous contrast describes the changes in perception of colours depending on the background against which they are viewed. Figure 19.5 shows two circles of the same grey with different backgrounds. It is simultaneous contrast that makes the circle surrounded by black appear lighter than the one surrounded by white.

19.6.3 Impaired colour vision

Colour blindness is caused by weak or deficient response from one of the cone cells communicating colour to the brain. It is a relatively common condition although more prevalent in men. In general it is more likely that the sufferer will confuse red and green shades since any deficiency in either long or medium responsive cones will result in this. It is possible, although rare, for colour blindness sufferers to confuse yellow and blue shades if the short responsive cones are affected. Another type of colour blindness in which all cone responses are missing results in the sufferer only being able to distinguish between light and dark from the rod response. Impaired colour vision can be tested by using Ishihara colour charts. It is important for those

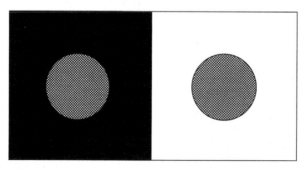

19.5 Simultaneous contrast effect.

working in an environment where visual colour appraisal is a requirement to understand if their perception of colour is affected and how it is affected by a visual defect.

19.6.4 Metamerism

Metamerism describes the effect of coloured objects that match under a specific light source no longer matching when viewed under different lighting conditions. This effect is of particular importance in printing since the choice of pigments or inks used to match a specific colour can often show metameric effects.

19.6.5 Ideal conditions for appraising colour

Based on what we now know, it is clear that in order to establish a reliable assessment of colour we must observe several specific rules (Nobbs, 1998):

- The level of illumination must be sufficient to produce cone vision (photopic).
- The illumination must be a good simulation of one of the CIE illuminants and comparison between colours should be made under the same illuminant (if possible it should match the end use illuminant).
- The nature of the background against which the sample is viewed must be controlled. A medium bluish grey (smoky) is to be preferred.
- The field of view should be controlled so that the image produced by the lens is focused roughly on the same area of the retina.

19.7 Colour printing

In 1931, based on the theories of human colour vision, the CIE developed a diagrammatic representation of all the colours able to be perceived by the average person. The chromaticity diagram in Fig. 19.6 (and Plate II between pages 460 and 461) shows the colour gamut of our visual system. It is possible to reproduce areas of this gamut of colour by additive (RGB) colour mixing or subtractive colour mixing. The colour gamut able to be produced by subtractive colour mixing reproduction techniques compared to the gamut of the human eye is approximately shown in Fig. 19.6 (and Plate II).

19.7.1 Printing process colours: subtractive colour mixing

Colour printing makes use of this method by printing one or more of four specific inks each designed to remove specific wavelengths from white light. These are cyan, magenta, yellow and black (CMYK). When cyan, magenta or yellow are printed, the transmission of light from the surface is altered, for example when cyan is printed, wavelengths of light above 580 nm are absorbed and the transmission of light back to the eye is focused in the blue-green part of the spectrum, 400–580 nm. The same can be said of magenta and yellow absorbing light between 490–580 nm and 380–490 nm, resulting in transmission of light from the blue-red and green-red parts of the spectrum, respectively.

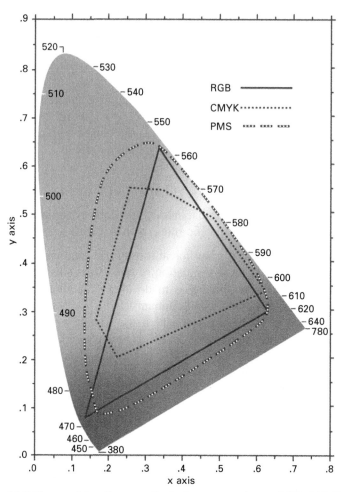

19.6 Chromaticity diagram (including approximate RGB, CMYK and PMS colour gamuts).

Each of these specifically designed inks is transparent so that combinations of two or more of these colours can absorb light through layers of one another. In an ideal case, a printed layer of all of these colours should absorb wavelengths across the whole of the visible spectrum and produce black. In practice, the black produced by a combination of CMY tends to be weak in sensation and results in a heavy coating weight made up of three colours which can be easily replaced by a single ink film of one. It is for this reason when printing process colours that a Black is used. The letter K denoting black refers to its use as the 'key colour.' It should be noted that black does not always have to be the key colour but is in most cases.

19.7.2 Tonal reproduction and halftone printing

Certain printing methods have only the ability to produce consistent thickness of ink. Tonal reproduction is achieved by printing in halftone. Halftone is created by printing dots of specific size in different areas. This was first demonstrated as a monotone or single colour reproduction and can be seen everyday in the black and white image reproduction of newspapers or this book. The effect is due to limitations in the resolution of the eye. By printing dots close together the eye blends them to give the impression of grey tones. Applying this same principle to the four process colours resulting in a mosaic of dots of each colour overprinting one another, it is possible to create a full colour image. The XY chromaticity diagram in Fig. 19.6 shows the plot of colour sensations the average human can perceive. Reproduction of colour using CMYK or subtractive colour mixing can achieve a gamut of colour within this plot. This is approximated in Fig. 19.6.

Screening and resolution

The image resolution for halftones is determined by the screen ruling. The term screen ruling is used to describe the frequency of dots in a given area of image. For conventional screening (amplitude modulated screening or AM), the halftone dots are arranged in a grid structure termed the screen. The quality of the halftone cell is determined by the overall cell size (lines per inch, lpi) and the dots (dots per inch, dpi) required to create the cell. The screen ruling (lpi) is the number of dpi used to produce the halftone image and the resolutions used will determine the screen rulings that can be produced. Course screen rulings tend to be used in low quality reproductions, e.g. newspapers, and fine screen rulings for high quality reproductions, e.g. fashion magazines or art prints (Kinyo, 2004).

Screen angles

The screen angle refers to the angle formed by the direction of the screen ruling in relation to the vertical finished position of the image (Kinyo, 2004). The role of screen angles is particularly important for multicolour image creation since each colour (CMYK) is required to have a different angle. Assigning the wrong angles to each colour can result in a compromised image appearance. If the wrong angles are applied to colours the image tends to produce objectional interference patterns (moiré) or screen clash. A moiré pattern can be seen in Fig. 19.7. Moiré patterns are produced when two similarly repetitive patterns are almost but not quite superimposed (Kipphan, 2001). Assigning the same angle to all colours can also be a cause of colour variation since each dot should print in the same position but any misalignment may result in variable transmission of light.

Typically in process printing the angles are assigned so that each colour is separated by an angle of 30°. Cyan, magenta and black are assigned angles furthest away (30° separation) from each other and yellow is assigned an angle 15° from the other colours. Yellow is assigned the 15° difference as it is less likely to visibly clash with the other colours. As a general rule, printing colours in halftone over or into other halftone colours should be assigned angles as far from each other as possible without being on the same axis. Difficulties can arise for complex artworks with four or more colours printing in halftone into or over each other, as there may not be enough angles to avoid a clash.

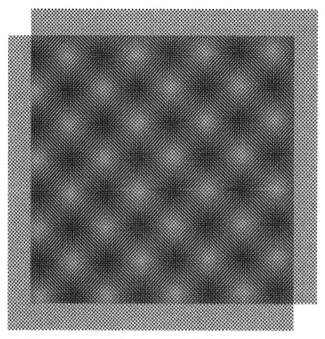

19.7 Moiré pattern (from Kipphan, 2001).

Dot shapes

Dots may be printed in a number of shapes, for example: round dot, square dot, elliptical dot and diamond dot. It has been reported that the elliptical dot gives smoother midtone (30–70% dots) particularly for lithographic techniques (DTP Tutorials, 2008). The effect of dot shape on print is a complicated one in that the choice of dot shape is influenced by the image to be printed. Elliptical dots may give a smoother transition to a vignette throughout the midtones; however, it is likely to be more susceptible to moiré patterns. Square dots tend to print sharper in the midtones but they too are more susceptible to moiré patterns. The round dot is least susceptible to moiré but more likely to show dot gain. The round dot tends to be the most common in packaging printing; however, for the screen printing technique it is the least favoured choice (Coudray, 2007). It is difficult to establish the ideal dot shape because applications and process techniques are often too diverse (Kipphan, 2001).

19.7.3 Hexachrome

The limitations in the colour gamut able to be produced using CMYK can be demonstrated by comparison of the Pantone swatch guide for special colours to process reproductions (Pantone solid to process guide or Color Bridge guide). The original Pantone guide produced in 1963 was introduced as a colour matching system (Pantone matching system, PMS) for use with special inks blended from 14 individual colours printed in solid and various tints. Its use as a colour matching tool is now widespread and has been adopted by most suppliers, converters and buyers of packaging.

Following soon after the original Pantone guide for special/spot colours was the colour bridge or solid to process guide which shows the closest match achievable for CMYK to the original special colour. When compared visually to special colour swatches achieved through mixing special inks to provide unique colours, the process reproductions appear noticeably less vibrant or do not match at all. Only around 30% of the Pantone spot colour library can be accurately reproduced using process colours. In 1994 Pantone introduced a six-colour system which addresses the limitations of the CMYK colour gamut in part. In addition to cleaner CMYK inks, orange and green (in some cases a strong blue shade of ink has been used to replace the orange or green) were added to complement the modified process set and to maximise simulating the original Pantone spot colour library (Reid, 2008).

In Fig. 19.6 the improvement of printable colours based on the Pantone system compared to the standard CMYK is approximated. In fact between 50% and 90% of the original Pantone library can be achieved using hexachrome. One potential issue with the use of hexachrome is that the increased brightness of the inks required to achieve the improved gamut can result in a loss of lightfastness properties. This increase in gamut is advantageous since it allows improved colour reproduction from an identical set of inks. This allows the printer to produce a range of work without costly wash-ups and the opportunity to produce composite reproductions of more than one item at the same time (Davey, 1999).

Hexachrome initially had limited success since it was not embraced by the industry in general; this may have been due to the cost of implementation, a lack of understanding of the print processes by design and marketing personnel, or the failure of pre-press systems to provide adequate conversion software. However, hexachrome popularity has steadily grown, particularly in certain areas of the packaging market and for specific production processes. Hexachrome does have its limitations, namely:

- its colour gamut still cannot reproduce many Pantone matching system (PMS) special colours
- colour control can be problematic in specific production situations requiring tight control of pre-press and production machinery, e.g. variations in dot gain from one printing press to another may result in unacceptable colour variation of the product
- the lightfastness of the finished product may be compromised
- existing colours/artworks/standards may be compromised or re-established when reproduced in hexachrome.

19.7.4 Alternative screening techniques

The use of conventional screening described in Section 19.7.3 could cause a problem when using hexachrome since the number of angles available is limited and may result in moiré patterns emerging in certain designs. The use of alternative screening types overcomes this potential issue and is used in other situations to impart specific advantages.

Stochastic screening (frequency modulated or FM)

Unlike conventional screening (AM) which alters the size of the dot in specific areas in an ordered pattern to create tonal variation in an image, stochastic screening puts dots of equal size, distributed randomly in varied volume to create the tonal variation in an image. Clearly the first advantage of this type of screening is the elimination of interference patterns (moiré) since the dots are randomly distributed without angle. Stochastic screening can also give the visual benefit of cleaner colour reproduction in certain tonal areas. This is a result of the reduced paper effect, since stochastic screens cover a greater percentage of the substrate than the conventional equivalent as shown in Fig. 19.8. It follows from this that ink usage can be reduced for the printer.

Stochastic screening also allows greater control of printed colour since variations in ink film weight on press result in a reduced colour change due to reduced tonal value increase (TVI) effects (see pp. 458–459). Another benefit of using stochastic screening is its ability to hide a certain amount of misregistration between colours. If colours in conventional screening are misregistered, the effect is more noticeable as a result of the ordered distribution of dots. The random patterns of stochastic screening

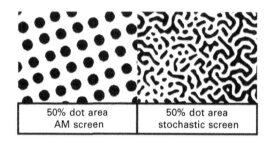

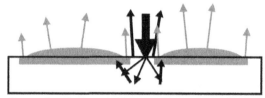

Conventional screen mid-tones allow paper reflection – paper colour directly affects printed colour

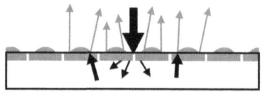

Stochastic mid-tones minimise the direct reflection of paper because of the refraction of light – minimises the effect of paper on printed colour

19.8 Reduced paper effect (from Braden Sutphin Ink Co., 2009).

disguise this to some extent which can lead to reduced wastage for the printer and a greater registration tolerance on press. The advantages of stochastic screening are clear: its use should be well controlled on the press and selected for the appropriate reproductions for the benefits to be realised. Its use is, however, not universal and the process may not prove beneficial for flesh tones, pastels, low contrast subjects, rough paper types or average process colour illustrations (Prince, 2009).

Hybrid screening

Hybrid screening is designed to make the most of both conventional and stochastic screening types, combining the advantages of both across the full tonal range, e.g. using AM screening in flat tints and flesh tones, whilst shifting to FM screening for fine detail or moiré sensitive areas. Many companies have developed variations of stochastic screening and hybrid screening each marketed to sell the benefits in the final printed result. Hybrid screening is commonly used in flexographic printing to disguise its inability to reproduce fine screens in comparison to lithography. In this case the hybrid will result in FM screening of some type in the 0–10% dot regions and the 90+% dot regions.

19.7.5 Special colours

The use of special or spot colours in printing is common; in general their use is to add something extra to the reproduction of an image or artwork. As already described, the reproduction of colour using CMYK or hexachrome does not provide a full colour gamut and reproduction of certain colours using these colour mixing techniques can be poor. In these cases spot colours can be introduced to overcome this problem.

Spot colours

A spot or special colour can be described as a single or blend of one of 14 (or more) base colours created from individual pigments. In the western world the most common method of description of these colours is the PMS guide created by Pantone Inc. (there are other guides, e.g. DIC, TOYO). The PMS guide shows the recipe and resulting colour swatch for combinations of the 14 basic colours. The results of these blends are cleaner and more vibrant than the CMYK representation. It is for this reason that special colours are often used for communicating important brand logos or corporate colours. Special colours are also often used in 1–3 colour designs as this may be more cost effective than using process colours. Typically these spot colours can be used in conjunction with process colours to add visual impact and in some cases can be used to replace one or more of the process colours to reduce the total number of colours.

The use of special colours can also be advantageous for a higher quality result not only as a result of the improved perception of colour. Consider the control of colour on a printing press; it is usually simpler for the printer to control a single ink film throughout a production run than to control all four process colours potentially reducing the likelihood of colour variation. Also in terms of registration of fine positive or negative type (e.g. instructions or ingredients listings) the reproduction in a single ink tends to be clearer than a combination of two of more colours which may exhibit some misregistration.

Special effects

Special effects can be relatively simple and designed into packaging. For example, the contrast between matt and gloss effect varnishes in specific areas can be striking. Special effect inks or inks containing special effect pigments can be categorised by the type of effect they impart to the final product. The use of special effect pigments has grown steadily in recent times, particularly in the packaging industry primarily to give products an exclusive edge.

Metallic effects

These inks, originally used in the automotive industry, impart a metallic lustre to the finished product. An ink is prepared with metallic pigments (generally aluminium or bronze flakes) within it. The metallic pigments act as tiny mirrors resulting in a bright metallic quality to the coating depending on the angle of viewing (Gilchrist, 2001). Standard metallic ink formulations can be viewed in the current PMS colour guide and an extended metallic PMS guide shows other shades that are available. In recent years the introduction of high lustre substrates, e.g. vacuum metallised films and paperboards, has further improved the quality of lustre achievable. A good example of the potential of metallic ink effects has been demonstrated by the Metal FX® technology (Metal FX® Technology Ltd, 2009). Although this is not a new technology, it demonstrates the principles of printing transparent inks (CMYK and other special colours) over a high lustre metallic ink to create metallic effect images.

Fluorescent effects

A series of fluorescing inks can also be printed which in daylight give rise to colours which possess a remarkable vivid brilliance as a result of the extra glow of fluorescent light (Christie, 1993). The pigments used in fluorescent inks tend to be used for their brilliance properties. However, they do have a drawback in that they tend to exhibit reduced lightfastness due to their dye-based composition. This can be a significant problem, particularly for packaging which may require high levels of lightfastness when on display.

Pearlescent effects

The original pearlescent inks/coatings which are still commonly used are based on mica flake pigments, typically coated with a thin layer of an inorganic oxide. These pigments reflect and partly transmit incident light leading to multiple reflections of light from the layered material. Interference between the reflected light beams results in specific colours at particular angles (Gilchrist, 2001). A commercial example of this type is Iriodin® by Merck. Improvements in this type of pigment technology have led to new improved inks which can exhibit more dramatic colour changes (often termed colour shift inks). Commercial examples of this type are: Colorstream® & Miraval® by Merck, Chromaflair® by Flex Products and Variochrome® from BASF. It should be noted that for the most dramatic effects close attention should be paid to the colour of the coating beneath the application of pearlescent pigmentation since the effect can be compromised considerably. To make the best use of pearlescent pigments, they should be applied over dark colours so that maximum incident light is absorbed allowing the pigment reflection to be the dominant response.

Other special effects

It is not only visual effects that can be printed to add value to a pack or product. In terms of our senses, colour is only one of many that we perceive when we examine an object. Table 19.1 describes other special effects used in the packaging industry.

19.8 Graphic design, reprographics and pre-press

In Section 19.7 the basics of image formation through printing were discussed in terms of tonal reproduction by subtractive colour mixing with process colours, screen ruling/angles and screening/dot types. The following section describes the processes of artwork or graphic design and the preparations for printing including modern platemaking techniques.

19.8.1 Graphic design

The discipline of graphic design is used in a wide range of industries for communication of information, an idea or concept. In the modern printing industry we can think

Table 19.1 Other sensory special effects used in packaging

Sense	Description
Touch	Tactile varnish effect used to engage the sense of touch, e.g. velvet type feel
	Braille applied to packaging for communication purposes to the blind
	Scratch off revealing hidden information beneath an opaque layer, e.g. lottery cards, contraceptive pill calendar
Smell	Encapsulated smells applied to a product which release the odour either by scratching or breaking a seal
Hearing	Smart packaging which can convey a message via sound, e.g. musical birthday card messages, prescription drug reminders or communications for the blind

of graphic design as construction of an artwork from a given brief. This may be something relatively simple such as the layout of text in a monotone printed book (the origin of graphic design, typesetting in the earliest reproductions of printed text) or more difficult graphic communications such as the setting of a magazine with multiple images or a packaging product required to meet legislative guidelines and promote a particular brand.

Artwork design

The graphic design industry is now dominated by computer-aided design or desktop publishing. The current types of software for graphic design are capable of performing most tasks the designer may wish to carry out enabling start to finish artwork creation, including image, type, solid colour and halftone or vignette manipulation. The computer-based nature enables the designer to immediately view the artwork on screen and the effect of alterations made to it. It is vital that graphic designers have at least a basic knowledge of print and the reproducible aesthetics of printing technologies so that designs can be readily reproduced with minimal additional work, making the most of the printing method intended to be used.

Computer-aided design (CAD), computer-aided manufacture (CAM)

CAD is typically used in the packaging industry to design the shape and style of a pack. The design of packaging construction is often used as a unique selling tool within the packaging industry. However, the use of CAD is not only a selling tool for novel packaging designs but is also used to create layouts/repeats for packaging manufacturers ensuring minimal waste. An example of CAM in the packaging industry would be the suitability and performance assessment of a specific design/layout/repeat manufactured on a small scale (e.g. plotting table) before large-scale manufacture.

19.8.2 Reprographics and pre-press

The term reprographics (repro) essentially refers to the reproduction of graphics in some form or another and pre-press refers to the work required for an artwork to be printable including platemaking. In terms of packaging printing this means the conversion of the graphic design artwork file to a file that is able to be reproduced by the chosen printing method. Each printing method has different requirements based on its individual properties and limitations. It is vital for the reprographics operations to be run in close communication with the print production, since the final product is highly dependent on not only the design itself but the way it has been set to print. The processes of graphic design and reprographics tend to be sold as a service to packaging manufacturers where in-house capabilities are not available and to packaging buyers seeking consistency across multiple items of a particular brand. The following sections show examples of typical repro operations.

Colour separations

Colour separation involves the creation of individual printed channels or separations from an artwork file and assigning the relevant parts of the artwork to print in the relevant areas in each chosen colour. In terms of artwork conversion, separation into process colours is now done at the click of a button using an artwork design or specific reprographics software. However, it is not always advisable to simply separate artworks into four-colour process or hexachrome images since the complexity of the resultant separations may be difficult to print. The introduction of special colours or additional colour separations is carried out at this stage and can reduce the printing difficulty and/or improve the overall printed effect.

Trapping (choking, spreading, gripping)

The trapping of two or more colours refers to the overprint (or underprint) at the interface between the two (or more) colours in the artwork. It is common to apply a degree of trapping taking into account the particular printing process to avoid gaps or remove overlap which may occur with misregistration on the printing press. Figure 19.9 shows a typical example of grip applied to two colours to prevent gaps

19.9 Trapping of two colours.

showing at an interface between two printing colours. The solid colour applied first has a letter 'T' in reversed out text and the overprinting colour is a slightly larger positive 'T' image. This results in the 'T' having a visibly darker border appearance around it where the two colours overlap. A negative grip could be used i.e. printing a smaller positive type 'T' into the reversed out text resulting in the substrate being visible between the two printing colours. This is often used to increase the clarity of text on some substrates that are difficult to print on such as metallised materials and plastics. Different printing processes require specific trapping rules to be applied.

The term 'kiss fit' is given to an interface when no trapping has been applied. In this situation, the reversed out 'T' and positive type 'T' would be identical in size. This prevents the border of either substrate (negative grip) or overprint (positive grip) from being visible around the text. However, this may be difficult to register in production printing resulting in an aesthetically unpleasing effect.

Dot control

The control of printed dots is also a reprographics function as much as it is a function of print production. The transfer of ink from plate to substrate will result in a tonal value increase (TVI) or more commonly termed dot gain. Essentially this is the increase in percentage dot printed when compared to the percentage dot on the printing plate/ repro artwork file. Each printed colour will exhibit individual properties in terms of printed dot and this increase varies depending on the percentage dot printed. We can split printed dots into three specific regions: highlights (low dot percentages, around 20%), midtones (around 50% tonal values) and shadows (high dot percentage, around 80%) (Nobbs, 2002). The extent of TVI is often measured for a particular printing press during a fingerprint exercise in which a full range of percentage dots (CMYK and/or special colours) are printed under production conditions and analysed to give an understanding of the dot growth that can be expected during production.

TVI can have a profound effect on the final printed product particularly if the design is reliant on full colour images and/or vignettes. For example, if the magenta separation is exhibiting a higher TVI than expected during printing, the image will appear to be redder than expected. This effect should be controlled at the reprographics or platemaking stages by applying compensation to the file based on the expected performance of a given printing press. Control of this phenomenon through reprographics has far-reaching implications particularly for consistency of final product produced on different machines, using different print processes and different materials. ISO 12647-2:2004 (Graphic Technology – Process control for the production of half-tone colour separations, proof and production prints) is a standard for offset lithography used to describe process parameters for printing and the results that should be achieved to ensure conformance (Jones, 2009).

By working to this standard often requiring reprographic alteration to artwork images, it is possible to recreate process coloured images and communicate the requirements to recreate colour images consistently. A disadvantage of this type of artwork control is that it does not fully take into account variable substrates which can have a significant effect, and it is designed only for process printed images, not spot colours. Since spot colours react differently on press it is common now to carry out and provide dynamic TVI results prior to printing. This allows repro to compensate for the variable dot gain effect of special colour in formulations. Reprographics also provide the screen ruling, screening type (AM, FM or hybrid) and the dot shape for a printable artwork all of which are specific to the design and print process being used.

Step and repeat

The process of step and repeat is again built into reprographics software and is vital for the packaging manufacturer producing many thousands of items. Individual artworks are stepped onto a production layout (sheet or web repeat) so that many units can be reproduced in a singe printed impression. The step and repeat function allows the single artwork/repro file to be duplicated into set positions on a layout file (Fig. 19.10). This layout file tends to be provided by the packaging manufacturer to fit the characteristics of the packaging machinery (e.g. repeat length, sheet size) and of the artwork itself, to avoid print defects. Some reprographics programs can take the individual layout and artwork files and construct an onscreen three-dimensional image to view.

Other useful additions

At the reprographics stage for artwork conversion, it is common for the repro operators and/or print providers to add useful print production aids. For example, it is common

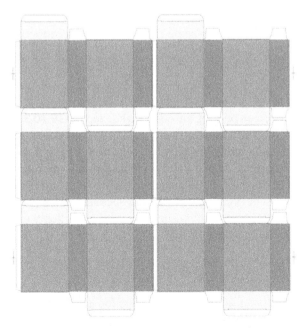

19.10 Step and repeat 6-up including artwork.

to add station numbers (providing traceability between positions on a given layout or repeat), company logo (confirmation of print provider), colour step wedges (used for monitoring TVI on special colours) and registration marks (to aid the printer in achieving registration quickly and consistently).

For some packaging applications, the use of deformation software may also be used at the reprographics stage and designed into the artwork. Take, for example, the shrink sleeve commonly used around plastic bottles. During the application of the sleeve to the bottle it is passed through a heating unit which causes the label/sleeve to shrink and conform to the size and shape of the bottle (Gates, 2002). At the reprographics stage the artwork is distorted to allow for shrinkage based on the bottle shape and expected shrinkage of the substrate. This results in the desired appearance of the artwork around the bottle after processing.

19.8.3 Image transfer

Once a file is ready for print, with the exception of digital non-impact printing techniques, an image carrier, usually a plate, must be manufactured. Each printing process uses different methods of image transfer and thus has a different technique for plate manufacture. Until the relatively recent digital revolution, platemaking was a labour intensive process. The development of CTP processes has improved image quality, speed of processing (thus reducing press downtime), flexibility for fast amendments and reduction in platemaking costs.

Once an artwork is ready for a digital platemaking process, the data/file is passed to a RIP (raster image processor) which transforms the various aspects of the file (text, screening, colour separations) into a format that the following plate production process can interpret. This file is then sent directly to the plate production output device. The plate production method for some of these devices is described in the following sections.

Computer to screen

The transfer technologies for screen printing are described in Section 19.10.3. Computer to screen technology works in the following way. Initially a full coverage layer of emulsion coating is applied to the screen (usually by inkjet printing) to block its mesh. Following this the image is printed, again usually by inkjet onto the relevant areas of the screen. Then the screen is exposed to an active light source and the unprinted areas of the screen are cured. The areas coated with the second printed image remain unexposed and the emulsion layer is then washed out with water leaving the mesh in the image area open and able to transfer ink.

Computer to gravure cylinder

The gravure printing process is an intaglio process (Section 19.10.3) and the image is transferred from an engraved cylinder. The image areas of the gravure cylinder are mostly mechanically engraved using a stylus and the engraving process is controlled

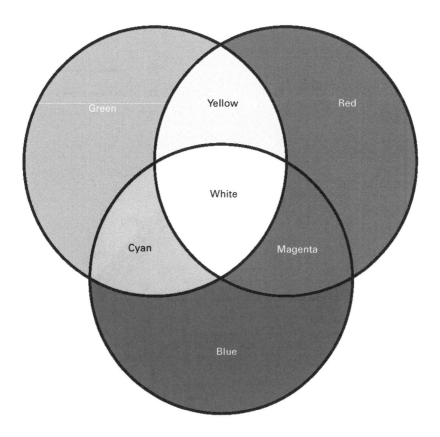

Plate I Additive colour mixing.

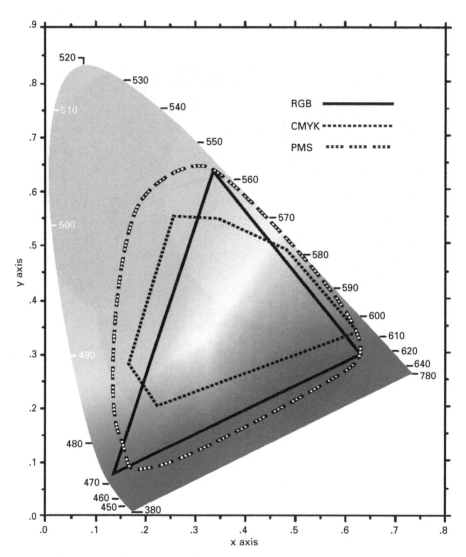

Plate II Chromaticity diagram (including approximate RGB, CMYK and PMS colour gamuts).

digitally. Direct laser engraving of gravure cylinders is also available, although it is less common.

Computer to relief plate

The relief printing processes of letterpress and flexography are described in Section 19.10.3. Although traditionally letterpress plates were metal and though they are still hard in comparison to flexographic plates, they are both now commonly manufactured from photopolymer materials. So called 'conventional' relief plate manufacture requires exposure through a digitally produced film from the artwork file. Digital relief platemaking removes the requirement for this film by having an ablation layer (black heat sensitive coating) applied to the surface of the photopolymer. A digitally controlled laser removes this layer in the image areas of the plate and the plate is then subjected to UV radiation to cure the exposed areas of the photopolymer material. The uncured photopolymer is removed, commonly using solvent wash although water wash and abrasion removal systems can be used, leaving raised image areas. The plate is usually finished with drying and a UV post exposure.

Computer to off-set lithographic plate

There are a number of different processes for digital production of lithographic printing plates, although the process by which they are imaged is similar. The basic process is one of digital exposure to a light or heat source. The plate itself is coated with a material reactive to the imaging head (laser, light/heat source) and the plate is developed and non-image areas are removed. The plate may be baked to increase its life on press although non-bake technology is now common and the plate is usually coated with a water-based gum or coating to protect the non-image areas from oxidation, which could result in them accepting/transferring ink on press.

19.9 Proofing options and approval processes

Proofing is a method of checking aspects of the artwork before going to press via single or small-scale reproduction of an artwork. Proofing can be used to provide a range of checks from relatively simple checks on copy and position to full colour reproduction simulation of the likely results from the print production process. When considering the legislative requirements and brand communication implications of packaging, proofing is an important part of the artwork conversion process. Examples of different proofing options and their uses are demonstrated in the following sections.

19.9.1 Digital proofing

Digital reproduction of an artwork is now a common tool used to assess artworks for the required qualities. The methods of digital proofing vary with each system in terms of the printing method and the properties the software capabilities can provide. The printing methods are the non-impact methods inkjet printing and electrophotography printing (see pp. 476–478). Relatively simple assessments of artwork can quickly be made by reproduction on desktop/low quality digital printers in terms of copy, content and aesthetic impact, although the colour reproduced is not accurate nor the dot structure used in most printing methods. The reproduction of artwork onto transparent film is also commonly used as a copy check tool to lay over an existing/altered file and even a production printed sample or product.

Application specific high quality inkjet or eletrophotographic proof printers are now common. The inclusion of software that can handle printing press profiles to build in production TVI and PMS colour control (including special colours) mean that digital proofing is now a popular method of assessment of artwork. The ease of production and low cost compared to traditional wet proofing methods mean that it has gained much popularity. Development of digital print proofing and its accuracy in terms of reproduction matching that of specific processes and machines is ongoing. Digital proofing is likely to be the dominant proofing assessment method of the future.

19.9.2 Wet proofing

Wet proofing is the traditional method of pre-approving an artwork and is used as a matching tool for a packaging converter to work with. Essentially wet proofing is a small-scale version of the printing process being used for full-scale manufacture. Wet proofing is available for all of the major print processes (lithography, flexography, gravure, screen). The advantage is that the actual components used in full-scale reproduction can be used to make the proof (substrate, inks, varnishes and plates) giving an accurate representation of what is achievable. It is popular with printers since the reproduction exactly replicates the dot structure and any variation from the proof on press can be spotted more easily. However, this proofing system relies on the proofing operator to match the supplied properties of the production press as any deviation from this will result in a difference between the printed product and the proof.

19.9.3 Sign-off and approval

If a proof has been made accurately (to the specific properties of the actual printing press) with either wet proof or digital proofing technology, then we can be confident that the actual printed product will be an exact or close representation of the proof supplied. This is particularly the case for wet proofs if relevant TVI information and actual production material have been supplied and adhered to. Given the importance of brand communication, it is not uncommon for representatives of the brand to attend a press approval or press pass at which sign off is achieved by approval of the actual printed package during production. If the requirements are not met, it also allows the opportunity for fine tuning artwork/colours to match expectations. In some instances the brand owners may prefer to proof their artwork on the production printing press although this is rare as it is a high cost option. Most packaging manufacturers will have procedures set in place for approval processes from the very first stage through to product manufacture and these should include common customer requirements relevant to their product.

19.10 Technological aspects of printing processes

The major print processes can be characterised in many ways; each process has a unique mechanism with advantages/disadvantages associated to its use. All print processes are essentially a method of communication via image transfer. Characterisation of these processes may be by their market dominance or common features such as ink type, drying method and press format. The processes in the following sections are described according to their mechanism of image transfer; however, other general characteristics and properties are illustrated first.

19.10.1 Ink types and drying

In any printing process, it is the purpose of the ink to impart the information by producing an image on a substrate. The desired properties of inks are as follows:

- They must be controllable during the application by the printing process both on the press and on contact with the substrate.
- They must dry or 'set' at a rate commensurate with high speed printing.
- They must convey the information as a thin film or halftone dots.
- They must be suited to a large range of substrates including paper, board, plastics, ceramics and metals (Thompson, 1998).

A traditional definition of an ink is a combination of pigment/dye, binder precursor and/or film former and/or curable monomer and solvent, assembled in such a way as to produce a near homogeneous composition (Guthrie and Lin, 1994). Inks are formulated with the requirements of the printing process in mind and they can be classified in terms of their application. Table 19.2 shows the general properties of inks designed for each print process (Leach and Pierce, 1988).

Two other useful classifications of inks are the nature of the ink formulation and mechanism of drying which can be dealt with simultaneously. The two dominant ink types by formulation are examined here and their methods of drying described. Firstly, there are so-called 'conventional' solvent-based inks in which the pigment, film forming and additive elements are carried by the solvent as a usable ink. The solvents tend to be organic solvents, such as petroleum distillates, alcohols, oils and resins or water in water-based formulations; they are an important component since they directly affect the properties of the ink in printing and the final film formed. The method of drying for these types of inks is by evaporation of the solvent from the ink film, although absorption and oxidation effects contribute also. This method of drying is shown in Fig. 19.11.

Secondly, in relatively recent times the printing industry has embraced ultraviolet (UV) drying technology. Although these inks are more costly and have certain health and safety implications, such as handling and product taint from residual monomers/photoinitiators, they are widely used throughout the industry. The main

Table 19.2 Properties of inks for various printing processes

Printing inks			
Printing process	Pigment (%)	Viscosity (Pa s)	Ink film thickness (µm)
Letterpress	20–30	50–150 (relatively high viscosity or paste inks)	0.5–1.5
Lithography	20–30	40–100 (relatively high viscosity or paste inks)	0.5–1.5
Gravure	10–30	0.05–0.2 (relatively low viscosity or liquid inks)	5–8
Flexography	10–40	0.05–0.5 (relatively low viscosity or liquid inks)	0.08–2.5
Screen	Highly variable	Dependent on film thickness and fineness of mesh	Up to 12
Ink-jet	1–5	0.05–20 (mPa s) (relatively low viscosity or liquid inks)	<0.05

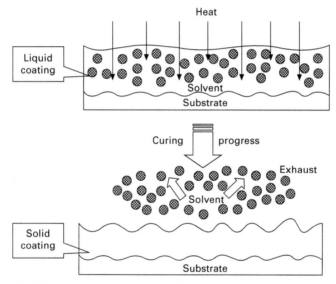

19.11 Conventional drying mechanism.

advantages of these types of inks are that they dry extremely rapidly on exposure to UV radiation, undergo no drying prior to and during printing, and they release no volatile organic compounds (VOCs) into the atmosphere since they do not contain conventional solvents. The inks contain photoactive species which when irradiated with UV radiation initiate a polymerisation chain reaction. Figure 19.12 shows the basic mechanism of drying in which 100% of the printed ink is converted into the solid film. Other methods of drying such as electron beam curing or infrared drying are used in the printing industry and in the case of infrared can be used in conjunction with conventional/evaporative drying and UV curing.

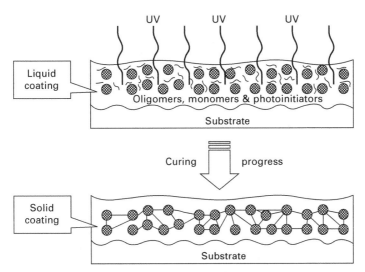

19.12 UV cure mechanism.

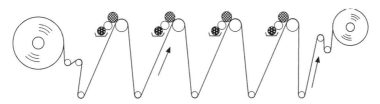

19.13 In-line printing configurations.

19.10.2 Printing press configurations

In general we can say that printing presses are designed in one of three formats. Each basic design is described in this section.

In-line presses

The configuration of this type of design is as expected, with each print unit following the next in a line. Figure 19.13 shows a typical example of this type of press design. The in-line press can run either sheeted material or reels. It is a popular set-up in the packaging market. These types of presses can be set up to print either side of the substrates, often termed 'perfecting'.

Central impression (CI) presses

A printing press with a design of this type has each unit distributed around one central impression cylinder as shown in Fig. 19.14. These types of presses tend to be web-fed. The advantages of CI presses are that they take up considerably less floor space than in-line presses and can hold tight registration as a result of the support from the

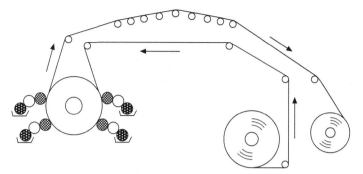

19.14 Central impression (CI) printing configuration.

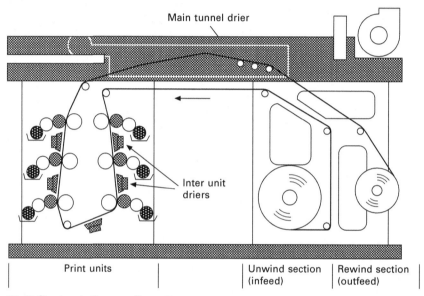

19.15 Stack printing configuration.

central cylinder, which can be of particular advantage when working with deformable substrates such as low density polyethylene (LDPE). However, they can be more expensive as a result of engineering the critical and large single impression cylinder. Also this type of configuration cannot print either side of the web simultaneously.

Stack presses

In stack press configurations the printing units are stacked on top of one another which reduces the floor space required. The space saving advantage permits more colour stations to be built in and, since the press can 'perfect', allows multi-colour printing on both sides of the substrates. A typical example of the stack configuration is shown in Fig. 19.15. Stack presses are reel fed and generally used for multi-colour double-sided work such as newsprint, magazine and catalogue work. In packaging they are popular configurations for flexible film conversion.

19.10.3 Image transfer mechanisms

In this section each of the major print processes is explained in terms of the method of image transfer. In addition, their general applications, advantages/disadvantages and recent developments are also discussed.

Screen printing

Screen printing is a type of stencil printing and a hugely versatile technique. It is used across a wide range of industries for image transfer due to the fact that it can print on a range of different substrates. It is used selectively for both small and large production orders. In this process, ink is transferred to the substrate through a stencil supported by a fine fabric mesh of silk (hence the term silk screen), synthetic fibres or metal threads, stretched tightly across a frame (Takai, 1999). The image areas of this screen or mesh are left open whilst the non-image areas of the screen are masked off, resulting in ink transfer through the open image areas onto the substrate (Peyskens, 1989).

Screen printing is carried out using either a flat-bed or a rotary system depending on the application. Flat-bed assemblies allow not only the printing of flat substrates such as paper or fabrics, but also of conical objects such as plastic bottles, ceramic mugs, etc. This process is known as the body printing method screen process, shown in Fig. 19.16. The rotary screen method also shown in Fig. 19.16 is generally used for longer print run jobs, typically for labels, wallpapers and textiles. It greatly improves the output efficiency for such products compared to the flat-bed method. The flat-bed method forces ink through the image areas of the screen by a moving squeegee. The screen and substrate remain stationary. The rotary screen method uses a rotating cylinder to pass the substrate into impression with the moveable screen. The squeegee remains stationary throughout.

Screen printing tends to be carried out with conventional drying inks, although UV is not uncommon. The ink film thicknesses screen printing can provide (Table 19.2) require more drying than other printing processes, often slowing the process and/or requiring a separate drying stage which adds costs to ink, energy and time. However, the large film weights have advantages, particularly for special effects such as glittering, metallic lustre, achievable opacity and tactile finishes. Also, good opacity can be obtained and this is especially useful when printing light colours on dark backgrounds, for example.

The biggest asset of screen printing is its ability to print on a wide range of substrates of all shapes and sizes; in fact it can be used to print on almost any item. Examples of screen printed items include: compact disk and DVD images, toys, traffic signs, display panels, electronic circuitry, textiles, textile transfers, ceramics, glass and wallpapers. In terms of packaging, screen printed items tend to be bottles, tubes, aerosols, labels and the large-scale advertisement posters and shelf-ready assemblies seen in shops. The speed of production, which may be considered slow when compared to other print processes, makes it less attractive/cost effective to use in certain areas of the packaging sector particularly for volume reproductions of low

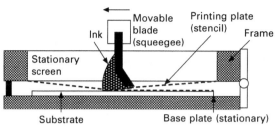

(a) Printing principle: flat-to-flat (flatbed)

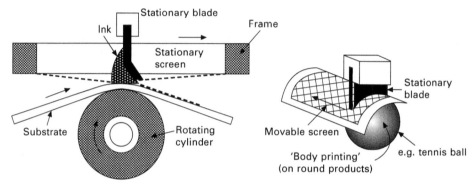

(b) Printing principle: flat-to-round (also for cylindrical products)

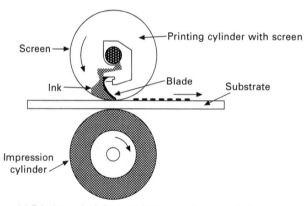

(c) Printing principle: round-to-round (rotary print)

19.16 Flat-bed and rotary screen printing (from Kipphan, 2001).

margin products. In recent years the focus for development in screen printing has been on improvements in the manufacture of the screen itself, namely the development of digital platemaking (see Section 19.8.3). Image quality has also been a focus since the achievable quality compared to other print processes is of relatively low resolution.

Letterpress

Letterpress printing is the oldest print process and is termed a relief printing process. The term relief refers to the fact that the printing is done by means of raised printing elements (image area). Originally the raised images would be set as a rigid solid plate made from metal (alloys of lead or tin) on a flat-bed press. These plates are now commonly made from a photopolymer material. Rotary press assemblies dominate the letterpress market. Letterpress printing requires the use of relatively high pressure, typically between 5 and 15 MPa which must be held uniformly across the relief plates on the substrate. The inks used for printing by the letterpress process are termed 'paste inks', due to their viscous nature. The inks are metered onto the hard plate by a series of metering rollers (Leach and Pierce, 1988). The plate then transfers the ink onto the substrate by impression as shown in Fig. 19.17.

In recent times the letterpress process has lost much of its market to lithography and flexography (based on the letterpress process). In general, it is now used for relatively low quality print such as newspapers, books, directories, and speciality products such as business cards and invitations. However, it is still widely used for some packaging applications particularly for label and self-adhesive label production (Kipphan, 2001), and, most commonly, for printing 3-dimensional objects such as tubes, cans and tubs, where it is known as *dry offset letterpress*. Here, instead of laying the colours down on the substrate sequentially, the image is assembled on a blanket and then transferred in total to the container (Fig. 19.18). Ink coating weights are low (compared with screen printing, for example) and line speeds are high.

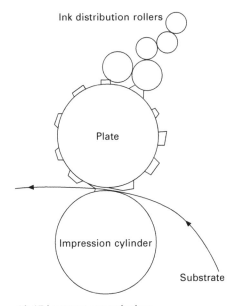

19.17 Letterpress printing.

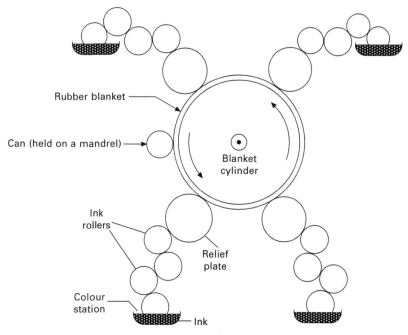

19.18 Dry offset letterpress.

Gravure

It is the simplicity of gravure printing that makes it a practical and favourable process to use. The gravure process is an intaglio process, with the image areas recessed below the non-image areas. This image carrier is known as the 'gravure cylinder'. The image is applied to the metal cylinder by means of an etching or engraving process. This image engraved/etched cylinder is flooded with ink and the excess ink is removed from the non-printing elements of the cylinder by means of a doctor blade assembly. The ink within the cells that form the image is transferred to the substrate via impression as shown in Fig. 19.19 (Leach and Pierce, 1988).

It is the gravure printing cylinder that imparts the qualities, advantages and disadvantages to the process. The high cost of these cylinders tends to promote large volume print with infrequent design changes such as wallpapers, wrapping papers, high quality magazine work as well as some areas of the packaging market. It is not cost effective for short run work. The process is capable of producing high quality image work, with lpi values of 150 as standard and higher values for high quality reproductions. This is a result of developments in the etching/engraving process allowing better tonal control through close control of cell structure and depth.

Gravure printing inks tend to be conventionally dried and are in general solvent based. The drying process is usually by hot air knife which essentially directs a stream of heated air at the printed substrate; conventional oven type drying systems are also used, as are UV curable systems. Typically, gravure printing presses are web fed (although there are some applications for sheet-fed gravure systems) and are

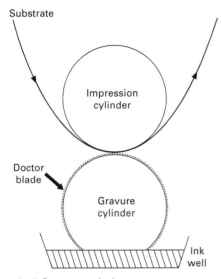

19.19 Gravure printing.

capable of printing on to film, metal foils, papers and board substrates. The versatility in substrates combined with high print quality and consistently reproducible results over long runs are the main advantages for gravure. It is for these reasons it is a popular process in packaging since it is very cost effective. Typical examples of gravure print in the packaging market are: plastic bags, film over-wraps, foils, labels, flexible packaging for food products, tear tapes, cartons and wrapping papers.

As shown in Table 19.2, the ink film thickness for gravure print is high; although this can cause drying problems, it also gives the advantages of vibrant special effects, particularly of high lustre metallic/bronzing and glitter or pearlescent varnishes. Much of the development in gravure in recent times has been concentrated on improving its costs and efficiency. Although the efficiency for high volume is high, it has lost popularity through not being able to compete for shorter run work. Cylinder engraving companies have worked on reducing the cost, time and effort required to produce cylinders (see Section 19.8.3). The main development for machine manufacturers has been the so-called 'wrap tooling' which replaces the traditional cylinder with an engraved thin plate or sleeve for quicker changeovers. Machine manufacturers and printers have also developed more elaborate press specifications allowing multi-colour printing on both sides of a substrate simultaneously and the addition of value adding/finishing processes in-line. Machine design has led manufacturers to build hybrid printing presses which incorporate units of different print process (i.e. flexography) into gravure presses and vice versa. These units can even be interchangeable depending on the printer's requirements. The result is that the benefits of the gravure process can be achieved whilst taking advantage of other process benefits.

Tampo printing has some similarities with gravure, but it is an indirect process. The design is cut/etched into a nylon, steel or photopolymer plate. The plate is inked, and the excess ink wiped off, leaving ink in the design of the image in the recessed areas. A flexible silicone pad then picks up the image from the inked plate and transfers it to the object to be printed. Dependent on its design and shape, this pad can transfer images to compound curved surfaces, smooth or rough, convex or concave, which is a key advantage of Tampo over other printing methods.

Flexography

Flexographic printing (commonly termed flexo) is a process derived from the letterpress process that was described above; it is a relief process. The main differences are the use of a flexible printing plate as the image carrier, the method of application of ink to the plate and the ink itself. In the flexographic printing process the ink is metered onto the surface of the plate by an engraved/etched roller of chrome or ceramic similar to a gravure cylinder but with a uniform distribution of cells (size, shape and depth). This roller is called an 'anilox'. The specification of the anilox determines the volume of ink transferred to the printing plate. The ink is taken into these cells and the excess ink is subsequently removed by a doctor blade assembly. The plate transfers the ink film from the anilox to the substrate by impression. This process is shown in Fig. 19.20.

The flexible printing plate, originally made from rubber, is now commonly made from photopolymeric materials. On early presses with rubber plates and no sophisticated duct control, quality was restricted to line and type work on corrugated board containers, envelopes, bread wrappers and sacks. However, with anilox metered ink feed and photopolymer plates, flexography has become a major process for flexible packaging giving high productivity and quality on filmic substrates. Doctor bladed

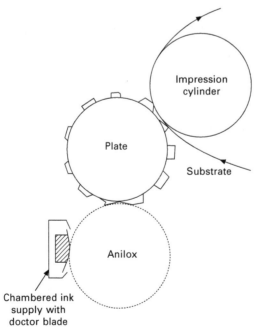

19.20 Flexography.

aniloxes with selected cylinder cell line counts (screen ruling) and cell volumes have raised the print quality beyond all earlier expectations (Gray, 2001). Typical lpi for flexography is now 133 lines although use of lpi values of 150 and higher have been demonstrated.

Flexography is now not only the dominant process for labelling, it is competing with gravure in the flexible market and also with other processes in the leaflets and cartons markets in packaging. Flexographic inks/drying tend to be similar to the gravure process for flexible packaging and are hot air dried solvent-based formulations. Water-based inks have gained in popularity and are widely used in specific markets such as corrugated carton board, plastic sacks and bags. The main advantages water-based formulations have are the improved environmental properties when compared to their solvent-based counterparts through reduction of VOC emissions and the health and safety impact on press operators. Water-based formulations can also be said to be more easily cleaned down and more stable on press due to low evaporation rates. However, the lower evaporation rate of water-based formulations can also be problematic particularly for non-absorbent substrates; also the use of water as solvent can result in reduced wetting of specific substrates. UV inks are also popular in the packaging market, particularly for flexibles, labels and cartons.

The flexographic plate is mounted onto the print cylinder, although it is now common to mount onto a sleeve which sits over the printing cylinder reducing set-up times. This has clear advantages over gravure printing since the cost of plate is much reduced. The most recent developments have been the introduction of digital photopolymer plates (see Section 19.8.3) and continuous sleeve systems. These advantages have won favour against the more established gravure process, particularly in the flexible market, as the plate cost is lower than the gravure cylinder allowing shorter runs and frequent design changes.

The anilox is the control method for ink application onto the plate and substrate. It is the cell volume and geometry that control the final ink film thickness on the substrate. The film thicknesses available in flexography can be advantageous for special effects as with gravure; however, the uniform distribution of cells can cause problems when printing solid colours from the same plate/anilox as halftone screens. As a result, the solid and screen work are often split at the repro stage, and printed using different plates, although plate and screening technologies are slowly removing the need for this (Galton, 2009).

As a general rule, the screen ruling of the anilox should be higher than that of the plate to avoid interference patterns or screen clash. The developments in gravure cylinder manufacture also apply to anilox production and have contributed greatly to the quality of flexography. Sleeved anilox rolls can be also used, reducing the set-up/changeover times further. Chambered ink application and reverse angled doctor blades have made a positive contribution to the control of ink in the flexographic process. Machine manufacturers have also developed gearless printing units allowing infinitely variable printing repeats, again reducing set-up times and costs for expensive machine gears. Flexographic units are now used in conjunction with gravure units in hybrid presses and are also common as coating units for application of varnishes on sheet-fed lithographic presses.

Lithography (off-set)

Lithography (litho) is unlike other printing processes in that it is planographic. By this it is meant that the image areas of the plate are neither raised nor recessed from the non-image areas, and are separated by the mutual repulsion between oil and water. Lithographic printing is currently the most popular of the printing processes worldwide. The success of lithography is largely due to its versatility in different markets and the achievable print quality which is high; typically 150 lpi is used although it is common for up to 200 lpi to be used in art reproductions and high quality magazine work. The term offset refers to the fact that ink is not transferred directly from the plate to the substrate but is offset onto an intermediate carrier or 'blanket', which then transfers the image onto the substrate. Lithography can be web-fed for longer run work such as newspapers, forms and magazines; or sheet-fed for shorter run length magazines, books and cartons and labels in the packaging sector.

The lithographic plate (almost exclusively computer to plate, see Section 19.8.3) has image areas that are oleophilic/hydrophobic (ink accepting/water repelling) and non-image areas which are hydrophilic/oleophobic (water accepting/ink repelling). This is a result of the surface chemistry designed into the plate. The non-image areas are covered with a dampening solution, which is essentially water with additives to lower its surface tension and thus aid more complete wetting of the non-image areas of the printing plate. Ink is then applied to the image areas of the plate via a series of distribution rollers, to ensure consistent film thickness. The areas of greatest application of ink onto these rollers are controlled by the operator and ink is then transferred to the blanket. The mutual repulsion between ink and water keeps the non-image areas free from ink allowing the ink to transfer only from the image areas of the plate. The relationship between the ink and the dampening or fount solution is a complex one. If the surface tension of the dampening solution is not sufficiently low, inadequate wetting of the plate may occur resulting in transfer of ink from non-image areas. Similarly if the surface tension of the dampening solution is too low then emulsification of the ink can occur. The ink is transferred from the printing plate onto the blanket and finally transferred by impression from the blanket to the substrate. A diagrammatic representation of this process is shown in Fig. 19.21.

The ink film thicknesses of lithographic printing are relatively low (Table 19.2) when compared to those of gravure, flexography and screen printing, limiting its success for printing special effects which rely on large film weights. However, the viscosity of lithographic inks (paste inks) allows for a high pigment loading which means the colour strength can be considered good for colour reproduction and hiding power on most substrates. The ink drying mechanisms for lithography tend to be either conventional solvent-based or UV curable. For web-fed applications drying time can be built into the process to allow for conventional drying inks to fully cure before the following processes. However, in sheet-fed applications the use of spray powder assemblies is common. A fine powder is sprayed onto the sheets as they are delivered into a stack. The powder effectively acts as a spacer between the sheets allowing air to oxidise the wet ink film and prevent the sheets from blocking (sticking together) in the stack. UV curable inks have gained huge popularity in the

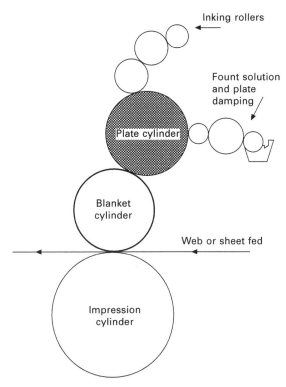

19.21 Lithography.

lithographic format due to the versatility within a multi-colour press, efficiency and improvement in the printing environment without the need for spray powder.

In terms of technology of printing presses, lithography has led the way such that lithographic printing presses are the most technologically advanced in many ways. Machine manufacturers have concentrated on improving the quality of the printed product achievable, the set-up times and the workflow/control technology. Advancements have been such that a good proportion of the processes required for changing from one job to another are fully automated (plate changes, size setting blanket and roller washes). Repeat job data can be stored on press or delivered to press from pre-press to improve set-up times. Colour can now be measured using spectrophotometry on printed colour control strips and automatic adjustments made in real time. This technology lends itself to sheet-fed lithography; although it is certainly used in the other processes, it is not generally an automatic feature of the presses. Machine configurations have also become more advanced with hybrid presses containing flexographic coating stations added particularly for overvarnishing. In fact the most recent additions to lithographic machinery have been cold foil blocking units and thin plate rotary cutting units.

Waterless lithography (dry offset) is another relatively recent technology which negates the use of a dampening solution. The technology is built into the printing plates. The non-image area of the plate has an oleophobic coating to repel the ink to replace the need for the dampening solution applied in standard offset. Its popularity has been limited due to the increased cost of the plates and their susceptibility to wear. Also the inks used can create problems since each ink is uniquely temperature dependent, requiring the press operator to closely control the printing temperature for each unit.

The other major development in the chemistry of lithography is the removal of alcohol (isopropanol, IPA) from the fount solution. IPA has been used in the fount solutions as an additive for the water, to reduce the surface tension and achieve better wetting of the plate. However, it is expensive and a VOC, therefore its removal has cost and environmental benefits. Fount chemistry is such now that IPA can be removed; in fact it is banned in some areas. Its removal can result in more vibrant colour reproductions since it acts as a solvent to inks used and removes colour. Also, it requires tight control of all aspects of the press to be successful.

Non-impact printing techniques

Non-impact technologies do not require an image carrier (plate) as do the more conventional methods of printing covered so far. This unique property allows the design to be changed simply and at any point. The current dominant technologies, electrophotographic printing and inkjet printing, are described in this section.

The process of electrophotography is made up of five simple steps as shown in Fig. 19.22. Initially, the image is created using a light source that impinges onto a suitable conductive surface by altering the surface's charge distribution. An appropriately charged powder or liquid toner ink is attracted to this charged image surface and the ink is then transferred to the substrate via a further charge applied in the printing nip. The toner is then fixed to the substrate generally with heat and pressure. Finally, the

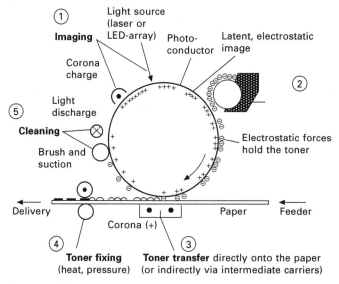

19.22 Electrophotography or xerography (from Kipphan, 2001).

image is cleaned from the photoconductive surface. The surface is now ready to be charged with the next image (Davidson, 1995). This process is familiar to all in the form of modern photocopiers and desktop laser printers and is sometimes known as xerography.

For inkjet technology the ink is transferred directly from an ink well to the substrate. This process is controlled digitally and can be classified into two main categories: continuous technology and drop on demand (DOD) technology. Figure 19.23 describes continuous inkjet technology; the main feature is that pumping ink through a nozzle on which a piezo crystal is constantly vibrating generates a continuous stream of ink droplets. An electric charge is applied to the droplets to enable their deflection onto the substrate. Ink droplets that are not deflected onto the substrate are collected and returned to the reservoir (PIRA International, 2009).

In DOD-type inkjet systems, a drop is produced as and when required on the substrate. In piezo-type inkjet print heads, the ink well has a piezo crystal wall that, when electronically charged, deforms and forces a droplet from the nozzle. Such a system is shown in Fig. 19.24. In thermal bubble-jet-type print heads, a localised area within the ink well is super heated, causing vaporisation of a section of ink. This vaporisation causes a change in pressure within the chamber and results in the expulsion of a droplet of ink. A schematic of a thermal bubble-jet printer is shown in Fig. 19.24. Either the superheating or charging of the piezo crystal is controlled digitally according to the image data that needs to be realised. Thermal bubble-jet printers tend to be restricted to desktop applications with piezo systems more popular for industrial printing use.

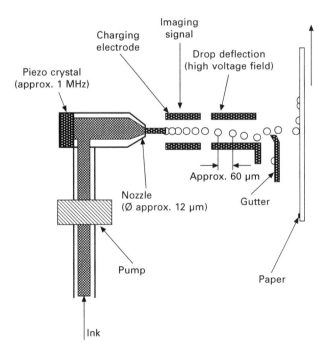

19.23 Continuous inkjet (from Kipphan, 2001).

478 Packaging technology

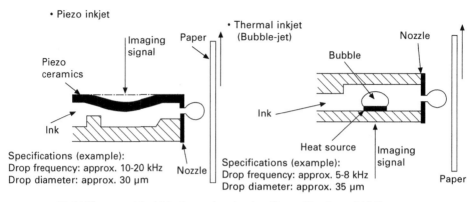

19.24 Piezo and bubble-jet technologies (from Kipphan, 2001).

The use of digital print systems for industrial print applications (outside proofing applications in Section 19.9.1) is a fast growing sector of the print industry. However, certainly in the packaging market, it has not made the impact initially expected. Digital packaging printing is expected to grow exponentially over the next 10 years; in fact the percentage growth between 2003 and 2007 in the packaging and label markets has been around 80–90% (Springford, 2007). It can be said that inkjet has been successful in making significant inroads into markets that were traditionally dominated by screen printing (printed textiles, ceramics, glass, large-scale advertisements and shelf-ready packaging) but it has failed to do so in others. The main reasons for this failure are the many requirements for specific packaging applications requiring tailored inkjet solutions. The speed of printing is considered slow compared to conventional processes, particularly for high volume work, and digital print is limited in its ability to produce large single colour areas with high colour density.

However, digital printing does have distinct advantages that with technological advancements will enable it to compete on a much greater scale in all markets. In recent times the packaging industry has seen two particular trends that suit digital printing systems. Firstly, the number of design changes and relaunches for brands is increasing in highly competitive markets; as the cost of pre-press for digital is vastly lower than other processes, it is well suited. Secondly, the reduction in batch sizes in the packaging market following just-in-time (JIT) principles also suits digital processes which can convert small run lengths quite effectively (Slembrouk, 2007). The recent developments in digital printing are clearly many, since it promises so much and is still in its infancy when compared to established processes. In terms of packaging the most significant developments are its use for individual coding of products often as an in-line process, its expansion into the flexible and carton markets, developments in the technology of the printing head allowing a vast array of ink systems to be printed and the technology designed into the inks.

19.10.4 Typical print defects

Although the printing processes discussed are markedly different, similarities exist in the type of defects that can occur, even if the cause of the defect may differ. Typical print defects are discussed in the following section, although the defects listed and typical causes are not exhaustive.

Colour variation

Colour variation is a common print fault particularly with lithographic printing. In general, flexographic and gravure have relatively stable colour since if the ink and print cylinder or anilox are in good condition, the repeatable print result should be the same; however, even though the colour consistency is more simply controlled, batch variation can be an issue as can process variation if close attention to the press is not taken by the print operator. Colour variation in lithography is typically a product of poor ink/water balance. If the balance between ink and water is not in a state of equilibrium, the tendency is for the colour (ink) and film weight to be unstable and since the film is only very thin, any minor changes can result in visible colour difference.

Hickies/splashing

'Hickies' is the industry term for dust or debris interfering with the ink transfer resulting in a small imperfection in the print (a spot with a halo around it). It is common in all printing processes and is particularly prevalent in carton production with coated boards which have inherent dust particles near or in the print surface. Other imperfections of this type may be caused by splashing of solvent, water or oil from the press itself; these are generally spots without a halo and the shape of the spot can often determine the source of the splash.

Misregistration

Misregistration is an incorrect arrangement of colours in relation to each other. It is a common fault with all processes. Areas of misalignment of colours can cause images/text to appear blurry with edges of visible single colours. The advantages of FM screening to hide misregistration are described in Section 19.7.4. There are several possible causes of misregistration, particularly in reel or web-fed applications, such as incorrect printing pressures, variable web tension, and poor plate mounting.

Scumming

Scumming tends to be an issue with gravure printing in particular and also lithographic printing. Scumming in both processes can be described as ink transfer in non-image areas, usually a patchy washed out appearance of a specific printed colour. In gravure printing the effect is a result of ink being held in the non-image sections of the cylinder and not being removed by the doctor blade. Cylinder manufacturers have done a great deal of work to reduce the likelihood of scumming at cylinder manufacture, and ink companies try to achieve the correct formulations to reduce this effect. Ink control, cylinder type and printing unit set-up are key to minimising the likelihood of scumming in gravure printing. In lithography it is an imbalance in the ink/fount relationship that is the cause of scumming. Insufficient wetting of the plate (too little fount or fount with incorrect surface tension) tends to be the cause.

Bleeding and feathering

Bleeding and feathering are two different defects, although they can appear similar on the printed product. Both phenomena appear as areas of ink transfer close to or part of a printed area. Bleeding tends to be associated with flexography and gravure printing systems and is generally a result of insufficient drying, ink film weight being too high and other ink faults such as wrong pigment choice and incorrect viscosity. Feathering is common in gravure and flexography and can generally be attributed to ink drying on the plate/cylinder prior to ink transfer, pressure settings and ink formulation issues. In lithographic printing, feathering is usually a result of ink/fount imbalance and is common with strong colour shade requiring a high level of ink.

Ghosting, repeating

Ghosting is described as a faint printed image or reduced transfer image which appears where not intended in a design. It is a defect that can occur in all of the major printing processes. It can generally be attributed to issues with the design or layout of the design, particularly in lithography. In lithography it may be a result of the blanket being embossed with a previous design and a similar effect can occur in the mesh for screen printing. With flexography and gravure processes the problem tends to be a mechanical issue (e.g. gear wear), doctor blade set-up or ink problem.

Screen clash

Screen clash is an interference pattern resulting from inappropriate screen angles being chosen in printed images or areas with vignettes running into one another. It can affect any of the printing processes which run conventional screening and is described in more detail in Section 19.7.2.

Dot gain or tonal value increase

Dot gain, as it is commonly known, is a deviation from the expected increase in tonal values for halftone printed dots. The process of fingerprinting presses for pre-press operations and proofing accuracy is described in Section 19.8.2. Deviation from the expected values can result in a reduction of detail and variation in expected colour and contrast. The effect is commonly a result of ink type and ink/fount balance in lithographic printing, although pressure, substrate and drying can also be potential causes. For flexographic processes the causes tend to be the plate, or mounting tape and pressure settings. Dot gain is less of an issue for gravure printing and tends to be a result of ink control. Errors in fingerprinting, variable unmeasured gains for special colours and errors in pre-press can also be a cause of variable dot gain.

19.11 Other processing techniques

Aside from the basic image transfer techniques described in Section 19.10.3, there are other processes often used in-line or as off-line processes to add further value and impact to the finished product. The mechanisms and uses of these processes are studied in the following sections.

19.11.1 Varnishing/lacquering

A varnish or lacquer can be described as a transparent coating applied to a printed product. Overvarnishing of printed products is a common process in the packaging industry and is carried out for a number of different reasons:

- to protect the printed product beneath, preventing rub off, abrasion marking, imparting colour protection (lightfastness) or provide a barrier to external forces and internal contact
- to aesthetically improve the appearance or add decorative effect to the product
- to provide required properties to the product for subsequent processing and its end use.

In the packaging industries varnishes tend to be applied as a liquid coating via gravure or flexographic processes. The protection properties offered are not only useful for the end user but also help to speed up production processes by reducing waiting time between handling.

Varnishes are formulated with specific properties in mind, whether they are visual or process requirements. Visual effects can simply be the level of gloss imparted on the product. A high gloss finish with incident light boundary reflection in a narrow set of directions imparts a smooth mirror-like surface and glossy appearance. In contrast, a matt finish reflects incident light in many directions. Varnishes can be formulated to have specific gloss levels. It should be noted that the gloss level can affect the appearance of colour (matt finishes tend to dilute the colour sensation) and that combinations of different levels of gloss can provide contrast in the finished product.

Other aesthetic improvements provided by varnishes are the so-called 'special effect', varnishes described in Section 19.7.4, such as pearlescent pigmented coatings. Security technologies have also been built into varnishes for anti-counterfeit purposes and tactile varnishes are now common to stimulate the sense of touch of the end user.

Property modifying varnishes may control simple processing characteristics such as slip, yet they may also be used to impart other product-specific properties such as mould inhibition in cosmetic packaging, barrier properties in primary packaging (direct contact packaging) and remelt sealing properties in blister packs and flexible packaging closures.

19.11.2 Foil blocking

Foil blocking is a decorative process used to transfer high lustre, mirror-like foil finishes or holographic images to a printed product. These transfer foils are available in a range of colours, and holograms can be designed and applied in the same way. Traditionally, foil blocking is a heated transfer process (Fig. 19.25) and can be either a rotary or flat-bed configuration. The heated die melts the transfer adhesive and releases lacquer in the area of impression, and the foil layer is transferred from its backing film onto the printed product.

Cold foil blocking has been introduced recently and has seen increased popularity particularly in the label market because of its reduced cost and increased speed of transfer compared to its heated counterpart. The general process described in Fig. 19.26 is for a UV curable cold foiling systems, although other drying formats can be used but the web path may be altered. The limitation of cold foiling tends to be that the quality is not as high as the hot foil method, particularly for clean straight edges and fine detail reproduction. However, relatively simple designs can be processed adequately and the production efficiency benefits are considerable. Hot and cold foil blocking processes are popular in packaging for their decorative properties. They are predominantly used in the label and carton markets.

19.11.3 Embossing

Embossing describes the process of raising areas of image/product so that they sit proud of the general surface of the product. Debossing, in which an area is recessed in specific areas, is also common and can be used in conjunction with embossing in the same product to create more enhanced three-dimensional qualities. This process

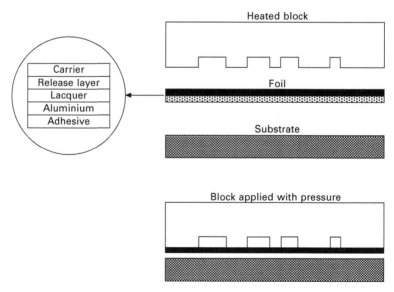

19.25 Hot foil blocking.

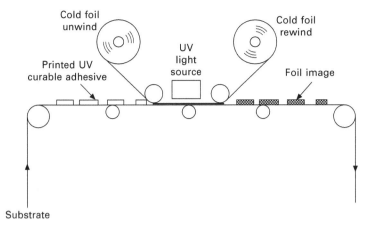

19.26 Cold foil blocking.

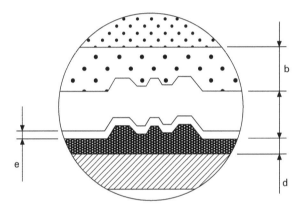

19.27 Embossing (from Bobst, 2002).

is particularly common in the carton and label packaging markets and can also be carried out in conjunction with the foil blocking process. The flat-bed method of embossing/debossing is depicted in Fig. 19.27, where the substrate (e) is depressed between male (b) and female (d) forces resulting in emboss. Rotary embossing is also common and the process is similar.

This method is also used to apply Braille to printed cartons. Embossing effects can also be created using special ink systems or with screen printing due to the large film weights it is capable of applying. Thermal or radiation expanding inks are currently not popular as they have two distinct limitations: the speed of expansion is not fast enough for packaging processing efficiencies, and they are less robust when compared to a mechanical emboss to withstand further processing. The mechanism of these systems is simply that an ink containing an active species which, when heated, expands is printed in the areas required to be embossed. At some time during or after processing, the package is heated or exposed to IR radiation to activate the expansion process. It is expected that this technology will improve to allow competition with the conventional embossing process.

19.11.4 Heat transfer printing and ceramic or glass decal

Heat transfer printing is similar to the hot foil blocking process in that it is a process by which an image is transferred by heat and impression. This method of image transfer has typically been used for the manufacture of special design clothing (e.g. T-shirts). However, more recently it has gained in popularity in the printing of large-scale fabrics, wood and metals. The process itself is a clean one since no liquid inks, solvents or drying powder are used. The increase in popularity is largely due to the improvements in manufacture of the transfers with the advent of digital print. Previously the transfers would be manufactured using the screen process which made short-run work not cost effective. Inkjet and electrophotography printing are now the dominant methods of transfer creation and have all of the benefits expected of the digital process: cost effective independent of run length, tight registration control and frequent design change flexibility.

Ceramic decal or transfer is a term given to the heat transfer process used for applying permanent images to glass and ceramics. The inks used contain reactive pigments which, when heated at high temperatures in contact with the glass or ceramic substrate, permanently bond to it. As with heat transfer prints, inkjet and electrophotographic printing are the print processes of choice for creating decals.

19.11.5 Metallising

Metallising or vacuum metallisation is the deposition of a thin film of pure aluminium (see Chapter 9) onto a substrate by vaporisation under vacuum. It is commonly used in the packaging industry to create the metallic/holographic foils used in hot and cold foil blocking processes and the metallised films applied to carton board to produce metallised/holographic substrates for printing. It can apply a metallic finish to almost any substrate. The process itself is not particularly cost effective since the scale of the process is large and it requires lots of energy to power aluminium vaporisation and vacuum enclosure.

19.12 Quality control in packaging

The importance of quality and quality control in the packaging industry should not be underestimated. The potential consequences not only in terms of lost revenue or additional costs but of consumer safety as a result of poor quality are severe. The following sections describe the requirements for production of a fit-for-purpose product, typical quality control measures and measurement techniques and details of quality control systems and their importance.

19.12.1 Specifying requirements and setting standards

The most significant aspect of setting standards for the packaging manufacturer is the communication of the requirements of the packaging. In order for a fit-for-purpose product to be supplied, every detail of the performance of the product should be investigated and understood for each manufacturing, packing, transit and end use process. In all likelihood the packaging supplier will be familiar with requirements and the relevant test procedures for their product and can both advise on specification and make provision for testing prior to and during manufacture. It is thus important to communicate all required properties for the packaging at an early stage in the development process. Table 19.3 lists some of the testing methods available for packaging products (ASTM, 2009; IOPP, 2009; PIRA, 2009). Certainly there are many other testing procedures which may be required depending on the processes and conditions the packaging is subjected to throughout its life. These tests should be an accurate reflection of performance, reliable, cost effective and, where possible, non-destructive (Stauffer, 2005).

Once the manufactured requirements of the packaging are fulfilled and relevant test procedures indicating a pass/fail response agreed, a standard for all future productions has been set. The manufacturer has the ability to ensure that this standard is upheld and product failure can be limited. Recording test results and maintaining batch samples allow for accurate retrospective analysis should a failure occur at any stage.

It is important for any packaging product that it fulfils the criteria specified and this is not simply a matter of fulfilling the technical requirements of the product's use. The Food Standards Agency (FSA) and Pharmaceutical Quality Group (PQG) are examples of UK bodies overseeing the rules and regulations regarding packaging for their specific markets. The requirement to also satisfy the EU Regulation REACH (Registration, Evaluation, Authorisation and restriction of CHemical substances) is

Table 19.3 Selection of common testing methods for packaging materials

Packaging type	Property	Test	Examples standards
All	Colour/image quality	Visual assessment, densitometry, tonal value increase testing, spectrophotometry	Often trained visual assessment only or numerical measured data comparison
All	Ink film	Tape tests, solvent rub testing, abrasion testing	ASTM F2252 ASTASTM D5264M D4752
All	Slip tolerance	Friction coefficient testing	ASTM D3108
All	Light fastness/ weatherability	Xenon arc lamp testing, daylight exposure tests	ASTM D3424 and 3794
Cartons	Carton assembly strength	Compression testing	ASTM D642, ISO 12048
Cartons	Carton erection	Opening forces for cartons	None specified
Cartons	Heat seal peel test	Seal peel test	ASTM F88
Flexibles	Flexible seal strength	Burst and seal decay testing	ASTM F2054–07 ASTM F1140
Flexibles (and others)	Barrier properties	Oxygen and water transmission rates	ASTM D–3985 and F1249
Labels	Adhesion, peel and repeel	Peel adhesion test	ASTM D3330

a legal one. The requirements set by these groups form standards that can be used to demonstrate compliance with a particular body (e.g. ISO 9001, PS 9000, GMP, PCOP).

19.12.2 Inspection processes

Inspection processes in the print production of packaging manufacturing tend either to be manual sampling and analysis or visual in-line systems for validating quality. The manual sampling technique requires a number of agreed samples to be taken throughout a production run (batch sampling); usually an initial pass must be achieved for the run to commence. The testing of these samples provides a level of confidence that the whole production run meets the required standard.

In-line visual inspection of printed material is not a new technology. It has been used in the packaging manufacturing industry for many years now and allows the quality of a production run to be monitored in real time. There are a number of technologies which have been adapted to printing processes. These are described below.

Basic photo-electric sensors

This type of sensor can be used to detect the presence of an object and is typically used in sheet-fed presses to warn the printer of missing sheets. Similarly, relatively simple contrast scanners are used to detect splices (joins) in reel to reel printing applications allowing the splice affected part to be labelled if it cannot be removed, thus acting as a warning for downstream processes. These types of sensors are also commonly used in registration systems of web-fed printing machines.

Laser sensors and scanning sensors

These more complicated systems can be used to measure incorrectly positioned product (web movement) and product outside specification tolerance (film thickness). They can also be used to scan simple codes such as bar codes. Bar codes and similar types of code are often used as a quality control for distinction of different products. Often they are added to artworks, packaging outers and pallets of product by both the manufacturer and customer to control batch traceability, remove substandard product and prevent mixing of similar products.

Machine vision systems

Simple versions of these types of systems take still images of printing material as it is running with the aid of stroboscopes (strobe lighted image capture) or from freeze frame continuous image capture. These systems allow the printer to monitor the quality during production at running speeds. They are popular in reel to reel applications and are often linked to labelling systems to allow defective areas of print to be marked up clearly so that they can be removed during downstream processing treatments.

Advancements in technology now allow for the more complex systems of optical character recognition/verification. These systems allow comparison of individual measurements against a given standard (Anon., 2009). Capable of spotting a whole host of print defects, these systems are being introduced/retrofitted to print machinery. They have gained popularity particularly in sheet-fed application with double delivery systems which allow product affected by defects to be segregated from good quality product whilst in full production in a separate delivery section of a press. However, the sensitivity of these systems may potentially add to increase wastage and labour costs for sorting acceptable defect levels from unacceptable defect levels.

19.12.3 Tracing, packing and tracking

The issue of traceability in packaging is clearly of great importance, particularly the requirement of batch traceability for food/beverage and pharmaceutical packaging. This may be performed in a number of different ways and the packaging user should have a procedure set in place to manage the traceability of their product and its packaging. Much of this can be carried out at the artwork stage of the packaging design by utilisation of individual item/product identity information that may be printed on the packaging at manufacture, e.g. product-specific barcoding, in-house item coding, formulation coding or simple printed logos. As such it is vital for packaging manufacturers to be able to manage the requirements for batch identification for their customer, the packaging buyer.

The traceability of individual batches of packaging supplied is generally also managed by similar methods depending on the customer requirements and/or the capabilities of the packaging supplier. This again may be relatively simple coding such as a bar code, date or job number via labels on outers containing a packaging product (cartons, corrugates) or labels attached to bulk packaging items themselves (label/flexible packaging reels). The relatively recent technology of radio frequency identification (RFID), which is essentially a tag that can communicate with a scanning antenna, has been used to great effect to monitor large-scale items in the packaging industry such as whole deliveries of product. In fact, this is more widespread and many products are being monitored using RFID, in particular individual products of high value such as cosmetics (perfumes) and multimedia products (CD, DVD). As a result, much research has gone into actually printing the receiving antennae of RFID tags directly onto packaging and reducing the cost of the tags themselves. It is expected in the near future that many commonplace items will include RFID technology.

In Section 19.12 fit for purpose and achieving standard requirements were discussed. The same principles also apply to the packing specifications of packaging materials, whether built into the packaging product or part of the delivery specification. It is important for the packaging buyer to specify requirements for packing and delivery of the packaging product. Also of importance is good stock management of packaging materials since packaging products may also exhibit deterioration in performance over a period of time. For example, the performance of carton board packaging assemblies tends to show a decrease in erection performance if unused for extended periods and adhesion in any glued components may be reduced or lost. Similarly self-adhesive label packaging may also show a reduction in adhesive performance over a period of time. The packaging manufacturer will almost certainly be able to provide a guide as to the effectiveness of their product over extended periods of time and are likely to specify a period of time in which they guarantee their product performance.

19.13 References

Anon. (2009) 'Automated Inspection Systems: A Brief History and Introduction', Next Generation Pharmaceutical (NGP) Summit Proceedings, available at: www.ngpsummit.eu.com

Antonine Education, 'What are the uses and hazards of waves that form the electromagnetic spectrum', available at: www.antonine-education.co.uk

ASTM (2009) *American Standards Worldwide*, available at: www.astm.org

Bobst (2002) *Autoplatine SP: Converting Tools and Production Guide*, Bobst SA.

Braden Sutphin Ink Company (2009) 'Stochastic screening – What is it and why is it such a hot issue ... today ... and tomorrow?', available at: www.bsink.com/tech/stochasticscreening

Christie RM (1993) 'Pigments: Structures and synthetic procedures', *Surface Coatings Reviews*, Oil and Colour Chemists Association.

Coudray M (2007) 'Understanding the behaviour of midtone dot gain', *Screen Printing Magazine*.

Davey A (1999) 'Printing inks for packaging applications', *Colour Science, Volume 4, Colour and Image Creation: Inks, paints and packaging materials*, Department of Colour Chemistry, University of Leeds.

Davidson JW (1995) 'Non-impact printing processes', *Surface Coatings International*, 5, 196–202.

DTP Tutorials (2008) 'Halftone Screens', www.dtp-aus.com.

Editure Education Technology, 'The human eye – structure and function', www.schools.net.au.

Galton D (2009) Discussions with D. Galton, Asahi Photochemical Products, April.

Gates GR (2002) 'Shrink sleeve perfection', *Flexo Magazine*, September, available at: www.flexography.org

Gilchrist A (2001) 'Characterising Special Colour Effects', Department of Colour Chemistry, University of Leeds, OCCA SURCON 2001 Chemistry and Application of Colour.

Gray P (2001) 'Ink Delivery Systems', Coates Lorrileux, OCCA SURCON 2001 Chemistry and Application of Colour.

Guthrie JT and Lin L (1994) 'Physical-chemical aspects of pigment applications', *Surface Coatings Review*, Oil and Colour Chemists Association.

IOPP (2009) *Package Label Qualification*, IOPP Medica Device Technical Committee, available at: www.iopp.org

Jones P (2009) 'An Investigation of ISO 12647-2 and the Characterisation of Tonal Value Increase of Special Colours', Final Year Project, IMPRESS, Department of Colour Chemistry, University of Leeds.

Kinyo (2004) 'Understanding pre-press: FM or stochastic screening', Technical Support Bulletin No. 26, January 2004.

Kipphan H (2001) *Handbook of Print Media: Technologies and Production Methods*. Springer Verlag, Berlin.

Leach RH and Pierce RH (1988) *The Printing Ink Manual*, Blueprint, London.

Metal FX® Technology Ltd website, http://www.metal-fx.com

Nobbs J (1998) 'Colour Assessment for Paints', COLO 5171 Paint Formulation and Characterisation. Department of Colour Chemistry, University of Leeds.

Nobbs J (2002) 'Fundamentals of Colour Science', COLO1241 IMPRESS Course. Department of Colour Chemistry, University of Leeds.

Peyskens A (1989) *The Technical Fundamentals of Screen Making*, SAATI S.p.A, Como, Italy.
PIRA (2009) *Materials and Primary Packaging: Packaging Testing*, available at: www.pira-testing.com
PIRA International (2009) 'Developments in Ink-jet Print Head Technology', available at: www.smitherspira.com
Prince R (2009) 'Stochastic has advantages but it's not for every print job', *NAPL Business Review*, 20. Available at: www.napl.org
Reid D (2008) 'Hexachrome print process: primary considerations for implementing hexachrome printing', *Digital Output*, August, available at: www.digitaloutput.net
Slembrouk S (2007) 'Challenges for Ink-jet printing in Packaging', www.profitthoughinnnovation.com
Springford C (2007) 'Future of Short-run Colour Printing', available at: www.profitthoughinnnovation.com
Stauffer T (2005) 'Package Testing: A Q&A with Tony Stauffer, President of PTI', *Packaging Technologies and Inspection*, available at: www.packagingdigest.com
Takai M (1999) US Patent 5,906,158.
Thompson B (1998) *Printing Materials: Science and Technology*, PIRA, Leatherhead.

20
Packaging machinery and line operations

G. CROMPTON, The Packaging Society, UK

Abstract: This chapter gives an introduction to help build the basic understanding of the principles of operation of the main types of packaging machine. The design and operation of a packaging filling line is a complex task, calling for a multidisciplinary team, within which each will require a broad understanding, in addition to their own specialist field. It is inevitable that the study of this subject will necessitate reference to the material-specific chapters throughout this text. In addition, reference to Chapters 4 and 21 is highly recommended.

Key words: packing, filling, line equipment and design, unscrambling, filling, counting, capping, labelling, cartonning.

20.1 Introduction

The packaging line is an application of materials technology and production engineering and developments in both disciplines have contributed to innovations in packaging. It is now commonplace to see flexible wrappers and bags with tear strips or laser cuts to make them easy for the consumer to open or ingenious tamper-evident devices to give the brand owner and the consumer assurance about the integrity of the product. Complete packaging lines are now operating where bottles are moulded from resin, filled, sealed, labelled, cartonned and palletised with a high level of automatic control and inspection. Such developments have brought the need for a very wide range of packaging machinery. This means that it is impracticable to explain them all in great detail. The aim of this chapter is to give an insight into the basic operating principles and to act as a springboard from which further information may be gathered.

20.2 The packaging line

A typical sequence of activities of a packaging line is to:

- bring the packaging components from the packaging warehouse
- place the containers onto a conveyor belt which will move them through the packaging line in the correct orientation
- clean containers (e.g. if filling with food products)
- fill the product into the containers
- close the containers
- if required at this stage, check product has been filled accurately and safely
- apply suitable identification such as a label
- apply any additional packaging (e.g. a carton with a leaflet)

- collate the primary packs into secondary packs and palletise the secondary packs for transfer to the finished goods warehouse.

The following sections discuss types of packaging line layout and some of the common technologies used in these operations.

20.2.1 Types of layout for packaging lines

The two most common configurations for a packaging line are straight-line or linear and rotary layouts. Since they depend on each operation happening in sequence, straight-line layouts have an intermittent operation, depending on the complexity of a particular operation in the line. Filling operations are often the slowest operation in the line since they require each container to be positioned for a period under a filling head for filling to occur before the container can move to the next stage. Multiple filling heads may be used and some filling machines move the filling heads with the container to allow more continuous filling as the containers move along the packaging line. More complex machines use a dual conveyor belt system with fill heads for each line operating in sequence to provide a smoother flow of filled product. In practice, to make the most of the available space, this kind of layout may have a 'U' or 'S' shape. Conveyor belt systems take product down one section of the packaging line and then loop round for the next section.

One way of minimising delays in straight-line packaging lines is to use a rotary layout. In these systems product is fed out of the main packaging line into a rotating circle (called a turret) where a more complex operation such as filling can take place before the product is fed back into the main packaging line (Fig. 20.1). The advantage of this configuration is that the turret can process more units at a slower

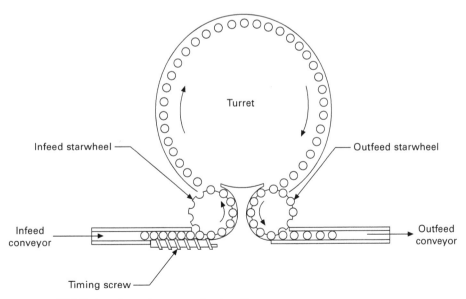

20.1 Rotary layout on a packaging line.

speed than the main packaging line, for example by incorporating a large number of filling heads, thus allowing a more rapid, continuous flow of product onto the next stage. Rotary machines typically use a timing screw to control the flow of product into a starwheel which diverts the product into the turret, whilst another starwheel redirects the product back into the main production flow.

20.3 Unscramblers

Typically the first operation in a packaging line is to transfer the containers to be used for the product to the packaging line. This requires moving them up to the packaging line and removing any packaging used to protect them during transport. They may also need to be sterilised before filling. The technology required depends on factors such as the type of container, how it is supplied, the degree of cleaning required (determined by the product type) and the speed at which the line is to run.

Metal cans for food and glass bottles for beer and spirits are usually filled on relatively high speed lines and the containers are delivered stacked in rows on pallets. Once any outer wrappings are removed, an automatic depalletiser takes an entire layer of containers from the pallet and places it on an in-feed table. Figure 20.2 shows a depalletiser with a sweep-off mechanism, although lift-off devices also exist. The containers are then marshalled into an orderly row. If cleaning is required, this may be done by inverting each container and blowing clean air into it, to remove any loose debris, or spraying with steam or hot water. Drying then follows (if necessary) and the containers are aligned in the correct orientation ready for filling. Rinsing with hot water is especially important for glass containers which are going to be filled with a hot product as it helps to prevent thermal shock. On the other hand, if the product is chilled at the point of filling, care must be taken that the temperature difference between the glass container and the product is not excessive. Plastic containers are normally much more durable and a mechanical means of removing containers from their transit packaging, cleaning them, orientating and supplying them to the packaging line is much easier to achieve. Cleaning may be done using hydrogen peroxide, especially if the containers are used for aseptically packed products.

Containers which are used in more modest quantities, for example glass bottles for perfumes, are usually supplied in corrugated board cases, possibly with internal divisions to provide protection against shock in transit. Such containers are likely to be loaded onto the packaging line manually, with the corrugated cases being opened as and when needed, which means that any unused stock can be returned to the warehouse and remain clean and contamination free. It may be necessary to invert each container and inject air into it to remove any loose debris, prior to filling. Whilst this means more manual labour than on the high speed lines mentioned above, this is invariably more cost effective than investing in automated handling equipment which would require several complex changeovers every time a new style of container is filled.

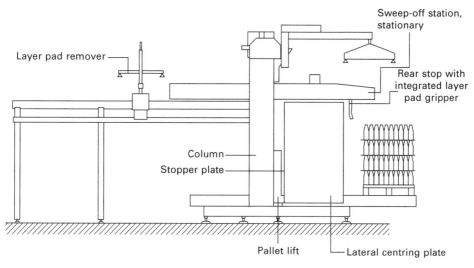
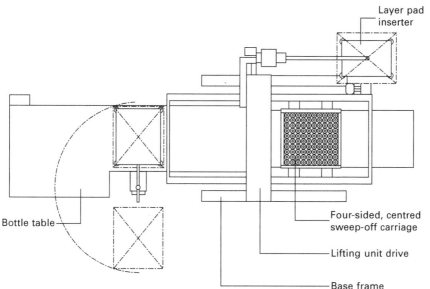

20.2 Automatic depalletiser with sweep-off mechanism.

20.4 Fillers and filling

Filling machines measure the product by weight, volume or count. There are several factors to consider when selecting which type of machine to use. These include:

- whether the product is a solid or a liquid
- the type of container required
- the level of accuracy needed in the filling operation.

The following paragraphs examine the various properties of solids, liquids and container types.

Solid products can be categorised in two ways:

- discrete solid items such as tablets which are often filled by counting a given number of items per container
- powders which can be further categorised by characteristics such as particle size, moisture content and bulk density.

Discrete items may vary in handling properties. Some may have varying shapes and sizes and may be fragile (e.g. crisps). They may trap air and have a tendency to settle over time. These characteristics make gentle handling necessary and filling by weight more accurate than filling by level or volume.

For the purposes of filling, powders can be sub-divided into the following types:

- free-flowing
- non-free-flowing/agglomerated.

Because individual particles move freely in relation to one another, free-flowing powders or granules have a consistent density, do not trap air and pour readily. As a result they typically form a flat cone with a shallow repose angle when poured onto a level surface (Fig. 20.3(a)). This type of powder is relatively easy to handle by a range of filling machines.

Whether because of characteristics such as irregular particle shape or high moisture content, which results in particles adhering to one another, non-free-flowing and agglomerated powders do not pour easily. They tend to form a steep cone when poured onto a level surface or agglomerate with particles clumping together (Fig. 20.3(b) and (c)). They are therefore prone to sticking to surfaces and clogging passages within filling machinery. They also tend to trap air within and between agglomerations of particles which results in variations in density in different areas of the product. Because this type of product may require more force to move it, filling machines based on gravity may not be suitable. Variations in density mean that filling by volume may not be accurate and that filling by weight is more appropriate.

Liquids can be categorised in a number of ways:

- by viscosity, e.g. low-viscosity, free-flowing liquids, viscous semi-liquids and highly viscous 'semi-solids'
- by other characteristics such as surface tension which may result in behaviour such as frothing or foaming when agitated.

These characteristics determine what type of filling operation is feasible and will

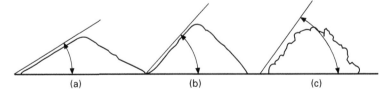

20.3 Angles of repose: (a) free-flowing solids; (b) non-free-flowing solids; (c) agglomerated products.

result in an accurately filled container. Some products (e.g. soups) may contain a mix of liquid and solid components and these will usually be filled separately with the solids filled first, followed by the liquids.

Another basic factor in selecting the type of filling machine is the type of container to be filled. Containers can be categorised as:

- rigid
- semi-rigid
- flexible.

Rigid containers made from glass, metal or thick plastic can withstand the application of a vacuum or high pressure during filling. Semi-rigid containers such as many plastic bottles or yoghurt pots are not able to withstand the same level of stress. They may bulge under pressure or collapse if a vacuum is applied. This will result in potential damage to the container, inaccurate filling and poor sealing which may have a major impact on product safety and quality. Flexible containers such as tubes for toothpaste may, on the other hand, require insertion of a feed line into the bottom of the tube and the application of pressure to push the product into the tube so that it is properly filled without trapping air.

Filling operations must meet the legislation and quality standards relevant within the country in which the product is to be marketed, with an adequate validation of the process capability of the filling system, and adequate sampling and measurement checks to ensure that each batch is correctly filled. See Chapter 4 for more information. Another regulatory consideration is the ingredients lists which may be required to be shown on the pack. If ingredients are sold by weight, the filling system must ensure an accurate weight is consistently delivered to each container. If a high level of accuracy is required (e.g. in pharmaceutical products where exact dosages may be critical to patient safety), the filling system must be capable of the precision required. In the case of high-value products, a high level of accuracy in filling may also be critical to the profitability of the business. Over-filling or a high level of wastage can be extremely costly. If a product consists of a mixture that may not be relied upon to remain consistent during the filling process, it is necessary to fill the various ingredients separately to ensure that accurate amounts of each ingredient are delivered to the container. This is known as sequential filling, and explains why all the cherries are at one end of a can of mixed fruit salad. In some products (e.g. bottles of milk), customers will judge the accuracy of filling by level and will expect that level to be consistent between different containers.

The maintenance of food quality over the required shelf-life of the product can depend largely on removal of air from the container and adequate sealing. The filling process needs to meet these objectives if it is not to reduce product quality. In hermetically sealed glass or metal containers used for heat-sterilised foods, for example, a headspace is needed above the food to form a partial vacuum. This reduces pressure changes inside the container during processing and oxidative deterioration of the product during storage. When filling more viscous liquids such as pastes, for example, it is very important to prevent air from becoming trapped in the product which would reduce the headspace vacuum.

20.4.1 Filling solid products

Solid products are filled into containers by:

- volume
- weight
- count (for unit items such as tablets).

Filling by volume

The simplest form of volume filling machine consists of a device for holding open cups which are filled in turn as they are passed under a hopper. A typical design involves a circular plate which revolves to present each cup under the product hopper in turn (Fig. 20.4). Scrapers or brushes (called 'doctor' brushes) wipe over the top of the cup to level off the amount of product in the cup. The cups then move to a discharge point where they tip their contents into the final container. An alternative system uses trap doors under each cup to release the product directly into the container. Some systems use adjustable telescopic cups which can be opened to accommodate a larger volume of product in each cup. These systems are suitable for free-flowing solids of consistent density. It is important to maintain a constant level in the product hopper feeding the cups since this maintains an even flow into each cup.

Some volume filling systems make use of a vacuum in addition to gravity. As the product flows from the hopper into the cup, a vacuum pump draws out air to

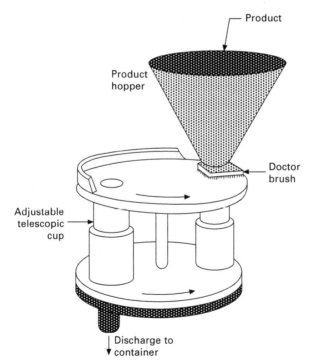

20.4 Solid products: filling by volume using a cup filler.

compact the product. A fine mesh prevents product escaping. The cup then delivers the compacted product to the container. This technique is suitable for light powders (e.g. cocoa powder) which might otherwise trap air and lead to inaccurate filling and reduce shelf life through oxidation.

An alternative to the use of cups is auger filling. A typical auger filler consists of a hopper with a funnel or tube at the bottom. An auger or screw runs through the middle of the hopper together with an agitation blade (Fig. 20.5). Towards the bottom of the auger each turn of the screw (or flight) is calibrated to a precise volume. As the auger rotates, the agitation blade rotates in the opposite direction to remove air, homogenise and then feed the powder into the flights. As the powder enters the lower flights of the auger it is divided into separate doses defined by the volume of the flight which can then be delivered to separate containers. Tapered augers can be used to compact finer powders. Auger filling is particularly suitable for non-free-flowing solids, e.g. moist brown sugar. It is not appropriate for solids with variations in bulk size or particle size and distribution, or where precise doses are required, where weigh filling may be more suitable.

Filling by weight

Weigh filling typically involves the use of weigh buckets. Product is fed from a hopper into the bucket. Once a certain weight is reached, the hopper is closed and the bucket tips the product into a discharge chute or directly into the container. In older machines this would be done using a mechanical balance system that would trip a mechanism to stop the feed of the product when the required weight had been supplied. The container supply mechanism would then release the filled container and position an empty one ready for filling.

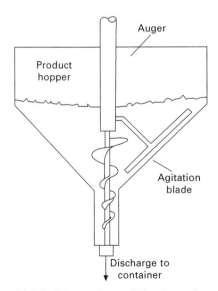

20.5 Solids products: filling by volume using an auger filler.

One potential problem with weighing is the risk of surplus product reaching the bucket after the bucket has reached the desired weight but before the flow of product has been halted. This can be dealt with by having separate bulk and 'dribble' feed lines. The bulk feed line supplies the product to the bucket until the required weight is nearly achieved. The bulk feed line is then closed, leaving the slow-flowing dribble line to top up the bucket. These systems ensure greater accuracy but make weigh filling a relatively slow process.

The development of electronically controlled magnetic force balances has now made it possible to have much more sophisticated, rapid and flexible systems. It is now possible to have a multi-head machine which fills products (e.g. items of fruit) into several containers simultaneously (see Fig. 20.6). These systems often use several weighing stations to fill each container, selecting which stations to use to fill the container to the exact weight. As with more traditional systems, they use bulk feeders to fill containers to near their final weight and fine feeders to top up the container to an exact weight. The machines weigh each container so that they can allow for slight variations in container weights. This allows them to weigh either the product without packaging (net weight) or the product plus packaging (gross weight). The machine can then calculate how much each container requires, weigh each piece being supplied, and select which container to place it in and calculate continuously how much more needs to go into each container as it fills. Any packs with the wrong weight are rejected. Free-flowing powders can be measured by volume using a telescopic measuring section of the machine which is filled before the product is transferred to the container.

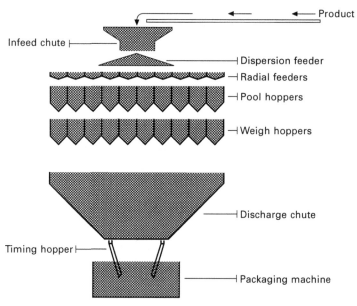

20.6 Solids products: multi-head filling by weight.

Filling by count

For some products, such as tablets or capsules, the amount of product is declared by count. Machines which fill by count normally use mechanical or photo-electronic systems. Mechanical counting systems usually place the product over a plate with an appropriate number of holes of suitable size. Once each hole is filled, the excess product is wiped off the plate and the product counted by the plate is then released into the feed system and into the container. The counting mechanism of a tablet blister-packing machine (see later) works in a similar way, using the thermoformed holes in the plastic film as the counting mechanism. Some systems consist of a series of thin perforated slats which pick up product as they move up through the product hopper. The slats then move down the front of the machine to enable inspection to ensure each hole contains a product. The product is then ejected into feed chutes which fill each container.

Photo-electronic counting systems usually involve tumbling or vibrating the product to create a flow of single items which move past the photocell on a conveyor belt for counting. The photocell counts each shape as it passes, typically by identifying its shadow. Once the required number is reached, a gate diverts the counted items into a feed chute which fills the container. These electronic systems can now also measure the size of the shadow to ensure that it is not too small (e.g. a broken tablet) or too large (two tablets together). In order to improve the detection capability of these systems, the wavelength of the light used may not be in the visible spectrum (i.e. ultraviolet or infrared). Other electronic detection systems (such as capacitance) may also be used. Modern systems can also be used to create sets of different items, e.g. mixes of nuts and bolts to be packed into sachets for furniture self-assembly kits. The machines are programmed to identify different shapes and sizes, separate out specific items and combine them with others of a different shape and/or size to create each set.

20.4.2 Filling liquid products

As has been noted, there are several criteria to consider in filling containers with liquids, including:

- the properties of the liquid, e.g. viscosity, foaming capacity, particulate size
- the conditions required for filling, e.g. temperature.

In any filling operation it is necessary to give some consideration to the differing handling properties of the product. At first thought, liquids may be considered to be easy, but variations in their properties may give rise to problems and these require solutions in the type of filling machine used and/or the design of the packaging. For instance, a viscous liquid such as jam may have to be heated to enable it to be filled. In this case the jars have to be manufactured and tested to ensure that they are able to withstand the thermal shock of being filled with the hot product. The extent of thermal shock has to be controlled (see previous discussion) and the pumps and other components handling the liquid and the container have to be constructed of materials capable of withstanding the heat. The machine has to exert sufficient

500 Packaging technology

force in its operation to move a more viscous liquid such as jam compared to a low viscosity liquid such as fruit juice.

Another product characteristic to consider is low surface tension in some liquids, causing the product to froth when filled. Some types of filler cause less turbulence to a liquid than others and are, therefore, better suited to liquids prone to foaming. Frothing or foaming may also be reduced by having nozzles which penetrate into the container and either cause the product to flow down the inner sides of the container, or direct the product below the filling level. These requirements can make the machines more complicated and more expensive both to build and maintain.

There are three main ways of filling liquids:

- by level
- by volume
- by weight.

Within these three types, filling can be done either from the top or the bottom of the container. Top filling involves inserting the filling tube into the neck of the container and either allowing the liquid to drop to the bottom or directing the liquid to run down the container sides. The latter will minimise turbulence and air entrapment. Bottom-up filling involves inserting the filling tube down to the bottom of the container and gradually withdrawing it as the container fills. This can be done either by moving the tube itself (Fig. 20.7(a)) or raising and lowering containers on the packaging line (Fig. 20.7(b)). Bottom-up filling is particularly effective in minimising air entrapment, limits frothing or vapourisation of more volatile liquids and is particularly suited to filling flexible containers such as sachets.

Level fillers

Level fillers use the container's volume to measure the amount of liquid filled. Since containers of the same design will have differing volumes due to slight variations in dimensions and wall thicknesses (Fig. 20.8(a)), this form of filling is less accurate than

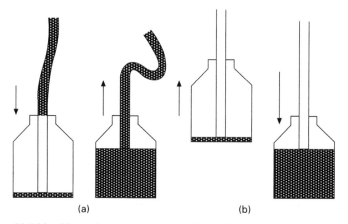

20.7 Liquid products: bottom-up filling to minimise foaming.

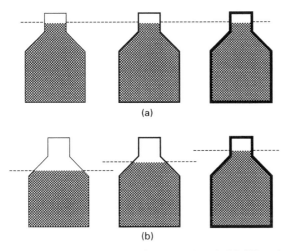

20.8 Liquid products: (a) filling by level; (b) filling by volume.

volume or weight filling. It is therefore used for lower cost liquid products such as soft drinks, beer and sauces where accurate volume is less important than a visually constant fill level. In many cases (e.g. milk, as already mentioned) customers will expect a container to be filled to a certain level and for that level to be consistent between containers even if that means there are slightly differing volumes in each container.

Modern level filling systems use sensors to identify when the right level of liquid has been supplied. Pneumatic systems use a flow of low pressure air in a tube next to the filling tube. As the liquid reaches the correct height, back pressure in this second tube triggers a valve to close off the supply of liquid. More advanced sonic systems use high-frequency sound waves which bounce off the surface of the liquid. The changing pattern of sound waves as the liquid reaches the desired height closes the valve to prevent the target level being exceeded.

There are three main types of level filler:

- gravity fillers
- vacuum fillers
- pressure fillers (also known as over-pressure or counter-pressure fillers).

In basic gravity fillers the liquid flows from a supply tank into the container below (Fig. 20.9). The height of the supply tank above the container determines the flow rate. A typical design involves a filling tube with a valve connected to a spring-loaded outer tube that fits over the container neck. As the container is raised, it activates the spring to open the valve to fill the container (Fig. 20.10). In some systems a sensor identifies when the liquid has reached the top of the container and closes the valve. In others each valve is independently timed by a control computer to deliver the target amount to the container. Any excess can be channelled into an overflow tank. Gravity filling is a relatively cheap process but is slower than vacuum filling. It is particularly suited to liquids prone to foaming since there is less agitation during filling. Bottom-up filling can be used for very foamy products. It is also suitable

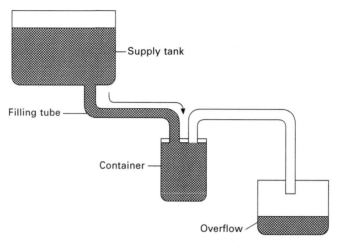

20.9 Liquid products: basic gravity filling.

for a wide range of container types. This type of filling is less suitable for viscous, slow-flowing products or products containing large particulates.

Vacuum fillers typically work by lowering a filling tube and a vacuum line (connected to a vacuum pump) into the neck of the container and then sealing the neck (see Fig. 20.11). Air is then drawn from the container to create a vacuum using the vacuum pump. Liquid is then drawn from a supply tank through the filling tube into the container. When the liquid reaches the vacuum line, suction draws it into an overflow tank, ensuring the desired level is not exceeded. The surplus liquid can then be returned to the supply tank. An alternative approach involves keeping the supply tank initially at low vacuum to draw the liquid in. The pressure then equalises, allowing the liquid to flow into the container by gravity. As with other systems, vacuum fillers also use sensing devices to identify when the desired level has been reached, halting further flow until the system is ready for the next container. Vacuum fillers are fast, flexible and relatively low cost. They are, however, limited to rigid containers (e.g. glass bottles) which are not distorted by creating vacuum conditions in the container and to liquids which are less susceptible to aeration.

Pressure filling uses a pump to move the liquid from a supply tank to the container (Fig. 20.12). In over-pressure or counter-pressure machines, the supply tank is kept at high pressure, forcing the liquid through to the container. The fill level is determined by the vent tube inserted into the container. When the liquid reaches the vent tube, the supply is interrupted by the difference in pressure. Alternatively, the difference in pressure can also be used to draw off any surplus liquid into an overflow tank. Pressure filling is relatively fast and is suited to viscous products which need minimum agitation.

Volume fillers

In volume filling, a measured volume of liquid is placed into the container. This means more accurate measuring than level filling but may result in variations in

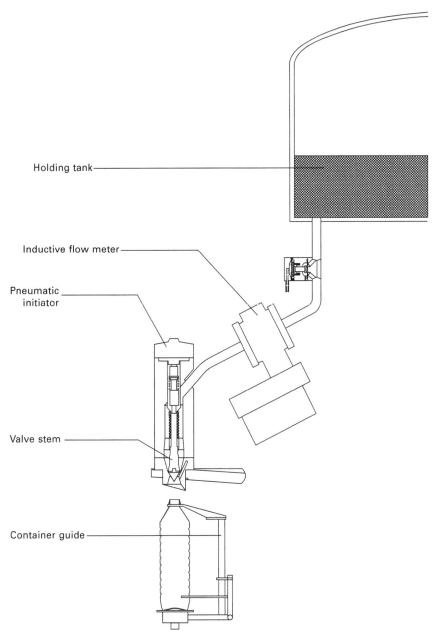

20.10 Liquid products: gravity filling using flow meter and valve.

the level of a liquid in containers, depending on variations in container size (see Fig. 20.8(b)). This will be obvious to the consumer if the container is transparent, although this can be obscured by the careful application of a neck label. Volume filling is used for high-value products and particularly for products sold by weight or where accurate weight or volume is important (e.g. pharmaceutical products where accurate dosage may be critical).

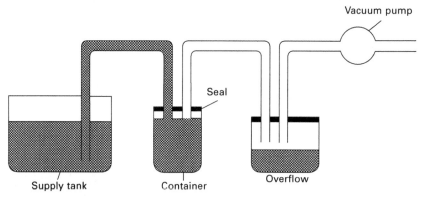

20.11 Liquid products: vacuum filling.

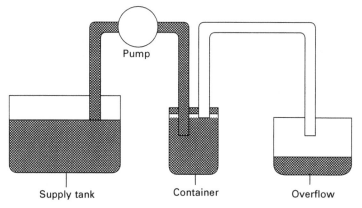

20.12 Liquid products: pressure filling.

The three main types of volume filler are:

- piston
- cup
- flow: time-pressure or flow meter.

The basic type of piston filler is the piston force-pump, which is connected to an adjustable crank, driven by a control system similar to the parking mechanism of a car windscreen wiper (Fig. 20.13). When the pump is activated, the crank makes one full turn, delivering the contents of the piston to the container, and then re-filling the piston in readiness for the next container. The amount of product supplied is controlled by the size of the piston cylinder and the stroke which is adjusted by the setting of the crank. Piston fillers typically have a control device which prevents the piston turning if there is no container ready to be filled.

Piston fillers are typically fully automated with groups of positive action rotary pumps working in sequence (Fig. 20.14), with electronically controlled drives which regulate the number of turns the pump makes to provide product to each container and which allow adjustments to be made without stopping the machine. A control

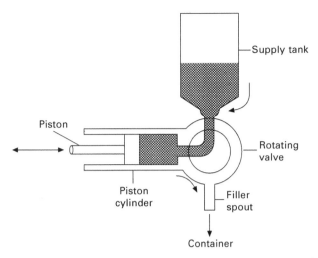

20.13 Liquid/paste products: filling by volume using a piston filler.

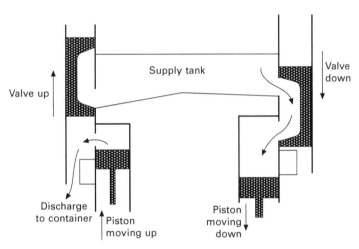

20.14 Piston filler using automated rotary pumps.

computer tracks the number of rotations of the pumps to determine precisely how much product has been filled. When the target fill volume is reached, the computer stops the pump. Piston fillers can be used for low viscosity liquids but there can be problems in leakage between the piston and the cylinder. Piston fillers are usually the most cost effective, rapid and accurate way of filling viscous products such as jams. As an example, pastes can be filled into tubes using piston fillers. The product first has to be drawn into the measuring cylinder. The machine nozzle fits into the tube so that the paste is pushed off the nozzle as it is filled, thereby minimising the amount of air trapped inside the tube.

In cup fillers the liquid product flows from a supply tank into a measuring cup. Once this is full, the measured quantity of liquid is emptied into the waiting container. Time-pressure controlled fillers divide the liquid into portions using a valve which are then fed to individual containers. It is important to keep a steady supply of liquid to ensure accurate measurement. This means maintaining a constant pressure and temperature as well as consistent viscosity in the liquid.

Flow meter filling machines use meters which control the opening and closing of measuring valves. Meters measure the liquid in a number of ways, including measuring its conductivity or mass. These machines typically use positive displacement pumps or constant output impellers.

Weighing fillers

Weighing fillers for liquids typically use scales which weigh the desired quantity and then open and close a valve for filling the container. Net weight liquid fillers are suited to liquids that are filled in bulk quantities or smaller amounts of products that have a high value and are sold by weight. Filling by weight is suitable for liquids of varying consistencies. The advantage of this type of filler is a high level of accuracy; the disadvantages are the high cost per filling head and the relatively slow rate of filling.

20.5 Closing and sealing of containers

As discussed in Chapter 15, closures for containers have a variety of functions and there is a wide range of types and methods of sealing which are covered in more detail there. This section considers some of the basic types and how they are applied to a container on an automatic packaging line.

20.5.1 Push-fit closures

Push-fit closures fall into two types: those in which the closure is pushed *in* the open neck of the container, e.g. a wine cork, and those in which the closure is pushed *on*, or over the outside of the top of the container, e.g. the metal lid on a tin of biscuits or the plastic overcap sometimes supplied as a resealable feature on metal containers for dry products such as ground coffee.

On the packaging line, the application of push-fit closures is a relatively simple process and typically involves positioning the closure on the container which then moves under an inclined belt which presses the closure into place as the container moves under the belt. For products such as wine, the corked neck may then be covered by thin metal or plastic to provide further protection as well as decoration.

20.5.2 Screw-threaded closures

Screw-threaded closures (which can be manufactured from metal or plastic) have a thread on the inside which must be matched up with the thread on the neck of the container. Closures are typically fed down a chute to land on filled containers which have been spaced out at regular intervals. A chuck then rotates the closure onto the neck of the container until a pre-programmed torque (tightness) is achieved. Some container designs include a rim which prevents the closure from being tightened beyond a certain point. The chuck then releases, allowing the sealed container to move to the next stage.

It is essential that the neck dimensions of the closure and container match to allow efficient engagement to provide an effective seal. Thread engagement refers to the number of turns required for the closure to fit completely on the neck of the container. The greater the thread engagement, the tighter the seal and the more effective the tightening torque in keeping the closure in place. The correct torque is also essential. If too tight, the closure and container might be damaged or customers might not be able to remove the closure. If too loose, the closure might loosen, allowing air in or it may come off completely due to vibration during transit.

A further consideration is the materials used. Thermoplastics such as polyethylenes are vulnerable to creep, i.e. they gradually deform under stress and will therefore gradually lose torque. This may mean increasing thread engagement and application torque as well as allowing for a potentially shorter shelf life. Polypropylene is widely used for multi-use screw-threaded closures due to its good resistance to creep.

Screw-threaded closures can be wadded or wadless (see Chapter 15) and the latter may be enhanced by the use of a membrane sealed across the neck of the container. An essential requirement for this type of seal to function correctly is that there must be a metal layer included in the membrane (usually aluminium foil). The cap, with the membrane inside, is screwed on to the filled container, which is then passed between induction heating coils. The aluminium foil absorbs heat, which softens the polyethylene layer, which in turn bonds to the neck of the container, thus providing secure containment and tamper evidence.

Induction sealing systems require careful attention in setting the correct operating parameters for the torque applied by the capping machine to the seal, the position of the electro-magnetic head and the amount of energy supplied, as well as the speed of the track (which controls the time during which the electro-magnetic energy is applied to the container). It is also necessary for operating personnel to remove metal items such as jewellery before working near such machines as these items act like a transformer and could cause serious burns to the fingers, ears, etc.

20.5.3 Roll-on pilfer-proof (ROPP) closures

This type of closure is typically made from soft-temper aluminium with a partially perforated ring (or skirt) at the lower edge of the closure, and is typically supplied fitted with a wad (see Chapter 15). The closure is positioned above the neck of the container which includes a thread and, at the bottom of the thread, a ridge (known as a ring grip) to take the perforated ring. The head of the capping machine is lowered, bringing the wad into contact with the top of the container. A thread is then formed from the outside of the closure using rollers which force the soft aluminium to take on the thread form of the container (see Fig. 20.15). The perforated ring is closed around the specially formed ring grip on the neck of the container and this provides the tamper-evident feature, as these perforations are broken when the closure is unscrewed. The pressure required limits this type of closure to use on containers

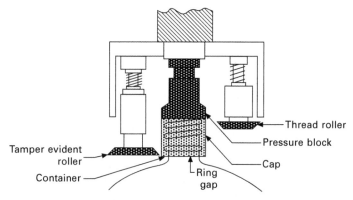

20.15 Application of a metal ROPP closure.

with rigid neck sections, e.g. glass bottles, or injection blow moulded plastic bottles which are specially designed for this use.

Plastic tamper-evident closures with breakable rings are applied as conventional screw-threaded closures. Extra pressure is required to push the perforated ring at the bottom over a ridge around the neck so that it snaps into a specially designed groove which holds it in place.

20.5.4 Lug, press-on twist-off and crimped or crown closures

These three types of closures are simpler than screw-threaded or ROPP closures. Lug closures include lugs or protrusions on the inner edge, designed to engage with an interrupted thread on the container neck (see Chapter 15). They are secured by placing on the container, exerting a downward pressure and twisting so that the lugs slide under the thread to hold the cap in place. Because less rotation is required, they can be applied at a higher speed than screw-threaded closures. Given the stress exerted on both closure and container, this type of closure is usually manufactured from heavy gauge steel and is restricted to rigid containers such as glass.

Press-on twist-off caps do not always require lugs but often include a thread. First used for baby foods, they commonly include a soft thermoplastic seal and rely on a partial vacuum in the container (developed when the hot-filled product cools down) to keep the lid in place. Crimped closures, also known as crown corks and widely used on bottles of beer, are made from heavy gauge metal and have a sealing material inside (commonly a flowed-in liner of soft plastic material around the inner circumference). The closure is placed over the neck of the container, often using magnets, and the outer circumference is crimped around the lip of the bottle, to secure it in place.

20.5.5 Can closing

Can manufacture (including closing) is discussed in detail in Chapter 8. Only a brief summary is given here. Can lids are sealed by a double seam. Lids are typically placed on cans immediately after filling. As the lid goes on, carbon dioxide may be blown over the can to displace the air at the top of the can if a product is likely to suffer from oxidative or microbial spoilage. Each can and its lid are then raised against a sealing head. One seaming roller bends the outer edge or hook of the lid round the can rim or hook (Fig. 20.16). An airtight double seam is obtained by pressing the two hooks together with a second seaming roller. A sealing compound (typically a water-based latex emulsion or synthetic rubber compound) incorporated on the lid rim completes a hermetic seal following tightening. Since the seam is the weakest part of the can, seams are routinely inspected, often using x-ray technology to identify defects.

20.6 Labelling

Whilst many containers are produced with decoration and identification requirements already pre-printed, many high volume products require the application of labels as part of the packaging line. Even when packaging has been pre-printed, there is often the need to apply a label to provide additional information such as shade or flavour variant. Chapter 17 provides information on the different types of labels available and the materials commonly used for labels, and an overview of label application methods. This section is confined to providing more explanation of the requirements of automatic label application. Chapter 16 on adhesives should also be referred to at this point.

20.6.1 Applying self-adhesive labels

Self-adhesive labels are provided already cut out and on rolls of silicone-coated backing paper (Fig. 20.17). Only one of the different possible orientations of the label on the roll will be suitable for the particular application and thus this must be

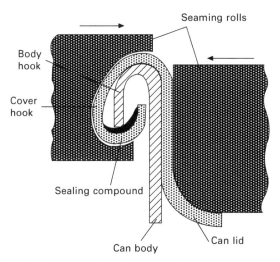

20.16 Seaming of metal can end.

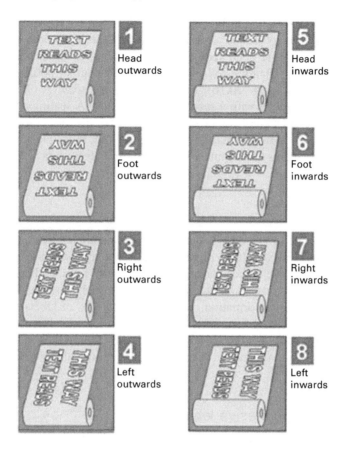

20.17 Layout options for self-adhesive labels.

clearly stated in the label specification. Other important features determined by the labelling machine include the diameter of the core around which the roll is wound and the maximum overall diameter of the roll.

A self-adhesive label applicator works by pulling on the backing paper using a rubber roller gripping onto a capstan. When a container is detected by a photoelectric device, the capstan begins turning and pulls the backing paper around a tight radius on a piece of metal known as the dispensing beak. The speed of the label and the speed of the container are matched so that the label comes off the backing paper and is wiped onto the container using a brush to press it home. If the container is cylindrical, it is turned and the label pressed down using a wrap-around mechanism (comprising a belt running at twice the speed of the track which presses the container against a soft rubber-faced stationary plate). Another photoelectric device detects when the end of the label has been reached and the capstan stops, awaiting the arrival of the next container. This stop signal is also used to trigger an on-line coding system, if required. In general, self-adhesive label applicators are flexible and relatively easy to adjust for different shapes and sizes of containers.

20.6.2 Applying ungummed labels

Ungummed labels are usually delivered to the packer-filler in stacks which fit into the appropriate magazine on the labelling machine. It is vital that the labels are easily separated so that only one is picked from the stack at a time. Defects in the cutting dies used in label production may lead to uneven edges which can hinder this picking-up operation, as will any edge damage incurred during transit to the packer-filler.

Applying ungummed labels, does, of course, require the packer-filler to select the most appropriate adhesive for the label substrate, applicator type and machine speed. When using water-based adhesives, it is important that the labels are printed onto the appropriate grade of paper and in the correct direction of the grain (the direction in which the fibres in the paper line up on the paper-making machine). If not, the moisture from the adhesive will cause the paper to curl with sufficient force to lift the edge of the label from the container before the adhesive has had time to set. The simplest way to test the direction of grain of a paper label is to wet it. The label will curl into a tube with the straight length of the tube indicating the grain (i.e. machine) direction. Usually, but not always, the grain direction is parallel to the base of the label.

Machines which handle ungummed labels include those which apply hot-melt adhesives, often used to label cans and cylindrical plastic bottles. Here the can (or bottle) is rolled along a set of rails and a vertical bead of hot-melt adhesive is applied. This picks up one edge of a label from a stack. The can rotates, wrapping the label around its circumference and a further bead of adhesive is applied to the trailing edge of the label to hold it in place. Where water-based adhesives are used, e.g. in applying labels to bottles in the drinks industry, there is often a complex series of pallets, adhesive applicators and pressure systems used to apply the labels (Fig. 20.18). Labels can be applied to the front, back and neck area of the bottles,

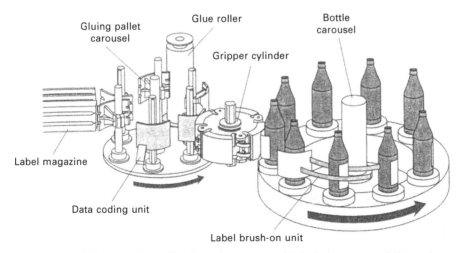

20.18 Automatic application of ungummed labels (courtesy of Krones).

all at the same time, ensuring excellent alignment of labels both to each bottle and to each other. Such machines tend to be bespoke systems which require extensive modification for each bottle shape, although modern developments have concentrated on making changeovers less complex and therefore less time consuming.

20.6.3 Applying sleeves

Shrink sleeves are typically made from heat-sensitive thermoplastic film. They are normally supplied as a flattened welded tube, printed on the inside of the sleeve, to prevent damage by scuffing, and wound onto a cardboard core. A register mark is incorporated into the print which indicates to the machine where to cut off the section required. The required length of sleeve is placed over the container and positioned correctly. It is then heated, either using jets of wet steam or hot air, to shrink the sleeve to the profile of the container. If necessary a strip of hot-melt adhesive may be included onto the inside of the sleeve which holds it on to the container to ensure that it does not slip during this process or later during transit, since shrink sleeve materials are prone to creep (the tendency to deform under prolonged stress) which may loosen them over time. In designing the graphics of the sleeve, it is important to compensate for the distortion which will result from the differences in shrinkage across the profile of the container, and to ensure that any bar code symbols are subject to the minimum of distortion. The selection of the grade of sleeve film with the appropriate shrink characteristics is critical to the successful operation of this process.

An alternative means of application involves printing and then applying the sleeve using the same technology as for a self-adhesive label. This sleeve is then shrunk to fit the container by passing the container through a heat tunnel. Another technique uses a polyethylene sleeve smaller than the container which is stretched over the container and then closes around it.

20.6.4 Applying neck collars and tags

Both neck collars and tags present a simple addition to the pack and provide promotional opportunities. Neck collars are normally made using a thin paperboard with a hole cut out to allow the collar to fit over the neck. They can be automatically presented to the container and located onto the neck using a robotic pick-and-place machine. If the line speed is slow enough, they may be applied by hand. Tags are normally constructed using a fine cord, which may be elasticated, and which is attached to the tag. It is very difficult to apply tags automatically as the cord has to be passed over the neck of the container. Both neck collars and tags are, unfortunately, susceptible to damage in end-of-line operations and in the opening of the transit packaging. They are also susceptible to pilferage. One solution is to build a neck collar into a multi-layer self-adhesive label and apply it from above using an adapted label application machine. Once situated over the neck the collar adheres to the container, making it more secure.

20.7 Cartonning

Cartons are used for a wide range of both solid and liquid products. See Chapter 10 for information on styles of cartons and grades of board used. Cartonning of liquids will be addressed later in the section on form, fill and seal machines. There are numerous different styles of cartons and cartonning systems and this section covers only the key principles.

Folding cartons are amongst the most common types used. They are delivered packed flat, usually with the side seam already glued, and placed in a magazine on the cartonning machine. Using suction caps, a carton is taken from the magazine and erected into its required shape. One end is then closed using adhesive, or by means of a tucked-in flap and the assembled carton is now ready for insertion of the product (which may have been sealed into a bag or sachet prior to this point, or may be unwrapped). Robust products can be dropped into the assembled carton from above, using gravity, whilst more fragile items (e.g. a decorated sponge cake) require more careful insertion by pushing in the horizontal direction. The carton is then fully closed, again using adhesive and/or an arrangement of tuck-in flaps and the filled cartons are moved to the next stage in the packaging line. An alternative is to use a flat, unglued carton blank, which is erected on the cartonning machine prior to filling and closing – see Fig. 20.19.

Sleeves are an alternative to cartons, and are also supplied folded flat and creased to allow automatic assembly. Both sleeves and cartons can be supplied unglued, in which case they are wrapped around the inner product and glued as part of the packaging line operation.

As mentioned, carton ends can be glued or tucked in place, the choice being decided by the degree of tamper evidence required and whether or not the carton is expected to be used more than once (a glued carton usually has to be torn to gain access to the contents). Another approach for one end is the 'crash-lock' design which uses

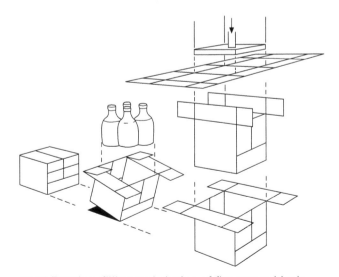

20.19 Erection, filling and closing of flat carton blank.

interlocking flaps and is easy and quick to assemble for filling, as well as being more robust than glued or tucked ends. However, this requires more complex folding and gluing at the carton production stage.

20.8 Form, fill and seal (FFS) packaging operations

Form, fill and seal (FFS) describes a packaging operation in which the 'container' is formed as part of the packaging line (rather than being made elsewhere, such as on a glass bottle forming machine) immediately prior to the product being filled into it, and then the filled container is closed, usually by heat sealing. The flexible packaging materials used are supplied in reel form and invariably include thermoplastics, often combined with paper, board and aluminium foil, according to the product's needs. FFS is used for a wide range of packs and products, including sachets for single portions of sauces, cartons of fruit juices and sacks for 25 kg of fertiliser and animal feed products.

The main types of FFS machine are:

- vertical (VFFS) machines, used for liquids such as fruit juices and soups, and solids such as frozen vegetables, sugar, crisps and wrapped sweets
- horizontal (HFFS) machines used for cakes, biscuits and bars of confectionery.

Variations on the above include:

- sachet forming and filling machines used for dried soups, moist tissues and trial sizes of products such as shampoo
- thermoforming machines used for meat, cheese and yoghurt
- blister packing machines used for tablets.

20.8.1 Vertical FFS machines

The basic operating principles of VFFS machines are shown in Fig. 20.20. The film is unwound from a reel and drawn over a forming collar to create a bag shape. It is then wrapped around the vertical filling tube and sealed along what will become the length of the bag. The film is pulled down by the draw down belt and sealed at the bottom. The product is loaded through the tube into the bag. Horizontal jaws come up to seal the top of the bag and cut it away so that the filled bag can fall down the outfeed chute of the machine.

The finished pack can be a simple pillow shape as shown. Alternatively, it can be formed in an appropriately shaped forming box to give a rigid rectangular cross section with excellent 'stand up' properties (e.g. for liquids) or the sides can be gusseted to improve shelf and pallet stacking. Opening and reclosing features such as plastic zippers can also be incorporated. Rotating the sealing bars can be used to create tetrahedron shapes.

An alternative type of VFFS machine does not require a forming shoulder and has a continuous action (see Fig. 20.21). The machine feeds two webs of film or laminate to form a vertical channel, using heated and crimping rollers. A horizontal seam is

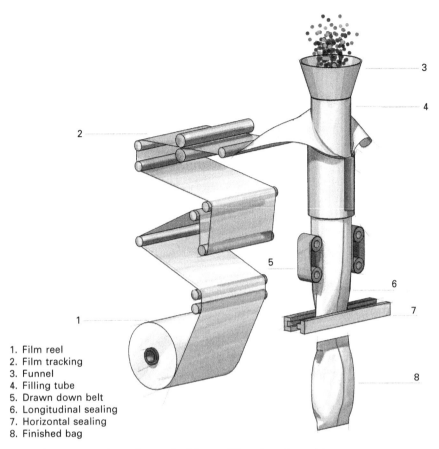

1. Film reel
2. Film tracking
3. Funnel
4. Filling tube
5. Drawn down belt
6. Longitudinal sealing
7. Horizontal sealing
8. Finished bag

20.20 Basic operation of a vertical form, fill and seal machine.

created to seal the base. Product is then filled into the newly-formed pouch and the top is sealed, at the same time forming the base of the next pouch. The strip can then be cut up into individual pouches. This type of machine is particularly suitable for filling liquids, and in some systems liquid flow is continuous, which means that the film or laminate must be suitable for sealing through the liquid product. Sealing and cutting of the pouches creates a pack with no trapped air, ideal for perishable foods and aseptic food processing.

20.8.2 Horizontal FFS machines

HFFS machines or 'flow wrappers' are used when the product is fragile and cannot withstand the drop down the filling chute of a vertical machine, e.g. bars of chocolate, cake bars and biscuits. The film is fed into a forming box where it is formed into the desired shape, continuously sealed along the sides and partially cut so that it starts to form an individual container (Fig. 20.22). At this point the product is fed into the container from a conveyor belt using push bars or 'flights' to separate and direct

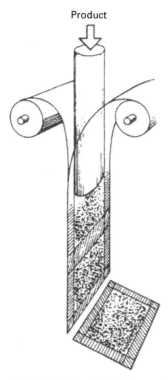

20.21 Variation of VFFS operation.

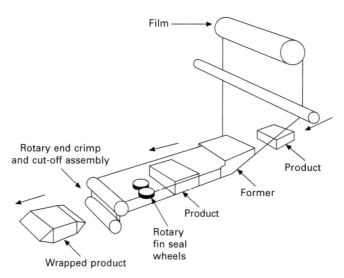

20.22 Basic operation of a horizontal form, fill and seal machine.

each product into a single container. Each container is then sealed at both ends and then separated into individual packs.

Most flexible films or laminates used on VFFS or HFFS machines are heat sealed.

A heat sealer heats the surfaces and applies pressure to fuse them. The strength of the seal is determined by the temperature, pressure and time of sealing as well as the type and thickness of the two films or the heat-sealable coating. The seal is weak until cool and should not be subjected to stress during cooling. There is a range of different types of sealer: hot-bar or jaw sealers hold the two films between heated jaws until the seal is formed, whereas impulse sealers clamp the films between cold jaws which are then heated to create the seal. Impulse sealers can reduce wrinkling or shrinking of films during sealing. Rotary or band sealers are used for higher filling speeds, working by passing films between two heated belts followed by cooling belts which clamp the films together until the seal develops. Other types of sealer include high frequency sealers, in which an alternating electric field induces molecular vibration in the film which heats and seals it, and ultrasonic sealers which induce high frequency vibrations to heat and seal the film.

The development of cold seal adhesives, requiring only pressure to seal the film wrap, has removed the need to apply heat. This has the benefit of not adversely affecting a heat-sensitive product (e.g. chocolate or ice cream). It also allows the line to operate at much higher filling speeds. However, cold seals are not generally as strong as heat seals. This is an advantage in terms of pack opening, as it is relatively easy to pull the seal open.

There are three common styles of seal:

- bead seals
- fin seals
- overlap seals.

A bead seal (Fig. 20.23(a)) is a narrow weld joining the two edges of the film. A fin seal involves folding one edge of the film over before it is sealed to the adjoining edge, leaving a double layer of material as a 'fin' which is often folded back on itself, to lie flat (Fig. 20.23(b)). It uses more film than a bead or overlap seal but is suitable for film that is only sealable on one side (e.g. because it has a sealable coating on one side only). In an overlap seal the front and back surfaces of the film are lapped

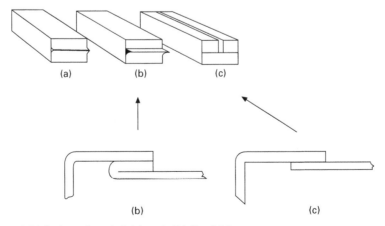

20.23 Styles of seal: (a) bead; (b) fin; (c) lap.

one over the other, giving a flat seal (Fig. 20.23(c)). This is very economical on material, but requires the front and back surfaces to be sealable to each other.

20.8.3 Sachet forming and filling machines

In this method of filling, the web is usually a printed laminate. It is unwound from the reel and folded in half along its length. Using the photoelectric cell registration marks on the web, the machine applies sealing and cutting bars at the appropriate locations, thus making open-ended pouches, which can then be filled with the product (Fig. 20.24). In the illustration shown there are two filling heads (1 and 2), one of which can be used to insert a folded towel and wad, and the second to dispense a liquid cleansing or sterilising agent. The filled sachets are then sealed at the top and moved to a collation station for further wrapping, e.g. using a film overwrap or cartonning.

20.8.4 Thermoforming packaging machines

Such machines usually operate on a horizontal bed. A base web is unwound and thermoformed into the required shape, e.g. rectangular cross section for stacks of sliced ham, circular cross section for pots to contain yoghurt. See Chapter 14 for information on thermoforming processes. Multiple packs can be formed across and along the length of the web. The product is dosed, by hand or machine, into the newly-made shapes and a top web applied to seal it in place. The finished packs are then cut out and the waste skeleton or matrix wound up and removed for disposal.

20.8.5 Blister packaging

Blister packaging is widely used for items such as tablets, which are contained in indentations or pockets in a plastic strip and sealed with a foil material. The base film which forms the strip is initially unwound and thermoformed to make the pockets. These are then filled with tablets using systems such as channel-feed (which can give a mix of different products on the same blister) or a relatively simple brush box (see previous section). After inspection (normally using video inspection techniques), a

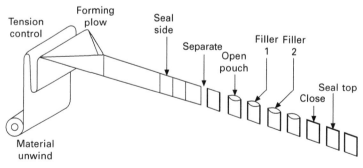

20.24 Sachet forming and filling machine.

top web of aluminium foil or a laminate is placed over the strip and heat sealed onto the filled film. The batch code and other details are embossed into the blister, which is then cut out of the film. The waste skeleton is removed and the blister travels on, normally to a horizontal cartonning machine.

An alternative form of blister packaging is used, for example, for hardware items such as small tools and decorative handles (see Chapter 10). The item(s) to be packed are placed on a perforated board which has been coated with a heat-sensitive adhesive varnish. A thermoplastic film is heated and draped over the items and the board, and a vacuum is applied to pull the film onto the products, which heat bonds the film to the board.

The construction of blister packaging machines depends on how fast they are required to run, and how often they need to be reconfigured to accommodate a different product specification. Intermittent operation flat-bed machines have an operating speed around 30 cycles per minute. They are relatively simple which makes changes easier to undertake. Higher speeds and longer runs of the same shape of product require machines with wider widths of film and rotary forming stations, coupled with more sophisticated filling, inspection and cutting devices. These machines are more expensive to build and reconfigure. This has to be balanced against the increased output available.

In more sophisticated systems, FFS machines can also be combined with a blow moulding system to develop more complex three-dimensional shapes from a thermoplastic web. Another variation of the form-fill-seal process is the unit-dose bottle blow-fill seal machine. This incorporates an ingenious combination of extrusion-blow, filling and sealing techniques. The small containers are formed using the extrusion blow process, but the necks are shaped to a small funnel. The liquid product is then filled into the containers. The containers are then heat sealed to close the pack. The bottles may be supplied singly or in strips. They are then finish-packed on to display cards or placed in cartons as required.

20.9 Direct product shrink-wrapping and stretch-wrapping

Flexible films such as LDPE are designed to be stretched over and then shrink around an object, for example food placed in a pre-formed tray. A small amount of shrinkage may be required to tighten around a simple shape requiring loose wrapping (e.g. greetings cards) whilst higher shrinkage is required for more contoured packs requiring tighter wrapping, e.g. frozen turkeys, to exclude air.

The degree of shrinkage of flexible films is measured according to two types of ratio: degree of shrinkage across one axis known as the machine direction (MD) compared to shrinkage across the crosswise or transverse direction (TD). Films are categorised as one of:

- preferentially balanced, e.g. MD = 50%; TD = 20% (i.e. high shrinkage in one direction)
- fully balanced, e.g. MD = 50%; TD = 50% (high shrinkage in both directions)
- low balanced, e.g. MD = 10%; TD = 10% (low shrinkage in both directions).

These different characteristics play a major role in selecting film, measuring and cutting it to shape and the means to complete closure around the product.

There are two ways of closing the film around the product and its container:

- shrink-wrapping
- stretch-wrapping.

In the former, the film is wrapped loosely over the product, which may be placed in a plastic tray, or the product may be placed in a bag, e.g. poultry or joints of meat. It is then passed through a hot-air tunnel, beneath heaters or subjected to a pulse of hot air, or dipped into a hot water bath, the heat causing the film to shrink around the product. In stretch-wrapping, the film is wrapped around the product (e.g. cuts of meat on a tray) under tension. As the tension is released, the film secures the product to the tray. The main advantages of stretch-wrapping over shrink-wrapping are lower energy and film use. This kind of packaging is often combined with, and needs to be compatible with, modified atmosphere packaging systems.

20.10 Modified atmosphere packaging

As mentioned in Chapter 3, the purpose of modified atmosphere packaging (MAP) is to extend the shelf life of a product, typically fresh foods. It inhibits spoilage by replacing the air in a package with a mixture of gases. The gas mixture used is usually selected from oxygen, nitrogen and carbon dioxide. The mixture of gases displaces the air in the headspace above the product fill level. For fragile products such as prepared salad and savoury snacks, the pressure of the gases in the package also helps to protect the product from damage during transport. Another way of modifying the atmosphere is to remove as much of the air as possible, in which case the process is described as vacuum packaging.

MAP works by limiting the microbial activity and chemical processes which cause spoilage. Maintaining suitable atmospheric conditions within a package is complicated because fresh produce continues to 'breathe' or respire within the package, absorbing oxygen and producing CO_2. Accounting for respiration requires the use of a permeable film which allows a certain amount oxygen and CO_2 to move in and out of the package to maintain the balance of gases in the package. Films may also need to allow the moisture generated during respiration to escape so that it does not build up in the pack. Typical films used are coextruded films or laminates of HDPE, LDPE, EVOH, PP and PET. MAP operations also require maintaining a low microbiological count in the product to limit the potential for microbial spoilage.

In a typical batch MAP processing system, air is removed from preformed bags which are then flushed with the gas mixture (typically using a tube inserted into the bag) before the package is heat sealed. In continuous MAP systems, food is packaged either in semi-rigid thermoformed trays which are covered with film or alternatively in pillow pouches or flow packs using either vertical or horizontal form-fill-seal equipment. FFS and shrink-wrapping systems are particularly suitable for MAP. The process of shrinking the film around the product under heat helps to sterilise the packaging film, limit microbial contamination and remove the air to allow an appropriate gas mixture to be added to the product headspace as the film closes around the product and is sealed.

20.11 Miscellaneous wrappers

Twist wrapping machines are mainly confined to the traditional confectionery industry for the wrapping of individual sweets. The sweet or other item is automatically placed onto the edge of the wrapping material (normally a crease-retaining film such as Cellulose™). This is rolled around the sweet whilst holding the ends of the film so that they are twisted together. Bunch wrapping is also used for confectionery and for wrapping soaps. The product is placed in the centre of a square of material which is brought up around it to give the required effect. For soaps, a decorative label is often applied to the bunched area.

Roll wrappers are used for some confectionery products and are widely used for biscuits. The items are collated into the required number per pack and placed on the edge of an oblong sheet of wrapping material such as polypropylene film. This is rolled around the product stack, sealed along its length and the ends folded in and sealed either with glue or through a combination of heat and pressure to fuse the film, using either an overlap or fin seal.

20.12 Coding systems

In any packaging operation it is essential to be able to identify when a batch was made and packed, and even which machinery was used and who operated it. In the event of a problem with the safety or quality of a product, it is essential to be able to identify which batch it came from. It will then be possible to identify if other products from the same batch are similarly affected, the likely source of the problem and whether other batches might also be affected. If the problem is serious, it may be necessary to recall all of the possibly affected batches from the market. These requirements can be met by giving each batch a code which can be applied to the packaging of each item in the batch during the packaging operations. As well as a unique code, dates and times are sometimes added to identify precisely when an item was packed. In the case of perishable items, it may also be necessary to add 'best before' or 'use by' dates for the benefit of consumers during the packaging process either to the package or to the label.

The techniques available for batch coding include:

- embossing
- hot foil printing
- thermal transfer
- inkjet printing
- laser marking.

The process of embossing is essentially letter press printing, with or without the ink, but using steel type to press into the substrate material. Embossing can be done by setting the type into a roller and using this to imprint the flaps of a carton as they pass through the machine. Another example is in tablet blister packing where the type may be built into the heat seal station.

In hot foil printing the type is heated to approximately 130°C and pressed against a film which carries a coating. When the hot type presses the film against the packaging, the heat seals the coating onto the material leaving the mark of the typeface. This method was the established means of coding self-adhesive labels until thermal transfer printing was introduced. Thermal transfer printing uses a similar film to that used for hot foil coding, but the print image is generated by wiping a heater bar across it. Within the bar there is a row of tiny ceramic heaters which are switched on and off at the appropriate moments to create the required print effect.

Whilst hot foil and thermal transfer require the product to be stationary during the print process, inkjet printing relies upon the movement of the product along the belt to give the horizontal component of the print matrix. This requires some form of encoder to enable the machine to respond to variations in the speed of the belt. The inkjet head incorporates a set of individual jets which are switched on and off at the appropriate moments to print a dot of ink to build up the matrix. Others use the electrostatic deflection of a stream of ink particles (see Chapter 19).

Similar systems may be used with low power lasers which are cheaper to operate because no ink is required. These may work by 'ink ablation' in which the surface of the material is removed to leave a white mark out of a coloured background, or by reactive colour change in which an appropriate light reactive pigment is included within the material and changes colour when exposed to the intensity of the laser beam. Lasers may also work on the 'dot-matrix' principle as described for inkjet printing, or may work using the 'pulse-mask' principle, which is similar to a slide projector but uses stainless steel masks to create the required image. Laser systems can be used to etch codes onto suitable surfaces such as glass.

20.13 End-of-line equipment

The equipment required at the end of the line depends on the speed at which the line is running, the weight of the product and the form of secondary or transit packaging being used. This is usually:

- a shrink-wrap (possibly utilising some form of collation tray)
- a corrugated case (with or without internal fitments)
- a combination of both, such as shrink-wrapped point-of-sale units.

20.13.1 Secondary packaging: shrink-wrapping

The shrink-wrapping operation is influenced by pack design and operating speed. If a collation tray is required to hold the primary packs together for ease of handling, the appropriate mechanisms are required to handle those trays and load the product into them prior to wrapping with film. The design of the mechanism will depend on whether the trays have precise locations into which each of the packs has to be placed. If a simple flat-based tray is to be used, a simple mechanism to group the packs together and slide them into the tray will probably be sufficient. If specific locations are required then a more complex mechanism, such as a robotic pick-and-place machine, may be required.

A fully automated system for collating primary packs and presenting them for shrink-wrapping is shown in Fig. 20.25. Once the film is applied it must be shrunk in place. This is normally carried out in a heated tunnel. Care should be taken to avoid overheating both the film and the product. The selection of the correct grade of film for strength and shrink characteristics is essential for the effective operation of these processes, as well as control of dimensions such as film gauge and reel width. In some instances, after shrink-wrapping, products may be collated by the application of a band which is heat sealed to complete the wrap around the pack. Banding may also be used as a means of sealing cases.

20.13.2 Secondary packaging: corrugated cases

Corrugated cases are discussed in more detail in Chapter 11. Corrugated cases are usually delivered packed flat and, like folding cartons, can be glued for ease of making up on the packaging line, or left unglued for wrapping around a collation of primary packs. These operations may be carried out manually or automatically, depending on line speed. Automatic case erecting machines normally demand a higher level of dimensional accuracy from the cartons supplied than those which are manually assembled. This may mean that the cases need to be die cut to provide the required accuracy and consistency of dimensions together with features such as tapered slots and offset creases. Cases may be sealed using adhesive tape, or by cold wet glue

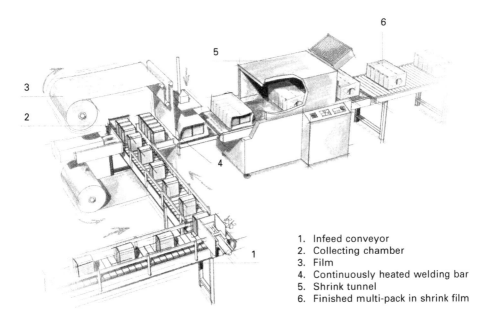

1. Infeed conveyor
2. Collecting chamber
3. Film
4. Continuously heated welding bar
5. Shrink tunnel
6. Finished multi-pack in shrink film

20.25 Shrink-wrapping of collated primary packs.

or hot-melt adhesives. It is normal to include a batch coding at the case sealing stage to enable product identification at subsequent stages in the warehousing and logistics chain.

20.13.3 Palletisation

It is normal to transport the secondary packs on pallets. The loading of the product onto the pallet may be manual, mechanically assisted or fully automatic, using robotic handling systems. Examples of mechanical assistance may include facilities such as:

- a roller conveyor to take the product as close to the pallet as practicable
- a vacuum-operated lift mechanism which uses a suction pad to lift the product and a swinging arm which operates like the jib of a crane to allow the operator to place the product onto the pallet without having to take its weight; the vacuum is then released so that the operator may direct the machine to pick up the next pack
- a moveable platform so that the pallet can be lifted and moved automatically to a designated storage slot.

An example of the latter is shown in Fig. 20.26. In a fully-automated pick-and-place robotic palletisation system:

- the layout of packs onto the pallet is programmed into a computer
- the computer controls the pick-up mechanism and the actuators
- the actuators control the movement of the pack
- the pack is lifted and placed in the next available location on the pallet.

If the packs are of a uniform shape, it is possible to use a mechanism which will:

- collate the packs into the required layout of the next pallet layer
- use a push-bar system to slide the whole layer onto the pallet
- re-position to the appropriate height for the next layer.

Palletised loads are normally stabilised using stretch-wrap film, which may be applied either manually or automatically. However, manual application requires the operator to hold a roll of film and walk around the pallet, bending up and down to guide the film to the required height, whilst pulling on the film to keep it tight. Clearly this is undesirable, both from a health and safety point of view, and in seeking to achieve a consistently effective wrap. Automatic machines normally work by placing the pallet onto a turntable and supplying the film through a mechanism which pre-stretches the film as it is applied to the pallet. The roll applicator mechanism moves up and down to apply the film to the required location on the filled pallet, as it is rotated by the turntable. Again, the selection of the correct grade of film for strength and stretch characteristics is essential.

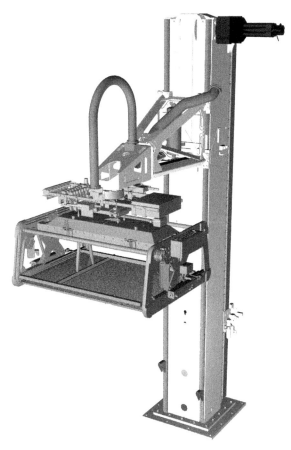

20.26 Automatic palletisation.

20.13.4 Pallet labelling

The palletised load should be labelled to facilitate its identification and destination. This may be:

- a simple handwritten slip of paper tucked into the stack (not ideal)
- a pre-printed label which is applied to the pallet, either manually or automatically
- a label onto which the product details are printed by the application machine (probably using either inkjet or thermal transfer methods, and possibly incorporating the information into a bar code symbol) and then applied automatically to the required location on the pallet.

The location and number of pallet labels depend upon requirements of later stages in the supply chain. A label may need to be visible on each face of the pallet, either by applying four identical labels, or by repeating the information on labels which are applied around diagonally opposite corners of the pallet.

20.14 Quality and efficiency aspects of packaging operations

There is a range of automatic techniques available to ensure that the correct materials are being packed, and that the product is being assembled correctly. Sensors include those using light, electromagnetic waves, infrared radiation, microwave or radio frequency waves, gamma rays or ultrasound. A summary of types of sensor which may be used to monitor aspects of filling and packing sensitive items such as food and beverage products is shown in Table 20.1.

An example of these inspection systems is video tracking of the components and video inspection of the lot coding process. A widely used technique uses a simplified version of bar codes which can be read at speed. These may be applied, for instance at the fold and glue stage of the carton-making process, to ensure that different prints on the same cutting profile do not become mixed in the same job. They may also be used along a filling line to ensure that the correct components are being assembled. To take this to a further stage, optical detectors may be used to ensure, for example, that a promotional item (e.g. a leaflet) is not only dispensed but is actually placed within the carton. These techniques require adequate control procedures to ensure that they operate effectively.

20.14.1 Storage of materials

The storage of packaging materials is important both to ensure they are safe to use (e.g. clean and contamination free) and that they are fit for purpose. As an example, creases in cartons or corrugated cases delivered flat for folding into their final shape will become ineffective if storage times are excessively long and are cold and damp. The items will be difficult to make up, resulting in high wastage on the packaging line. Such materials should ideally be stored in the following conditions:

- ~20°C, 45–55% relative humidity (RH)
- no double stacking of pallets (which might put too much weight on materials at the bottom and could constitute a health and safety hazard)
- away from direct sunlight or cool air.

Other examples of packaging materials which may be adversely affected on the packaging line, due to poor storage are:

- Self-adhesive labels: adhesive failure may occur, or adhesive bleed around the die-cut edges of the label, which can cause labels to adhere to the underside of the carrier web.
- Glass bottles: labelling problems will occur if the bottles are brought into a warm, moist environment whilst still cold enough (from the warehouse) to cause condensation to form on the glass surface.

For maximum performance, all packaging materials, and especially those based on cellulose, are best used after they have acclimatised to the temperature of the packaging line environment, a practice known as 'conditioning'. Sudden changes

Table 20.1 Examples of measured parameters and types of sensors used in food processes

Parameter	Sensor/instrument type	Examples of applications
Bulk density	Radiowave detector	Granules, powders
Caffeine	Near infrared detector	Coffee processing
Colour	Ultraviolet, visible, near infrared light detector	Colour sorting, optical imaging to identify foods or measure dimensions
Conductivity	Capacitance gauge	Cleaning solution strength
Counting food packs	Ultrasound, visible light	Most applications
Density	Mechanical resonance dipstick, gamma-rays	Solid or liquid foods
Dispersed droplets or bubbles	Ultrasound	Foams
Fat, protein, carbohydrate content	Near infrared, microwave detectors	Wide variety of foods
Fill level	Ultrasound, mechanical resonance, capacitance	Most processes
Flowrate (mass or volumetric)	Mechanical or electromagnetic flowmeters, magnetic vortex meter, turbine, meter, ultrasound	Most processes
Foreign body detection	X-rays, imaging techniques, electromagnetic induction (for metal objects)	Most processes
Headspace volatiles	Near infrared detector	Canning, MAP
Humidity	Hygrometer, capacitance	Drying, freezing, chill storage
Interface – foam/liquid	Ultrasound	Foams
Level	Capacitance, nucleonic, mechanical float, vibronic, strain gauge, conductivity switch, static pressure, ultrasound	Automatic filling of tanks and process vessels
Packaging film thickness	Near infared detector	Packaging, laminates
Particle size/shape distribution	Radiowave detector	Dehydration
pH	Electrometric	Most liquid applications
Powder flow	Acoustic emission monitoring	Dehydration, blending
Pressure or vacuum	Bourdon gauge, strain gauge, diaphragm sensor	Evaporation, extrusion, canning
Pump/motor speed	Tachometer	Most processes
Refractometric solids	Refractometer	Sugar processing, preserves
Salt content	Radiowave detector	Pickle brines
Solid/liquid ratio	Nuclear magnetic resonance (NMR)	In development
Solute content	Ultrasound, electrical conductivity	Liquid processing, cleaning solutions
Specific micro-organisms	Immunosensors	Pathogens in high-risk foods
Specific sugars, alcohols, amines	Biosensors	Spoilage of high-risk foods

Table 20.1 Continued

Parameter	Sensor/instrument type	Examples of applications
Specific toxins	Immunosensors	High-risk foods
Suspended solids	Ultrasound	Wastewater streams
Temperature	Thermocouples, resistance thermometers, near infrared detector (remote sensing and thermal imaging), fibre-optic sensor	Most heat processes and refrigeration
Turbidity	Absorption meter	Fermentations
Valve position	Proximity switch	Most processes
Viscosity	Mechanical resonance dipstick	Dairy products, blending
Water content	Near infrared detector, microwaves (for powders), radiowaves, NMR	Baking, drying, etc.
Water quality	Electrical conductivity	Beverage manufacture
Weight	Strain gauge	Weighing tank contents, checkweighing

of temperature and humidity should always be avoided, e.g. taking cartons from a cold warehouse or, worse still, from a cold delivery vehicle onto an automated packaging line will almost certainly result in poor performance. The technical ideal would be to have controlled atmospheric conditions in warehouses but this is likely to be economically prohibitive. The alternative is to aim for minimal storage time of packaging materials (by having close stock controls and just-in-time deliveries) and to hold the materials in a conditioning area for 48 hours prior to use.

20.14.2 Maintenance and training

The correct setting up, operation and adequate maintenance of the packaging line equipment is essential to ensure that product of the correct quality is produced as economically as possible. Thorough training of line technicians and machine operators is required to ensure that machines are operated as efficiently and safely as possible (for both the operators and the machines). Health and safety regulations and the ISO 9001 quality standard both require that accurate records are kept and reviewed to ensure that the staff and equipment are capable of the processes required of them to manufacture the product. It is good practice to have clear, written operating and maintenance procedures as well as records of what training staff have undertaken and what maintenance has been done.

20.14.3 Calculating line efficiency

Line efficiency is defined and calculated in different ways by different companies, and this section is not intended to debate the pros and cons of the various methods used. It is, however, important to try to be clear at the outset about what is being measured. Some key terms and their common definitions are defined below.

- *Station*: the term commonly used to describe each machine serving a specific function in the packaging line.
- *Running speed*: the time taken for a station to complete its cycle (e.g. for a filling machine to fill one container), sometimes defined as the number of cycles the station can complete in a given time.
- *Design cycle rate*: the speed of a station running empty (set by the machine manufacturer after testing).
- *Design speed*: the theoretical running speed of a station under perfect operating conditions (slower than the design cycle rate). This is often defined as the number of containers the station can process in a minute (cpm).
- *Input*: the numbers or volume of product (e.g. the number of containers to be filled) entering the station at a given point in time.
- *Output*: the quantity of product leaving the station at a given point in time under realistic operating conditions.
- *Efficiency*: usually defined as a ratio of output over input.

It is important to understand that the design cycle rate or design speed are not the same as output or efficiency. The latter are (or should be) calculated based on real operating conditions over time as experienced by other manufacturers using the equipment. No machine runs continuously at its design speed. A machine may be marketed with a design speed of 50 cpm which means that, in theory it should process 3,000 containers each hour. In practice, however, it may be affected by common operating problems such as variations in power supply, variability in supply or quality of product, containers or other packaging materials required to complete its operations, blockages, component problems, the need for maintenance and cleaning as well as wear. This may reduce average output in real operating conditions to, for example, 2,500 containers each hour (41 cpm) or 83% efficiency.

In working out the specification required for a machine, it is best to start from the required output, select a machine with a higher design speed specification and look for reliable output data which will match the output required. Reputable machinery suppliers will have this data based on customer experience. It is always advisable to speak to other users. It is important to check potential differences in other users' operations such as the product being processed, levels of maintenance and operator training to assess whether data is comparable.

In setting a required output, it is essential to bear in mind the output of each station in the packaging line since the overall output can never be greater than that of the slowest station. Furthermore, variations in station efficiency have a cumulative effect on overall line efficiency, as the following example shows:

Unscrambler	→	Filler	→	Sealing machine
Design speed: 150 cpm		Design speed: 100 cpm		Design speed: 125 cpm
Efficiency: 95%		Efficiency: 90%		Efficiency: 98%

In this case the line cannot theoretically run any faster than 100 cpm, the filler's design speed. However, its output is 90 cpm (90% efficiency). The overall efficiency of the entire line would be:

100 cpm × 90% efficiency (filling) × 95% efficiency (unscrambling)

× 98% efficiency (sealing) = 84 cpm

In this example, the output of the entire line would be 84% of the maximum theoretical design speed of the line. As a further example:

- a station with a design speed of 100 cpm might run at 95% efficiency (i.e. an output of 95 cpm or packs per minute)
- coupled with a second station also at 95% efficiency, this becomes 90 packs per minute coming out of the second station
- coupled with a third machine running at 95% efficiency, overall efficiency becomes 85 packs per minute.

By the time there are six machines in the line, overall output has been reduced to 73 packs per minute, i.e. overall line efficiency has been reduced by over a quarter from the theoretical design speed of the entire line.

20.14.4 Ways of optimising line efficiency

There are various ways of optimising line efficiency. As a rule, the most critical station in the packaging line is taken as the benchmark. This is usually the filling operation which is often the most complex operation and usually critical to product quality and safety. All other stations are, if possible, specified to run at a higher design speed so that they will always be able to match the output of the filler. Post-filling operations in particular are often designed with a higher design speed. Modern control systems then use this extra capacity to regulate the speed of stations to ensure a smooth movement of packaging to and filled packs from the filler. They also include stop/start controls to close down a station if a serious problem occurs.

In practice, of course, individual stations do not all operate continuously at the same level of efficiency and may vary in performance at different times, creating potential bottlenecks. One engineering answer to this problem is to have accumulating devices between each machine which store up product. Accumulators effectively isolate parts of the line from each other to allow production to continue. This means that, if one of the machines slows down, the machines feeding it may continue feeding the product into the accumulator. At the other end, subsequent operations can also continue by using up the accumulated stock produced by the problem machine. When the problem machine recovers, it also has a stock of accumulated product to process. It is possible to assess the size of accumulators by estimating the likely duration of any interruptions in supply. If it typically takes one minute to resolve most common production problems in a filler with an output of 90 cpm, a following accumulator would need to hold around 90 containers to ensure a continued supply of containers to the following machine. At low line speeds it may even be possible to remove the product from the line into temporary containers and feed it back on again when the line returns to full capacity. It is important to note that, whilst an integral element in many packaging lines, accumulators should not be a substitute for maximising the efficiency of individual machines and the packaging line as a whole.

A typical response to a problem with a particular station is to slow the line down until the problem is resolved and then increase the speed in stages back up to the desired level. It is important to strike the right balance between overall line speed and efficiency. As a general rule, the faster a machine is, the less able it is to cope with variations in inputs such as product and packaging materials and the more vulnerable to problems which reduce efficiency or product quality or which cause stoppages. It may be better to have a slower machine with higher efficiency, better output and more consistent product quality than a fast machine operating at a low level of efficiency.

It is also important to strike the right balance between standardisation and flexibility in optimising production. In general terms, the more dedicated a machine is to process a particular product and container type, the more efficient and reliable it is likely to be. There are often pressures to create innovative types of product and packaging format, or to have a broad range of product offerings (e.g. in differing sizes or weights), as a way of establishing a distinctive position in the market. It is unlikely that a single existing packaging machine will be sufficiently versatile to encompass all these needs. In these circumstances, it is essential for a business to factor in the capital costs of commissioning and acquiring new machinery as well as the operating costs of running a more complex packaging line or lines with potential delays for changeovers (see below). It may be better for overall product quality and cost to consider adapting existing packaging formats, for example, and settling on a small number of standard sizes which can be more easily handled by modifying existing technology with fewer changeovers.

20.14.5 Changeovers

Many packaging lines are designed to accommodate some variations in product, for example by filling different sizes of containers with the appropriate volume or weight of product. Some sort of changeover is then required. In general, machines using a cup or other container to measure by volume have a relatively fixed range compared to weighing machines or machines using metering devices (e.g. pumps) which can accommodate a wider range of variations. As a general rule, the faster a machine, the more complex the changeover is likely to be.

Where frequent changeovers are required, the ease and speed with which each changeover can be achieved may be more important than machine speed. A fast machine with complex, time-consuming changeovers may be much less efficient overall than a much slower, more easily adjusted machine. Ideally, changeovers should involve simple recalibration procedures (e.g. programming instructions for new weights) or simple component changes. It is important to factor in the higher staff costs involved in managing even relatively simple changeovers.

To avoid high inventory costs, many industries operate just-in-time (JIT) manufacturing systems which require maintaining continuity of supply if the rest of the supply chain is not to be compromised. Rapid changeovers may be critical in these circumstances. High volume production of standardised product will drive down unit costs but potentially increase inventory costs. On the other hand, frequent changeovers will increase unit costs for each run but keep inventory costs down. There is a point for an individual business at which each cost is minimised, commonly known as the economic order quantity (EOQ) (Fig. 20.27).

A changeover can be divided into several stages:

- *Preparation*: having appropriate procedures, components and tools together with properly trained staff familiar with the technology and the steps involved
- *Changeover*: replacing components; recalibrating machines
- *Trial run*: testing, final adjustments and run-in time, i.e. the time to bring the machine and packaging line back up to full speed.

Changeover times can be improved by measuring what actions take up most time in practice and ways in which these actions can be made more efficient, for example by:

- improving procedures (e.g. by simplifying and standardising steps and, where possible, having precise, quantified settings to follow)
- improving the quality of documentation (e.g. by breaking down the process into a series of simple, easy-to-follow procedures)
- improving training
- managing as much of the changeover as possible off-line
- using as few components and tools as possible to effect changes
- performing changeovers from one access point.

20.14.6 Using existing or new machinery to increase production

There are several options for increasing production:

- use a third party, e.g. a contract packer

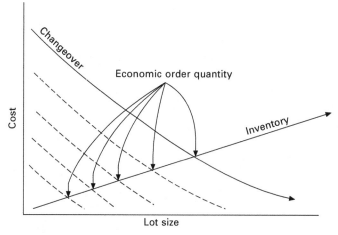

20.27 Economic order quantity.

- increase the efficiency of the existing equipment
- expand production with additional or larger machines (either new or refurbished).

A contract packer is a specialist company offering additional packing services to businesses. Temporary fluctuations in demand may best be met by using reputable contract packers rather than investing in expanding existing capacity which may then be unused when demand returns to normal. Contract packers can also be used in the short term if there is a more sustained increase in demand whilst a business is building up additional capacity, or if a new product line is required which may take a business time to plan and implement. If a contract packer is used, it is important to use one familiar with the product and able to meet the packaging specification, quality, legal and delivery requirements of the customer. These requirements will differ significantly for products such as nuts and bolts vs. food/pharmaceuticals.

Adapting, increasing the capacity or improving the efficiency of existing machinery to accommodate new lines or to increase production has obvious advantages over purchasing additional machinery. The technology is proven and familiar, capital costs will be lower, development time is shorter and less training will be needed. It may also be possible to add extra capacity by buying second-hand machines at reduced cost from companies specialising in selling reconditioned technology. However, these companies may not specialise in a particular product area and may have a limited understanding both of the specific requirements of a customer and of the detailed characteristics and history of a particular machine. Using existing technology will involve higher maintenance costs and may result in reduced efficiency due to wear and increased down time. Adapting an existing packaging line can also become counter-productive if the adaptations push it too far beyond its original specification. Older machinery may be incapable of dealing with new products or types of packaging, and may not be able to achieve increased quality standards in the market. A point may come when it is advisable to plan for a redesign of the packaging line using new machinery, particularly if a substantial new pack design or product line is agreed or a significant and sustained increase in production is anticipated.

20.14.7 Specifying and purchasing of new packaging machinery

Most packaging machines are built within a standard framework or configuration based around a standard pack format. Within this framework, they can then be customised in a number of ways:

- to process a range of sizes of pack
- to achieve a certain range of weights or volumes of product
- to include extra functions such as date coding, leaflet insertion, etc.
- to be compatible with other packaging machinery in the packaging line.

If a manufacturer wishes to avoid the risk and cost often associated with designing a completely new type of packaging machine, it is best to base any new design on a modification of an existing format that could be accommodated by adapting existing technology.

It is normal to begin the purchasing process by drawing up a detailed specification of what is required, possibly by discussion with suppliers. The specification should cover:

- product type, characteristics and requirements (e.g. liquid or solid)
- product processing requirements (e.g. temperature ranges machinery will need to operate in)
- packaging and sealing materials and requirements
- labelling materials and requirements
- basic pack shape and dimensions (including stability – simple shapes such as cylinders are easier to handle – and dimensions such as neck size)
- range of pack weights or volumes (and degree of accuracy) required
- number, type and speed of changeovers required
- output required
- ancillary items required (e.g. promotional leaflets)
- legal requirements (i.e. legislation applicable to the machinery and the product)
- hygiene requirements (e.g. for aseptic processing of food products)
- level of compatibility with other machinery in the packaging line
- potential hazards
- power and installation requirements
- maintenance, staff and training requirements.

Formal quotations are then sought. Other considerations include:

- the capital investment (a line could cost in excess of a million pounds)
- the quality and availability of support (e.g. access to spares and service engineers, quality of documentation and training for in-house staff)
- the timescale for design, construction installation and commissioning
- the operation validation requirements.

Some companies offer a 'turn-key' project management service. It is important to ensure that the requirements and scope of the service are clearly specified.

Once the contract has been agreed, the processes of design and construction may commence. These lead to the manufacturing site acceptance trials, which would include a full validation of all the operational requirements of the machine. The final stage is delivery, installation and commissioning, at which the performance of the machine is assessed to ensure that the required operational performance has been achieved.

20.14.8 Installing new packaging machinery

A packaging line consists of a series of linked machines with separate functions. Typically each will have come from a different supplier and will have its own requirements. To understand how each machine fits into the whole, the line has to be split into a series of clearly logical steps. This requires an understanding of the implications of making a change at any one stage. For example, the addition of an induction heat seal wad into a cap, with the associated equipment required, would call for close control of the torque applied by the capping machine, and of the speed of the belt through the induction tunnel. Failure to control either of these aspects would cause variations in the effectiveness of the induction field applied to seal the membrane across the opening of the container.

In planning the lay-out of a packaging line, it is important to leave sufficient space around the equipment for the delivery of bulk product (and how this feeds the filling machine) and components, through to the removal of the finished stock. It is also necessary to allow space for access by the line team to operate the machines, and to make it possible to clean and maintain the entire line, which may mean removing machines for an upgrade or major repair.

Health and safety of personnel is always of paramount importance in a packaging line. Moving parts must be adequately guarded and factors such as the total weight of material being handled by people must be considered to avoid repetitive strain injury. Professional advice must always be sought here. Of equal importance, packaging lines for products which are to be consumed must be designed and operated to an appropriate level of hygiene, such that the quality, safety and legislative requirements of the finished packed product are not compromised.

A general requirement is that the machine has been fully validated for the product it is to handle, and the materials it is to use. Not only is this in the interests of both health and safety, and of the economical use of resources, but it also has legal implications under material minimisation requirements of the EU Packaging Directive. If the filling line operation is being used as one of the criteria restricting further minimisation, it is necessary to have undertaken trials to confirm that the process may not be operated effectively using less substantial grades of packaging materials. This is addressed further in Chapter 4. In addition to the consideration of the material/machinery interface and implications for the selection of appropriate packaging materials indicated within this chapter, reference must also be made to the chapters relating to those particular materials.

Installing new machinery or packaging lines requires consideration of the services required to operate machinery, including power sources such as:

- electricity
- steam
- compressed air.

It also requires allowing for:

- ventilation
- means for the provision of packaging to the line
- the removal of any wrappings used to supply the packaging (e.g. film wraps and core bungs used on reeled materials, or corrugated cases used for supplying folded cartons)
- the removal of finished product and its secure storage.

Accommodation will also be required for:

- the line staff, who will require to change into appropriate work clothing, and may also be required to have washing and changing facilities
- the supporting engineering services require stores for equipment and change parts
- facilities for the cleaning of machines, especially the parts in contact with the bulk product, and equipment for the cleaning of the production facility
- the space needed to carry out quality inspection processes and store any inspection equipment
- the warehousing and materials handling facilities
- if the line is not being installed solely as a production unit, there would also be the need to provide accommodation for the other aspects of the company.

20.15 Problem-solving on the packaging line

There are several factors which may contribute towards problems on a packaging line, including:

- the design and specification of the machines (which may have been modified from their original design to handle this product)
- the setting adjustment of the machines
- the specification of the packaging components
- the condition of the packaging components (as indicated earlier these may be adversely affected by their storage)
- the adequacy of training given to the line personnel
- the standards of quality of output required.

Consequently it is necessary to take a careful methodical approach to establish the contributory factors, in order to determine what steps would be appropriate to deal with the problems. This is a science in itself, known as operational research. Thankfully, the development of computerised controls on the machines has made the diagnostics of machine operation easier. However, it will not help if there is an issue with the quality of the packaging components. These need to be carefully assessed as the source of the problem may not lie with the supplier.

As an example, a problem of the backing paper of self-adhesive labels snapping in the application machine may be caused by:

- incorrect manufacture of the face material of the label
- incomplete coverage of the silicone coating of the backing paper
- damage to the backing paper at the die-cutting stage of the label printing
- damage to the labels such as inadequate slitting of the rolls or failure of splices introduced during re-wind/inspection of the labels
- poor storage conditions leading to damp labels
- careless handling of rolls of self-adhesive labels may damage the edges of the backing paper and cause tiny cuts to occur which spread under tension
- damage by a hot-foil coder unit which is incorrectly set and creates cuts in the backing paper
- sharp edges in the machine which may damage the backing paper

- an engineering fault within the mechanism of the capstan causing it not to pull evenly
- a worn dispensing beak which is no longer straight, thereby causing uneven tension within the backing paper
- or a combination of some of the above.

Each of these possible causes has further tests that may be undertaken to identify the sources of the problem. As some causes go beyond the packaging components, it is important that the investigation is done in collaboration with the other members of the production team.

20.16 Sources of further information and advice

Details of packaging equipment may be obtained from specific suppliers, which are too numerous to list here. Contact may be made via the appropriate trade magazines and exhibitions, or the Process and Packaging Machinery Association (www.ppma.co.uk).

Other sources of information include: British Contract Manufacturers and Packers Association (www.bcmpa.org.uk) and the Packaging Machinery Manufacturers Institute (www.pmmi.org).

Reference should also be made to the BRC/IOP Global Standard for Packaging and Packaging Materials, Issue 4, London, TSO.

21
Hazard and risk management in packaging

M. EWART, Authenta Consulting, UK

Abstract: This chapter deals with hazard and risk management (HARM), as it applies to the packaging industry. The chapter describes the use of specific methodologies for the development of effective HARM systems. Industry technical standards and associated certification schemes which also have a part to play are reviewed.

Key words: hazard and risk management (HARM), hazard analysis critical control point (HACCP), prerequisites, packaging industry standards.

21.1 Introduction

Hazard and risk management (HARM), as it applies to the packaging industry, has emerged in recent years as a result of two main drivers: cost control and avoidance of litigation, which are to some extent linked. It is possible to see the evolution of HARM as starting from 'failure mode effects analysis' (FMEA) (Beauregard et al., 2008) and hazard analysis critical control point (HACCP) (United Nations Food and Agriculture Organisation, 2003). Each of these techniques was based on a preventative approach to the assurance of reliable and safe products, mechanical and food, respectively.

More recently the term 'risk management' has come into use, describing an approach to providing a means of protecting companies, their stakeholders and their brands from changes, events or incidents with the potential to cause damage (Kaye, 2008; Smith and Politowski, 2008). One reason why risk management has increased in importance is recognition that many business operations have greater dependence on third parties. This is a consequence of a trend towards outsourcing of components (or complete products) as well as services such as design, development, marketing and distribution. Control of risks associated with remote, supplier-dependent aspects of a business has become an inherent part of normal operating arrangements for many companies.

A reliance on suppliers and service providers characterises most of the packaging industry. Not only does the manufacture of packaging have its own, often extended, supply chain, but its products are destined to be part of a complex and demanding chain of supply for virtually every manufactured product, and a wide range of agricultural produce, as well as raw materials. Many of the raw materials used in packaging manufacture are sourced globally. These included paper, carton board, polymers and coating chemicals, each of which may be used in contact with hygiene-sensitive products such as food. This is why, in some circumstances, HARM may be applied backwards in the supply chain, as well as forwards. Industry technical standards and associated certification schemes also have a part to play. This is dealt with in Section 21.5.

Such is the diversity of packaging types and uses that the application of HARM is generally specific to each circumstance. This is why hazard and risk analysis is a vital tool for packaging professionals. Whilst it has a wide application across all of commerce, it is mainly industry sectors with links to the consumer that have applied HARM. There are several reasons why this is so. Major among these has been the perceived need amongst UK retailers to protect their brand value from damage. In the UK in particular, concerns over threats to the consumer from foodstuffs have had a very high profile from time to time. Each of the above influences has played a part in raising the importance of HARM as a management technique. There have also been social and societal changes resulting from increasing prosperity and the expectations and aspirations that follow. Caution is a phenomenon that thrives on affluence.

The terms 'hazard' and 'risk' are themselves problematical. In some languages they have the same or very similar meaning. It is essential to clarify this so that a consistent use of these words, with a common understanding of meaning, is achieved. For the purpose of this text the preferred meanings are as follows.

- 'Hazard' is the *potential* to cause harm/damage. Note that actual cases of harm need not have occurred for the potential to be recognised. Hazards are normally judged in terms of *severity*. This can be seen simply as estimating how dangerous an entity or event might be.
- 'Risk' is the *probability* that a given hazard will occur. The term 'risk' is often used interchangeably with hazard; for example commercial risk management as discussed above has to deal with hazards as well as risks. Risks are normally assessed in terms of *likelihood*.

Arriving at a rational position in these matters requires that hazard and risk are recognised as being different and must be dealt with as distinct phenomena that are nonetheless linked in any given case. The practical implications of this are dealt with in Section 21.4.

21.2 Packaging life-cycles in the supply chain

21.2.1 Background

Present-day developed countries, with high urban population densities, rely on complex systems for the distribution of consumer goods and food. The balance between supply and demand can be delicate. For food, an extension of shelf life greatly reduces the fragility of supply management. This is where packaging can offer one of its important contributions. Foods with extended shelf life are almost invariably given this characteristic by a combination of processing, packaging and, in some cases, temperature control. Many perishable foodstuffs can also have greater longevity where appropriate protection is provided, normally by packaging.

It follows that the performance and sustained integrity of packaging are vital to food safety. Similarly, high-value consumer goods must complete the journey to the consumer in good condition. Two categories of packaging are employed in containment, storage and distribution. These are single-use (one trip), and returnable (often referred to as returnable transit packaging or RTP). They have different characteristics and functions, summarised in Table 21.1.

The life-cycles of these packaging types are illustrated in Figs 21.1 and 21.2. At each stage of manufacture, distribution and use, a range of differing hazards to the

Table 21.1 Comparison of single-use and returnable packaging

Single-use packaging	Returnable transit packaging
Often decorated for consumer appeal	Generally functional in design
Wide range of base materials used	Mainly plastics or corrugated board
Constructed for one trip only	Multi-trip capability
Re-cycling can be problematical	Long life and typically recyclable
May be crucial to food safety	Unlikely to compromise food safety

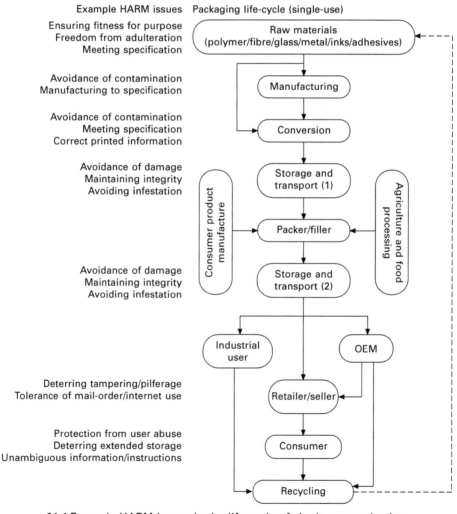

21.1 Example HARM issues in the life-cycle of single-use packaging.

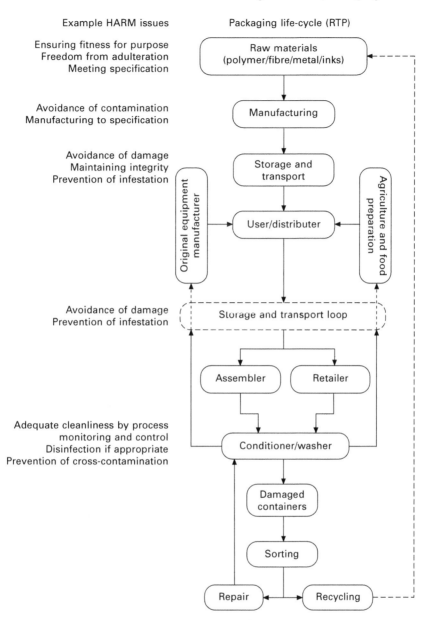

21.2 Example HARM issues in the life-cycle of RTP.

packaging and its contents may apply. Examples that illustrate this are shown in the figures. These examples are not exhaustive but are intended to provide an indication of issues that need to be considered. The use to which packaging is put determines the design, materials and the hazards and risks it will be exposed to. It is not possible to list every combination of packaging type, product packaged, life-cycle challenge and end-use in this chapter. Table 21.2 list examples for a selection of packaging types, uses, associated hazards and possible consequences.

Table 21.2 Packaging types, example uses, hazards and consequences

Type	Use	Hazards	Example consequence(s)
Flexible polymer	Frozen food	Puncture	Contamination, freezer burn
		Seal failure	As above + product loss
	Mail order outer	Rupture	Damage to or loss of contents
		Abrasion	Weakening, loss of print/decoration
Rigid polymer	Consumer products	Weld or adhesive failure	Product damage or loss, pilferage
		Fracture	As above + fragment release
	Food and drink	As above	Product leakage and/or microbiological compromise or physical contamination
Carton board	Consumer products	Physical deformations	Loss of shelf appeal Package/product tampering
	Food	Adhesive failure	Product loss or contamination
Glass	Bottles and jars	Broken glass fragments	Consumer distress; high likelihood of claim and adverse publicity
		Closure seal failure	Product leakage or microbiological ingress post process
Metal	Cans and tins	Pinhole	Product leakage and/or pressure loss (carbonated drinks)
		Sharp edges	Consumer incident leading to claim

21.2.2 Single-use (one trip) packaging

Packaging intended for single-use may be made from any of the materials or combinations of materials employed by the industry. Whilst suitability for recycling is not a question appropriate to this chapter, it is worth noting that ultimately brand value for brand owners, retailers and consumers may be affected, for ill or good, by choice of material. This is especially true of single-use packaging which inevitably raises the spectre of land-fill disposal where recycling is not possible.

There may be occasions when a conflict arises between product protection or shelf life, and environmental impact. This is further complicated by the reduction in the use of packaging being sought by manufacturers, retailers and legislators alike. Similarly the use of materials that are readily recycled is being encouraged. This trend makes the role of HARM more important than ever. Consumers are often cited as the drivers of such developments but the greatest pressure probably comes from organised sources in conjunction with the media. Even at its most simple, packaging can extend shelf life and control potential spoilage. For example, shrink-wrapping a cucumber with polyethylene greatly increases its life in-store and at home yet there have been moves to prevent this on so-called environmental grounds. The cost/benefit case for such packaging use may be overwhelming but often perceptions outweigh facts. A similar case can be made for other more sophisticated techniques that extend shelf life, reduce spoilage losses and improve food safety such as vacuum packing and modified atmosphere packaging (MAP).

The contribution that HARM can make is obvious. It can help in finding the balance between protecting the product, protecting the consumer, and protecting brand value. Some compromise is inevitable. When the performance of packaging is linked to food safety there is less scope for debate. Package types such as cans and retortable flexible pouches, where total integrity is mandatory, have an exemplary record of safety. It is in everyone's interest that this reliability be sustained. There will always be a need for those responsible for commissioning and developing new packaging designs, materials or uses to consider all of the implications. Carrying out a hazard and risk analysis must be part of the process.

21.2.3 Returnable transit packaging

Returnable (or reusable) packaging makes an environmental case for itself. It is mainly used as part of the supply chain for ingredients, components, produce and products. There are few examples of consumer packaging that are reused. Examples of RTP range from the ubiquitous timber pallet to complex multi-material constructions for fragile and high-value manufactured components and assemblies. The largest numbers of RTP types are injection moulded plastic crates used by retailers for fresh produce and meat. A typical retailer pool of these can exceed 10 million units. Figure 21.2 gives a representation of the life-cycle of RTP packaging. Successive storage, transport, recovery and maintenance are common to all types of RTP. Ruggedness is an essential characteristic and in applying HARM is often the primary consideration.

It has already been noted that most types of RTP do not have an interface with the consumer. Generally a hazard and risk analysis will deal with fitness for purpose and protection of contents. The exception is types that are used to contain food. For these, part of each 'trip' includes a wash/disinfection process that is comparable in its thoroughness to food industry practice. An example of a hazard and risk analysis based on a food crate wash line is given in Section 21.4.

21.2.4 Packaging performance

It is essential that packaging is able to sustain its characteristics throughout its life-cycle. The physical challenges that need to be taken into account are often numerous and/or unpredictable. A range of tests is available dealing with structural and material adequacy. Structural tests include drop, vibration, compression and stacking. For materials, adhesive and weld strength, tensile strength, abrasion resistance, ink adhesion, etc., are applicable. More information about the nature and use of such tests is given in Chapter 18. HARM can make a useful contribution to this by identifying likely challenges and thus suggesting which testing methods should be applied. Factors that might be taken into account include:

- product value
- physical characteristics
- product fragility
- mode of transportation
- single-use or RTP

- budgetary or cost considerations
- packaging materials choice/limitation.

Again, the importance of a team approach is clear. The team must be able to draw on market experience and technical knowledge. Customer requirements will be part of the mix, especially when decorative and presentation aspects are important.

21.3 Prerequisite systems and controls

21.3.1 The prerequisites

Management systems that make use of formal HARM or HACCP must ensure that a number of prerequisite requirements have been satisfied (United States National Advisory Committee on Microbiological Criteria for Foods, 1997; the British Retail Consortium/IoP the Packaging Society *Global Standard for Packaging and Packaging Materials* – hereafter the 'Global Packaging Standard' – British Retail Consortium/IoP the Packaging Society 2011). Prerequisite measures are those controls that deal with hazards and sources of hazard that may affect the product at several or all of the stages of production and delivery. HARM is more focused; it deals mainly with definable points in the manufacturing process upon which the safety, legality and integrity of the product may rely. There is no firm dividing line between prerequisites and HARM. Most industry Codes of Practice and Standards provide guidance in this respect, but each case must be dealt with in its own right. Table 21.3 provides a guide

Table 21.3 Prerequisites to HARM

Prerequisite	Example requirements
Premises	Buildings shall be located, constructed and maintained according to appropriate design principles to assure the maintenance of standards that enable product safety and legality to be achieved
Suppliers	The company shall ensure that suppliers have in place effective quality and product safety management systems
Specifications	There shall be written specifications for all raw materials, agreed with suppliers
Equipment	Equipment shall be constructed, installed, calibrated and operated so as to ensure that product will consistently meet specification
Maintenance	Planned preventative maintenance procedures shall be in place
Cleaning	Cleaning of equipment and premises shall be the subject of written procedures, schedules and records
Personal hygiene	Employees and visitors/contractors shall follow written requirements for personal hygiene
Training	Employees shall receive documented training in all relevant aspects of their work
Receiving, storage and dispatch	Raw materials and products shall be stored under clean conditions with documented controls for location and disposition
Traceability and recall	Materials and products shall be recorded in a system which allows rapid and accurate traceability to source and recall from customers
Pest control	An effective pest control programme shall be in place

to aspects of the manufacturing environment, facilities and operational arrangements that are generally dealt with as prerequisite to HARM.

21.3.2 Management systems

Well developed management systems supported by authoritative third-party certification are the most convincing way to demonstrate compliance with legal requirements and industry standards. There are two determinants for the adequacy of such systems: firstly, a basis in good practice for the relevant industry sectors often sourced from trade associations or government bodies; secondly, the robustness or 'auditability' of the system, and the records it generates as evidence of effectiveness. The structure of management systems has now reached an almost classic status with a hierarchy of documents having specific functions at each level. Figure 21.3 illustrates basic structure and functions of documents that form parts of a management system.

Integrated management systems, whereby the requirements of several technical standards and industry sector-specific criteria are documented (on paper or electronically) and managed as a single entity, have become commonplace. In such cases a policy statement will be very general, referring to company objectives. It is at the procedure level that integration is most focused. For example, many of the process monitoring and controls called for by HARM will be common to those of the quality management system. Health and safety procedures will also overlap with HARM to some extent. As an example of this we can consider prevention of product contamination during manufacture. Contaminants of concern are normally considered as falling into three categories: physical, chemical and (micro)biological. Such contaminants, if poorly controlled, can have quality and operator safety implications, as well being potentially hazardous to the consumer. Examples of procedures that might be developed to control potential contaminants are given in Tables 21.4, 21.5 and 21.6.

More detail of typical prerequisite requirements is given in the BRC Global Packaging Standard. It should be noted that both HARM-based management systems and the associated prerequisite controls may have significant legal implications. The defence of 'due diligence' was given prominence as a result of its incorporation in the UK Food Safety Act 1990 and has been cited frequently since. The essential aspect of this is that any manufacturer or supplier may have a defence in law should he be able to convincingly demonstrate that all reasonable precautions were observed prior to

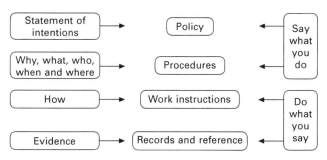

21.3 Structure and functions of a management system.

Table 21.4 Physical contaminants and procedural controls

Potential contaminant	Typical procedure
Glass (e.g. windows, lights, screens, watches)	Broken glass and brittle materials Internal hygiene inspection Personal belongings
Timber (e.g. tables, internal structures, tools, pallets)	Pallet condition Internal hygiene inspection Transport, storage and distribution
Dust and dirt (e.g. machinery, building, lint materials, carried from exterior)	Cleaning Maintenance hygiene Internal hygiene inspection
Paper, card, adhesive tape (e.g. board and tape engineering, documentation)	Maintenance hygiene Internal hygiene inspection
Blades and sharps (e.g. knives, needles)	Control of blades and sharps
Maintenance debris (e.g. wire insulation, machine parts, metal cuttings)	Cleaning Maintenance hygiene Internal hygiene inspection
Items from incoming goods (e.g. metal and plastic items)	Inspection of incoming goods
Personnel (e.g. hair, fibres from clothing, jewellery)	Personnel Personal belongings Training
Pests (e.g. rodents, flying and crawling insects, birds)	Pest control Internal hygiene inspection Inspection of incoming goods

Table 21.5 Chemical contaminants and procedural controls

Potential contaminant	Typical procedure
Cleaning chemicals (e.g. detergents, solvents)	Cleaning
Lubricants (e.g. oil, grease, aerosol sprays)	Cleaning Maintenance hygiene
Odour and taint (e.g. solvents, perfume, disinfectants, preservatives, vehicles)	Cleaning Inspection of incoming goods Personnel Personal belongings Internal hygiene inspection
Non-approved materials (e.g. agents and chemicals not on approved list)	Inspection of incoming goods Supplier assurance Product recall
Water (e.g. from roof leaks, spillages)	Internal hygiene inspections Inspection of incoming goods

any occurrence and a resulting alleged offence. A HARM-based management system would thus constitute a contribution to such a defence. Chapter 4 on legislation gives information relating to legal requirements that impact on packaging manufacture.

Table 21.6 Biological contaminants and procedural controls

Potential contaminant	Typical procedure
Personnel (e.g. incorrect use of protective clothing, dirty hands)	Personnel Training
Pests (e.g. contamination by droppings or dead bodies)	Pest control Internal hygiene inspection
Dust and dirt (e.g. contamination from equipment and environment)	Cleaning Maintenance hygiene Internal hygiene inspection
Materials (e.g. contaminated packing and raw materials)	Inspection of incoming goods Supplier assurance
Damp (mould growth) (e.g. goods affected by water during storage/transport)	Internal hygiene inspection Inspection of incoming goods

The ability to demonstrate due diligence is one of the fundamental reasons for adopting the HARM approach. It has become a routine customer requirement and is incorporated in several technical standards including those that apply to packaging, storage and distribution. The contents of such standards are dealt with in more detail in Section 21.5.

21.4 Hazard identification and risk assessment

21.4.1 Hazard and risk analysis

It is often convenient to consider hazards in relation to the packaging/product, and/or the consumer as distinct issues. This distinction is most important when the product packed may compromise the consumer as a result of contamination during manufacture, or due to a defect. In most instances the manufacturing controls that deal primarily with quality parameters also have relevance to fitness for purpose in general, and consumer protection in particular. It is normal for a hazard and risk analysis to be carried out in circumstances where a mature quality management system already exists. Herein lays a dichotomy. The existing, proven, controls might be assumed to provide sufficient assurance of product integrity and safety. The system of which they form a part will probably be based on many years of production and customer experience. On the other hand, testing that system by rigorously applying hazard and risk analysis is as likely to uncover deficiencies, as to confirm adequacy.

The process of hazard and risk analysis must be carried out in a rational and disciplined way. It is vital that each stage is documented to an extent that confirms competence of the individuals involved, and credibility of outcome. To this end there are a number of requirements that must be met. The commitment of senior management is a fundamental requirement. A properly conducted analysis will require resources of people and time which also equates to cost. Such is the complexity of the task that a multi-disciplinary team will be required. If a sufficient level of skill and knowledge is to be available, then several senior members of staff will be required to commit their time. In some circumstances the use of external experts is desirable or necessary. For example, a manufacturer of packaging destined for the food industry is unlikely to have a staff microbiologist. It should be remembered that whilst an external expert may contribute to the analysis and development of associated management controls, subsequent operations will be the responsibility of the company.

As with any teamwork, it is necessary for a leader to be identified. This individual will have overall responsibility for the progress and outcome of the analysis. Such responsibility must be supported by adequate authority delegated from the highest level in the company. The authenticity of support and authority given to the team leader is one measure of management commitment. The team leader and the team should receive appropriate training in HARM techniques. There are numerous organisations offering such training. Any worthwhile prospective training provider will be able to offer several reference clients for whom they have worked and this simple test should be applied.

The formation of a HARM team cannot be regarded as a one-off exercise for the sole purpose of carrying out the hazard and risk analysis. Like any component of a management system, the controls developed and introduced through HARM will require continuing review and revision. Consequently the system must include reliable arrangements for keeping the team informed of changes to the factory, equipment, products and customer requirements so that these can be reflected in the HARM controls. The general approach to hazard and risk analysis is similar for all applications of the methodology. The working examples used in this text are applicable to hazards to the consumer that might arise from failures in the manufacturing process. They are equally relevant to hazards to the manufactured product during the product life-cycle as discussed in Section 21.2.

21.4.2 HARM development

Records of the initial HARM development work should include a list of the team members and their job titles. Each stage of the process thereafter must be recorded. The process itself has been outlined in several published versions which are, in essence, the same. The following list of stages is based on these. These steps are:

(a) Describe the product(s), intended use, and consumer/user.
(b) Decide the scope of the HARM system and draw up an accurate diagram of the process flow(s).
(c) Identify and assess the possible hazards due to process failure at each process step.
(d) Introduce or confirm monitoring and control at critical process steps.
(e) Validation and verification.
(f) HARM review.

Describe the product(s), intended use, and consumer/user

Where a number of products that differ in materials, production methods or intended use are manufactured, it may be necessary to conduct more than one analysis. The decision as to the significance of any differences and whether these merit separate consideration should be made by the team. It may be that, in the light of subsequent stages of the HARM development, this decision will merit review. The team leader should note this and call for review as appropriate. The intended use of the product and any potential consumer impact should take account of possible consumer vulnerability. For example, a label for infant food, or one where confusion over composition (ingredients) of contents could arise, may require special attention to text layout and wording.

Decide the scope of the HARM system and draw up an accurate diagram of the process flow(s)

Scope in this context is the proportion of the product life-cycle to be covered by the HARM system. This needs careful consideration as some parts of the product life-cycle may not be completely under the control of the company. As an example of scope, it could include the manufacturing process from reception of raw materials to dispatch of finished products. In some cases the scope may be extended beyond these limits. The effectiveness of prerequisite measures in dealing with possible hazards not under the company's direct control should also be taken into account when establishing scope. Processes vary in complexity and in some cases aspects normally dealt with as prerequisites may require full analysis.

Diagrams or flowcharts that describe a process vary in the amount of detail they include. This will be influenced by the complexity of the process as well as the sensitivity of the end use of the product. Generally, a flowchart to be used as part of HARM in packaging manufacture need not be highly detailed. The number of process steps identified will normally be between eight and fifteen and the flowchart should include ancillary processes in addition to the main line of flow, appropriately linked.

Hazard and risk analysis allows for more than one hazard to be identified at each step; hazards should not be considered at this stage. The objective should be to accurately and adequately describe the process or processes without consideration of hazards. It can prove helpful to conduct an on-site comparison of the flow diagram with the actual factory arrangements. The involvement of personnel not familiar with the process can help to avoid assumptions about the validity of the diagram. An example of a flow diagram is given in Fig. 21.4.

Identify and assess the possible hazards due to process failure at each process step

This is the most problematic aspect of HARM development. The challenge is to decide which of the hazards is of sufficient severity and likelihood to merit special attention. Process steps where such hazards are identified are often described as 'critical control points' (CCPs) or 'critical process steps' (CPSs). A number of methodologies that aid this decision have been developed. It is important to recognise that such methodologies are best used for guidance in reaching the 'criticality' decision.

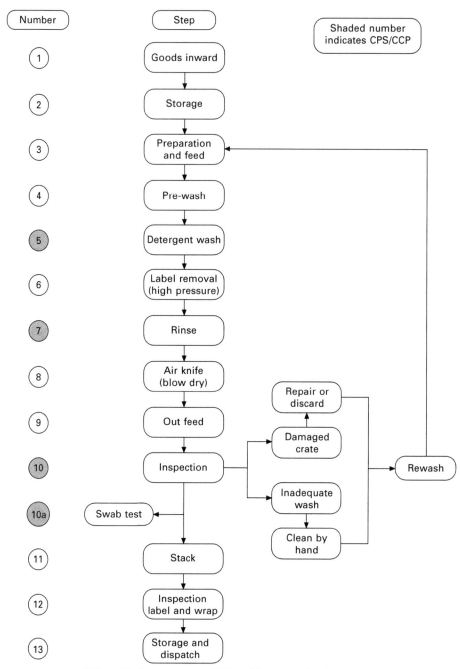

21.4 Example process flowchart (RTP washing).

Ultimately that decision will be a matter of judgement as well as analysis, again emphasising the importance of using a team in arriving at conclusions. A number of 'decision trees' have been available, derived from that first published by Codex Alimentarius for use in HACCP development. Codex Alimentarius have recommended caution in the application of their decision tree, implicitly recognising that it may have been misapplied in the past. Such misapplication can result in large numbers of CCPs being identified resulting in excessively large and complex and potentially unworkable HARM/HACCP systems.

HACCP decision trees were devised to deal with process steps in the food industry. Given the inherent biological stability of packaging materials, it is possible to simplify the decision tree without compromising its validity. An example of a simplified decision tree is given in Fig. 21.5. The simple decision tree adds another consideration for inclusion in the analysis. If loss of process control will render the product recognisably out of specification to an extent that it will be unsuitable for sale, then that point in the process may not be critical.

In addition to the use of the 'decision tree' for guidance in determining critical process steps, a 'scoring' system is often adopted. One such system is set out below. Note that the essentially subjective nature of the criticality decision is not made objective by the adoption of such a scoring system. Critical process step decisions are always to some extent judgemental.

What would be the consequence of the hazard occurring (the Severity)?

High   3 Death or serious injury
Medium 2 Temporary disability/minor injury
Low   1 Discomfort
None   0 Negligible consequence

Similarly, the likelihood of occurrence can be given a rating as follows:

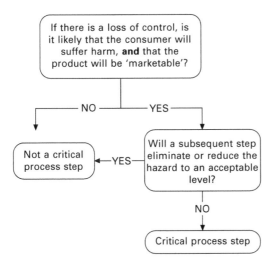

21.5 Simplified 'decision-tree'.

How likely is the hazard to occur (the Risk)?

> High   3 Likely to happen
> Medium 2 Could happen
> Low   1 Unlikely to happen
> None   0 Cannot happen

The score for each identified hazard/control is calculated using the following formula:

$$\text{Score} = \text{Severity rating} \times \text{Risk rating}$$

The range of resultant scores is 0 to 9. A process step should be considered to be potentially critical where the score is *higher* than 2. If a score of '3' is a result of the estimate of risk being '3' and severity '1', then the answer to the question 'Has this ever happened?' should form part of the judgement. Documentation of the decision process can be by tabulation of process steps and each aspect of the analysis. An example of this is given in Table 21.7.

Introduce or confirm monitoring and control at critical process steps

For each process step identified as critical, it is necessary to confirm that current system controls are adequate where they exist, or to take action to achieve adequacy. This might be done by improvement of existing controls or development of new procedures and work instructions that will lead to effective control. It is normal to record this. Table 21.8 is an example of how this may be presented.

In order to demonstrate that all of the requirements of HARM have been fully met, it is necessary to state for each critical point the method and frequency of monitoring,

Table 21.7 Example hazard analysis chart, RTP (retail crates) washing

Process step #	Hazard(s)	Type[a]	Hazard analysis — Severity	Hazard analysis — Likelihood	CCP decision	Reason (Decision Tree?)
1	None identified				N	Pre-process of incoming soiled crates
2	None identified				N	Pre-process of incoming soiled crates
3	Potential contaminants not removed	M P C	2	0	N	Subsequent steps are specifically intended to deal with this hazard
4	Potential residues due to process failure	M P	2	1	N	Subsequent steps are specifically intended to deal with this hazard
5	Potential residues due to process failure	M P	2	2	Y	Score of 4; process parameters under direct control at this step

[a]M (microbiological), P (physical) or C (chemical) as appropriate.

Table 21.8 Example HARM control chart, RTP (retail crates) washing

Process step #	Process monitoring	Frequency	Target	Tolerance	Corrective action	Document[a] References/ records
5a	Temperature measurement	4 times per shift	55°C	± 5°C	Check m/c parameters and functionality. Re-process	TP011 TF004 WI004
5b	Detergent concentration measurement	Continuous by probe. Manual check if step 10 or 10a dictates	½ % in water	± 10%	Check dosing equipment function. Re-process	Instrument calibration TP011 TF004 WI001
7	Temperature measurement	4 times per shift	75°C	± 5°C	Check m/c parameters and functionality. Re-process	TP011 TF004 WI004
10	Physical inspection	Constant during process	No visible residue	Nil	Reject and re-process	TP011
10a	ATP Swab testing	4 times per shift	<150 rlu	+150 rlu	Re-process as necessary	TP011 TF004 WI003

[a]TP = technical procedure, TF = technical form, WI = work instruction.

target and tolerance values, corrective actions that will be taken if tolerances are exceeded, and the identity of related documents in the management system.

Validation and verification

In this context validation is a test of fitness for purpose of the documented system, its procedures, work instructions and records. It is the process of confirming that they are capable of achieving the intended control of hazards. Validation should be undertaken prior to the implementation of the control measures. It is best if validation is at least overseen by a person or persons who were not directly involved in system development although this is not essential. Validation can include carrying out tests to the product to confirm that the parameters set are being met. This might include microbiological testing or chemical analysis as well as dimensional checks, where these could have a safety-related effect on product performance.

Verification is the process for confirming that the system is meeting its objectives in practice. This is carrying out mainly through internal audits. An audit programme must take account of HARM, with verification one of its specific activities. Frequency of audit of critical process steps may be greater than that of other parts of the system where this is deemed necessary. Verification seeks to establish the extent to which the management system is correctly implemented, and meeting its intentions.

HARM review

Management systems are subject to periodic review usually by report to a review group including senior management. The HARM system is part of this, taking account of the results of internal and external audits, and customer concerns and complaints. The reports should identify any need to update or improve the HARM system, and any trends in the occurrence of incidents linked to product safety. Actions taken to address these issues should also be reported. Incidents where critical limits have been exceeded merit special attention and corrections applied and corrective and preventative actions must be fully documented. It is the role of the review group to confirm that satisfactory outcomes have been achieved.

21.5 Industry technical standards

21.5.1 The standards

There are three technical standards relevant to packaging manufacture, storage and distribution that require a HARM or HACCP system as part of compliance and certification. Hazard and risk analysis, and management, can be applied to other standards. For example, the BS *Occupational Health and Safety Standard* (BS OHSAS 18001:2007) calls for analysis of hazard and risk (although the methodology differs from that set out here). HARM methodologies can also be of value when establishing controls for compliance with quality and environmental standards.

Table 21.9 lists the relevant standards and gives an indication of their scope. The BRC/IoP *Global Packaging Standard* and the BRC *Global Standard for Storage and*

Table 21.9 Industry technical standards

Standard	Scope	HARM requirements
BRC/IOP *Global Standard for Packaging and Packaging Materials*, Issue 4, February 2011	Manufacture of packaging and packaging materials for food and 'hygiene sensitive' products (includes other food-contact items)	Mandatory for all categories of packaging/materials
BRC *Global Standard for Storage and Distribution*, Issue 2, September 2010	Pre-packaged and loose food products, packaging materials and consumer products (includes wholesale and contracted services such as cleaning of returnable packaging)	Mandatory with variations for consumer products. Some technical guidance is appended
BS EN ISO 22000:2005, 'Food safety management systems – Requirements for any organisation in the food chain'	All organisations, regardless of size, that are involved in any aspect of the food chain	HACCP is a mandatory requirement

EU Regulations (No. 178/2002, No. 852/2004) legally oblige food storage and distribution companies to ensure systems are in place to supply safe and legal products.

Distribution (British Retail Consortium, 2010) contain detailed requirements for the conduct and recording of hazard and risk analysis for the establishment of HARM. In addition the BRC standards require the use of hazard and risk analysis in deciding the level of control appropriate for meeting some of the standards' requirements, according to the degree of risk encountered. Examples of this include the frequency of pest contractor visits and the identification and control of risk of microbiological contamination.

The BRC *Global Packaging Standard* was written so as to give sufficient flexibility to embrace the wide range of packaging types and potential hazards and risks. This was achieved by enabling companies to seek exemptions from some of the standard's prerequisite requirements by using hazard and risk analysis. For an exemption to be valid, a company must demonstrate that hazard severity and risk likelihood are such as to make compliance with the specified requirement inappropriate or unnecessary in given circumstances.

Although ISO 22000 (British Standards Institution, 2005) is nominally a food safety standard, its title does indicate broad scope and some sectors of the packaging industry have adopted it. The washing of retail crates is an example of this. RTP conditioning companies do not manufacture packaging and thus fall outside of the scope of the BRC/IoP Packaging Standard. Many packaging manufacturing companies have adopted ISO 22000 and become certificated to it. There is a perception that there are benefits in it being international in its derivation and acceptance.

ISO 22000, like ISO 9001 with which it has been devised to be compatible, is technically non-prescriptive. In this respect it differs from the BRC Standards which are technically detailed and prescriptive. In order to give direction to companies seeking to comply, ISO 22000 requires that reference be made to appropriate sources of guidance. These include the Codex Alimentarius Commission (Codex) principles and codes of practice (United Nations Food and Agriculture Organisation, 2003), and national or sector-specific standards. There is a paradox here in that the obvious sector-specific standard for packaging manufacture is the BRC/IoP *Global Packaging Standard*. It may be that customers from the food industry will ultimately prefer their suppliers to be certificated to ISO 22000 because of its clear food remit. Some companies have adopted certification to both standards in order to meet differing requirements of customers.

21.5.2 Using the standards

The practical impact of adopting any standard is much affected by interpretation and implementation. This can result in the management systems and operational circumstances of companies certificated to the same standard appearing to differ. Whilst the development and introduction of industry standards has always been promoted as a way of achieving adequate control and consistency, many customers continue to carry out technical visits. Some of the variability between certificated companies is a reflection of customer preferences. Often a customer will require more demanding site standards or production controls in specific respects than are judged to be necessary for compliance with a standard.

Certification to a standard implies complete compliance. Since all certificated companies are in that position, compliance is often seen as meeting the minimum that is acceptable. Many companies choose to go beyond that, as a result of a determination to be seen to be leading their sector, or to demonstrate excellence to customers.

Packaging companies may be in one of three circumstances when working with a technical standard. They may already be certificated, working towards certification, or in the process of changing from one standard to another. These alternatives are shown in Figure 21.6.

All technical standards require the commitment of management. Organisations working towards certification, using a standard for the first time, must support their objective with adequate resources. Unless the individual designated to lead the exercise has demonstrable previous experience then training may be necessary. Often the use of a consultant will provide an experiential route to developing the required knowledge and skills without the need for formal training. The best approach can only be decided by the management team and initial informal discussions with certification bodies and consultants will guide this. Familiarity with the contents of the chosen standard and the process of certification can only be achieved in the light of experience. Training will help this. A properly conducted training course, based on interactivity as well as the imparting of knowledge, and shared with others with a common interest, can be invaluable.

The size of the task is best assessed by carrying out a gap analysis. That is, comparing the current position with what has to be achieved. This enables an action plan to be drawn up. There are two main areas of activity that need to be planned. Firstly, some management system documents may require revision. Additional documents are also likely to be needed. Secondly, the site, factory, equipment and personnel management will have to be reviewed to ensure compliance. Most certification bodies will be seeking evidence that staff at all levels have received training commensurate with their responsibilities.

When the above work is complete or nearing completion, it is normal to have a pre-audit carried out. Many certification bodies are able to offer this, as are independent consultants. The pre-audit is a final critique of the system and site prior to the certification audit. If an independent consultant is used, it can also serve as the first internal audit, evidence of which is a requirement of certification. The system documentation, especially that concerned with the non-conformance, corrective action and review processes, should be in use at this stage to subsequently provide objective evidence of functionality to the certification auditor. Similarly, records of process parameters, cleaning activities, pest control visits and so on will be essential.

21.5.3 Certification

The first certification schemes for hygiene and product safety in packaging manufacture were launched in 1990. During the following years a number of competing schemes were offered. None of these early certifications had links to accreditation bodies such as the United Kingdom Accreditation Service (UKAS). These schemes relied on being accepted by the packaging industry and their customers, and the credibility of the

Hazard and risk management in packaging 557

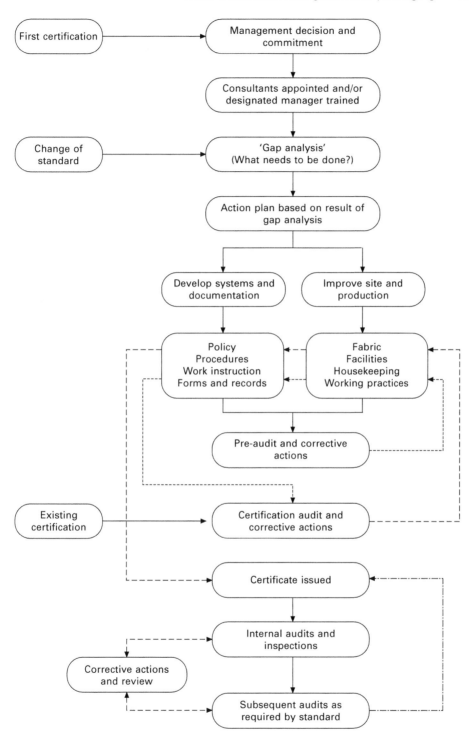

21.6 Certification process.

organisations that operated them. The first issue of the BRC/IoP Packaging Standard was published in 2001, following development by a committee with representation from the packaging industry, their customers and the certification organisations. Among the reasons for the development of the BRC/IoP Standard was drawing together the various competing certification schemes into a single common position. In addition, certifications were required to be operated by accredited certification bodies. This continues to be the position for all the BRC Standards and ISO 22000.

The terms accreditation and certification are often confused. Stated simply, a company is *certificated* by an accredited certification body. A certification body (of which there are several in most countries) is *accredited* by a National Accreditation Body (in the UK this is the United Kingdom Accreditation Service – UKAS). These relationships are conducted according to internationally accepted standards and guides. They are illustrated in Fig. 21.7.

There can be no doubt that during almost 20 years since the introduction of industry standards dealing with hygiene, and hazard and risk management, there has been a transformation in operational arrangements in the packaging industry. Sites in the UK and throughout the world have developed, in some cases transformed, their working environments and production controls. Factories have become more pleasant places to work, to the benefit of all involved. Improvements to product integrity have been an inevitable consequence.

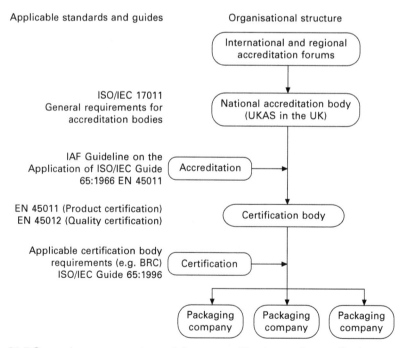

21.7 General arrangements applying to certification and accreditation.

21.6 References

Beauregard, M. R., McDermott, R. E. and Mikulak, R. J. (2008). *The Basics of FMEA*, 2nd edn. New York, Productivity Press.

British Retail Consortium (2010). *Global Standard for Storage and Distribution*, Issue 2. London, TSO.

British Retail Consortium/IoP the Packaging Society (2011). *Global Standard for Packaging and Packaging Materials*, Issue 4. London, TSO.

British Standards Institution (2005). BS EN ISO 22000:2005. *Food Safety Management Systems. Requirements for any organisation in the food chain*. London, TSO.

British Standards Institution (2007). BS OHSAS 18001:2007. *Occupational Health and Safety Management Systems. Requirements*. London, BSI.

Kaye, D. J. (2008). *Managing Risks and Resilience in the Supply Chain*, London, BSI.

Smith, D. and Politowski, R. (2008). *Good Governance – A Risk-based Management Systems Approach to Internal Control*. London, BSI.

United Nations Food and Agriculture Organisation (2003). *Codex Alimentarius Food Hygiene Basic Texts*, 4th rev. Available from http://www.fao.org/docrep/006/y5307e/y5307e03.htm#bm3 (accessed 4 April 2009).

United States National Advisory Committee on Microbiological Criteria for Foods (1997). *Hazard Analysis Critical Control Point Principles and Application Guidelines*. Available from http://www.cfsan.fda.gov/~comm/nacmcfp.html (accessed 4 April 2009).

Index

abiotic spoilage, 42–3
acid sulphite process, 193
acidity, 42
Aclar, 303
acrylic, 387
acrylonitrile butadiene styrene (ABS), 301–2
adaptation, 84
additives, 199–200, 283–4
　nano fillers for improved gas barrier, 284
adhesive labels, 376
adhesives, 245
　adhesion theories, 382–4
　adhesive selection, 389–90
　application methods, 390
　packaging, 381–2
　performance evaluation, 391
　terminology, 392–3
　troubleshooting adhesive problems, 391–2
　types, 384–9
　　100% solid, 388–9
　　solvent-based, 388
　　water-based, 384–7
aerosol safety, 55
alkaline sulphate process, 193
alternating copolymers, 271
aluminium foil, 163–77
　container types, 177
　end uses as packaging material, 176
　finishes, coatings and lacquers, 170–2
　　chemical resistance of coating materials, 171
　　general properties and characteristics of compounds, 171
　　standard treatments, 170
　laminate, 172–4
　metallised films, 174–6
　overview, 163–5

properties of aluminium as a packaging material, 164
printing and embossing, 172
processing, 165–6
　aluminium manufacturing process, 166–7
　production, 168–70
　continuous casting or hot-strip casting, 169
　conventional rolling-mill method, 169
　refining, 167
　smelting, 167–8
aluminium metallised films, 174–6
　laminate samples, 174
　vacuum metallising, 175
aluminium oxide, 128
American Society for Testing and Materials (ASTM) International, 391
analysis, 427–8
　assessment chart of design concepts, 428
anilox, 472
animal glues, 386
annealing, 115
atactic configuration, 268
auger filler, 497
autoclave, 266
average fill, 56

Barex, 302, 303
basis weight, 325
bauxite, 128, 165
Bayer process *see* refining
beating, 196–8
　effect on paper and paperboard properties, 198
　fibres before and after, 197
　illustration, 196
Bendtsen Test, 260

bio-based plastic, 79
bio-based polymers, 306–8
biodegradable plastic, 79–80
biodegradation, 78–80, 307
biotic spoilage, 42
bleaching, 198–9
bleeding, 480
block copolymers, 271
blow-and-blow, 113–14
blow moulding, 350–9
  design specifications, blow moulded
      bottles, 358–9
    carbonated soft drinks bottle
        specification, 358
    standard bottle terminology, 359
  extrusion blow moulding, 352–8
    alternating and rotary blow moulds, 358
    moulding cooling, 356
    multilayer extrusion blow mouldings,
        357
    parison thickness control, 355
    parison thickness control problems, 355
    process, 354
  injection, 350–2
    multilayer injection moulded PET
        bottles, 353
    process, 351
    stretch process, 352
blow up ratio, 318
blown film, 318–21
  annular blown-film die cross-section, 320
  bubble process, 322
  standard blown-film process, 319
bore gauger, 116
borosilicate glass, 110
bottle spacer, 115
Box Compression Test (BCT), 258, 259
brand
  consistency and complex information, 99
  marketing matrix, 95–7
    place, 97
    price, 95–6
    product, 95
    promotion, 96
  product promotion and advertising, 93–5
    coherence, 94
    distinctiveness, 94
    market appeal, 94–5
    protection, 95
    relevance, 94
    promotion, 97–8
    value determination, 92
brand name, 4
brand values, 417, 431
  top-branded products, 418
break elongation, 326
bright finish, 170
BS 4816:1972, 258
BS 4817: 1972, 258
BS 7325, 258
BS 6538-2:1992, 260
BS EN 13432, 78, 79
BS EN 23035:1994, 258
bubble film method, 320
burst strength, 182

calcination, 167
can body, 158
can decorating, 159–60
  cylindrical decorating machine, 160
can end, 158
  making process, 144–8
  process performance, 126–7
cannibalism, 421
carbon dioxide, 83
carbon footprint, 68, 85
  calculation, 85
Carbon Trust, 68, 85
carded blister, 236–8
cartonning, 513–14
  erection, filling and closing of flat carton
      blank, 513
casein, 385–6
cast sheet, 315–18
cellophane, 305, 311
cellulose, 521
cellulose acetate, 306
cellulose acetate butyrate, 306
cellulose film, 321–3
  casting process producing regenerated
      cellulose film, 323
cellulose materials, 305–6
certification, 556–8
  general arrangements, 558
  process, 557
changeovers, 531–2
  economic order quantity, 532
Chartered Institute of Marketing, 87
chemical adhesion, 383
chemical pulp, 193–5

Index

production of chemically separated and bleached pulp, 193–5
pulp preparation, 195
pulp production using a hydrapulper, 195
chemical stability, 290
chemical thermo-mechanical pulping (CTMP), 191, 194–5
child-resistance, 376–7
child-resistant closures, 55
chilled air, 318–19
Chromaflair, 455
clarification, 167
clay coated Kraft back (CCKB), 185–6
clay coated news back (CCNB), 185
closing time, 392–3
closures, 506–9
    can closing, 508–9
        seaming of metal can end, 509
    lug, press-on twist-off and crimped or crown closures, 508
    push-fit closures, 506
    roll-on pilfer-proof (ROPP) closures, 507–8
        application of metal closure, 508
    screw-threaded closures, 506–7
coating weight, 393
Cobb Test, 260
Cobb value, 180
Code of Federal Regulations (CFR), 131
coding systems, 521–2
coefficient of friction (CoF), 326
Coefficient of Friction Test, 260
coextrusion
    cast and blown films, 323–5
        cast film coextruder feeding into feed box and die, 325
    coextrusion process, 324
cohesive strength, 393
cold foil techniques, 406
cold-seal temperature, 326
collation shrink film, 294
Colorstream, 455
colour, 339
    factors affecting perception, 445–7
        ideal conditions for appraising colour, 447
        illuminants, 446
        impaired colour vision, 446–7
        metamerism, 447
        simultaneous contrast, 446

colour mixing, 444–5
    additive colour mixing, 445
    subtractive colour mixing, 445
colour printing, 447–55
    alternative screening techniques, 451–3
        hybrid screening, 453
        stochastic screening, 452–3
    chromaticity diagram, 448
    hexachrome, 450–1
    special colours, 453–5
        fluorescent effects, 454
        metallic effects, 454
        other special effects, 455
        pearlescent effects, 454–5
        sensory special effects used in packaging, 455
        special effects, 454
        spot colour, 453–4
    subtractive colour mixing, 447–8
    tonal reproduction and halftone printing, 448–50
        dot shapes, 450
        screen angles, 449
        screening and resolution, 449
colour specification, 405
colour variation, 479
colour vision, 443–4
    human eye, 444
coloured lacquers, 172
Combibloc, 8
competitor research, 104–5
composite closures, 154
compostable materials, 78
compostable plastic, 79
composting, 78–80
compression, 32–6
    dynamic and static compression correlation, 33
compression strength, 182
conceptual design, 423–31
    generating concepts, 424–31
        analysis, 427–8
        design development, 428–31
        glass jar project sketch, 425
        graphic design, 427
        structural design, 425–7
        working ways, 424–5
    inspiration sources, 424
Concora Medium Test, 260
condensation polymerisation, 269–70

PET polymerisation, 270
consumer needs, 100–1
consumer panels, 103–4
Consumer Protection Act (1987), 58
consumer requirements
  benefits, 420
    packaging, 421
consumer research, 101–2, 103–4
consumers, 13–14
container base, 120
container body, 119
container finish, 119
container heel, 120
container neck, 119
container shoulder, 119
containment
  function evaluation, 25–6
    failure samples, 26
contract packer, 533
control of substances hazardous to health (COSHH), 56
convenience, 46–7
conventional soda glass, 110
convolute single wrap containers, 218–19
  can design, 219
coordination polymerisation, 267
copolymer, 280
copolymerisation, 271–2
copyright, 58
Copyright, Designs and Patents Act, 58
corporate identity manuals, 99
corporate social responsibility (CSR), 85
corrugated cases, 523–4
corrugated packaging, 240–61
  container design types, 249–53
    case dimensions, 255
    case styles, 249, 250
    common types of fitment, 254
    flat sheet laid out with the resulting folded dimension, 255
    multi-component case designs, 252
    regular slotted container (RSC) blank parts, 251
    scoring allowances for board of various thickness, 256
    sizing a container, 256
    typical die-cut case design, 252
  decoration and printing, 253–6
  manufacturing, 245–9
    corrugating machine, 246
    gap tolerances for B- and C-flute single-wall cases, 248
    gap tolerances for out-of-square and fishtail (U joint) configurations, 248
    length and width of glue tabs, 249
    overlap identification, 248
    scored and slotted sheet tolerances, 247
    three-point and five-point scoring wheels profile, 247
  materials, 242–5
  materials and containers testing, 257–61
    equilibrium moisture content, 258
    mechanical strength, 258–60
    paper properties, 257–8
    properties testing, 260–1
  overview, 240–2
    single-wall and double-wall configuration, 240
  special board treatments, 256–7
cost, 65
cost effective packaging, 14–19, 75–6
counter-pressure fillers *see* pressure filler
Courtauld Commitment, 81–2, 84
crack detector, 116
cradle-to-grave packaging, 11
crimped crown cork closures, 372–3
  image, 373
cross direction (CD), 316
crosslinking polymerisation, 271–2
  crosslinked polymer, 272
crown cork closures, 155
crystallinity, 276–8
  LDPE to HDPE progression, 277
  polyethylene homopolymer main form, 277
cup fillers, 505–6
cyclic olefin copolymers, 304
cylinder method, 206–7
  uniflo vat cylinder, 207
  vat paperboard machine, 207

damage, 20
Danner process, 121
data errors, 20
deceptive packaging, 58
decoration, 425
degradable materials, 78
degradable plastic, 79–80
delivery errors, 20
demographics, 102–3

UK socio-economic status classifications, 102
density, 290, 292, 325
design cycle rate, 529
design development, 428–31
　Gantt chart, 429
　highly finished model package, 430
design rights, 58
design speed, 529
dextrin adhesives, 385
die-cut containers, 249, 251
diffusion theory, 383–4
digestion, 167
digital technology, 98
DIN 19303, 186
DIN EN 20535:1994, 260
dirt, 38
distress purchases, 416
distribution logistics, 13
dot gain, 480
double-wall configuration, 240
dry bond adhesion, 332
dry bonding, 172–3
dry offset, 339
dry offset letterpress, 469
drying method, 463–5
　conventional drying mechanism, 464
　UV cure mechanism, 465
dust, 38

easy-open ends
　end shells conversion
　　plain end/shell into easy-open end, 147
　　plain food can ends and shells, 145–6
　　　end lining compound application, 146
　　　food end and beverage end shell forming, 146
Edgewise Compression Test (ECT), 258–9
efficiency, 529
electro chrome coated steel (ECCS), 128, 157
electrophotography, 476–7
elemental chlorine free (ECF), 198–9
elongation, 289, 326
embossing, 172, 482–3, 521
　flat-bed method, 483
EN 1230.2-2001, 210
end-of-line equipment, 522–5
　corrugated cases, 523–4
　pallet labelling, 525
　palletisation, 524–5

　shrink-wrapping, 522–3
energy recovery, 67–8, 77–8
environment, 66
environmental aspect, 66
environmental impact, 66
　packaging and life cycle analysis, 84–5
environmental stress cracking, 290
environmental sustainability
　climate disruption, 82–5
　definitions, 66–8
　environmentally responsible packaging, 80–1
　law compliance, 68–76
　overview, 65–6
　　future generation, 66
　packaging, 65–85
　packaging re-use and recovery, 76–80
　voluntary agreement compliance, 81–2
　waste, 68
Essential Requirements, 59, 60, 70, 72–6, 84
ethylene acrylic acid, 296
ethylene vinyl acetate (EVA), 295, 386–7
ethylene vinyl alcohol (EVOH), 295, 333
European Chemicals Agency (ECHA), 54
European Committee for Standardisation (CEN), 73
exhaustive panel tests, 55
expanded polystyrene packaging, 239
extruded-film adhesives, 172
extrusion bonding, 173

feathering, 480
fibre length, 189
fibre tear, 393
filling machine, 493–506
　angles of repose, 494
　filling liquid products, 499–506
　　bottom-up filling to minimise foaming, 500
　　level fillers, 500–2
　　volume fillers, 502–6
　　weighing fillers, 506
　filling solid products, 496–9
　　by count, 499
　　by volume, 496–7
　　by weight, 497–8
　　multi-head filling by weight, 498
　　using a auger filler, 497
　　using a cup filler, 496
film extrusion, 314–27

cast-film extrusion line, 316
coated overwrap film for tobacco
    packaging, 318
orienting cast film, 317
slot-orifice die cross-section, 315
typical PP/EVOH/PP barrier layer
    coextrusion, 315
film treatments, 327–35
    carton containing liquid laminate example,
        328
    film metallising chamber, 330
    flexible materials lamination and coating,
        328–35
    typical packaging laminate example, 327
Flat Crush Test, 258
flexible materials
    lamination and coating, 328–35
        coating process, 331
        dry bond laminator, 332
        extrusion coating process, 330
        gravure coating process, 334
        laminates examples, 329
        wax bond laminator, 332
        wet bond laminator, 333
flexographic printing, 253–4
flexography, 406, 472–3
    schematic diagram, 472
flow meter, 506
fluoropolymers, 303
flute materials, 243–5
    characteristics of the main flute sizes, 244
    most commonly used grammages for liners
        and flutings, 245
    relative flute flat crush values, 244
    standard flute configurations, 243
foil blocking, 482
    cold foil blocking, 483
    hot foil blocking, 482
folding boxboard (FBB), 180, 184
    illustration, 185
folding cartons, 222–36
    making, 229–36
        carton with concora opening, 235
        crease depth and width, 233
        cutting and creasing forme with make-
            ready counter, 231
        die-cutting rule with a notch, 232
        glueing operation, 236
        nicks position, 231
    styles, 222–9

carton grain direction, 224
dimensions, 225
end load carton, 225
end load carton aeroplane tuck, 228
end load carton basic design, 226
end load carton locking mechanism,
    228
end load carton reverse tuck, 227
glue flaps and working creases
    relationship, 223
length and width determine the location
    of opening, 223
top load carton heat-sealed, web-
    cornered, tapered tray design, 229
top load carton lock tab design, 229
top load carton multi-pack for dairy
    packs and carbonated drinks, 230
top load carton six-corner, glued, folded
    design, 230
top load carton with locked corners,
    229
Food and Drugs Administration (FDA), 131
food can ends
    peelable membrane, 147–8
        illustration, 148
    plain and shells for food/drink easy-open
        ends, 145–6
food contact, 129–31
    EU National regulations, 130–1
    Harmonised European regulations, 130
    USA regulations, 131
Food Labelling Regulations (1996), 58
food packaging
    regulations, 52–5
        EC-approved symbol denote food
            contact materials and articles, 53
Food Safety Act (1990), 52
form, fill and seal (FFS) packaging, 182,
    514–19
    blister packaging, 518–19
    horizontal FFS machines, 515–18
        basic operation, 516
        styles of seal, 517
    sachet forming and filling machines, 518
    thermoforming packaging machines, 518
    vertical FFS machines, 514–15
        basic operation, 515
        variation of VFFS operation, 516
Fourdrinier method, 201–6
    basic wire process, 201

blade coating, 204
calendering, 204
extrusion coating and lamination, 205
headbox and slice process, 203
multi-formers process, 202
size press, 203
vertiformer, 206
free-flowing powder, 494
free radical polymerisation, 265

gap analysis, 101, 556
gas barrier, 290, 325
gas phase polymerisation, 267
gaseous liquids, 118
general line decorative cans
  mechanical lock seam, 143–4
    forming and lock seaming general line can, 144
    general line lock seam process flow, 145
    general line non-performance can, 144
ghosting, 480
glass containers
  decoration and labelling, 120
  design, 117–20
    nomenclature, 117
    relative strengths of different shaped container, 118
    relative weights of cylindrical and irregularly shaped container, 118
    shingling causing toppling and breakage on filling line, 118
  formation, 113–14
    blow-and-blow process, 114
    press-and-blow process, 114
glass inspection, 115–16
glass manufacture, 112–13
  glass furnace and gob-forming process, 112
glass packaging, 7, 109–21
  advantages and disadvantages, 110–11
  annealing, 115
  container design, 117–20
  forming glass containers, 113–14
  glass container decoration and labelling, 120
  glass inspection, 115–16
  glass-making processes, 121
  glass manufacture, 112–13
  overview, 109–10

composition and properties of soda lime glass, 109
surface coating, 115
tolerance, 116–17
glass tolerance, 116–17
global brand management, 98
Global Standard for Packaging and Packaging Materials, 63
global warming, 82
globalisation, 5–6
grab, 393
graft copolymers, 271–2
granular solids, 118–19
graphic design, 455–6
  artwork design, 456
  CAD/CAM, 456
gravity fillers, 501
gravure, 470–2
  printing, 471
grease barrier, 325–6
grease resistance, 183
green bond, 393
greenhouse gases (GHG), 82–4
  global warming potential, 83
greenwash, 80–1
Gurley Porosity Test, 260

halftone, 448–9
hazard, 539
hazard and risk management
  HARM development, 548–54
    HARM control chart (RTP), 553
    hazard analysis chart (RTP), 552
    process flowchart, 550
    simplified 'decision-tree,' 551
  hazard identification and risk assessment, 547–54
    hazard and risk analysis, 547–8
  industry technical standards, 554–8
    certification, 556–8
    relevant standards and scope, 554
    standards, 554–5
    using the standards, 555–6
  life-cycles in supply chain, 539–44
    HARM issues in RTP, 541
    HARM issues in single-use packaging, 540
    overview, 539–42
    packaging performance, 543–4
    packaging types, example uses, hazards

and consequences, 542
returnable transit packaging, 543
single-use packaging, 542–3
single-use vs returnable packaging, 540
packaging, 538–58
prerequisite systems and controls, 544–7
management systems, 545–7
prerequisites, 544–5
hazardous substance, 75
heal-seal temperature, 326
heat sealability, 290
heat-sealable adhesives, 172
heat transfer printing, 484
heating, 150
hexachrome, 450–1
hickies, 479
high density polyethylene (HDPE), 8, 267–9
Home Authority Trading Standards Officer, 57
home compostable packaging, 79
hot foil printing, 521, 522
hot-melt bonding, 173
hot melts, 172, 389
hue, 443
humidity, 42
hybrid screening, 453
hydrapulper, 191
hydraulic pressure tester, 116
hydrofluorocarbons, 83

image transfer, 460–1
computer to gravure cylinder, 460–1
computer to off-set lithographic plate, 461
computer to relief plate, 461
computer to screen, 460
mechanisms, 467–78
flexography, 472–3
gravure, 470–2
letterpress, 469–70
lithography, 474–6
non-impact printing techniques, 476–8
screen printing, 467–8
impact strength, 289
in-mould labelling, 339, 401–2
independent section machine, 113
Industrial Revolution, 4
Industry Council for Packaging and the Environment (INCPEN), 61
industry guides, 62
injection moulding, 339–45

decoration of injection moulded parts, 344
hot runner moulds, 344
image, 345
injection points, 342
mould designs, 343–4
parts with undercuts, 344
multi-cavity mould, 341
reciprocating and ram type, 340
simple injection mould, 342
thermoforming comparison, 348–50
ink, 463–5
properties for various printing processes, 464
inkjet printing, 403, 521, 522
inkjet technology, 477
insects, 38
intensity of colour, 443
inter-atomic forces, 280–3
aromatic benzene, 281
effect on polymer properties, 281
styrene block copolymer TPE, 283
thermosetting and thermoplastic polymers, 282
Intergovernmental Panel on Climate Change (IPCC), 83–4
intermediate bulk containers (IBC), 7, 312
international trading, 60–1
trade conventions between Britain/Europe, the United States and Japan, 61
Internet Advertising Bureau (IAB), 97
ionomers, 296
Iriodin, 455
ISO 535:1991, 260
ISO 1294, 210
ISO 2759:2001, 258
ISO 3034:19, 258
ISO 3035:1982, 258
ISO 3036, 258
ISO 3037:2007, 258
ISO 5628, 210
ISO 5636, 209
ISO 7263:2008, 258
ISO 8296, 209
ISO 9001, 63
ISO 12048:1994, 33
ISO 14001, 66
ISO 14021, 60
ISO 22000, 555
ISO 287-1985, 210
ISO 534-2005, 210

ISO 535-1991, 210
ISO 536-1995, 210
ISO 1974-1990, 210
ISO 1994.2-2008, 210
ISO 2470.2-2008, 209
ISO 2471-2008, 209
ISO 2493-1992, 210
ISO 2758-2003, 210
ISO 3783-200, 209
ISO 5636-5:2003, 260
ISO 8254.3-2004, 209
ISO 8295-1995, 210
ISO 8791.3-2005, 209
ISO 9895-2008, 210
ISO 11476-2000, 209
ISO 12647-2:2004, 458
ISO 14000/140, 60
ISO 15754-2009, 210
ISO 105-X12-2001, 209
isotactic configuration, 268

jute bags, 8

key performance indicator (KPI), 15, 17, 18, 22
Kraft, 193
Kyoto Protocol, 83

label claims, 58
label manufacturer, 398–400
  adhesive selection, 399
  label application, 399
  label substrates, 398–9
  storage requirements, 399–400
label market, 405–6
labelling, 509–12
  neck collars and tags, 512
  self-adhesive labels, 509–10
  sleeves, 512
  ungummed labels, 511–12
laminate aluminium foil, 172–4
  wet-bonding foil lamination process, 173
Landfill Directive, 69
Landfill Regulation (2002), 69
laser marking, 521, 522
laser sensors, 486
latex, 386
law
  compliance, 68–76
  enforcement, 50–1

  trading conventions, 51
leak testing, 26
leakage, 25–6
legislation, 51–2
  compliance consistency, 63
  consequence of compliance failure, 63
  environment protection regulations, 59–60
    EU Packaging Directive essential requirements, 59
    factory operations, 60
    green claims and recycling information, 60
    Producer Responsibility Obligations, 59–60
  good practice, 62–3
    industry guides, 62
    trade conventions, 62–3
  international trading, 60–1
  overview, 50–1
    law and trading conventions, 51
    law enforcement, 50–1
    role of law within civilised society, 50
  packaging, 50–63
  product quality, safety regulation during manufacture, distribution, storage and use, 52–6
  trade associations, 62
  trade regulations, 56–8
    deceptive packaging and label claims, 58
    design rights, patents and copyright, 58
    e mark, 57
    filling of products, 56–8
    permitted tolerable negative error changes, 57
letterpress, 406, 469–70
  dry offset letter press, 470
  schematics, 469
level fillers, 500–2
  basic gravity filling, 502
  filling by level and volume, 501
  gravity filling using flow meter and valve, 503
  pressure filling, 504
life cycle analysis, 84–5
lifestyle changes, 5
light, 37–8
lightness, 443
line efficiency, 528–9
linear forming, 218

linear low density polyethylene (LLDPE), 271
liner/facing materials, 242–3
linerless closure, 368–9
liquid crystal polymers, 304
liquid packaging cartons, 219–21
  brick design, 221
  laminates used, 220
  Tetrapak closure designs, 221
lithographic printing, 254
lithography, 474–6
  schematic diagram, 475
long softwood fibres, 190
low density polyethylene (LDPE), 265–7
  autoclave and tubular reactors for processing polyethylene, 266
lug closures, 371–2
  lug neck finish and closure, 371
  metal lug tamper-evident pop-up button, 372

machine direction (MD), 316
machine vision systems, 486–7
management information system (MIS), 406
management systems, 545–7
  biological contaminants and procedural controls, 547
  chemical contaminants and procedural controls, 546
  physical contaminants and procedural controls, 546
  structure and functions, 545
manufacturing, 13
market, 417
market segmentation, 101
marketing, 13
  brand promotion, 97–8
  branding and product promotion and advertising, 93–5
  communication consistency, 98–100
  definition, 87–9
  market research tools and techniques to identify customer needs, 100–5
  packaging, 87–105
  product branding, 95–7
  role, 89–92
    brand value determination, 92
    distribution strategy determination, 90–2
    identifying market sector, 89–90
    identifying product need, 89
    packaging relationship, 92
    price determination, 90
marketing communications agency, 100
Marketing Department, 88
marketing-led companies, 88–9
Mater Bi, 307
matte finish, 170
mechanical adhesion, 382–3
mechanical lock seam
  general line decorative cans, 143–4
    forming and lock seaming general line can, 144
    general line lock seam process flow, 145
    general line non-performance can, 144
mechanical pulp, 193
mechanical seaming ends
  can bodies, 148–9
  operations, 149
  overview, 149
mechanical strength, 258–60
  testing the strength of corrugated board, 259
medicine packaging, 55
medicine safety, 55
melt flow index (MFI), 314
merchandising, 13
metal can
  end process performance, 126–7
  formats, 123–6
    aerosol prior to fitting of value mechanism, 126
    body manufacture, 124
    bottle, 125
    non-round food, 125
    processed food and drink, 124
metal closures, 153–5
  outlines, 153
metal containers
  manufacture, 133–53
  specifications, 157–9
Metal FX, 454
metal packaging, 7
metallisation, 174
metallocene, 294
metamerism, 447
methane, 83
minimum net weight, 56
Miraval, 455

Misleading Advertising Regulations (1998), 60
misregistration, 479
mitigation, 84
modern packaging, 4–5
modern retailing, 5–6
modified atmosphere packaging, 520–1
modified molten aluminium, 168
moisture, 290
moisture barrier, 325
moisture vapour transfer (MVT), 257
moisture vapour transmission rate, 44–5, 325
molecular weight, 278
mono-orientation, 316
monomer
  different number, 278–80
    linear low density polyethylene, 279
    orientation in a copolymer, 280
    polymer formation, 273–5
      substitution with hydrogen atom in ethylene, 273
moulded pulp boxes, 238–9
moulded pulp trays, 238–9
Mullen Burst Test, 258
multi-injection moulding, 345–8
  single-stage coinjection, 347–8
    coinjected PP and PET preforms, 349
    process, 348
  two-stage injection moulding and overmoulding, 345–7
    process, 346–7
multifunctional monomers, 272
multipacks, 222–36
multiple drawn cans, 136–8
  can drawing from flat metal, 137
  cup making for DRD and DWI cans, 136
  DRD can process flow, 138
  redraw can forming, 138

narrow neck press-and-blow, 113, 114
NatureFlex, 305–6, 307
neck collars, 512
non-free-flowing powder, 494
non-impact printing techniques, 476–8
  continuous inkjet, 477
  electrophotography or xerography, 476
  piezo and bubble-jet technologies, 478
noxious substance, 75
nutrient source, 42

open mouth sacks, 211
open time, 393
optical properties, 290
optimised pallet fill, 20
output, 529
over-pressure fillers *see* pressure filler
oxodegradable, 307
oxygen, 42
oxygen chemical bleaching (OCB), 198

pack size, 91
package top-load strength, 20
packaging, 66
  adhesives, 381–93
    adhesion theories, 382–4
    adhesive selection, 389–90
    adhesive types, 384–9
    application methods, 390
    performance evaluation, 391
    terminology, 392–3
    troubleshooting adhesive problems, 391–2
  brief, 414–15
    checklist, 415
  containment, 25–6
  convenience, 46–7
  design and development, 411–39
    brief, 414–15
    case study, yoghurt for children, 431–9
    conceptual design, 423–31
    design process, 412–13
    research, 415–23
  design process, 412–13
    chart, 413
  different level of packaging, 6–7
  environmental sustainability, 65–85
    climate disruption, 82–5
    definitions, 66–8
    environmentally responsible packaging, 80–1
    law compliance, 68–76
    overview, 65–6
    re-use and recovery, 76–80
    voluntary agreement compliance, 81–2
    waste, 68
  functions, 24–49
  globalisation and modern retailing, 5–6
  hazard and risk management, 538–58
    hazard identification and risk assessment, 547–54

industry technical standards, 554–8
life-cycles in supply chain, 539–44
prerequisite systems and controls, 544–7
history, 3
legislation, 50–63
  compliance consistency, 63
  consequence of compliance failure, 63
  environment protection regulations, 59–60
  good practice, 62–3
  international trading, 60–1
  overview, 50–1
  product quality, safety regulation during manufacture, distribution, storage and use, 52–6
  trade associations, 62
  trade regulations, 56–8
machinery and line operations, 490–537
  cartonning, 513–14
  closing and sealing of containers, 506–9
  coding systems, 521–2
  direct product shrink- and stretch-wrapping, 519–20
  end-of-line equipment, 522–5
  fillers and filling, 493–506
  form, fill and seal (FFS) packaging operations, 514–19
  labelling, 509–12
  miscellaneous wrappers, 521
  modified atmosphere packaging, 520–1
  packaging line, 490–2
  problem-solving on the packaging line, 536–7
  unscramblers, 492–3
marketing, 87–105
  brand promotion, 97–8
  branding and product promotion and advertising, 93–5
  communication consistency, 98–100
  definition, 87–9
  market research tools and techniques to identify customer needs, 100–5
  product branding, 95–7
  role, 89–92
materials polymer chemistry, 262–86
  addition polymerisation, 265–9
  condensation polymerisation, 269–70
  copolymerisation and crosslinking polymerisation, 271–2
  factors affecting polymer characteristics, 272–86
  overview, 262–3
  polymerisation principles, 264
materials usage and development, 7–8
patterns and impact of consumption, 4–5
preservation, 41–6
printing, 441–88
  additive and subtractive colour mixing, 444–5
  colour description, 442–3
  colour printing, 447–55
  colour vision, 443–4
  graphic design, reprographics and pre-press, 455–61
  light and colour, 442
  other factors affecting colour, 445–7
  other processing techniques, 481–4
  proofing options and approval processes, 461–3
  quality control, 484–8
  technological aspects, 463–80
product information source, 47–8
product selling, 48
protection, 26–40
quality and efficiency aspects of operation, 526–36
  calculating line efficiency, 528–30
  changeovers, 531–2
  installing new packaging machinery, 534–6
  maintenance and training, 528
  measured parameters and types of sensors, 527–8
  optimising line efficiency, 530–1
  specifying and purchasing of new packaging machinery, 533–4
  storage of materials, 526, 528
  using existing or new machinery to increase production, 532–3
responsible use of resource, 9
society, 3–9
supply chain, 10–22
  challenges, 19–22
  delivering cost effective solutions, 14–19
  segments development, structure and interdependence, 12–14
  training importance, 22
Packaging and Packaging Waste Directive, 70

packaging closures, 361–80
  child resistance, 376–7
  crimped crown cork closures, 372–3
  dispensing and metering, 377–9
    dispensing pumps, 378–9
    flip-top closures, 377–8
  lug closures, 371–2
  peelable seal lids, 373
  push-fit closures, 362–5
    push-in styles, 363–4
    push-on styles, 364–5
  role, 361–2
  screw-threaded closures, 365–71
  tamper evidence, 373–6
    adhesive labels and tapes, 376
    plastic tamper-evident closures, 375
    roll-on pilfer-proof (ROPP) metal closures, 374–5
    ROPP closure, 374
    shrink seals or bands, 375
  testing closure performance, 379–80
  types, 362
Packaging Compliance Scheme, 59
Packaging Export Recovery Notes (PERN), 72
packaging labels, 395–406
  challenges, 405
  digital revolution, 406
  future trends, 406
  in-mould, 401–2
  label manufacturer, 398–400
  main end user sectors, 396
  market, 405–6
  printing process, 403–4
    process used, 404
  self-adhesive (pressure sensitive), 396–8
  sleeves, 402–3
  specifications, 404–5
  trends, 396
    label types produced by European label converters, 397
  wet glue (gummed labels), 400–1
packaging line, 490–2
  problem-solving, 536–7
  types of layout, 491–2
    rotary layout, 491
packaging materials
  plastic properties, 287–308
    bio-based polymers, 306–8
    common packaging plastics, 292–300
    key properties for packaging applications, 289–92
    market overview, 287–9
    specialist polymers, 301–6
  plastics manufacturing processes, 310–60
    blow moulding, 350–9
    environmental considerations, 359–60
    film treatments after forming, 327–35
    injection moulding, 339–45
    injection moulding vs thermoforming, 348–50
    multi-injection moulding, 345–8
    plasticating extruder, 311–14
    sheet and film extrusion, 314–27
    thermoforming process, 335–39
packaging minimisation, 73–4
packaging optimisation, 14, 17
Packaging Recovery Notes (PRN), 71
packaging reduction projects, 9
Packaging Regulations, 59, 60, 70, 72–6
packaging research, 415–23
  marketing, 415–21
    brand values, 417
    competitor products, 420–1
    consumer requirements and benefits, 420
    market, 417
    product positioning, 416
    target audience, 419–20
  technical consideration, 421–3
    containment, protection, preservation and compatibility, 422
    environmental factors, 423
    production, distribution and point-of-sale, 422–3
packaging technical development, 13
Packaging Waste, 59
Packaging Waste Directive, 70, 72
pallet construction, 33–4
  compressive forces when using single-side pallets, 34
pallet labelling, 525
pallet shrink wrap film, 294
pallet stacking, 34–6
  0201 style corrugated case, 35
  load-bearing ability around the perimeter of a case, 35
  overhang effect on compression stack strength, 34
  stacking factors guidance, 36

palletisation, 524–5
  automatic palletisation, 525
palletisation line, 21
Pantone, 405
Pantone swatch guide, 450–1
paper and paperboard packaging, 178–239
  manufacture, 200–8
  overview, 178–9
    matted fibres, 179
  paper conversion process, 208–21
  paperboard conversion process, 222–39
  post-pulping of fibres, 196–200
  properties, 179–89
    burst strength principle, 182
    Cobb test for water absorbency, 181
    combination of whiteness and strength for various pulp, 188
    combining paper with other materials, 188–9
    compression strength testing, 183
    different types comparison, 186–8
    paperboard grades, 184–6
    properties of changing types of pulp, 189
    SBB, FBB and WLC, 181
    total energy absorption (TEA) levels, 182
    warping, 181
    whiteness and strength for different paperboard grades, 190
    whiteness and tensile strength difference, 183
  pulping process, 191–6
  raw materials, 189–91
paper bags, 213, 215–16
  flat bag design, 215
  satchel bag design, 215
  satchel bag design with window, 216
  storage, 216
paper packaging, 7
paper sacks, 208–13
  specifications for multi-wall sacks, 214
  types of open mouth sack design, 212
  valve sack types, 213
paperboard grades, 184–6
  DIN 19303 European classification for qualities of paperboard, 187
patents, 58
peel strength, 391
peelable seal lids, 373

perfluorocarbons, 83
pester power, 435
pests, 38
photo-electric sensors, 486
pilferage, 38–9
piston filler, 504–5
plastic sacks, 294
plastic tamper-evident closures, 375
plasticating extruder, 311–14
  carbon black nanofillers producing anti-static polyethylene carbon nanocomposite, 313
  illustration, 312
  plastic conversion process, 314
plastics, 8
  bio-based polymers, 306–8
  common packaging plastics, 292–300
    polyethylene family of plastics, 293–6
    polyethylene terephthalate, 299–300
    polypropylene, 296–8
    polystyrene, 299
    polyvinyl chloride, 298–9
  key properties for packaging applications, 289–92
    chemical resistance, 291
    density, 292
    moisture and oxygen barrier properties, 292
  market overview, 287–9
    development of plastics, 288
    European demand vs other markets, 288
  properties for packaging materials, 287–308
  specialist polymers, 301–6
    cellulose materials, 305–6
    cyclic olefin copolymers, 304
    fluoropolymers, 303
    liquid crystal polymers, 304
    polyamide, 303
    polycarbonate, 303
    polyethylene naphthalate (PEN), 302
    polymers used in packaging, 301
    polyvinylidene chloride (PVDC) copolymers, 302
    styrene copolymers, 301–2
    thermoplastic elastomers, 304
    thermosets, 304–5
plastics manufacturing processes
  packaging materials, 310–60
    blow moulding, 350–9

environmental considerations, 359–60
film treatments after forming, 327–35
injection moulding, 339–45
injection moulding vs thermoforming, 348–50
material selection, 310–11
multi-injection moulding, 345–8
plasticating extruder, 311–14
sheet and film extrusion, 314–27
thermoforming process, 335–39
Ply Separation Test, 260
point-of-sale (POS), 423
polyamide, 269–70, 303
polycarbonate, 270, 303
polyethylene, 293–6
  ethylene acrylic acid, 296
  ethylene copolymers, 295
  ethylene vinyl acetate, 295
  ethylene vinyl alcohol, 295
  high density polyethylene, 294–5
  ionomers, 296
  linear low density polyethylene, 293–4
  low density polyethylene, 293
  polyvinyl acetate, 296
  polyvinyl alcohol, 296
polyethylene naphthalate (PEN), 302
polyethylene terephthalate (PET), 8, 299–300
polylactic acid, 270
polymer, 312
  addition polymerisation, 265–9
  condensation polymerisation, 269–70
  copolymerisation and crosslinking polymerisation, 271–2
  factors affecting characteristics, 272–86
    addition vacuum deposition chamber, 285
    additives, 283–4
    branched polymer showing short and long chain branching, 276
    crystallinity, 276–8
    forces, 280–3
    linear polymer, 276
    mixing, 285
    molecular weight and molecular weight distribution, 278
    monomer, 273–5
    monomer orientation in a copolymer, 280
    number of different monomer, 278–80
    physical orientation, 285–6

polymerisation methods, initiators and catalyst, 276
special material treatments, 284–5
tacticity, 275
overview, 262–3
  common packaging plastics, 263
  packaging material chemistry, 262–86
  polymerisation principles, 264
polymer melting, 313
polymeric film, 325–7
polymerisation
  addition, 265–9
  principles, 264
polypropylene, 296–8
  packaging applications, 297
polystyrene, 299
polyurethane, 387
polyvinyl acetate (PVA), 296, 386–7
polyvinyl alcohol (PVOH), 296
polyvinyl chloride (PVC), 298–9
  packaging applications, 298
polyvinylidene chloride (PVDC) copolymers, 302
precipitation, 167
prerequisite systems and controls, 544–7
  HARM, 544
preservation, 41–6
  packaging characteristics for foodstuffs, 43–6
  order of importance of specific deterioration indices, 45
  process, 43
    methods and associated packaging requirements, 44
    packaging requirements to reduce/prevent abiotic spoilage, 45
  shelf life, 41
  spoilage, 41–3
press-and-blow, 113, 114
press-on closure see push-on closure
press-twist closures, 155
pressure control, 313–14
pressure filler, 502, 504
Pricewaterhouse Coopers, 97
primary packaging, 6, 10, 15, 16, 74
printing, 172
  colour description, 442–3
    hue, 443
    intensity of colour, 443
    lightness, 443

light and colour, 442
    electromagnetic radiation spectrum, 442
  other processing techniques, 481–4
    embossing, 482–3
    foil blocking, 482
    heat transfer printing and ceramic or glass decal, 484
    metallising, 484
    varnishing/lacquering, 481
  packaging, 441–88
    requirements for viewing colour, 441
  technological aspects, 463–80
    image transfer mechanisms, 467–78
    ink types and drying, 463–5
    printing press configurations, 465–6
    typical print defects, 478–80
printing press, 465–6
  central impression presses, 465–6
    printing configuration, 466
  in-line presses, 465
  stack presses, 466
Producer Responsibility Obligations, 59–60, 70–2
product advertising, 93–5
product information, 47–8
  formats, 48
  types, 47–8
product/packaging design, 13
product positioning, 416
product promotion, 93–5
product type, 28
product value, 28
product viability, 100
production-led companies, 88
proofing, 461–3
  digital proofing, 461–2
  sign-off and approval, 462–3
  wet proofing, 462
protection, 26–40
  defining the environment, 28–39
  defining the product, 28
  function evaluation, 27–8
  function evaluation process completeness, 39–40
    typical data journey mapping, 40
  hazards in the supply chain, their causes and effects, 27
pry-off lever lid, 363
psychographics, 103
puncture, 36

puncture resistance, 326
Puncture Resistance Test, 258, 260
purchasing, 13
purified molten aluminium, 168
push-in closure, 363–4
  cork in glass bottle, 363
  pry-off lever lid design, 363
push-in plastic closures, 364
push-on closure, 363, 364–5
  plastic design, 364
  twist-off, 365
push-on plastic closure, 364
push-on twist-off closure, 365

quality assurance
  finished and semi-finished components, 151–3
    three-piece can air testing principles, 151
    two-piece can light testing principles, 152
quality control
  packaging, 484–8
    common testing methods, 485
    inspection processes, 486–7
    specifying requirements and setting standards, 484–6
    tracing, packing and tracking, 487–8
  paper and paperboard, 207–8
    tests carried out for surface appearance, 209
    tests carried out to control performance properties, 210–11

radio frequency identification (RFID), 395, 426, 487
random copolymers, 271
raw material suppliers, 13
raw materials, 127–33, 189–91
  coating materials and application, 131–3
    coater for applying lacquer to sheets, 132
    DWI food can internal lacquering, 132
  corrosion management, 128–9
  food contact, 129–31
  forming processes, 155–7
  hardwood and softwood, 191
  properties of some paperboard grades made from recycled waste, 192
  steel and aluminium, 127–8

re-use, 76–7
recovery, 67–8
recycled fibres, 191
    production, 196
recycling, 67–8, 77, 161
refining, 167, 196–8
    rotating disc refiner, 197
Registration, Evaluation, Authorisation and Restriction of Chemicals (REACH), 54
regular slotted containers (RSC), 249, 251
Regulation (EC) No. 1169/2011, 58
Regulation (EC) No. 1907/2006, 54
Regulation (EC) No. 1935/2004, 52, 130
reprographics, 456–60
    colour separations, 457
    dot control, 458–9
    other additions, 459–60
    step and repeat, 459
        set positions, 459
    trapping, 457–8
    two colours, 457
reroll stock, 168
resin, 313
retail packaging, 15
retail ready packs (RRP), 241–2
retorting, 149–51
returnable transit packaging, 543
reversed epsilon, 55
rigid boxes, 236–8
    designs, 237
rigid metal packaging, 122–62
    containers manufacture, 133–53
    containers specifications, 157–9
    decorating processes, 159–60
    environmental factor, 160–2
    metal closures, 153–5
    overview, 122–7
        history, 122
        market, 123
        metal can and end performance, 126–7
        metal can formats, 123–6
    raw materials, 127–33
    raw materials and forming processes, 155–7
risk, 539
risk management, 538
roll-on pilfer proof (ROPP) caps, 154
roll-on pilfer-proof (ROPP) metal closures, 374–5
roll wrappers, 521
rolling ingots, 168
running speed, 529

sales-led companies, 88
scanning errors, 20
screen angles, 449
    Moiré pattern, 450
screen clash, 480
screen printing, 339, 467–8
    flat-bed and rotary screen printing, 468
screw-threaded closures, 365–71
    container neck dimensions standard nomenclature, 367
    H dimensions accuracy, 367
    induction sealed combined wad and membrane seal, 370
    linerless closure with bore seal, 369
    linerless closure with top seal, 370
    matching thread profiles importance, 366
    thread profiles, 366
    wadded closure, 368
scumming, 479–80
secondary chlorine free (SCF), 199
secondary packaging, 6–7, 10, 74
self-adhesive labels, 396–8, 509–10
    label production, 397–8
    layout options, 510
semisolids, 118
sequential filling, 495
setting time, 392–3
sheet extrusion
    film extrusion, 314–27
        blown film, 318–21
        cast, 315–18
        cast and blown film coextrusion, 323–5
        cellulose film, 321–3
        polymeric films performance parameters, 325–7
shelf life, 17, 41
shock, 29–30
    cushioning effect, 31
    effect minimising ways, 30
short hardwood fibres, 190
short span compression strength (SCT), 242
Short Span Compression Test, 258
shrink film, 320–1
shrink seals, 375
shrink sleeve, 512
    labelling, 402–3

shrink wrap, 320, 519–20, 522
    collated primary packs, 523
shrinkage, 19–20
single-use packaging, 542–3
single-wall configuration, 240
skin packs, 236–8
    carded construction, 238
skiving, 217
sleeve labelling, 402–3
smelting, 167–8
smelting cell, 168
society
    packaging, 3–9
        different level of packaging, 6–7
        globalisation and modern retailing, 5–6
        history, 3
        materials usage and development, 7–8
        patterns and impact of consumption, 4–5
        responsible use of resource, 9
soda lime glass, 110
solid bleached board (SBB)
    illustration, 184
solid phase pressure forming (SPPF), 336
solid unbleached board (SUB), 184
    illustration, 185
solids (non-volatile) content, 393
solution phase polymerisation, 267
solvent-based adhesives, 388
solventless adhesives, 388–9
spiral wound containers, 216–18
    illustration, 217
    skiving, hemming and anaconda fold, 218
spoilage, 19, 41–3
squeeze tester, 115
starch, 384–5
station, 529
Statutory Instrument, 51, 52
stay-on-tab (SOT), 145
steric hindrance, 274
Stern Review, 84
stiffness, 180, 326
stochastic screening, 452–3
    reduced paper effect, 452
storage logistics, 13
stretch blow moulding, 351
stretch hooding, 294
stretch wrap, 320, 519–20
structural design, 425–7
    French pack for sugar, 426

styrene acrylonitrile (SAN), 301
styrene copolymers, 301–2
subtractive colour mixing, 445, 447–8
sulphur hexafluoride, 83
supply chain
    challenges, 19–22
    delivering cost effective solutions, 14–19
    packaging, 10–22
    segments development, structure and interdependence, 12–14
    training importance, 22
surface coating, 115
surface energy, 326
surface finishes, 339
surface friction, 289
sustainable development, 66–7
SWOT analysis, 101
syndiotactic configuration, 268

tablets, 119
tack, 393
tacticity, 267–8, 275
tags, 512
tampering, 38–9
Tampo printing, 471–2
TAPPI T803, 258
TAPPI T826, 258
tear resistance, 182
tear strength, 289
Teflon, 303
temperature, 42, 290
    humidity and, 36–7
        corrugated board compression strength vs moisture content, 38
temperature control, 313
tensile strength, 182, 289, 326
tertiary packaging, 7, 10, 11, 74
test markets, 104
testing closure performance, 379–80
Tetra Pak, 8
thermal transfer, 521, 522
thermoforming process
    plastic packaging, 335–9
        billow forming over a plug and cavity, 338
        plug assist vacuum forming, 337
        solid phase pressure forming, 339
        vacuum forming over cavity and plug moulds, 337
        vacuum snap back forming, 337

thermomechanical pulping (TMP), 191, 194–5
thermoplastic elastomer, 304, 345
thermoplastics, 172, 262, 288
thermoset resins, 256–7
thermosets, 304–5
thermosetting, 262
thickness control, 325
Thickness of Corrugated Board Test, 258, 260
three-piece can, 123–5
three-piece welded cans, 133–5, 156
    coil cutting operation, 133
    forming and welding, 134
    process flow, 135
    wall beading, 135
titratable alkalis, 110
tolerable negative error, 56
tonal reproduction, 448
tonal value increase, 458, 480
total energy absorption (TEA), 182
totally chlorine free (TCF), 198–9
traceability, 487–8
trade associations, 62
trade conventions, 62–3
Trade Descriptions Act, 58, 60
Trade Marks Act, 58
trading conventions, 51
trays, 222–36
triple-wall configuration, 240–1
turret, 491
twist-off closures, 154–5
twist-off metal lug closures, 371
twist wrapping machines, 521
two-piece can, 123
two-piece drawn cans, 138–41, 156
    DWI can process flow, 140
    wall ironing through one ring, 141
two-piece impact extruded cans, 141–3
    impact extrusion process, 142
    swage top of aerosol can, 143
two-piece impact extruded tubes, 141–3
    impact extrusion process, 142
two-piece single drawn cans, 136–8
    can drawing from flat metal, 137
    cup making for DRD and DWI cans, 136
    DRD can process flow, 138
    redraw can forming, 138
two-piece wall ironed cans, 138–41, 156
    DWI can process flow, 140
    wall ironing through one ring, 141

UK emissions total, 83–4
ungummed labels, 511–12
    automatic application, 511
United Nations Environment Programme (UNEP), 84
United Nations Framework Convention on Climate Change (UNFCCC), 83
unscramblers, 492–3
    automatic depalletiser with sweep-off mechanism, 493
UV flexo, 406

vacuum button, 371–2
vacuum fillers, 502, 504
vacuum metallising, 174
vacuum snap back, 336, 337
validation, 553
valve sack, 211–12
van der Waals force, 383
Variochrome, 455
varnishing, 481
vat method *see* cylinder method
Velio process, 121
verification, 553
vibration, 31–2
    sources in road vehicles, 31
    vehicle stack resonance, 32
vinyl heat-seal coatings, 171
viscoelasticity, 289–90
viscosity, 393
viscous liquids, 118
visual check, 116
volatile organic compounds (VOCs), 388
volume fillers, 502–6
voluntary agreement, 81–2

wadded closure, 368
wall thickness detector, 116
waste, 67
    scale, 68
waste stream, 79–80
water-based adhesives, 384–7
water-soluble dispersions, 172
water-soluble emulsions, 172
water vapour transmission rate *see* moisture vapour transmission rate
waterless lithography, 475–6
weighing fillers, 506
Weights and Measures Act (1963), 56
wet bond lamination, 333

wet bonding, 172–3
wet glue, 400–1
   product label application, 400
   substrate choice, 400–1
      material types, printed or converted, 402
wetting out, 393
white caps *see* twist-off metal lug closures
white lined chipboard (WLC), 184
   illustration, 186
workforce, 56
World Meteorological Organisation (WMO), 84

yoghurt, 431–9
   design study, 432–9
      design analysis, 438–9
      design concept, 435–8
      initial sketches, packaging concepts, 436
      leading companies and brands, 433
      mock-up, paperboard yoghurt pot, 438
      pouch and tube formats, 435
      research, 432–5
      rough graphic concepts, packaging design, 437
      target audience, 434
   overview, 431–2
      advertising, 431–2
      brand values, 431
      competitors, 432

Lightning Source UK Ltd.
Milton Keynes UK
UKOW07n1352050417
298400UK00005B/37/P